# PERSPECTIVES IN MEMBRANE BIOLOGY

*ACADEMIC PRESS RAPID MANUSCRIPT REPRODUCTION*

*Proceedings of an International Symposium*
*held at Oaxaca, Mexico on January 14-18, 1974*

FIRST MEXICAN SOCIETY OF BIOCHEMISTRY SYMPOSIUM

# PERSPECTIVES IN MEMBRANE BIOLOGY

*Edited by*

**SERGIO ESTRADA-O.**
**CARLOS GITLER**

*Departamento de Bioquimica*
*Centro de Investigacion y de*
*Estudios Avanzados del I.P.N.*
*Mexico, D. F.*

**ACADEMIC PRESS New York San Francisco London 1974**

A Subsidiary of Harcourt Brace Jovanovich, Publishers

ACADEMIC PRESS, INC.
111 Fifth Avenue, New York, New York 10003

*United Kingdom Edition published by*
ACADEMIC PRESS, INC. (LONDON) LTD.
24/28 Oval Road, London NW1

**Library of Congress Cataloging in Publication Data**

Meeting on Perspectives in Membrane Biology, 1st, Oaxaca, Mexico, 1974.
Perspectives in membrane biology.

Bibliography: p.
Includes index.
1. Membranes (Biology)–Congresses. I. Estrada-O., Sergio, ed. II. Gitler, Carlos, (date) ed. III. Sociedad Mexicana de Bioquímica. IV. Title.
[DNLM: 1. Cell membrane–Congresses. W3 ME427U 1974p / QH601 M495 1974p]
QH601.M44 1974 574.8′75 74-26692
ISBN 0-12-243650-4

PRINTED IN THE UNITED STATES OF AMERICA

# CONTENTS

CONTENTS

# PARTICIPANTS

Baltscheffksy, Margareta, *Department of Biochemistry, Arrhenius Laboratory, University of Stockholm, S-104-05 Stockholm, Sweden.*

Bennett, Vann, *Department of Pharmacology and Experimental Therapeutics and Department of Medicine, The Johns Hopkins University, School of Medicine, Baltimore, Maryland 21205.*

Bretscher, Mark S., *Medical Research Council, Laboratory of Molecular Biology, Hills Road, Cambridge, CB2 2QH, England.*

Burger, Max M., *Department of Biochemistry. Biocenter of the University of Basel, Klingelbergstrasse 70, CH 4056 Basel, Switzerland.*

Cerbón, Jorge, *Departamento de Bioquímica, Centro de Investigación y de Estudios Avanzados del Instituto Politécnico Nacional. Apartado Postal 14-740. México 14, D. F.*

Cone, Richard A., *Department of Biophysics, The Johns Hopkins University, Baltimore, Maryland 21218.*

Crofts, Anthony R., *Department of Biochemistry, Medical School, University of Bristol, University Walk, Bristol, BS8 1TD, England.*

Cuatrecasas, Pedro, *Department of Pharmacology and Experimental Therapeutics and Department of Medicine, The Johns Hopkins University, School of Medicine, Baltimore, Maryland 21205.*

Chance, Britton, *Johnson Research Foundation, School of Medicine, University of Pennsylvania, Philadelphia, Pennsylvania 19174.*

Cheng, Wen, *Johnson Research Foundation, School of Medicine, University of Pennsylvania, Philadelphia, Pennsylvania 19174.*

de Petris, Steffanello, *Medical Research Council, Neuroimmunology Project, Zoology Department, University College London, London WC1E 6BT, England.*

Erlij, David, *Departamento de Biología Celular, Centro de Investigación y de Estudios Avanzados del Instituto Politécnico Nacional. Apartado Postal 14-740. México, D. F.*

Estrada-O., Sergio, *Departamento de Bioquímica, Centro de Investigación y de Estudios Avanzados del Instituto Politécnico Nacional. Apartado Postal 14-740. México, D. F.*

Garoff, Henrik, *Department of Serology and Bacteriology, University of Helsinki, Haartmansgata 3, 00290 Helsinki 29, Finland.*

Gitler, Carlos, *Departamento de Bioquímica, Centro de Investigación y de Estudios Avanzados del Instituto Politécnico Nacional. Apartado Postal 14-740. México 14, D. F.*

Gómez-Puyou, Armando, *Departamento de Biología Experimental, Instituto de Biología, Universidad Nacional Autónoma de México. Ciudad Universitaria. México 20, D. F.*

Gómez-Puyou, Marieta Tuena de, *Departamento de Biología Experimental, Instituto de Biología, Universidad Nacional Autónoma de México. Ciudad Universitaria, México 20, D. F.*

Gray, G. Maurice, *MRC Unit on the Experimental Pathology of Skin, The Medical School, The University, Birmingham, England.*

Helenius, Ari, *Department of Serology and Bacteriology, University of Helsinki, Haartmansgatan 3, 00290 Helsinki 29, Finland.*

Kääriäinen, Leevi, *Department of Virology, University of Helsinki, Haartmansgatan 3, 00290 Helsinki 29, Finland.*

Kaback, H. Ronald, *The Roche Institute of Molecular Biology, Nutley, New Jersey 07110.*

Klip, Amira, *Departamento de Bioquímica, Centro de Investigación y de Estudios Avanzados del Instituto Politécnico Nacional. Apartado Postal 14-740. México 14, D. F.*

Koretz, S., *Department of Biochemistry University of Rochester, School of Medicine and Dentistry, Rochester, New York 14642.*

Kretsinger, Robert H., *Department of Biology, University of Virginia, Charlottesville, Virginia 22901.*

Lardy, Henry, *The Institute for Enzyme Research, The University of Wisconsin, 1702 University Avenue, Madison 53706, Wisconsin.*

Läuger, Peter, *Fachbereich Biologie, Universitat Konstanz, 775 Konstanz, Germany.*

Lawson, Durward, *Medical Research Council, Neuroimmunology Project, Zoology Department, University College London, London WC1E 6BT, England.*

Lesko, L., *Department of Biochemistry, University of Rochester, School of Medicine and Dentistry, Rochester, New York 14642.*

Luzzati, Vittorio, *Centre de Génétique Moléculaire, CNRS, 91190 Gif-Sur-Yvette, France.*

Machen, Terry E., *Department of Physiology and Anatomy, University of California, Berkeley, Berkeley, California 94720.*

Marinetti, Guido V., *Department of Biochemistry, University of Rochester, School of Medicine and Dentistry, Rochester, New York 14642.*

Martínez-Palomo, Adolfo, *Departamento de Biología Celular, Centro de Investigación y de Estudios Avanzados del Instituto Politécnico Nacional, Apartado Postal 14-740. México, D. F.*

Melchers, Fritz, *Basel Institute for Immunology, Grenzacherstrasse 487, CH-4058 Basel, Switzerland.*

Montal, Mauricio, *Departamento de Bioquímica. Centro de Investigación y de Estudios Avanzados del Instituto Politécnico Nacional, Apartado Postal 14-740. México, D. F.*

Mueller, Paul, *Eastern Pennsylvania, Psychiatric Institute, Henry Avenue and Abottsford Road, Philadelphia, Pennsylvania 19129.*

Nathenson, Stanley G., *Department of Microbiology and Immunology, and Cell Biology, Albert Einstein College of Medicine of Yeshiva University. Eastchester Road and Morris Park Avenue, Bronx, New York 10461.*

Peña, Antonio, *Departamento de Biología Experimental, Instituto de Biología, Universidad Nacional Autónoma de México. Ciudad Universitaria, México 20, D. F.*

Pethica, B. A., *Unilever Research, Port Sunlight Laboratory, Unilever Limited, Port Sunlight, Wirral, Cheshire, England L62 4XN.*

Pinto da Silva, Pedro, *The Salk Institute for Biological Studies, San Diego, California.*

Pohl, G. W., *Department of Biology, University of Konstanz, 775 Konstanz, Germany.*

Racker, Efraim, *Section of Biochemistry, Molecular and Cell Biology, Cornell University, Ithaca, New York 14850.*

Raff, Martin C., *Medical Research Council, Neuroimmunology Project, Zoology Department, University College London, London WC1E 6 BT.*

Renkonen, O., *Department of Biochemistry, University of Helsinki, Haartmansgatan 3, 00290 Helsinki 29, Finland.*

Singer, S. J., *Department of Biology, University of California, San Diego, La Jolla, California 92037.*

Simons, K., *Department of Serology and Bacteriology, University of Helsinki, Haaitmanikatu 3, 00290 Helsinki 29, Finland.*

Skou, Jens Christian, *Institute of Physiology, University of Aarhus, 8000 Aarhus C, Denmark.*

Steinemann, A., *Department of Biology, University of Konstanz, 775 Konstanz, Germany.*

Tiddy, G.J.T., *Unilever Research, Port Sunlight Laboratory, Unilever Limited, Port Sunlight, Wirral, Cheshire, England, L64 4XN.*

Tosteson, D. C., *Duke University Medical Center, Department of Physiology and Pharmacology, Durham, North Carolina 27710.*

Trissl, H.-W., *Department of Biology, University of Konstanz, 775 Konstanz, Germany.*

Vanderkooi, Jane, *Johnson Research Foundation, School of Medicine, University of Pennsylvania, Philadelphia, Pennsylvania 19174.*

Witt, H. T., *Max-Volmer-Institut für Physikalische Chemie und Molekularbiologie, Technische Universität Berlin.*

# PREFACE

The development of disciplines of research in Mexico and Latin America depends not only on the establishment of local groups actively pursuing a facet of science, but also on their accessibility to the new findings which are foremost in that science. In surveying those fields which are beginning to flourish in Mexico as well as in Latin America it was apparent that one of the most important is that of membrane research. It seems, in view of the fact that there is a significant scientific community devoted to membrane research, that the establishment of a periodic meeting in which these scientists can gather to exchange viewpoints and to delineate the perspectives of the field with leading exponents from other parts of the world would be most important.

The Mexican Society of Biochemistry therefore, established a recurrent Meeting on Perspectives in Membrane Biology to be held approximately every third year. The first Meeting was held at Oaxaca, Mexico on January 14 to 18, 1974. The proceedings of this Meeting are the subject matter of this volume.

The Editors would like to acknowledge that without the warmth and enthusiasm shown by all participants this venture would not have been initiated. It is our hope that future Meetings share the excellence of the scientific contributions which has characterized the present one.

The Editors are particularly grateful to Dr. Guillermo Soberón, Dr. José Laguna and Dr. Guillermo Massieu for their inspiring encouragement and support. The success of the Meeting was a product of their efforts in stimulating the development of our scientific community. The Editors also wish to express their appreciation to the Multinational Project of Biochemistry (National Scientific and Technological Program, O.A.S.), the National Council of Science and Technology of Mexico, the National Chamber of Chemical-Pharmaceutical Industry, Mexico, the International Union of Biochemistry, the International Union of Pure and Applied Biophysics and Laboratories Hoechst, Miles and La Roche of Mexico for their financial support which allowed the participation of our foreign colleagues and the Mexican faculty and graduate student group.

Appreciation and thanks are also due to Drs. Mauricio Montal and Armando Gómez-Puyou for assisting the Editors in organizing the International School which followed the Symposium. Also to Mrs. Josefina Quiroga and Miss Marina Peral for their kind help in the organization of the Meeting. A special note of gratitude goes to Mrs. Josefina Quiroga for her devotion and skill in producing the present volume.

# I

# STRUCTURE AND PLASTICITY OF BIOLOGICAL MEMBRANES

# SOME ASPECTS OF MEMBRANE STRUCTURE

Mark S. Bretscher

Medical Research Council,
Laboratory of Molecular Biology
Hills Road
Cambridge, CB2 2QH
England

Before setting out to describe what we believe is a working model for the structure of membranes, it might be useful to start with a few general remarks about (1) what is a membrane, and (2) what are we aiming to do in describing its "general structure".

## What is a membrane?

The conventional view is that a membrane is a permeability barrier which acts to compartmentalise a cell (in the case of the plasma membrane), or to separate one region within a cell from another region (in the cases of the endoplasmic reticulum, Golgi apparatus, secretory vesicles and so on). The main requirement for a membrane then is, in general, that it prevents small molecules and large ones from crossing it. The membrane is then made selectively permeable to specific molecules, as Nature requires, by specific proteins embedded in the membrane. These protein permeases can either function as passive "filters", or actively pump molecules across the membrane and against a concentration gradient.

The fundamental structure of a membrane is a lipid bilayer (1), (2): it provides the permeability barrier. We should note, however, that many molecules, especially organic ones, can pass freely across the lipid bilayer. This means that there is a great premium in Nature for enzymes to utilise and produce polar molecules which cannot spontaneously diffuse away from the cell. For example, it would not be helpful to a bacterium to produce and attempt to utilise acetic anhydride as its general acetylating agent. The acetic anhydride would pass straight out of the bacterium. Nature overcomes this problem by using either acetyl-phosphate or acetyl-CoA as acetylating agents. We can therefore conclude that the small molecule components of intermediary metabolism are hydrophilic and must not be capable of rapidly traversing a lipid bilayer.

There is another general property of membranes which I think it is helpful to bear in mind: it is concerned with what is "inside" a cell and what is not. The reason for caution is best illustrated by an example. A pancreatic exocrine cell contains many vesicles awaiting discharge from the cell. The process of discharge presumably occurs by fusion of the granule membrane with the plasma membrane of the cell; thereby its contents, which include the zymogens of several proteolytic enzymes, are released from the cell, or are externalised. The internal contents of these vesicles are therefore regions which are "potentially connected" with the outside of the cell. In general, one can divide a higher cell up into just two, separate, regions which are not potentially connected. These are the cytoplasmic region and the outside. The cytoplasmic region includes the internal contents of the nucleus, whereas the external region includes the internal contents of secretory vesicles, Golgi apparatus, lumen of the endoplasmic reticulum and the space between the nuclear membranes. If we define the cytoplasm as a region which contains ATP, then the zymogen granules, Golgi apparatus and lumen of the endoplasmic reticulum are external since there is (presumably) no ATP there. In general if we cross a membrane, we go from cytoplasm (ATP-rich) to external (no ATP) or vice-versa.

We therefore see that each membrane faces a cytoplasmic region on one side, and an external region on the other. However, caution is needed not to push the presence-of-ATP argument too far. For example, the secretory granules (chromaffin particles) of the adrenal medulla contain adrenalin complexed with ATP: the adrenalin and ATP are both external and the ATP must therefore be pumped there.

This generalisation may be useful for distinguishing different types of membrane. Certain bacteria have not only a plasma or inner membrane, but also an "outer membrane". This outer membrane is no membrane at all, in the sense that it has large pores in it which enable salts or sugars and even small proteins to pass through it. Nevertheless, the outer membrane of E. coli does contain large amounts of phospholipid.

There seems to be one exception to the above generalisation that, on crossing a membrane, one goes from high ATP to no-ATP, or vice-versa. This exception arises with mitochondria and probably chloroplasts which have ATP inside them, bounded by a plasma membrane and an outer membrane, but situated in the cellular cytoplasm. Assuming that their outer membranes are wholly porous, we then find that there exists high-ATP on either side of the mitochondrial or chloroplastic membrane. This is not hard to understand if we regard each of these organelles as degenerate bacteria living in a rich medium (namely, the cell's cytoplasm).

## What do we expect to learn about the "General Structure" of a membrane?

It is well known that membranes are not rigid structures, but that both lipid and protein molecules are free, in some cases, to rotate about an axis perpendicular to the membrane (3-4), or even to move in the plane of the membrane (5-8). In other words, such a membrane cannot be given a complete structure - it can be described as fluid (9). But not all proteins in membranes have this property - the best example is the bacterial rhodopsin which exists as a crystalline lattice lodged in the purple bacterium's membrane (10). The aim, then, is to provide a set of rules which govern the manner in which proteins, carbohydrates and lipid molecules are found in membranes. It is, perhaps, analogous to attempting to define what type of arrangement different segments of a polypeptide chain can adopt in a protein ($\alpha$-helix, $\beta$-pleated sheet, etc.), except on a much coarser scale. In addition, one would hope that such a general picture, once it had emerged, would lead to a better understanding of how membrane components function and how membranes are assembled.

It is evident that any single membrane studied could be labelled as atypical. I have therefore worked on the plasma membrane of the erythrocyte, which is the most complex of the simple membranes yet studied. Naturally, the membrane of any cell is likely to differ from the membrane of any other cell - the cytoplasms of two different cell types are also likely to be different. I have therefore not studied the biochemical activities of the different enzymes in the red cell ghost, but have rather tried to provide a crude spatial description of the relationship of the proteins present to the bilayer. I assume that the phospholipid molecules are arranged in a bimolecular leaflet, as originally proposed by Gorter and Grendel in 1925 (1), (11).

## Proteins

When the proteins present in the erythrocyte ghost are dissolved in sodium dodecyl sulphate (SDS) and fractionated according to their molecular weights on SDS-gels, one finds about a dozen main protein bands (12-17). I have tried to find out which of these proteins is exposed on the external surface of the erythrocyte by labelling intact cells with a membrane-impermeable reagent which is highly radioactive - to tag the proteins on the outer surface (15). The reagent I used for this purpose was $^{35}S$ formyl-methionyl sulphone methyl phosphate (FMMP) - a powerful acylating reagent. Alongside such an experiment one could look at which proteins are exposed on either surface by labelling fragmented erythrocyte membranes; in this case proteins on both surfaces of the membrane should be labelled. When applied to the human erythrocyte, we find the following: there are only two main components which become labelled from outside the cell (Figure 1, gel a). There must, of course, be lots of other proteins on the outside, but they are presumably either present in very much smaller amounts or else are quite unreactive. So, on the outside, this evidence indicates just two main components: one is a protein with a molecular weight of some 105,000 daltons which I call component _a_. The other is the major glycoprotein so extensively studied by the late Dr. R.J. Winzler and his colleagues (18, 19) and, more recently, by Segrest and his colleagues (21). On these SDS-gels it migrates just ahead of component _a_ (22).

By contrast, when ghosts are labelled, every protein visible in an SDS-polyacrylamide gel becomes labelled - all proteins are exposed on at least one surface (Figure 1, gel c). No protein seems to be hidden away inside the bilayer (not that a bilayer would be big enough to accommodate it anyway). This experiment shows that much more protein is associated with the inner surface of the erythrocyte membrane than with the external surface - a result which I think is important in understanding membranes, and to which I shall return later.

Perhaps the most interesting question to ask at this stage is whether either of the two proteins labelled from the external surface extends across the membrane. That is, is more of either labelled from both sides of the membrane (in ghosts) than is labelled from the outer surface (in the intact cell). I have tried to examine this possibility in the cases of both component _a_ and the glycoprotein. Because of the relative simplicity of the glycoprotein experiments, I should like to restrict myself here to that molecule.

The molecular orientation of the glycoprotein molecule in the membrane has, in part, been elucidated by Winzler (18-20). He showed that all the carbohydrate is attached at around ten different sites along the backbone of the polypeptide chain; the polypeptide is small, being around 90 residues in length. All the carbohydrate moieties are found external to the bilayer, attached to the N-terminal segment of the polypeptide. The N-terminal region can be cleaved by treating erythrocytes with proteolytic enzymes. Winzler also found that this small protein, when isolated in pure form and digested with proteolytic enzymes, produced a very insoluble core - presumably that part of the molecule associated with the hydrophobic interior of the membrane. But where is the rest of the glycoprotein - is it all in the membrane, or does it traverse it?

As already shown (Figure 1, gel a), the glycoprotein is labelled when the membrane-impermeable reagent, $^{35}$S-FMMP, is added to intact erythrocytes. A fingerprint of the labelled peptides obtained by digestion of this molecule with trypsin and chymotrypsin contains no discrete, radioactive peptides. Apart from a large smear of radioactive material around the origin (similar to that seen in Figure 2a), the fingerprint is a blank.

When the region of an SDS-gel which contains the glycoprotein, labelled in cell ghosts, is eluted and fingerprinted, many labelled peptides are found (Figure 2a). Most of these radioactive peptides have nothing to do with the glycoprotein since the glycoprotein comigrates with another major polypeptide (which is located on the inner surface of the membrane) which contributes most of these labelled peptides. Those peptides derived from the glycoprotein can be deduced by fingerprinting the material from the same region of a gel upon which has been fractionated labelled ghosts from which the glycoprotein has been removed. This latter condition is easily achieved by prior treatment of intac cells with pronase, followed by cell lysis and labelling of the ghosts with FMMP. This shows that, of all the strongly labelled peptides seen in Figure 2a, only two are derived from the glycoprotein molecule, peptides G1 and G2. That both these two peptides do arise from the glycoprotein is shown by purification of the glycoprotein from labelled ghosts, and fingerprinting it. The resulting map, containing just two discrete labelled peptides, G1 and G2, is shown in Figure 3.

We can thus deduce that peptides G1 and G2 have the following properties:

(1) They are not labelled when intact cells are labelled, but only when ghosts are labelled. This shows that they come from the cytoplasmic surface of the erythrocyte membrane.

(2) Proteolysis of the external surface of the erythrocyte, which causes cleavage of the glycoprotein, leads to loss of the protein carrying the two peptides, G1 and G2, from its normal position on an SDS-polyacrylamide gel. This shows that the peptides G1 and G2 are derived from a molecule which is partly exposed on the outer surface of the cell.

(3) These two peptides, G1 and G2, are a part of the glycoprotein molecule.

These data, therefore, reveal that the glycoprotein molecule extends across the erythrocyte membrane (22). The carbohydrate is located near the N-terminus, as shown by Winzler, and this is on the external surface of the cell. Because of the small size of the protein (some 90 amino acids long) (19), it is difficult to arrange it in any way other than placing its C-terminal end at the cytoplasmic surface. This would require a hydrophobic segment of some 25 residues, to extend across the bilayer, for which good evidence now exists (20, 21). Peptides G1 and G2 therefore must arise from the C-terminal end of the glycoprotein, which is in the red cell cytoplasm.

In experiments which are similar in principle, I have also shown that the other major erythrocyte protein, component a, also extends across the bilayer (23). All the remaining proteins of the erythrocyte membrane which we can see on an SDS-gel are presumably associated with its inner surface.

## Lipids

As I mentioned earlier, the phospholipids are arranged in a bilayer; this is the basic structure of membranes. But is there any greater order superimposed on this liquid-crystalline state ? There are several lines of evidence which indicate that in the erythrocyte there is a greater degree of organisation. There are four common kinds of phospholipids which occur in erythrocytes from most species (24): these are shown in Figure 4. Two of them, phosphatidyl-choline and sphingomyelin, have phosphoryl-choline as their head groups; two, phosphatidyl-ethanolamine and phosphatidyl-serine, have amino-alkyl phosphoryl head groups. The possibility of an asymmetry in the phospholipid composition of the two halves of a bilayer arose when I found that very little phospholipid could be acylated by FMMP in intact cells (25-27); much more was reactive in cell ghosts. Since only phosphatidyl-serine and phosphatidyl-ethanolamine have free amino groups which could, in principle, react with FMMP, it suggested that both these phospholipids are located primarily at the cytoplasmic side of the bilayer. (Figure 1 also shows that labelled phospholipid, which migrates at the SDS front, is labelled more heavily in ghosts gels c, d than in intact cells gels a, b). This would leave both sphingomyelin and phosphatidyl-choline - somewhat similar lipids as far as their headgroups are concerned - at the external surface. There is much other chemical data which support this conclusion (28-30), as well as enzymatic evidence based on the susceptibility of erythrocytes from different animals to lysis in the presence of phospholipases (31-34). The most striking enzymatic evidence is that of Turner (31,32), who found that, unlike most other erythrocytes, sheep erythrocytes have no phosphatidylcholine but a much larger-than-usual proportion of sphingomyelin. Again, unlike other erythrocytes, he found that these sheep cells are resistant to the action of crude cobra-venom phospholipase A. It is tempting to put these two observations in parallel, as did Turner, and suggest that the replacement by sphingomyelin of phosphatidylcholine is responsible for the difference in susceptibility to lysis by the venom. Obviously, the chemical basis for this would be that, unlike phosphatidyl choline, sphingomyelin has no 2-acyl linkage to a glycerol moiety and so cannot be hydrolysed. Furthermore, it suggests (26) that the sphingomyelin in sheep erythrocytes, (and therefore also the replaced phosphatidylcholine in other erythrocytes), makes up a large portion of the accessible lipid surface on the exterior of these cells.

The existence of an asymmetric lipid bilayer implies that phospholipid molecules are not able to flip from one side of the bilayer to the other very readily. The actual rate of migration -or flip-flop - of a spin-labelled phospholipid in a synthetic phospholipid vesicle has been determined by Kornberg (35). It was found that the absolute rate is very slow -a half-life of many hours at 30°C. Two points should be emphasised. First, the presence of either cholesterol or proteins which extend across the bilayer would be expected to slow the rate of flip-flop. In other words, the actual rate of flip-flop in an erythrocyte membrane, where both these restraints are acting, may be very much slower still than is suggested above. Second, it is possible that natural membranes have enzymes which catalyse the trans-membrane motion of phospholipids -a "flippase". The reason for this emerges if we consider a hypothetical scheme of how asymmetrical bilayers may be synthesised -and here, of course, I am assuming that membranes of all higher cells have an asymmetry of phospholipid distribution similar to that found in the erythrocyte. Since it is likely that all phospholipid synthesis occurs at the cytoplasmic surface of a membrane, there must be a flippase to catalyse the transfer of some phospholipids across the membrane to form the outer half of the bilayer (see Figure 5). If this flippase had a specificity for phosphorylcholine head groups, the suggested asymmetric bilayer would automatically follow: choline phospholipids outside -amino phospholipids inside. This means that great caution must be exercised in interpreting flip-flop measurements in real membranes. We can expect some of the rates to be much faster than in synthetic bilayers and therefore presumably enzymatically catalysis Recent experiments in which the rate of flip-flop has been estimated in different membranes indicate that this rate appears to be faster than in synthetic bilayers (36). The above arguments predict that an enzyme could be isolated from these membranes which would catalyse the specific flip-flop of choline-phospholipids.

## General Conclusions

Our picture of the erythrocyte membrane is based, then, on a lipid bilayer having compositional asymmetry in its phospholipid (and possibly also glycolipid) components. The glycoprotein and another major protein (component a) are located in a fixed orientation across the membrane; many more proteins are associated with the inner surface of the bilayer (Figure 6). We do not know if, and to what extent, some of these latter proteins replace, or indeed penetrate, the inner half of the bilayer, although the observed difficulty in removing most of them by washing the ghosts with strong salt solutions (17) implies that they are probably inserted into the bilayer. If this is so, there are two points to note.

First, the erythrocyte membrane also has compositional asymmetry within its hydrophobic phase.

Phosphatidylcholine, sphingomyelin and glycolipid each contain very few polyunsaturated fatty acid residues: by contrast, phosphatidylethanolamine and phosphatidylserine are rich in these components, particularly arachidonic acid (see reference 24 ). These polyunsaturated residues in the inner half of the bilayer may provide a less ordered phase which may in turn be a better solvent for accommodating proteins than the outer half of the bilayer.

Second, the composition of erythrocyte lipids shows that there is usually more external (choline) phospholipid plus glycolipid than internal (amino) phospholipid. The volume deficit in the inner half of the bilayer caused by this disparity may well be filled in by protein. That is, a substantial portion (maybe as high as 50%) of the cytoplasmic half of the bilayer may be protein. Alternatively,the suggested asymmetry of phospholipids may be incomplete: some choline phospholipids may be located within the cytoplasmic half of the bilayer. The observation that there seems to be much more protein associated with the cytoplasmic side of a bilayer raises the question of what all this protein is doing. It might, in part, contain all the enzymes for lipid synthesis, whose substrates are likely to partition into the bilayer. An interesting possibility has been suggested by Adam and Delbrück (37). They draw attention to the fact that diffusion along two dimensions is faster than in three dimensions and that therefore there may be advantages in having an enzyme whose substrate is somewhat lipophilic, remain attached to a membrane. The substrate, once it had loosely associated with the bilayer, could move around in two dimensions until it had found its enzyme. The cytoplasmic side of a bilayer could therefore be used extensively for this purpose (about a quarter of the total protein of *E. coli* is in their membranes) and may explain why so much protein is found on the cytoplasmic side of most membranes.

The compositional asymmetry of a bilayer means that membranes always have a polarity, even they are synthesised as cytoplasmic products which then diffuse into the membrane and become lodged there. At a simple minded level, consider a protein which only partially penetrates the cytoplasmic surface of a bilayer. Were it synthesised by a membrane-bound ribosome, and, let us suppose, "fed" into the membrane as it is assembled, how would it find its way to the cytoplasmic - rather than external - side of the bilayer? The presence of an asymmetric lipid bilayer might help, but bearing in mind that phospholipids do not flip-flop and that all the available evidence suggests the same is true for proteins, this seems unlikely. Rather, it is more plausible to imagine that such a protein arrived there by solution in the membrane from the cytoplasmic side. Then, consider a protein which extends across the bilayer. Two problems arise. First, how does it find its correct orientation in the bilayer if it is synthesised by membrane-bound ribosomes? This is a similar question to that above for a partially penetrating protein. Second, at a different level, I find it difficult to see how it could *fold up* properly in an environment which at once is very hydrophobic and is bounded by very hydrophilic (aqueous) regions. The main forces which are usually ascribed to hold proteins together are hydrophobic bonds and ionic interactions. If a protein were in the process of folding-up *in* a membrane, I believe the forces which would determine its folding would not be the usual peptide side-chain interactions, but that the whole process would be dominated by the very discontinuous environment provided by the lipid bilayer. If one bears in mind the fact that a transmembrane protein, like component *a*, is probably larger in each dimension (say 50 x 50 x 100 Å) than a lipid bilayer (say 40 Å), one can see the implausibility of its folding up inside the membrane. I therefore believe that, for membrane proteins, synthesis is by cytoplasmic ribosomes, followed by solution in the membrane. Whether a cytoplasmically synthesised protein remains as a soluble component, or whether it partially dis-

solves in the inner surface of the bilayer, or
whether it dissolves in the membrane so that it
traverses the bilayer, is determined by the nature of the protein. For a transmembrane protein, the nature of the protein will ensure its
correct polarity of insertion into the membrane;
once inserted, it seems likely that glycoscyla-
tion from the external side of the membrane will
ensure a permanent abode for that protein across
the membrane. In other words, the carbohydrate
may act as an irreversible lock on the protein to
hold it fast in the membrane, thereby preventing
its diffusion back into the cytoplasm.

There are two pieces of evidence which argue
strongly in favour of this hypothesis.

(1) The *lactose* operon of *E. coli* has three
genes which code for β-galactosidase (*z*), lactose
permease (*y*), and thiogalactoside transacetylase
(*a*). These three genes are transcribed and trans
lated as a single messenger RNA (39). It is then
very likely that each of these genes is translated by the same class of ribosome (especially
in view of the polar nature of some *z*-gene mutants on the *y*-gene). Since both the first and
last genes on the messenger RNA (β-galactosidase
and the transacetylase) are cytoplasmic proteins,
the middle gene, coding for lactose permease
(which probably spans the bacterial plasma membrane), must also be synthesised as a cytoplasmic protein.

(2) Sindbis Virus contains three main proteins, E1, E2 and a nucleocapsid protein (see,
for example ref. 40). The latter protein is associated with the RNA genome inside a membrane
coat. E1 and E2 are glycoproteins which largely
reside outside the virus surface as spikes. They
are membrane proteins and, since part of each resides on the external side of the bilayer, each
must traverse the membrane. On the other hand,
the nucleocapsid protein is clearly a cytoplasmic
protein: it binds to the RNA (itself synthesised
in the cytoplasm), and the resulting nucleoprotein particle can be seen to bud from the cyto-

plasm out of the cell becoming coated with a layer of plasmic membrane as it does so. The point of interest here is that a precursor protein has just been identified (41) which contains the polypeptide parts of E1, E2 and the nucleocapsid protein joined covalently as one polypeptide chain. Since there can be no doubt that the nucleocapsid protein is a cytoplasmic product this proves that E1 and E2 - membrane proteins - were also originally cytoplasmic proteins. This general conclusion, that a protein which traverses the membrane has arrived in this position by diffusion from the cytoplasm, could explain why the major proteins exposed on the outer surface of the erythrocyte seem so chemically unreactive: both the glycoprotein and component a appear to have very few amino groups on their external surfaces. We also know that that part of the glycoprotein which is on the cell exterior is built from rather neutral amino acids (almost one half is accounted for by serine plus threonine (19)). This N-terminal segment of the protein (without its carbohydrate) would thus be expected to diffuse spontaneously across the membrane and then become fixed there by the addition of carbohydrate from the external side of the membrane.

## Summary

The arrangement of lipids and the major proteins of the erythrocyte membrane has been presented. The conclusions from this I list here as a set of guide-lines (42) for the general structure of membranes of higher organisms: some of these rules may be wrong. But it seems useful to me that a scheme which is close to the molecular level should be precisely stated so that attention can be focussed on specific points.

1. The basic structure of a membrane is a lipid bilayer with (a) choline phospholipids and glycolipids in the external half and (b) amino (and possibly some choline) phospholipids in the cytoplasmic half. There is effectively no lipid exchange across the bilayer (unless enzymatically catalysed).

in the absence of any associated proteins. Cellular membranes can usually be given a gross polarity: one side is adjacent to the cytoplasm (high ATP) the other side is remote from it (no ATP). The usual model for membrane fusion requires that this topological polarity be maintained. Thus, when secretory vesicles or membrane viruses fuse with the plasma membrane, or when two cells fuse, the cytoplasmic side of each membrane always remains on the cytoplasmic side of the fused membrane, whereas the external side is always kept external. The asymmetry described here simply provides a molecular basis for defining each side of the bilayer.

Insertion into the bilayer of proteins which extend across the membrane provides another -more easily measured- polarity. How do these proteins come to be in the membrane? There seem to be two general classes of protein synthesised by a cell: cytoplasmic and secreted proteins. Cytoplasmic ribosomes synthesise cytoplasmic proteins such as haemoglobin, RNA polymerase, or β-galactosidase. Membrane-bound ribosomes synthesise proteins for secretion and eventual export from the cell (38). The decision of whether a protein is to be secreted or not - that is, on which class of ribosome it will be synthesised - is presumably genetically defined. A messenger RNA must therefore have information coded in it which determines the class of ribosome to which it attaches. Selection at this level determines the fate of the protein product. If the messenger RNA attaches to membrane-bound ribosomes, it seems probable that the secreted protein is extruded through the membrane as it is synthesised. This general scheme is consistent with recent studies of the specificity of protein synthesis by membrane-bound ribosomes.

It is widely assumed that membrane proteins are synthesised by membrane-bound ribosomes -that they are a special class of secreted protein. This view, I believe, should be discarded. There are several reasons why I believe that, rather,

2. Some proteins extend across the bilayer. Where this is so, they will in general have carbohydrate on that surface remote from the cytoplasm. This carbohydrate may prevent the protein diffusing out of the membrane back into the cytoplasm: it acts as a lock on the protein.

3. Just as lipids do not flip-flop, proteins do not rotate across the membrane. Lateral motion or rotation of lipids and proteins in the plane of the bilayer may be expected.

4. Most membrane protein is associated with the inner, cytoplasmic, surface of the membrane. Proteins are not usually associated exclusively with the outer half of the lipid bilayer.

5. Membrane proteins are a special class of cytoplasmic proteins, not of secreted proteins.

## References

1. Gorter, E. and Grendel, F. J. Exp. Med. 41 (1957) 439.
2. Danielli, J. F. and Davson, H. J. Cell Physiol. 5 (1936) 495.
3. Brown, P.K. Nature New Biology 236 (1972) 35.
4. Cone, R.A. Nature New Biology 236 (1972) 39.
5. Kornberg, R.D. and McConnell, H.M. Proc. Nat. Acad. Sci. U.S.A. 68 (1971) 2564.
6. Devaux, P. and McConnell, H.M. J.Am.Chem.Soc. 94 (1972) 4475.
7. Frye, L.D. and Edidin, M. J. Cell Sci. 7 (1970) 319.
8. Pinto da Silva, P. J. Cell Biol. 53 (1972) 777.
9. Singer, S.J. and Nicolson, G.L. Science 175 (1972) 720.
10. Oesterhelt, D. and Stoeckenius, W. Nature New Biology 233 (1971) 151.
11. Danielli, J.F. and Davson, H. J. Cell Physiol. 5 (1936) 495.

12. Berg, H.C. Biochim. Biophys. Acta 183 (1969) 65.

13. Lenard, J. Biochemistry 9 (1970) 5037.

14. Phillips, D.R. and Morrison, M. Biochem. Biophys. Res. Comm. 40 (1970) 284.

15. Bretscher, M.S. J. Mol. Biol. 58 (1971a) 775.

16. Bender, W.W., Garan, H. and Berg, H.C. J.Mol. Biol. 58 (1971) 783.

17. Fairbanks, G., Steck, T.L. and Wallach, D.F. H. Biochemistry 10 (1971) 2606.

18. Kathan, R.H., Winzler, R.J. and Johnson, C.A. J. Exp. Med. 113 (1961) 37.

19. Kathan, R.H. and Winzler, R.J. J. Biol. Chem. 238 (1963) 21.

20. Winzler, R.J. 1969. In: Red Cell Membrane, p. 157, eds. Jamieson, G.A. and T.J. Greenwalt. J. B. Lippimistt Co.

21. Segrest, J.P., Jackson, R.L., Marchesi, V.T., Guyer, R.B. and Terry, W. Biochem. Biophys. Res. Comm. 49 (1972) 964.

22. Bretscher, M.S. Nature New Biology 231 (1971c) 229.

23. Bretscher, M.S. J. Mol. Biol. 59 (1971b) 351.

24. Rouser, G., Nelson, G.J., Fleischer, S. and Simon, G. In: Biological Membranes, p. 5, ed. Chapman, D. Academic Press: New York.

25. Bretscher, M.S. Biochem. J. 122 (1971d) 40.

26. Bretscher, M.S. Nature New Biology 236 (1972a) 11.

27. Bretscher, M.S. J. Mol. Biol. 71 (1972b) 523

28. Maddy, A.H. Biochim. Biophys. Acta 88 (1964) 390.

29. Knauf, P.A. and Rothstein, A. J. Gen. Physiol. 58 (1971) 190.

30. Bangham, A.D., Pettica, B.A. and Seaman, G.V. F. Biochem. J. 69 (1958) 12.

31. Turner, J. C. J. Exp. Med. 105 (1957) 189.

32. Turner, J.C., Anderson, H.M. and Gandal, C.P. Biochim. Biophys. Acta 30 (1958) 130.

33. Ibrahim. S.A. and Thompson, R.H.S. Biochim. Biophys. Acta 99 (1965) 331.

34. Verkleij, A.J., Zwaal, R.F.A., Roelofsen, B. Comfurius, P., Kastelijn, D. and Van Deenen, L. L. M. Biochim. Biophys. Acta 323 (1973)178.

35. Kornberg, R.D. and McConnell, H.M. Biochemistry 10 (1971) 1111.

36. McNamee, M.G. and McConnell, H.M. Biochemistry 12 (1973) 2951.

37. Adam. G. and Delbrück, M. (1968) In: Structural Chemistry and Molecular Biology, p. 198, eds. Rich. A. and Davidson, N. W.H.Freeman & Co.

38. Siekevitz, P. and Palade, G.E. J. Biophys. Biochem. Cytol. 7 (1960) 619.

39. Kepes, A. Prog. in Biophys. and Mol. Biol. 19 (1969) 201.

40. Sefton, B. and Burge, B. J. Virology 11 (1973) 730.

41. Schlesinger, M.J. and Schlesinger, S. J. Virology 11(1973) 1013.

42. Bretscher, M.S. Science 181 (1973) 622

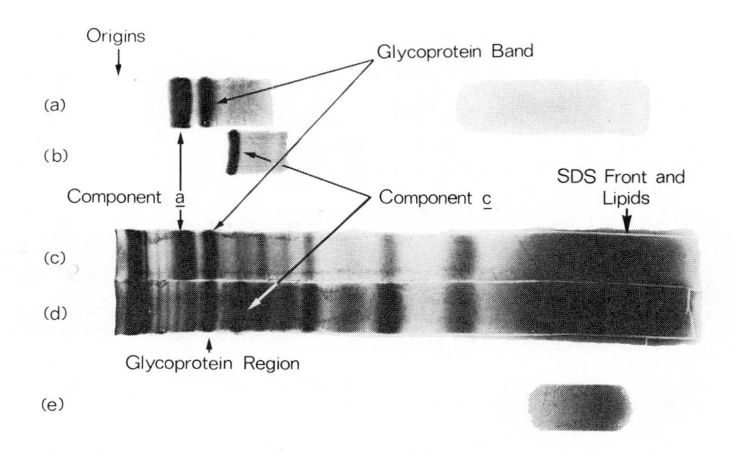


Figure 1. Autoradiogram of SDS-polyacrylamide gels of erythrocyte membranes (gels a-d) or synthetic phosphatidylethanolamine (e) labelled with FMMP. The erythrocytes were labbeled as follows: (a) Labelled intact cells. (b) Cells treated with pronase and then labelled. (c) Labelled ghosts. (d) Labelled ghosts derived from cells which had been digested with pronase.

Cells (A, $Rh^+$) were washed in 0.15 M NaCl. Pronase digestion of intact cells was in 0.15 M NaCl, 0.1 M $NH_4$ $HCO_3$ and 0.5 mg/ml pronase at 37°C for 1 hour. Cells or ghosts (0.2ml) were labelled in 0.05 M $NaHCO_3$, 0.05 M $Na_2CO_3$, 0.05 M NaCl buffer with about $10^8$ cpm FMMP at 0°C for 10 minutes. They were then washed in 0.15 M NaCl. Cells were lysed by dilution into 0.02 M tris-Cl pH 7.4 containing 3 x $10^{-5}$M $CaCl_2$. Ghosts were dissolved in SDS and fractionated on 7.5% acrylamide gels in phosphate buffer in standard conditions. After running, they were sliced and a centre section autoradiographed. The gels were not washed or stained prior to autoradiography.

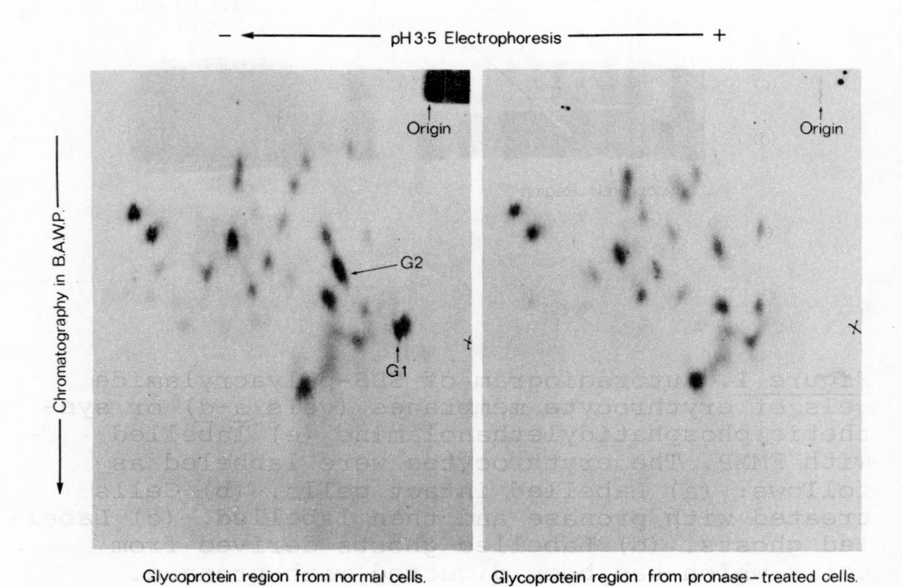


Figure 2. Crude Glycoprotein Fingerprint: Autoradiogram of FMMP labelled peptides. The glycoprotein region of gels similar to those in Fig. 1 (gels c and d) were cut out of a gel, the labelled material eluted, freed of SDS and digested with trypsin and chymotrypsin. The digest was fractionated by paper electrophoresis at pH 3.5 and by chromatography in butanol:acetic acid:water: pyridine = 30:6:24:20. Autoradiogram of material from the band off one gel took about two weeks.

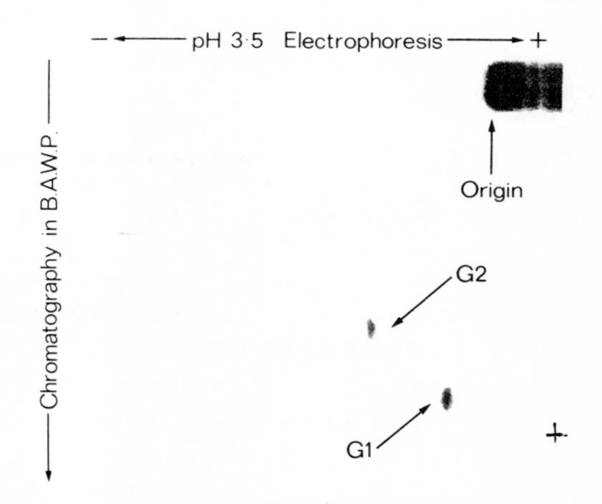


Figure 3. Pure Glycoprotein Fingerprint: Autoradiogram of FMMP-labelled peptides. The labelled glycoprotein, isolated from a 5% SDS-acrylamide gel, was rerun on a 7.5% gel, taking advantage of the anamalous mobility of glycoproteins on SDS gels (22). The pure glycoprotein was fingerprinted as described in Figure 2.

---

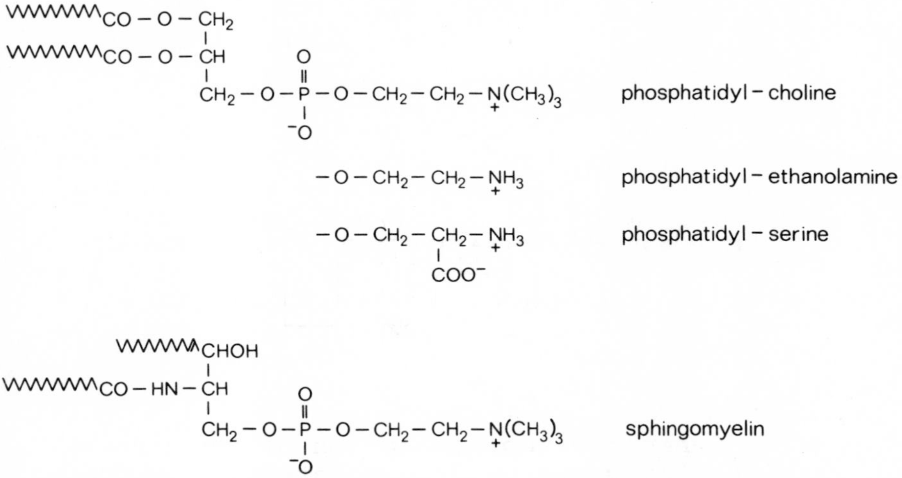

~~~~~~CO − O − $CH_2$
~~~~~~CO − O − CH
$CH_2$ − O − P(=O)($O^-$) − O − $CH_2$ − $CH_2$ − $\overset{+}{N}(CH_3)_3$ phosphatidyl − choline

− O − $CH_2$ − $CH_2$ − $\overset{+}{N}H_3$ phosphatidyl − ethanolamine

− O − $CH_2$ − CH($COO^-$) − $\overset{+}{N}H_3$ phosphatidyl − serine

~~~~~~CHOH
~~~~~~CO − HN − CH
$CH_2$ − O − P(=O)($O^-$) − O − $CH_2$ − $CH_2$ − $\overset{+}{N}(CH_3)_3$ sphingomyelin

Figure 4. The Main Phospholipids found in Higher Organisms, (not drawn to scale). In bacteria, the main phospholipids are usually phosphatidylethanolamine and phosphatidylglycerol.

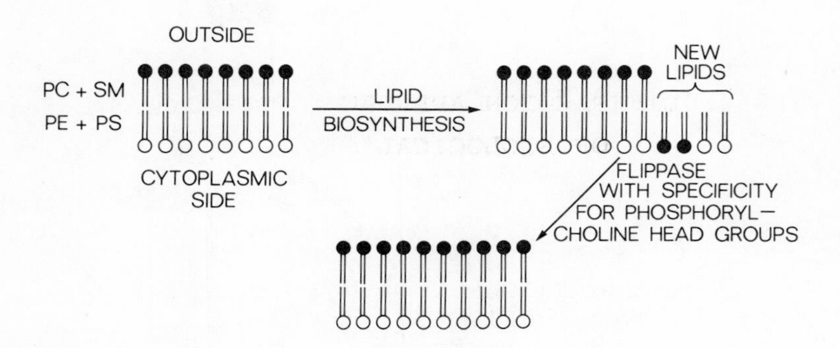


Figure 5. Schematic representation of a hypothetical mechanism for the biosynthesis of an asymmetric bilayer. A flippase embedded in the membrane acts to equalise the pressure in each half of the bilayer by catalysing the transfer of phosphorylacholine phospholipids from one side to the other.

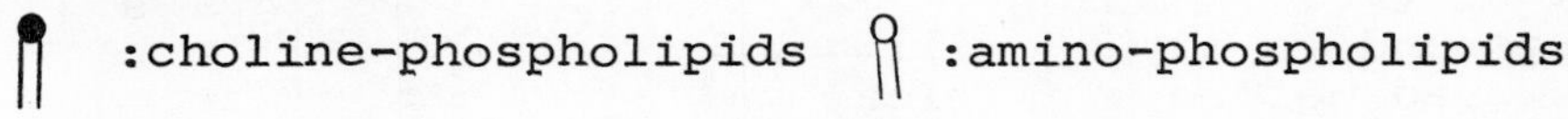
:choline-phospholipids :amino-phospholipids

---

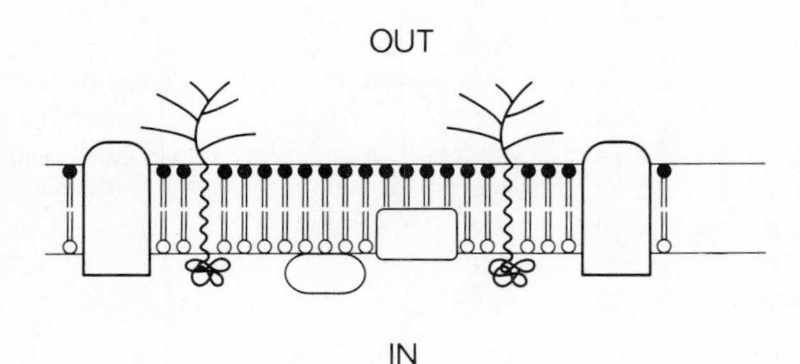


Figure 6. Schematic representation of the manner in which proteins may be accommodated in an asymmetric lipid bilayer in the erythrocyte membrane. (Not drawn to any particular scale).

# X-RAY DIFFRACTION APPROACH TO THE STRUCTURE OF BIOLOGICAL MEMBRANES

V. Luzzati

Centre de Génétique Moléculaire,
CNRS, 91190 Gif-Sur-Yvette,
France.

## Introduction.

In membranes, as in other biological systems, an understanding of the physiological mechanisms at the molecular level is heavily dependent upon structural information; questions like "how do ions cross membranes" or "how do retinal discs respond to light" will not be satisfactorily answered before a chemical and physical structure is associated with the operational notions of "pores", "channels" or "carriers".

Intact membranes as well as model systems derived from membranes - lipid-water and protein-lipid-water phases - have been the object of active X-ray diffraction studies over the last few years. I intend to survey here the results of these studies, to compare the structural information obtained with membranes and model systems, and to discuss future perspectives of the X-ray diffraction approach to the structure and function of membranes.

## Intact Membranes.

Several years of chemical and physical studies of membranes have allowed us to establish a few structural concepts (see Bretscher and Singer, this volume):

- membranes are thin two-dimensional objects, composed mainly of proteins and lipids,

- the thickness appears to be fairly constant, and is probably defined by the length of the lipid molecules,

- the hydrocarbon chains are concentrated near the centre of the membranes.

- membranes are chemically asymmetric,

- some of the proteins sit on the surface, others are embedded in the lipid matrix, others probably span the membrane,

- under physiological conditions the hydrocarbon chains of the lipid molecules are predominantly, if not entirely, in a liquid-like conformation,

- the lipid and protein molecules appear to move rather freely in the plane of the membrane,

- the movements of the lipid and protein molecules from one side to the other appear to be extremely slow.

Thus, to a first approximation, the organization of membranes may be visualized as that of two-dimensional liquid; this situation is far from ideal for the X-ray diffraction techniques, which lose most of their strength in the absence of periodic order. In fact membranes can often be stacked, and preparations can be obtained which display a one-dimensional periodic repeat; in this case the X-ray diffraction study can yield information on the average electron density distribution in the direction normal to the plane of the membrane. Several membranes have been studied under these conditions, some naturally (myelin, retinal rods) others artificially stacked. We can discuss the results of these studies.

Some authors have greatly emphasized the "bilayer structure" of membranes (1,2). From the operational standpoint the diagnosis of the "bilayer structure" is based upon the presence in the electron density profiles of a low density through sorrounded by high density peaks (see figs. 1, 2, 4, 5). This observations does indeed indicate that the hydrocarbon chains are located preferentially near the centre of the membranes, but by no means excludes the presence of other components, proteins for example, in the interior of the membrane (see below the example of retinal rods). In principle the important question of the penetration of proteins in the hydrocarbon layer could be clarified by determining the electron density profiles on an absolute scale: yet this operation is difficult to perform with X-rays (it is within easier reach of neutrons). Another structural problem, solved elegantly by Bretscher (see this volume) is that of chemical asymmetry. Caspar and Kirschner (3) have tackled this problem by the X-ray diffraction study of nerve myelin, and have put forward electron density profiles in excellent agreement with an asymmetric distribution of cholesterol (see fig. 1); nevertheless it has not been possible to prove that their choice of phases is unique, and thus to disregard another set of phases advocated by Worthington and King (4), which leads to a more symmetric profile (see fig. 2). This problem also is awaiting solution from neutron scattering experiments. Aside from these limitations, the diffraction study of stacked membranes is subject to a more general objection, namely that from the biological standpoint the organization of the proteins <u>in</u> the plane of the membrane is probably more relevant than the average distribution of the scattering elements <u>across</u> the membrane. In this regard it is rather disappointing to note that in some of the rare cases of membranes in which the exceptionally high concentration of one protein species appeared to justify hopes for an ordered distribution of the protein molecules, the structure has turned out to be disordered.The most striking example is that of retinal rods, in which rhodopsin has been shown by Richard Cone to be highly mobile and to be organized more like a liquid that like a solid, in spite of its extremely high concentration. Although future perspectives do not always look so dim - I can mention purple membranes (5) and gap junctions (6) - it appears unlikely that the X-ray diffraction studies of intact membranes will ever reach the degree of sophistication required for a sound molecular approach to the physiological events.

Lipid-water Systems.

Extensive studies of the lipid-water phases carried out over the last few years have revealed a remarkable polymorphism; no other class of compounds besides lipids is known to display such a number and variety of phases over such a narrow range of parameters (temperature, water content, chemical composition) and so near to physiological conditions (7,8). The possible biological relevance of these polymorphic transitions has been often discussed in the past (9, 10): it has been suggested that in membranes the structure of the lipid moiety is likely to undergo local transitions and fluctuations, some of which may well be involved in the physiological events. I regard the fact of neglecting even the possibility of such structural inhomogeneities as the most severe limitation of the "bilayer" concept (see above).

Another phenomenon discovered in the X-ray diffraction study of lipid-water phases, since thoroughly analyzed in membranes as well as in model systems, is the temperature-induced conformational transition of the hydrocarbon chains (see in refs. 11 and 12 a review of the extensive literature on this subject). The interest for this phenomenon has been greatly stimulated by the observation of similar transitions in intacts membranes and of stricking correlations between these transitions and some functional properties. The conformational transition has been studied by a variety of techniques, but X-ray diffraction appears to provide the most direct mean for a quantitative evaluation of the fraction of the chains in the ordered and in the disordered conformations; moreover recent technical developments have allowed the undertaking of a kinetic study of the transition by X-ray diffraction techniques (13). Although the biological significance of the conformational transition is still an open question, it has been suggested that some of the hydrocarbon chains might undergo conformational transitions under physiological conditions, and that these transitions could induce a segregation of the lipid molecules, alter the lipid composition around the different proteins and thereby exert a regulation over different functions of membranes (11, 12).

## Protein - Lipid - Water Systems.

*Structure and Polymorphism.* The X-ray diffraction studies of protein-lipid-water systems are far less advanced than those of lipid-water; nevertheless several systems have been explored, a variety of phases have been identified, and the structure of some of these has been determined. The number and variety of structures are quite large, and the polymorphism is at least as extensive as in lipid-water systems. The influence of various parameters-nature of the proteins and of the polar groups of the lipids, degree of saturation of the hydrocarbon chains, presence of cholesterol, temperature, ionic strength, redox potential and pH - on the structure of the phases has been investigated (14, 15).

I shall present here a few examples, chosen for the sake of illustrating how protein-lipid-water systems can be used to approach structural problems of membranes. All the examples are of lamellar structure since the only phases analyzed so far belong to this class; yet phases with a two- or three-dimensional repeat are commonly observed in protein-lipid-water systems.

*Electrostatic and hydrophobic protein-lipid interactions.* The X-ray diffraction analysis of lamellar phases containing proteins and lipids can give an insight into the nature of the protein-lipid interactions (14,15). Lamellar phases consist of planar lamellae, all parallel and equally spaced, without other correlations in position and orientation. The presence of "liquid" hydrocarbons, and the fact that the same type of structure is preserved when the water is removed, suggest that the seat of the disorder is a continuous layer of "liquid" hydrocarbons. Thus the structures appear to consist of lipid bilayers, surrounded by water and proteins. With this model in mind the partial thickness of the protein, lipid and water layers (defined as the thickness of the ideal planar slabs occupied by each of the components in the one dimensional unit cell) can be determined if the repeat distance, the chemical composition and the partial specific volumes are known (14). An important parameter in these phases is the thickness of the lipid layer, especially when this is compared to the thickness observed in the absence of protein. Two cases can be distinguished:

a). The thickness of the lipid layer is the same in the presence and in the absence of protein. In this case the protein-lipid interactions appear to be weak, and to involve mainly the polar regions of the structure: we presume in this case that the interactions are electrostatic in nature (Fig. 3, a and b).

b). The thickness of the lipid layer is substantially smaller in the protein-lipid-water phase than in the lipid-water phase. It is clear in this case that some of the hydrocarbon chains are exposed to the protein-water layer; this may be due either to the chains protruding into the hydrophobic core of the protein (Fig. 3 c) or the proteins penetrating the hydrocarbon layer of the lipid (Fig. 3 d). In any event this type of interaction may be presumed to be hydrophobic in nature.

The electrostatic and the hydrophobic nature of the interactions is confirmed by the effects of ionic strength (14, 15): when according to the partial thickness of the lipid layer the interaction is electrostatic, it is found that the lipids can be dissociated from the proteins by raising the ionic strength; on the contrary ionic strength has no dissociating effect on the phases with hydrophobic interactions. Other observa tions - fluorescence (16, 17), the effect of unsaturation of the hydrocarbon chains (15, 18) electron spin resonance of nitroxyde labeled lipids (18) - are also consistent with this operational distinction of the two types of interactions.

These concepts are applied hereafter to protein-lipid-water phase containing rhodopsin.

*Rhodopsin-containing phases*. Rhodopsin is one of the rare examples of a genuine membrane protein which has been extracted from membranes and incorporated into lipid-containing phases (19, 20). Moreover intact rhodopsin-containing membranes -retinal discs and rods- have been the object of thorough X-ray diffraction studies (21 to 25). Rhodopsin containing phases thus provide a particularly clear illustration of the use of model systems in the structural studies of membranes.

Rhodopsin can be extracted from retinal discs by detergents; subsequently the detergent can be eliminated and replaced by lipids (19, 20). In this way a rhodopsin-lipid-water lamellar phase can be obtained, whose protein-lipid ratio is fixed, and whose water content varies from 70 to 35%: this phase has been studied by X-ray diffraction and freeze-fracture electron microscopy (19). As discussed above, in the case of a lamellar phase whose repeat distance, chemical composition and partial specific volumes are known, the partial thickness of the protein and lipid layers can be determined; the result for the rhodopsin-lipid-water phase is approximately 18Å for the lipid and 20 Å for the rhodopsin layers (19). On the other hand the partial thickness of the lipid layer in lipid-water lamellar phases is close to 45 Å (7, 26), more than twice the thickness observed in the rhodopsin-lipid-water phase. Such a conspicuous shrinkage of the lipid layer indicates that the rhodopsin-lipid interactions are hydrophobic in nature (see above), and that a sub stantial fraction of the hydrocarbon chains is in contact with hydrophobic regions of the protein. The question of whether the rhodopsin molecules penetrate into the lipid layer or some of the hydrocarbon chains protrude inside the proteins can be clarified by an inspection of the electron density profiles. The profiles obtained with intact retinae, rhodopsin-lipid-water and lipid-water phases, all at the same resolution and properly (and arbitrarily) scaled, are represented in fig. 4. It is clear that the three curves are almost identical in the region of the low-density trough and high-density maxima, and thus that the apparent thickness of the lipid layer is the same. This sharp difference between the apparent thickness measured on the projection of the electron density distribution on the normal to the lamellae, and the partial thickness defined above, indicates that the rhodopsin molecules are deeply embedded in the lipid layer; more precisely, at least one half of the volume of the "bilayer" appears to be occupied by the protein.

These results suggest two comments. First the measurement of the penetration of rhodopsin into the lipid layer is far more accurate in the model system than in the intact discs, mainly because the chemical composition of the rod preparations used in the X-ray diffraction studies is not accurately known (27) (and the electron density profiles are not known on an absolute scale). Secondly some of the limitations of the "bilayer" model are clearly illustrated here by the fact that all the electron density profiles of Fig. 4 display the features generally accepted as typical of "bilayers" and yet in one case the low density region contains only hydrocarbons, in another at least 50% protein.

*Phases containing the basic proteins of myelin.* The so called "basic proteins" are the major protein components of myelin; one type, A1, is found in the central nervous system (28), two types, P1 and P2, in the peripheral nervous system (29). The discovery of the involvement of these proteins in the induction of experimental allergic encephalitis has stimulated a large number of chemical and immunological studies (30); the three proteins have been purified the molecular weight is 18,400 for A1, 14,000 for P1, 12,300 for P2 (29, 31, 32) - and the amino acid sequence of A1 has been determined (33).

Several protein-lipid-water systems formed with one of the basic proteins and different types of myelin extracted lipids have been studied by X-ray scattering techniques, and several phases have been identified (34): one of these turned out to be particularly interesting. This phase contains any one of the basic proteins (provided their chemical integrity is preserved) and the whole of the acidic lipids -sulphatides, phosphatidyl serine, phosphatidic acid. The structure is lamellar, with repeat distance 154 Å in the case of A1, 175 Å in the case of P1, and the conformation of the hydrocarbon chains is partly liquid-like, partly more ordered (type β, see 12). The structure was determined by taking into account the chemical composition and the partial thickness of the lipid and protein layers, and by selecting the electron density profiles according to a pattern recognition procedure (34). Fig. 5 represents the structure of the phase, with a schematic distribution of the lipids and proteins in the unit cell; the electron density profile is shown in the same figure. The structure consists of two types of lipid bilayers, one containing mainly the phospholipids with the hydrocarbon chains in a liquid-like conformation, the other the sulphatides with most of their chains stiff and hexagonally packed. The polar layer is asymmetric, and thus the protein molecules may be presumed to interact differently with the two classes of lipids; in fact the shoulder associated with the thick lipid layer suggests a specific affinity of that face of the protein for the polar groups of the sulphatides.

This phase displays striking analogies with myelin: the one-dimensional unit cell contains <u>two</u> lipid bilayers; the repeat distances of the native myelin (150-160 Å for central nervous system; 170-180 Å for peripheral nervous system) are very similar to those of the protein-lipid-water phase made with the corresponding protein (154 Å with A1, 175 Å with P1). These analogies are all the more interesting since the most conspicuous chemical difference existent between the myelin from the central and the peripheral nervous systems is precisely the nature of the basic protein. Some differences should also be pointed out: the chemical composition of the phase is highly simplified with respect to myelin; myelin contains one type of asymmetric membrane, the protein-lipid-water phase two types of symmetric lipid layers (see fig. 1 of ref. 34). However the segregation of the different lipid molecules and the protein-lipid interactions revealed in the protein-lipid-water phase are likely to be relevant to myelin. It is clear that further studies of model systems involving the components of myelin may well help to understand the molecular organization of that membrane.

## Discussion

Let me stress again how unrealistic it is to expect any satisfactory understanding of the molecular events underlying the physiological functions of membranes without the support of detailed structural information; membranes will remain "black boxes" as long as this information is lacking.

I have not concealed the opinion that the overall balance of the X-ray diffraction studies of intact membranes looks disappointing. Moreover, and in spite of some recent promising developments, it appears unlikely to me that the X-ray diffraction analysis of intact membranes will ever reach the degree of sophistication attained in other areas of biology. I think that a more promising approach is to isolate some of the membrane components, and to attempt to reassociate them into model systems of simple and well defined chemical composition and of highly regular periodic structure, thus favourable to to X-ray diffraction analysis. I could point out precedents of this approach in other areas of biology: for example the use of DNA fibers in establishing the molecular bases of genetic replication. The use of model systems is at an early stage in the field of membranes; the limiting factor is still the purification of membrane proteins and their incorporation in model systems. This explains why the examples I have chosen are fairly crude, and some even rather remote from membranes.

It would perhaps be worth trying to specify more precisely which structural aspects of membranes may be expected to become amenable to an experimental approach in the near future. One can take the viewpoint, rather widespread among biochemists, of regarding proteins as the only functional elements of membranes. In this case the molecular approach to membrane functions is reduced to the usual concepts of protein chemistry, and the purpose of X-ray diffraction studies is limited to the structural analysis of membrane proteins. Yet, without denying the obvious importance of proteins, and awaiting the crystallization of membrane proteins, one can take the alternative attitude of focussing attention on those properties of the membranes which seem to be characteristic of the whole intact organelles, and not a simple and direct consequence of the properties of the separate components. The evidence presently available would hardly justify the assumption that membranes "conceal a secret" - for example something akin to base pairing in nucleic acids. Nevertheless in at least one respect membranes differ from other cellular organelles, namely the high lipid content. If indeed membranes display any specific and characteristic property, it may be presumed to be related to the protein-lipid interactions. The X-ray diffraction study of model systems may well provide the most fruitful approach to this and to other aspects of the structure of membranes, at least for the time being. The experience acquired in the structural analysis of lipid-water and protein-lipid-water phases, illustrated by the examples given above, shows that low resolution diffraction studies (which do not require the use of single crystals) are capable of yielding information on the size and shape of protein molecules, on the distribution of the polar and apolar regions and on the nature and geometry of the protein-lipid interactions. In the long run however there will be no substitute for high resolution X-ray diffraction studies of membrane proteins.

Acknowledgements.

I wish to thank many colleagues - especially Annette Tardieu, Leonardo Mateu and Tadeusz Gulik - for often and patiently listening to the views expressed in this paper, and for keeping me alert with their wit and criticism.

References

1. Stoeckenius, W. and Engelman, D.M. J. Cell. Biol. 42 (1969) 613.
2. Wilkins, M.H.F., Blaurock, A.E. and Engelman, D.M. Nature New Biol. 230 (1971) 72.
3. Caspar, D.L.D. and Kirschner, D.A. Nature New Biol. 231 (1971) 46
4. Worthington, C.R. and King, G.I. Nature 234 (1971) 143.
5. Blauroch, A.E. and Stoeckenius, W. Nature New Biol. 233 (1971) 152.
6. Goodenough, D.A. and Stoeckenius, W. J. Cell. Biol. 54 (2972) 646.
7. Luzzati, V. (1968) In: Biological Membranes. p. 71, Vol. 1., ed. D. Chapman, Academic Press London and New York.
8. Shipley, G.G. (1973) In: Biological Membranes. p. 1, Vol. 2. , eds. D. Chapman and D.F.H. Wallach. Academic Press, London and New York.
9. Luzzati, V. and Husson, F. J. Cell Biol. 12 (1962) 207.

10. Luzzati, V., Reiss-Husson, F., Rivas, E. and Gulik-Krzywicki, T. Ann. N.Y. Acad. Sci. 137 (1966) 409.

11. Overath, P. and Trauble, H. Biochemistry 12 (1973) 2625.

12. Ranck, J.L., Mateu, L., Sadler, D.M., Tardieu, A., Gulik-Krzywicki, T. and Luzzati, V. J. Mol. Biol. (1974) In press.

13. Dupont, Y., Gabriel, A., Chabre, M., Gulik-Krzywicki, T. and Shechter, E. Nature 238 (1972) 331.

14. Gulik-Krzywicki, T., Shechter, E., Luzzati,V. and Faure, M. Nature 223 (1969) 1116.

15. Gulik-Krzywicki, T., Shechter, E., Luzzati,V. and Faure, M. (1972) In: Biochemistry and Biophysics of Mitochondrial Membranes. eds. G.F. Azzone, E. Carafoli, A.L. Lehninger, E. Quagliariello and N. Siliprandi. Acad. Press. New York, and London.

16. Gulik-Krzywicki, T., Shechter, E., Iwatsubo, M., Ranck, J.L. and Luzzati, V. Biochim. Biophys. Acta 219 (1970) 1.

17. Shechter, E., Gulik-Krzywicki, T., Azerad, R. and Gros, F. Biochim. Biophys. Acta. 241 (1971) 431.

18. Caron, F., Mateu, L., Rigny, P. and Azerad, R. J. Mol. Biol. (1974) In press.

19. Chabre, M., Cavaggioni, A., Osborne, H.B. and Gulik-Krzywicki, T. FEBS Letters 26(1972) 197.

20. Hong, K. and Hubbell, W. L. Proc. Nat. Acad. Sci. USA 69 (1972) 2617.

21. Blauroch, A.E. and Wilkins, M.H.F. Nature 223 (1969) 906.

22. Gras, W.J. and Worthington, C.R. Proc. Nat. Acad. Sci. USA 63 (1969) 223.

23. Webb, N.G. Nature 235 (1972) 44.

24. Corless, G.M. Nature 237 (1972) 229.

25. Chabre, M. and Cavaggioni, A. Nature New Biology. 244 (1973) 118.

26. Huynh, S. Biochimie, 55 (1973) 431.

27. Blauroch, A.E. and Wilkins, M.H.F. Nature 236 (1972) 313.

28. Nakao, A., Davis, W. and Einstein, E. Biochim. Biophys. Acta 130 (1966) 163.

29. London, Y. Biochim. Biophys. Acta 245 (1971) 188.

30. Eylar, E.H., Salk, J., Beveridge, G.C. and Brown, L. V. Arch. Biochem. Biophys. 132 (1969) 34.

31. Eylar, E.H. and Thompson, M. Arch. Biochem. Biophys. 129 (1969) 468.

32. Oshiro, Y. and Eylar, E.H. Arch. Biochem. Biophys. 136 (1970) 606.

33. Eylar, E.H. Proc. Nat. Acad. Sci. U.S.A. 67 (1970) 1425.

34. Mateu, L., Luzzati, V., London, Y., Gould, R.M., Vosseberg, F.G.A. and Olive, J. J. Mol. Biol. 75 (1973) 697.

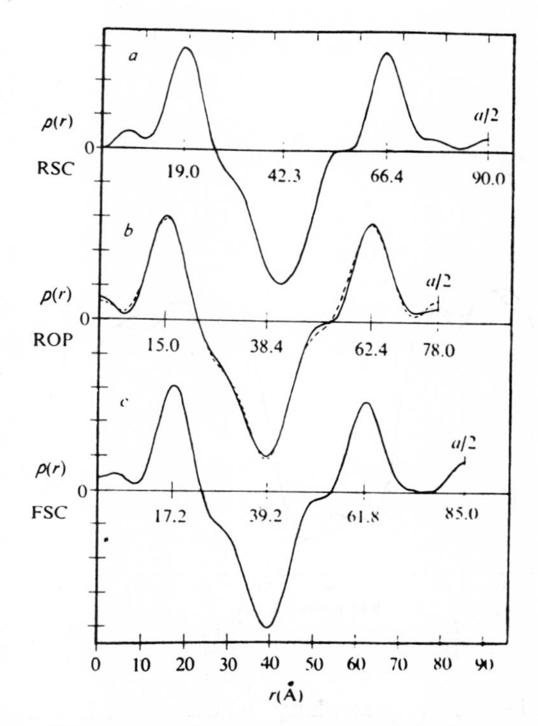


Figure 1. Electron density profiles of myelin membrane units as a function of distance r from the cytoplasmic boundary, at 10 Å resolution. See the original paper for a justification of the choice of phases. Note the asymmetric shape of the electron density trough, interpreted to be due to an asymmetric distribution of cholesterol: RSC: rabbit sciatic - ROP: rabbit optic - FSC: frog sciatic. (from ref. 3).

---

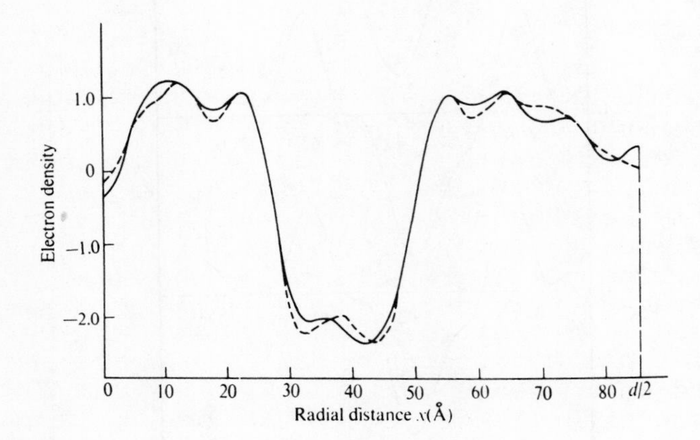


Figure 2. Electron density profile of frog sciatic myelin, with a different choice of phases. Note the symmetric profile of the low density trough. (From ref. 4).

---

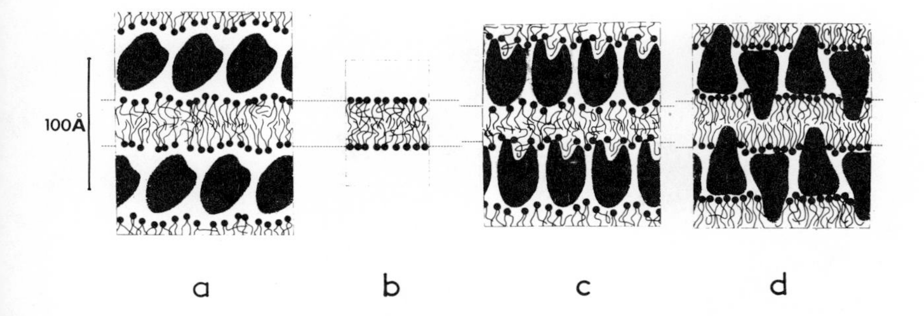


Figure 3. The structure of the lamellar protein-lipid-water phases with electrostatic and hydrophobic interactions. The sections are perpendicular to the plane of the lamellae. The densely and lightly hatched areas represent the protein molecules and the water regions (both of arbitrary shape). The polar groups of the lipid molecules are represented by a dot, the hydrocarbon chains in a liquid like conformation by a wriggle. The lipid-water phase is represented in b, the protein-lipid-water phase in a, c and d. a represents a structure with electrostatic protein-lipid interactions: note that the partial thickness of the lipid layer is the same in a as in b. c and d represent structures with hydrophobic interactions: note that the partial thickness of the lipid layer is smaller than in a and b. In c the hydrophobic interaction involves hydrophobic chains penetrating the hydrophobic core of the protein molecules, in d the protein molecules protruding into the lipid layer. (from Sadler & Gulik-Krzywicki, unpublished).

---

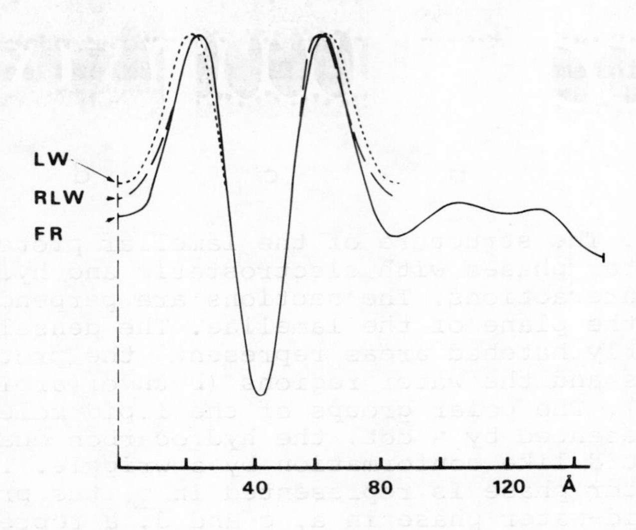


Figure 4. Electron density profiles of frog retinae outer segments (FR) (from ref. 25), of the lamellar phase of a rhodopsin-lipid-water system (RLW) (from ref. 19), and of a lipid-water lamellar phase (LW). The lipid-water phase contains phosphatidic acid (see table I of ref. 12); in fact at this resolution the profiles of the lipid-water lamellar phases are barely dependent on the nature of the lipid. The three profiles were computed at the same resolution, 25-30 Å. The three curves are arbitrarily scaled by bringing the maxima and minima into coincidence. Note that the three curves are almost indentical in the region of the low density trough and of the two surrounding maxima.

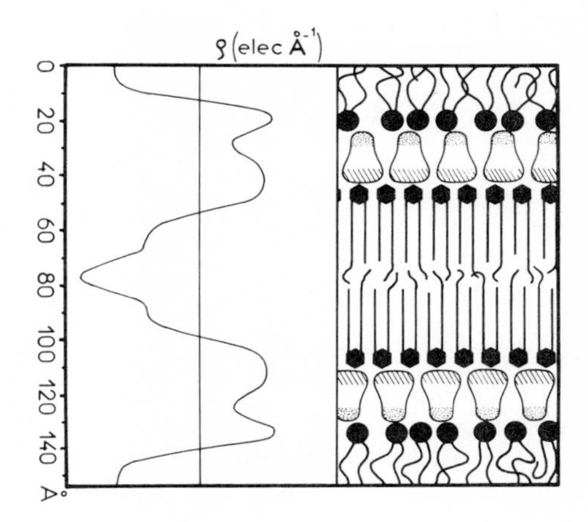


Figure 5. The structure of one protein-lipid-water lamellar phase containing basic protein and acidic lipids extracted from myelin. The polar head of the phospholipids is represented by a circle, that of the sulphatides by a hexagon. The hydrocarbon chains in a liquid-like conformation are represented by a wriggle, those which are stiff and hexagonally packed by a straight line. The protein molecules are represented as objects of irregular shape. The lower part shows the electron density profile.

Note the presence of two hydrocarbon layers of different thickness, one with a well defined minimum in its centre; note also the asymmetric peak in the polar region suggesting preferential association of the protein with the sulphatides. The segregation of the phospholipids from the sulphatides does not have to be as complete nor the chain conformation as extensively of type β, as suggested in the figure. (from ref. 34).

---

# STRUCTURE AND ASSEMBLY OF VIRUS MEMBRANES

Kai Simons, Henrik Garoff, Ari Helenius,[1]
Leevi Kääriäinen[2] and Ossi Renkonen[3]

Department of Serology and Bacteriology[1]
Department of Virology[2]. Department of
Biochemistry[3]. University of Helsinki,
Haartmansgatan 3, Helsinki 29.
Finland

## Introduction.

There are a number of animal, plant and bacterial viruses that have membranes. The viral membranes or envelopes have in recent years been used as useful experimental model systems to study the structure and assembly of biological membranes.

The most widely studied enveloped viruses and some of their properties are presented in Table I. These viruses contain on the inside of their membranes a nucleoprotein forming the nucleocapsid. Most of the viruses acquire their membrane by a budding process through a cellular membrane, usually the plasma membrane (Fig. 1). The envelopes of poxviruses and bacteriophage PM 2 are unique in the sense that these seem to be formed *de novo* free in the cytoplasm without connection to preformed cellular membranes.

Our intention is not to review the structure and assembly of all these different viruses. For such reviews the reader is referred to Choppin et al (1), Klenk (2), Lenard and Compans (3), and Franklin (4). We will mainly deal with the virus we have been working with, Semliki Forest virus (SFV) (Fig. 1) and try to illustrate in a rather speculative and subjective manner the present and the future of viral envelopes in molecular membranology.

## Viral envelope constituents

*Lipids.* It is now generally accepted that enveloped viruses acquire their lipids by utilization of host cell lipids (2, 3, 5). So far not a single enzyme involved in lipid metabolism has been found to be coded for by the viral genome. Klenk and Choppin (6, 7) have in a detailed study analyzed the lipid composition of a paramyxo (SV 5) virus grown in four different cell types. They found that the composition of the virus lipids resembled that of its host cell plasma membrane. Our studies show that the lipids in SFV also closely reflect the lipids in the host cell plasma membrane (8, 9). Similar observations on other enveloped viruses have been published (see ref. 2 and 3). The most dramatic difference between the same virus grown in different host cells has been reported by Renkonen et al. (10). They showed that SFV grown in animal (BHK21) cells and in insect (Aedes albopictus) cells have only 36% of their phospholipids in common.

The question whether the viral membrane proteins have any selectivity in determing the lipid composition of the viral envelopes is still unclear. More refined methods are needed than those thus far used to resolve the nature and the specificity of the lipid-protein interactions in these membranes. In the bacteriophage PM 2, Franklin et al. have proposed that electrostatic interactions between a basic viral membrane protein (isoelectric point 12.5) and the acidic phospholipid, phosphatidyl-glycerol, may be important in membrane assembly (11). The lipid composition of this virus, in which the virus membrane is not formed by a budding process, is clearly different from that of the cell membranes of the bacterial host.

*Viral membrane proteins and their properties.* In contrast to the lipids all the major viral structural proteins are specified by the viral genome. Most of these proteins contain carbohydrate.

*Viral glycoproteins*. The glycoproteins form projections or spikes on the external surface of the virus membrane. These spike glycoproteins can be removed by proteases. In SFV there are two medium sized ($E_1$ and $E_2$) and a small ($E_3$) glycopolypeptide (12) which are present in equimolar ratios in the membrane (Table II). The carbohydrate portions of these proteins contain N-acetylglucosamine, mannose, galactose, fucose and sialic acid but no galactosamine, thus suggesting that the carbohydrate units are linked by N-glycosidic linkages to asparagines in the polypeptide chains. The SFV glycoproteins can be solubilized by disrupting the membrane with the non-ionic detergent Triton X-100 (13). The proteins thus released are lipid-free and form a complex (see Table III) with the detergent so that about 75 molecules of Triton X-100 are associated with a trimer of $E_1$, $E_2$, and $E_3$. This complex shows a sedimentation constant of about 4S. The 4S "trimers" easily aggregate into a 23S complex (Table III) which contains 8 trimers of $E_1$, $E_2$, and $E_3$ and about 250 molecules of detergent. The 23S complex has a rosette-like structure when examined in the electron microscope and it has the hemmagglutinating activity of the virus (13).

The ability to bind appreciable quantities (20-50 % by weight) of Triton X-100 (15-18) appears to be a general characteristic of the class of membrane proteins that Singer and Nicolson call integral proteins (14). These proteins cannot be released from the membrane without disrupting the lipid matrix. Integral proteins have been postulated to have an amphipathic structure with a hydrophilic part in the aqueous phase, and a hydrophobic part which is interacting with the lipids in the membrane interior (14). The other major class of membrane proteins, the peripheral proteins (14), can be detached from membranes into water-soluble form with mild treatment such as raising the ionic strength of the medium. They bind only negligible amounts of Triton X-100 when released (15, 19). In this respect they resemble normal water-soluble proteins like albumin.

Further studies have indeed shown that the SFV glycoproteins $E_1$ and $E_2$ have an amphipathic structure (20). If SFV is treated with the protease thermolysin the external parts of the spikes are cleaved off leaving within the membrane peptide segments which derive from both $E_1$ and $E_2$. These segments have a molecular weight of about 5000 (about 10% of the protein molecule), and they are enriched in hydrophobic amino acids. The polarity index calculated as described by Capaldi and Vanderkooi (21) from the amino acid composition is as low as 28.0 which is even lower than for brain proteolipids (22). The hydrophobic nature of these peptides is also evident from their solubility properties (20). They are, like phospholipids and proteolipids soluble in neutral chloroform-methanol. The intact membrane glycoproteins are insoluble in this solvent.

In the influenza virus there are two different spike glycoproteins, one carrying the hemagglutinating and the other the neuraminidase activity (3). The hemagglutinin spikes are composed of two glycopolypeptides $HA_1$ (molecular weight 50,000) and $HA_2$ (molecular weight 25,000) which can be isolated from the virus using detergent solubilization (23). If the protease bromelain is used to release these proteins from the virus particle, components are obtained which correspond in size to polypeptide $HA_1$ and to a polypeptide of slightly lower molecular weight than $HA_2$ (24). Thus it is possible that a small segment of the $HA_2$ polypeptide is left in the membrane which perhaps is involved in the attachment of the hemagglutinin to the viral membrane. The remainder of the spikes is resistant to digestion. The protease-released spike glycoproteins do not form rosette-like aggregates and unlike the detergent-released spike glycoproteins they lack the ability to agglutinate erythrocytes (24).

The neuraminidase spike glycoprotein has a molecular weight of about 60,000 if released from the virus using detergent treatment (25). Also these spike glycoproteins form rosette-like aggregates when examined in the electron microscope. However, if trypsin is used to release the neuraminidase has a molecular weight of about 48,000 (26). This water-soluble form has lost the ability to form rosette-like agregates.

Rosette-like aggregates are also formed by the spike glycoproteins (released by detergent) of other enveloped animal viruses (3). The inability of the protease-released protein to form these aggregates could be due to the cleavage of hydrophobic peptide segments involved in attaching the spike glycoproteins to the membrane. It is therefore likely that an amphipathic structure may be a general characteristic of viral spike glycoproteins, and hydrophobic peptide segments similar to those found in the SFV glycoproteins should be searched for.

*Internal membrane proteins*. Most enveloped viruses contain at least one non-glycosilated protein in addition to the proteins of the nucleocapsid. The only exception appears to be the toga virus group to which SFV belongs. In the myxo-, paramyxo- and rhabdovirus groups the polypeptide with the lowest molecular weight is a major nonglycosylated polypeptide (3). This protein is not present in the nucleocapsid, and is thought to be associated with the internal surface of the viral membrane. Several observations indicate that these polypeptides form a shell of protein beneath the lipid membrane. Whether these polypeptides interact with the hydrocarbon interior of the lipid membrane or whether they are more peripheral to the membrane structure is not known. These proteins seem to be sensitive to the ionic strenght in the medium, forming aggregates at low ionic strength. It would be interesting to know whether the internal membrane proteins bind Triton X-100. In SFV the nucleocapsid proteins form the layer beneath the viral lipid; these proteins bind no Triton X-100 (27).

Gregoriades has shown that the internal membrane protein of influenza virus is soluble in acid chloroform-methanol (28). However, this does not necessarily classify it as an integral protein. Gitler and Montal (29) have shown that cytochrome c which is a peripheral protein of the inner mitochondrial membrane can be extracted into decane as a proteolipid. This they found to be due to the formation of ion pairs presumably between basic protein groups and acidic phospholipids. The requirements for the formation of such soluble complexes between lipids and proteins is not yet understood but may well give us a handle on molecular interactions which are important for membrane structure and function (cf. PM 2).

## The structure of viral membranes

The most extensive studies to date on viral membrane structure have been carried out with bacteriophage PM 2. These investigations mainly based on low angle x-ray diffraction indicate that the viral lipids are arranged into a bilayer structure 40 Å thick (30). The compositional and the structural data indicate that the bilayer may be fenestrated or separated into patches covering only about 65% of the area around the nucleocapsid. Electron microscopic studies show that this virus is clearly different in shape compared to animal membrane viruses. PM 2 appears to be icosahedral and has brushlike spikes at the icosahedral vertices. It is conceivable that these spikes are bridging the lipid bilayer (4).

X-ray diffraction studies of Sindbis and SFV indicate that the lipids in these membranes also form a bilayer (31). Spin labelling results are also compatible with a bilayer structure in other animal virus membranes (32-34). It therefore is likely that lipid bilayers are generally present in enveloped viruses as in other biological membranes. This suggests, although proof is yet lacking, that the lipid structure of the host cell membrane is preserved as the lipid leaves the cell membrane and becomes part of the viral envelope.

Recent studies using surface labelling with $^{35}$S-formyl methionyl sulphone methyl phosphate ($^{35}$S-FMMP) and phospholipase digestion have indicated that the lipids in the erythrocyte membrane are asymmetrically distributed with the choline-containing phospholipids predominating in the outer half of the bilayer leaflet of the bilayer while the amino-containing phospholipids are concentrated in the inner, cytoplasmic, half of the bilayer (35). When labelling of intact SFV with $^{35}$S-FMMP was compared to labelling after the virus was disrupted with Triton X-100, eight times more of the label was found in phospatidylethanolamine and phosphatidylserine after detergent disruption (36). This difference could arise from asymmetric distribution of the aminophospholipids. However, from data available on the lipid composition of viruses grown in different host cells it is clear that all the choline phospholipids cannot in all cases be located in the outer half and the amino phospholipids in the inner half of the bilayer. For instance, in SFV the ratio of amino phospholipids to choline phospholipids is 0.6:1 for virus grown in BHK21 cells (8) and 5.4:1 for virus from insect cells (10). In SV5 virus this ratio varies from 1.6:1 when grown in MK cells to 0.3:1 when grown in BHK21 cells (6). These very different molar ratios occur in viral membranes which contain for each virus the same set of proteins. If these viruses reflect the distribution of lipids in the host cell plasma membrane from which they bud then it is evident that the distribution of the different phospholipids between the outer and inner halfs of the bilayer vary greatly in plasma membranes from different cells. Viruses grown in different host cells may indeed be useful experimental models to study this aspect of membrane structure.

Physical methods have not yet given much direct information on how the proteins are organized in the lipid matrix of biological membranes. Most experimental data on how proteins are located in membranes have been obtained with biochemical methods. The MN-Glycoprotein (glycophorin) in the erythrocyte membrane is perhaps that best studied in this respect. Protease dissection and labelling studies have shown that the MN-glycoprotein is attached to the lipid bilayer in such a way that three different regions of the protein molecule can be differentiated (35, 37, 38). The protein has two hydrophilic parts located on either side of the membrane and these are connected by an intramembranous hydrophobic region. The hydrophobic region has been sequenced and is 23 amino acid residues in length i.e. long enough to span a lipid bilayer as an α-helix (38). All the carbohydrate is attached to the external hydrophilic part of the glycoprotein (37).

Our recent work shows that the SFV glycopolypeptides $E_1$ and $E_2$ not only contain hydrophobic segments which attach the proteins to the lipid membrane but that one or both of these polypeptides span the membrane (39). This has been shown by two different experimental approaches. The first employed the radioactive surface label $^{35}$S-FMMP and the second the bifunctional crosslinking reagent, dimethylsuberimidate (DMS). Both of these reagents react with amino groups in proteins (and lipids). The glycoproteins were labelled with $^{35}$S-FMMP from the outside in intact SFV or from both sides in membrane preparations which were obtained from SFV by treatment with small amounts of Triton X-100 (Stage II, see Fig. 2). When the membranes were labelled more of the glycoprotein became accessible to the reagent. This was shown by two dimensional peptide patterns by the presence of two additional basic, heavily labelled peptides which were not seen in the peptide patterns derived from the glycoproteins labelled in intact SFV. One explanation to this result is that segments of the viral glycoproteins extend through the membrane and that the two additional peptides seen when labelling was performed from both sides are derived from those parts of the proteins which are exposed on the inner (nucleocapsid) surface of the membrane. The crosslinking studies confirmed this interpretation. In contrast to FMMP, DMS is known to diffuse through cell membranes (40). DMS crosslinks amino groups within about 10 Å from each other (41). This is only about one fourth of the bilayer thickness in SFV. Using this reagent we were able to crosslink 70% of the viral glycopolypeptides $E_1$ and $E_2$ to the nucleocapsid in <u>intact</u> SFV. No significant crosslinking of the phospholipids to the nucleocapsid could be detected.

Previous interpretations on how the viral glycoproteins are located in the membrane have been based on X-ray and freeze fracture microscopy. Harrison et al. concluded from the chemical composition and the radial electron density distribution profile of Sindbis virus that the lipid bilayer is not bridged by protein (31). However, our calculation show that if both of the viral polypeptides $E_1$ and $E_2$ would span the membrane in an α-helical form, less than 15% of the area in the middle of the viral bilayer would be occupied by protein (20). Such penetration would be difficult to detect by low angle X-ray diffraction. Intramembranous particles are not observed in the viral membrane when examined in freeze fracture electron microscopy (42). As these particles are assumed to derive from spanning proteins, this finding has been used to support the view that viral spike glycoproteins do not span the bilayer. However, this result may also be due to limitations in resolution as structures smaller than 20 Å are not easily resolved by the freeze fracture microscopic method, and an α-helical protein segment is less than that in diameter. It should also be pointed out that the exact chemical nature of the intramembranous particles seen in membranes after freeze fracturing is not settled, and all spanning proteins may not give rise to such particles (see both Martínez-Palomo and Raff in this volume).

## Dissociation and reassociation studies

Triton X-100 is widely used to dissociate biological membranes into their lipid and protein components because many membrane proteins preserve their biological activities when solubilized with this detergent (43). However, not much is known of its mode of action. Virus membranes are suitable experimental models to study its action because they are simple in structure and their components can be conveniently labelled by radioactive precursors. Furthermore careful studies of the dissociation process can be expected to give us information on the molecular organisation of the viral membrane.

From our studies on the effects of Triton X-100 on SFV it is evident that the solubilization proceeds through definable intermediate stages (13, 27). The solubilization process is summarized in schematic form in Fig. 2. Binding of Triton X-100 to the virus begins below the critical micellar concentration of the detergent. Binding increases with increasing detergent concentration, but the concentration of the unbound detergent remains such that micelles are not formed. This is in agreement with the conclusions of Tanford (44) who has shown that micelle formation and detergent binding to proteins and membranes are competitive phenomena. If the free energy of binding is greater than that of micelle formation then the detergent binds to the membrane rather than forming micelles. The detergent is bound in monomer form.

When about 9,000 moles of Triton X-100 are bound per mole of virus, which corresponds to about 16,000 moles of cholesterol-phospholipid pairs, the SFV membrane ruptures (Stage II), releasing the nucleocapsid. When about 40,000 molecules of Triton X-100 are bound per viral membrane, breakdown into homogenous lipoprotein-detergent particles occurs and part of the lipid forms lipid-detergent mixed micelles (Stage III). This breakdown from a lamellar to basically micellar structure resembles the phase transition observed in simple detergent: lipid: $H_2O$ systems (45). When a moderate amount of Triton X-100 is added to relatively dilute aqueous suspensions of phospholipids two phases are observed; a lamellar phospholipid structure with incorporated detergent and an isotropic detergent solution. When on further addition of detergent, a detergent-lipid ratio of about 2 is reached, the lamellar phase disappears and both the lipid and the detergent are included in an isotropic micellar solution (45). The dissociation of the viral membrane by Triton X-100 may proceed in an analogous fashion except that the protein present gives rise to two populations of micelles; micelles containing protein, lipid and detergent and micelles with only lipid and detergent.

The amount of Triton X-100 needed to reach stage IV of the dissociation scheme is at least 20 times the amount of viral lipid. The protein and the lipids of the SFV membrane could then be separated from each other by density gradient centrifugation in the presence of 0.05% Triton X-100 (13). Removal of the lipids from the membrane protein of SFV by Triton X-100 resulted in the association of substantial amounts of detergent with the protein as described in section II. We have previously proposed that Triton X-100 displaces the lipid from the protein and is bound to the hydrophobic region of the protein which is interacting with the lipid in the membrane (15). This interpretation has recently been given additional support by studies using protease dissection.

Both the lipids in the membrane and Triton X-100 bound to the solubilized SFV glycoproteins protected the same (hydrophobic) region of the protein molecule from digestion (20).

Figure 3 shows the removal of the phospholipids from the protein during the solubilization process, which takes place during Stage III. Phosphatidylserine and phosphatidylethanolamine are slightly more difficult to extract from their association with the protein than the phosphatidylcholine. This may suggest that these amino phospholipids are more tightly bound to the proteins than phosphatidylcholine. However, such studies are as difficult to interpret as previous work using organic solvents of different polarity to extract "loosely or strongly bound" lipids from the erythrocyte membrane (46). It remains to be verified by more direct means whether these studies only reflect the nature of the extracting agent, asymmetric lipid distribution or whether the extraction really show that some phospholipids are more tightly bound to the protein than others.

Also the detergent can be removed from the lipid-free protein (Stage V). This is done by adding enough Triton X-100 to dissociate the lipids from the protein where after this mixture is sedimented through a band of Triton X-100 into a sucrose gradient free of Triton X-100. In this one-step procedure the lipids stay in the Triton band and the nucleocapsids sediment to the bottom of the centrifuge tube. Attempts to remove the detergent by dialysis or column procedures (ion exchange or gel filtration) have led to macroscopic aggregation. It is thus important how the detergent is removed to obtain water-soluble aggregates of these integral membrane proteins. The lipid- and detergent-free glycoprotein has a sedimentation coefficient of about 26 S and a molecular weight of about 900,000. The 26 S complex has a similar rosette-like structure as the 23 S complex in the electron microscope (Fig. 4 A). Apparently the hydrophobic segments of $E_1$ and $E_2$ are sufficiently shielded from contact with the aqueous medium so that the complex does not form larger aggregates.

Summarizing the events after delipidation of the SFV membrane, it appears that eight 4 S complexes associated to form one 23 S complex with extrusion of about 350 molecules of Triton X-100. The remaining 250 molecules of Triton X-100 are removed in the formation of the 26 S complex. The properties of the 4 S, 23 S and 26 S complexes are summarized in Table III.

One additional feature of our dissociation scheme should be noted. Triton X-100 is apparently very inefficient in breaking protein - protein interactions between subunits in oligomeric proteins. Our results based on crosslinking studies and stochiometry suggest that the structural protein unit in the SFV membrane may be a trimer of $E_1$, $E_2$ and $E_3$ (12, 48). So is the 4 S complex (13). It is therefore quite likely that the polypeptides in the protein-detergent complexes obtained after Triton X-100 solubilization also associate with each other in the native membrane. However, as also shown in our studies, aggregation may take place after solubilization (formation of 23 S complexes) which in more complex membranes than the SFV envelope may lead to association of proteins which are not interacting with each other in the undisrupted membrane.

No functional biological membrane has yet been reconstituted from its lipid and delipidated protein components. It is obvious that controlled dissociation studies are a prerequisite for such attempts. We have tried three different ways to associate the lipid with the protein in SFV membranes. In the first method SFV was solubilized into Triton X-100 and reassociation was attempted by simply removing the detergent using dialysis. After three days of dialysis through Neflex tubing small vesicles with surface projections were obtained. However, these still contained about 10% of their weight of the detergent. No virus-like particles were seen. The nucleocapsids appeared to be disrupted by this treatment. In the second method, liposomes were made by sonication of phospholipids. The liposomes were mixed with the lipid- and detergent-free 26 S complex. After long incubation at 37° (> 4 hours) reassociation began to take place. Large macroscopic aggregates were usually obtained which were difficult to analyze.

The most successful method we used was the sonication method introduced by Racker (49). If phospholipids and the 26 S complex were sonicated together at 40° rapid reassociation was observed (Fig. 4 B). In this way it may be possible to reconstitute an infective virus particle from nucleocapsids, lipids and the 26 S complex. The virus needs its membrane with its spikes for infectivity, which can be quantitatively assayed by plaque titration.

Hosaka and Shimizu recently demonstrated that a glycoprotein fraction and lipids solubilized from a paramyxo (Sendai) virus using Nonidet P40 (the same detergent as Triton X-100) can be recombined by gradual removal of the detergent by dialysis to yield functional lipoprotein particles (50). All of the assayable properties of the viral surface (hemagglutination, neuraminidase, and cell fusion) were associated with the reassociated particles. In a further study it was shown that purified phosphatidylethanolamine, sphingomyelin and phosphatidylcholine, but not phosphatidylserine were effective in reconstituting the membrane activities with the same Sendai virus glycoprotein fraction(51) . Such studies combined with physico-chemical methodology offer a promising approach for investigating the nature of the lipid-protein interactions in biological membranes.

## Assembly of viral membranes.

Most enveloped animal viruses are released from their host cells by budding out through the plasma membrane (Fig. 1). There are some general features in the assembly process which are common to these groups of viruses. The nucleocapsids and the viral glycoproteins apparently move to the plasma membrane through different routes (2, 3). The nucleocapsids are assembled in the cytoplasm whereas the glycoproteins become membrane-bound soon after translation and are apparently glycosylated by host cell enzymes in the intracellular membranes. The viral glycoproteins are then inserted into the plasma membrane by a mechanism so far unknown. As no significant amount of host cell proteins are found in the viral membrane in the mature virus particles (RNA tumor viruses apparently contain host membrane proteins (2)), they must be derived from segments of the plasma membrane from which the host cell proteins are excluded during the budding process. How such patches of viral membrane can be generated in the host cell has been an enigma.

Figure 5 shows a speculative scheme for toga virus biogenesis. These viruses shut off host cell protein synthesis early after infection (52), whereafter all the proteins made are virus proteins, which therefore can easily be labelled and followed in pulse-chase experiments. Schlesinger and Schlesinger have found a temperature-sensitive virus mutant in which a precursor protein accumulates (53). This molecule contains the viral structural proteins($E_3$ has not yet been demonstrated in this molecule). This finding indicates that these proteins are translated as a monocistronic message as has been shown earlier to be the case for poliovirus (54). After the nucleocapsid protein is split from the precursor, it rapidly binds to the 42 S RNA to form the nucleocapsid (55). The envelope glycoprotein precursors are incorporated into intracellular membranes where they are cleaved and glycosylated to form $E_1$, $E_2$ and $E_3$. The sequence and sites for these proteolytic cleavages have not been established, but it appears that $E_1$ is cleaved off first before the linkage between $E_2$ and $E_3$ is split. The proteases involved in this cleavage process have not been identified and these could be of both host cell and viral origins.

Several observations (for other enveloped virus groups) suggest that the first viral components to arrive at the plasma membrane are the envelope glycoproteins (3). There are also indirect observations which indicate that these proteins can undergo lateral diffusion in the plasma membrane (56, 57) and that they have a tendency to associate with each other (13). We propose that the next event in assembly is the binding of the nucleocapsid to the plasma membrane by attachment to those parts of the viral glycoproteins penetrating to the cytoplasmic side. This of course, implies that the spike glycoproteins are inserted in such a way into the plasma membrane that their membrane-attaching segments extend through the membrane as they do in the mature SFV particle. The nucleocapsid may act as a a nucleation site for additional viral glycoproteins moving into the growing patch by lateral diffusion. The association of the envelope glyco proteins with the nucleocapsid proceeds until all the binding sites are filled which leads to the release of the virus particles into the extracellular medium. The stochiometry of the assembly (Table II) could be explained if we assume that one trimer of $E_1$, $E_2$ and $E_3$ binds to one protein subunit in the nucleocapsid. We assume that the host proteins are excluded from the patch for steric reasons. Although the viral proteins are not closely packed in the hydrocarbon interior of the membrane the proteins may be quite close to each other on either side of the membrane.

It is interesting in this context to note that a lowering of the pH to about 6 or glutaraldehyde fixation has been observed to cause a contraction of the SFV particle with a decrease in diameter of about 100 Å (58). This effect appears to be due to a size decrease of the nucleocapsid (59). The viral membrane apparently adheres to the nucleocapsid during the contraction and excess membrane is extruded from the contracted virus particle into blebs. Spikes are not seen on the surface of these blebs suggesting that these are mainly formed by viral lipids and that the viral spike glycoproteins remain bound to the nucleocapsid.

This speculative model for the budding mechanism of virus assembly is analogous to the patch formation induced by antibody to surface antigens in lymphocytes (see Raff this volume), or more closely to the aggregation of anionic external sites induced by anti-spectrin reacting with the internal surface of erythrocyte membranes (see Singer this volume). It is possible that other enveloped viruses utilize similar mechanisms during budding.

## References.

1. Choppin, P.W., Klenk, H.D., Compans, R.W. and Caliquiri, L.A. 1971. Persp. Virol. p.127, ed. Pollard, M. London: Academic Press.

2. Klenk, H.D. 1973. Biological Membranes, Vol. 2, p. 145, ed. Chapman, D. and Wallach, D. London: Academic Press.
3. Lenard, J. and Compans, R.W. Biochim.Biophys.. Acta. (Review of Biomembranes) In press.
4. Franklin, R.M. 1973. Membrane Mediated Information. Ed. Kent, P.W. Lancaster: Medical and Technical Publishing Co., Ltd.
5. Blough, H.A. and Tiffany, J.M. Advances in Lipid Res. 11 (1973) 267.
6. Klenk, H.D. and Choppin, P.W. Virol. 40 (1970) 939.
7. Klenk, H.D. and Choppin, P. W. Virol. 38 (1969) 255.
8. Renkonen, O., Kääriäinen, L., Simons, K. and Gahmberg, C.G. Virol. 46 (1971) 318.
9. Laine, R., Kettunen, M., Gahmberg, C.G. Kääriäinen, L. and Renkonen, O. J. Virol. 10 (1972) 433.
10. Renkonen, O., Luukkonen, A., Brotherus, J. and Kääriäinen, L. 1974. Control of Proliferation in Animal Cells. Ed. Clarkson, B. and Baserga, R. Cold Spring Harbor Laboratory. In press.
11. Franklin, R.M. (personal communication).
12. Garoff, H., Simons, K. and Renkonen, O. (submitted for publication).
13. Simons, K., Helenius, A. and Garoff, H. J. Mol. Biol. 80 (1973) 119.
14. Singer, S.J. and Nicolson, G.L. Science 175 (1972) 720.
15. Helenius, A. and Simons, K. J. Biol. Chem. 247 (1972) 3656.
16. Meunier, J.C., Olsen, R.W. and Changeux, J.P. FEBS Letters 24 (1972) 63
17. Rubin, M.S. and Tzagoloff, A. J. Biol. Chem. 248 (1973) 4269.

18. Walter, H. and Hasselbach, W. Eur. J. Biochem. 36 (1973) 110.

19. Makino, S., Reynolds, J.A. and Tanford, C. J. Biol. Chem. 248 (1973) 4926.

20. Uterman, G. and Simons, K. J. Mol. Biol. In press.

21. Capaldi, R.A. and Vanderkooi, G. Proc. Nat. Acad. Sci., U.S.A. 69 (1972) 930.

22. Eng, L.P., Chao, F.C., Geostl, B., Pratt, D. and Tavastjerna, M.G. Biochemistry 7 (1968) 4455.

23. Laver, W.G. Virology 45 (1971) 275.

24. Brand, C.M. and Skehel, J.J. Nature New Biol. 238 (1972) 145.

25. Lazdins, I., Haslam, E.A. and White, D. O. Virology 49 (1972) 758

26. Wrigley, N.G., Skehel, J.J., Charlwood, P.A. and Brand, C.M. Virology 51 (1973) 525.

27. Helenius, A. and Söderlund, H. Biochim. Biophys. Acta 307 (1973) 287.

28. Gregoriades, A. Virology 54 (1973) 369

29. Gitler, C. and Montal, M. Biochem. Biophys. Res. Commun. 47 (1972) 1486.

30. Harrison, S.C., Caspar, D.L.D., Camerini-Otero, R.D. and Franklin, R.M. Nature New Biol. 229 (1971) 197.

31. Harrison, S.C., David, A., Jumblatt, J. and Darnell, J.E. J. Mol. Biol. 60 (1971) 523.

32. Landsberger, F.R., Lenard, J., Paxton, J. and Compans, R.W. Proc. Nat. Acad. Sci. U.S.A. 68 (1971) 2579.

33. Landsberger, F.R., Compans, R.W., Paxton, J. and Lenard, J. J. Supramol. Struct 1 (1972) 50.

34. Landsberger, F.R., Compans, R.W., Choppin, P. W. and Lenard, J. Biochemistry. In press.

35. Bretscher, M. Science 181 (1973) 622.

36. Gahmberg, C.G., Simons, K., Renkonen, O. and Kääriäinen, L. Virology 50 (1972) 259.

37. Marchesi, V.T., Tillack. T.W., Jackson, R.L. Segrest, J.P. and Scott, R.E. Proc. Nat. Acad. Sci. U.S.A. 69 (1972) 1445.

38. Segrest, J.P., Jackson, R.L., Marchesi, V.T. Guyer, R.B. and Terry, W. Biochem. Biophys. Res. Commun. 49 (1972) 964.

39. Garoff, H. and Simons, K. (Submitted for publication).

40. Niehaus, W.G., Jr. and Wold, F. Biochim. Bio phys. Acta 196 (1970) 170.

41. Bickle, T.A., Hershey, J.W.B. and Traut, R.R. Proc. Nat. Acad. Sci., U.S.A. 69 (1972) 1327.

42. Brown, D.T., Waite, M.R.F. and Pfefferkorn, E.R. J. Virol. 10 (1972) 524.

43. Razin, S. Biochim. Biophys. Acta 265 (1972) 241.

44. Tanford, C. J. Mol. Biol. 69 (1972) 59.

45. Ribeiro, A.A. and Dennis, E.A. Biochim. Biophys. Acta 332 (1973) 26.

46. van Deenen, L.L.M. (1968) Regulatory Functions of Biological Membranes. Ed. Järnefelt, J. Amsterdam: Elsevier Publishing.

47. Helenius, A., von Bonsdorff, C. -H. and Simons, K. (In preparation)

48. Garoff, H. (Submitted for publication)

49. Racker, E. Biochem. Biophys. Res. Commun. 55 (1973) 224.

50. Hosaka, Y. and Shimizu, Y.K. Virol. 49 (1972) 627.

51. Hosaka, Y. and Shimizu, Y.K. Virol. 49 (1972) 640.

52. Strauss, J.H., Burge, B.W. and Darnell, J.B. Virology 37 (1969) 367.

53. Schlesinger, M.J. and Schlesinger, S. J. Virol. 11(1973) 1013.

54. Summers, D.E., Roumiantzeff, M. and Maizel, J.V., Jr. Ciba Foundation Symp. on Strategy of the Viral Genome, p. 111, ed. Wolstenholme, B.W. and O'Connor, M.

55. Söderlund, H. Intervirology 1 (1973) 354.

56. Marcus, P.I. Cold Spring Harbor Symp. Quant. Biol. 27 (1962) 351.

57. Rutter, G. and Mannweiler, G. Arch Ges. Virusforsch. 43 (1973) 169.

58. Bonsdorff, C. -H., von, Commentationes Biologiae Societas Scientiarum Fennica. 74 (1973) 1.

59. Söderlund, H., Kääriäinen, L., von Bonsdorff, C. -H. and Weckstrom, P. Virology 47 (1972) 753.

---

TABLE I

Major Groups of Enveloped Viruses

| Group | Virus | | Nucleic Acid | |
|---|---|---|---|---|
| | Shape | Size (Å) | Type | MW x $10^6$ |
| *Animal Viruses* | | | | |
| Togaviruses<br>- Semliki Forest (SFV)<br>- Sindbis | Spherical | 700 | RNA | 4 |
| Myxoviruses<br>- Influenza | Spherical or filamentous | 800-1200 | RNA | 3-4 |
| Paramyxoviruses<br>- Simian 5 (SV5)<br>- Sendai | Spherical or filamentous | 1200-2000 | RNA | 6-7 |
| Rhabdoviruses<br>-Vesicular stomatitis (VSV) | Bullet shaped | 650-1750 | RNA | 3.5-4 |
| RNA tumor viruses | Spherical | 1200 | RNA | 10-12 |
| Herpesviruses | Spherical | 1800-2000 | DNA | 70-100 |
| Poxviruses<br>-Vaccinia | Brick shaped | 2200x2200 x2800 | DNA | 160 |
| *Bacterial Viruses* | | | | |
| Marine bacteriophage PM 2 | Icosahedral | 600 | DNA | 6 |

TABLE II

The Polypeptides in Semliki Forest Virus

| Polypeptide | MW x $10^3$ | Carbohydrate content (%) | Molar Ratio in SFV |
|---|---|---|---|
| *Membrane* | | | |
| $E_1$ | 49 | 8 | 1 |
| $E_2$ | 52 | 11 | 1 |
| $E_3$ | 10 | 40 | 1 |
| *Nucleocapsid* | | | |
| NC | 33 | - | 1 |

TABLE III

Properties of the Delipidated SFV Membrane Glycoprotein Complexes

| | 4 S | 23 S | 26 S |
|---|---|---|---|
| Stokes radius (Å) | 53 | 94 | 87 |
| $S_{20,W}$ | 4.5 | 23.5 | 26 |
| Molecular weight | $1.5 \times 10^5$ | $1.1 \times 10^6$ | $9 \times 10^5$ |
| Polypeptide composition | $(E_1E_2E_3)$ | $8(E_1E_2E_3)$ | $8(E_1E_2E_3)$ |
| Number of Triton X-100 Molecules | ∿75 | ∿260 | - |

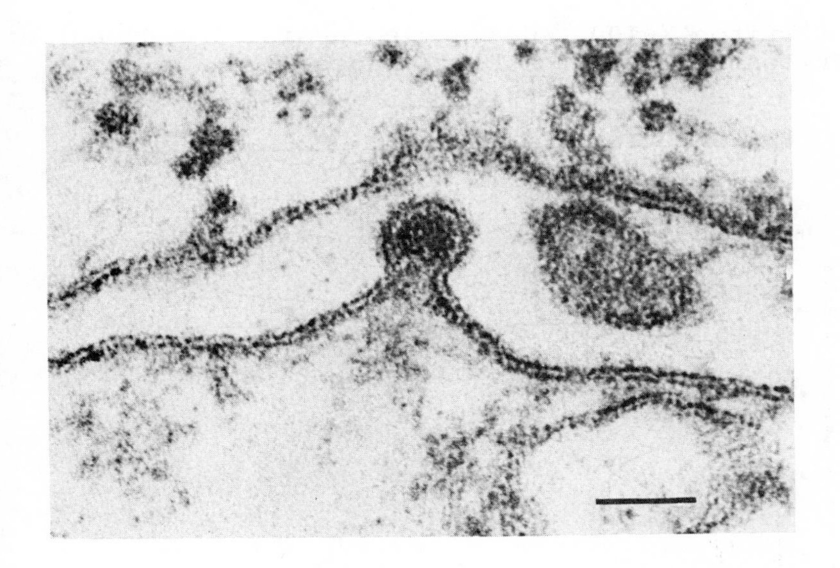

Figure 1. Semliki Forest virus assembly by budding from the plasma membrane. A nucleocapsid is seen enveloped by cell membrane. Bar 100 nm. The electron micrograph was taken by Dr. C.-H von Bonsdorff.

---

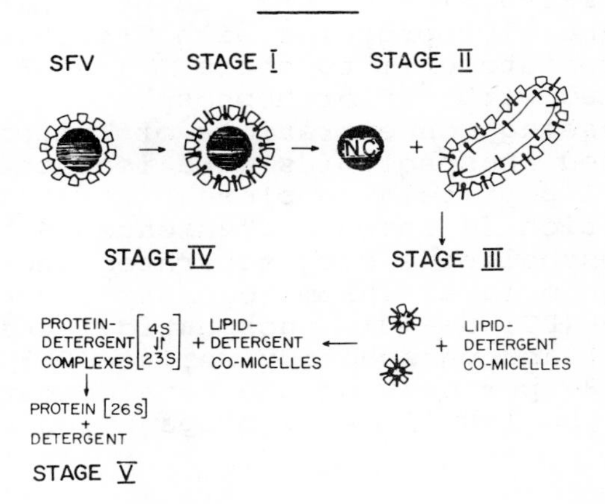


Figure 2. A schematical figure of the SFV membrane dissociation into soluble protein and lipid complexes caused by increasing concentrations

of Triton X-100. NC = nucleocapsid. The symbol of Triton X-100 is the black match-like rod.

---

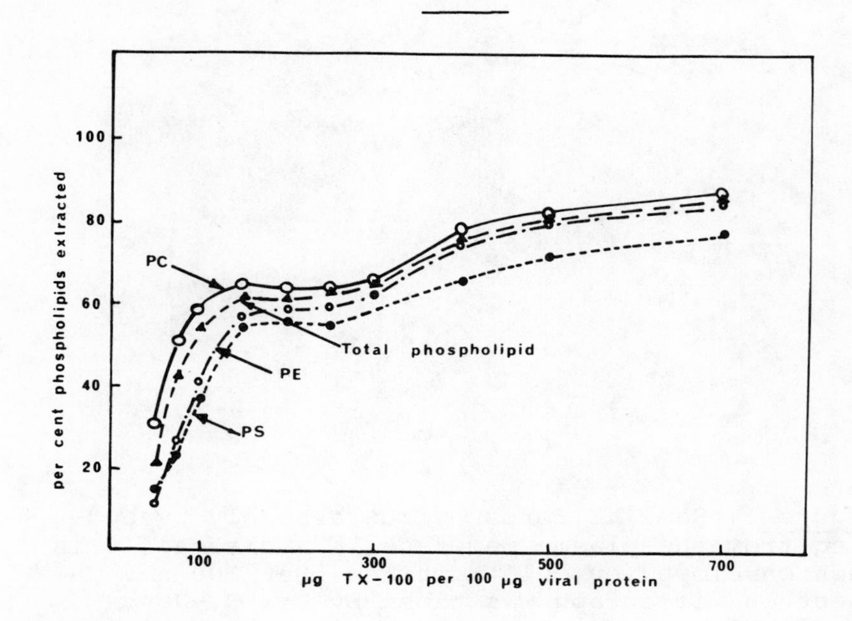


Figure 3. Extraction of the phospholipids from SFV membrane glycoproteins with Triton X-100 proceeding from stage II to stage IV (see Fig. 2). SFV labelled with $^{32}$P-orthophosphate was treated with increasing concentrations of Triton X-100. The released phospholipids were isolated from the detergent-lipoprotein complexes (stage III) by centrifugation in sucrose gradientes [27]. The different phospholipids were separated and quantitated by thin layer chromatography[8]. Phosphatidylcholine (PC; O——O), phosphatidylethanolamine (PE; o--•--o) and phosphatidylserine (PS; ●----●) comprised 86 per cent of the total phospholipid (▲--▲) in the labelled SFV preparation used.

---

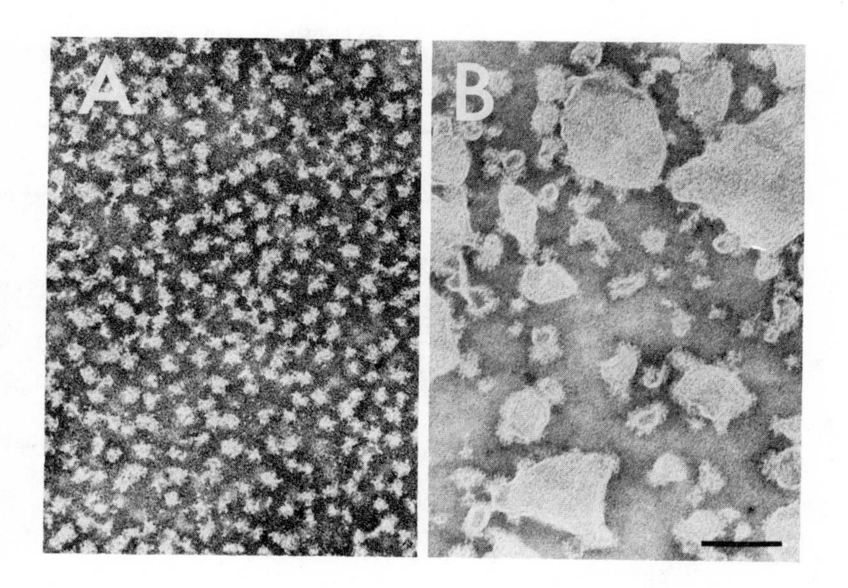


Figure 4. Reassociation of the SFV spike glycoproteins with BHK21 cell phospholipids. A. Electron micrograph of the 26 S complex containing less than 1 per cent phospholipid and detergent. B. Electron micrograph of reassociated SFV membrane. The 26 S complex was mixed with liposomes prepared with BHK21 cell phospholipids, and the mixture was sonicated for 10 minutes at 39°C. Negative staining with potassium phosphotungstate, pH 7.4. Bar 100 nm.

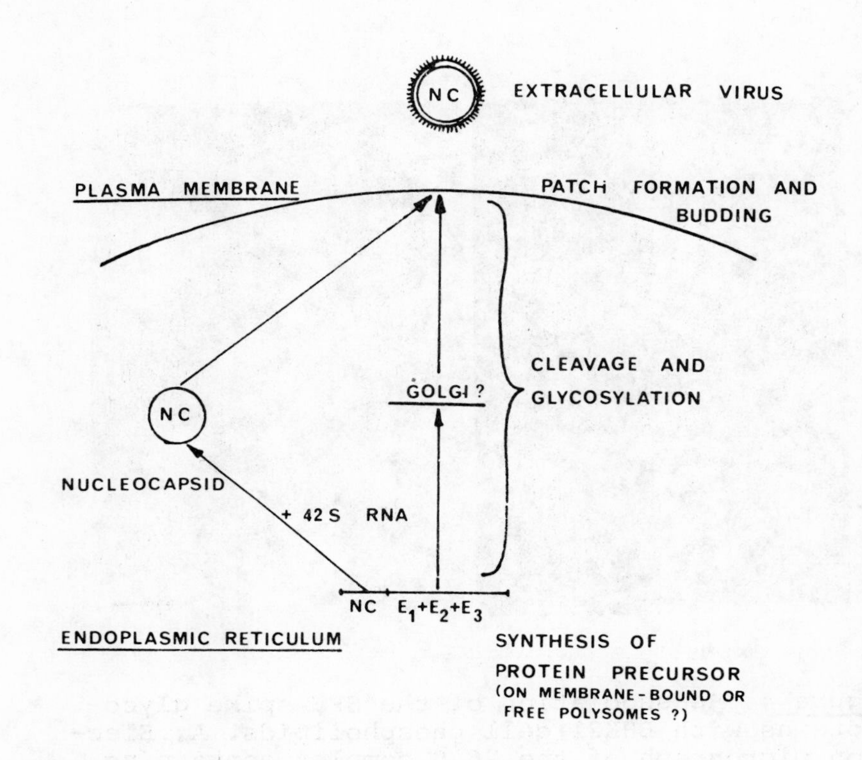


Figure 5. A scheme for togavirus biogenesis.

# WATER IN THE STRUCTURE AND FUNCTION OF CELL MEMBRANES

Jorge Cerbón

Departamento de Bioquímica
Centro de Investigación y de Estudios
Avanzados del I.P.N.
Apartado Postal 14-740
México, D. F.

The development of various NMR techniques has made it possible to directly examine the bonding and molecular motion of water protons and the application of these methods to living cells has now made considerable progress. However, there is no agreement on the interpretation of the NMR data obtained and the physical state of cellular water has become a topic of lively controversy. In spite of the experimental evidence that has been provided on the importance of water to the structure and function of the cell membrane (1-4), this component has not been taken into account in most of the studies and reviews devoted to this topic. If we consider that lipid bilayers with similar properties to those of cell membranes can only be formed in water, that many physiological properties are modified or abolished when water is substituted even by deuterated water, and that polypeptide conformation and stability depend to a large extent on the structure of the solvent (5), the very important role of water in structure and function of cell membranes cannot be ignored. Membrane components are: proteins, about 50-60%; lipids, about 30-40% and usually a small amount of carbohydrates (5-10%) and traces of RNA. Interpretation of the NMR data can be aided if the measurements of the different components available are taken as a guide to understanding the molecular behaviour of both the water and the membrane. Attempts to explain the state of water in living systems have taken into consideration only the major component of the cells, without realizing that the influence of proteins on water structure is only evident when proteins are in solution, and that even so, their effect is less marked than that of lipids and that of nucleic acids (3).

The purpose of this presentation is not a review of the present status of the state of water in living cells and well-defined model systems, but to analize and further document on the following aspects: 1) which are the components of the living cell and of their membranes responsible for the state of water? 2) what is their order of importance? 3) what is the effect of anionic, cationic and non-charged groups. Furthermore, the importance of the structure of the solvent at the cell surface on permeability and other membrane phenomena including membrane fluidity, will be discussed as well as a perspective of how boundary conditions can be further studied.

Walter and Hope have written an excellent review that includes a brief description of the important concepts and techniques involved, of the evidence that NMR studies have provided on the state of water in living and, in well-defined model systems.

## Properties of Water Bound to Proteins

Proton mobilities have been studied in hydrated systems of gelatin and conclusions about the state of water have been based on the frequency dependance of the NMR relaxation processes (7). This method allows the investigator to determine the distribution of proton mobility by keeping the temperature constant. This represents an advantage since the possible temperature range is somewhat limited with living cells or heavily hydrated specimens. The characteristics of the modified proton states in the gelatin-water systems are markedly dependent on water content. They are tentatively attributed to gelatin protons coupled for spin-lattice relaxation to those of the bulk water by exchange and spin diffusion. At 75% water a single exponential decay was observed that was indicative of rapid exchange between the bound water and the bulk water; but at 15% water or less,two exponentials appear.At this point, 10-30% water protons appear in a fast-decaying fraction of T2 less than 4 msec. The presence of two exponentials is indicative of a water fraction that does not exchange or exchanges very slowly with the bulk water. The authors present a model of distribution of correlation times. For gelatin samples containing 0.4 to 1.0 g water/g gelatin, the limit for (X) the fraction of modified water varies from 20 up to 100 % and the corresponding correlation times from log-9 to -11, taking log -12 for liquid water at 25°C. So, a variation ranging from water with a mobility only ten times slower than that of the bulk water up to water 300 times less mobile (if only 20% of it is affected by the protein) can explain the relaxation times obtained. For a gelatin sample containing more water 4.1g water/g of protein, that is 20% protein in water, a concentration similar to the value of proteins in a living cell, the value of (X) the water fraction modified by the protein varies only from 0.8 to 10% and corresponding correlation times range from log-9 to -11.

In order to obviate problems of interpretation that arise when studying nuclei with spins = 1/2 in which spin-lattice relaxation is governed both by intramolecular (rotational) motions and intermolecular motions (relative motions of other neighbouring nuclear or electron magnetic moments), studies of $^{17}O$ NMR relaxation in a system of 5% ribonuclease A in $H_2\,^{17}O$ (11% isotopically enriched) have been performed (8). The relaxation times for $^{17}O$ compounds are governed entirely by the rotational reorientation times of water molecules because of the very large electric quadrupole moment of the $^{17}O$ nucleus. Also, $^{17}O$ relaxation is very much less sensitive to paramagnetic impurities than $^{1}H$ relaxation. The presence of protein nearly doubles the relaxation rate of $H_2\,^{17}O$ when the temperature is varying and it is clear that the macroscopic viscosity has very little to do with the interpretation of rotational motion of water molecules in these systems. With this methodology, the magnitude of hydration in proteins appears again to be only from 0.15g. water/g. protein. In the case of RNAse A, this is equivalent to 150 molecules of water/mole of protein. This is well under a reasonable estimate for a monolayer. Studies on Lysozime (9), indicate a coverage of one water molecule per polar side chain group, taking into account the initial linear segment of the plot of $T_2$ versus water coverage.

Other studies of proton relaxation in protein solutions and gels of beef serum albumin, egg albumin, and RNAse (10), gave the following conclusions: There was only one $T_1$ time and only one $T_2$ time for the protein solutions and gels examined with all concentrations (5 to 10%) and at all temperatures (10 to 80°C). $T_1$ was greater than $T_2$ and the absolute values for $T_1$ and $T_2$ were reduced and the ratio $T_1/T_2$ was increased. The reported values were between 0.5 to 1.0 seconds for $T_1$ and around 300-400 msec for $T_2$. Since similar characteristics hold for maize leaves, bean leaves, frog liver and frog gastrocnemius, it is reasonable to conclude that the main component responsible for the state of water in living cells is that of the soluble proteins.

## Water in Macromolecular Systems of High Negative Charge Density

In this case, we will concentrate on information pertaining to nucleic acids and lipid membranes (liposomes). In 1967 the existance of immobilized water in lipid systems was reported both in vitro and in vivo by CW NMR studies (3). The viscosity contribution to the broadening of the resonance line was determined by plotting $T_2$ versus $\eta r$ values and it was established that both polar lipid liposomes and RNA solutions were able to restrict the water mobility on their highly-charged surfaces, while dextran and bovine serum albumin taken as examples of polysaccharydes and proteins were unable to show this phenomenon. When the systems were exposed to the well-known salt-breaking effect of calcium, the existance of irrotationally bound water was confirmed by the peak height increments in the water resonance signal of the lipids and RNA systems but not in these containing protein or polysaccharyde; these data indicated that a higher amount of water is affected in the first two systems as reflected by their NMR properties. Last year these systems were restudied by Spin Echo NMR techniques, since these techniques obviate to a large extent the inconveniences of magnetic inhomogenities, the results can be analyzed with more confidence. A simple analysis of the results obtained by Spin Echo NMR of the water in yeast-cell pellets, liposomes, RNA solutions etc. under different environmental conditions (detailed results to be reported elsewhere), again shows that the role played by the proteins in structuring the water in living cells is of minor importance when compared to that played by nucleic acids and phospholipids. Therefore, in the particular case of the cell membrane in which 40% by weight are lipids and the great majority of the protein molecules seems to be buried into the bilayer, the properties of water at the interphase membrane-water are mainly dictated by the lipid phase.

Let us now take a yeast cell which has about 10% nucleic acids, 65% proteins, 20% lipids and 5% polysaccharydes. A yeast pellet obtained by centrifugation of a yeast suspension at 3000 g during 10 min. has 60% intracellular water plus 40% extracellular water. Taking hydration values of 1.7g water/g RNA; 0.1-0.2 g water/g protein., 0.2-0.4 g water/g lipid and 0.1 g water/g polysaccharyde, we obtained a hydration value for the whole yeast of about 35%. A single exponential decay containing about 90% of the total water of the pellet was observed and a value of $T_1$=399 msec calculated (Fig. 1). The $T_1$ value corrected by the water that does not participate in the exchange (small fast exponential decay*) was=272 msec. Because of this we can be sure of rapid exchange between the major part of the intracellular water and the extracellular water and therefore the appropiate formula can be utilized to calculate $T_1$ for the bound water.

$$\frac{1}{T_1 \text{ observed}} = \frac{X}{T_1 \text{ bound}} = \frac{1-X}{T_1 \text{ free}}$$

$T_1$ observed = 0.399 sec.; $T_1$ free = 3.0 sec. and X= 0.234, since the water of hydration 35% corrected by the total water in the sample is equal to 23.4%. A value of $T_1$ for the water bound of 0.104 sec. is obtained.

If we apply the same calculations considering only thw water bound to proteins, assuming all the proteins are in solution and with a value of 0.2 g water/g protein, we still obtain a value of $T_1$ for the bound water of only 3.5 msec. This value is too low for water bound to proteins. The values reported in the literature on protein water systems at concentrations equivalent (similar) to that existing in the yeast sample (20% protein) are in general much larger 0.400 to 0.600 sec, or more than 10 times as high. On the other hand, if a 2% yeast RNA solution is dialyzed against EDTA for 24 hours and first studied by ESR to show the absence of paramagnetic cations, then utilized for Spin Echo NMR studies of water relaxation and the results obtained used to calculate the relaxation rate of water, a value of $T_1$ (water bound) = 0.191 sec is obtained. In spite of the high dilution employed, the magnitude of this value is more in accordance with values obtained for water in living systems when these are corrected by the amount of extracellular water present.

Similar studies with lecithin liposomes (5% egg lecithin in water) gave values of $T_1$ = 300 msec. Therefore, the order of importance of the components of a living cell in contributing to define the state of the water is as follows: RNA, DNA, Lipids, Proteins and polysaccharydes. Information obtained mainly by NMR techniques for natural and artificial products including detergents showed that the order of importance is also anionic>cationic>neutral. The weak influence of the proteins can be seen very clearly when we consider that most of the protein in the cells is particulated; for example, 1 gram of wet liver has only 40 mg of soluble protein. Since liver has 70 to 80% water, 1g. of liver has 200 to 300 mg dry weight of which 130 to 182 mg are protein (∿ 65%); therefore, 40 mg of soluble protein represent 22 to 32% of the dry weight and about 5-7% of the wet weight. A 5-7% aqueous solution of protein or gel has a $T_1$ of about one second at room temperature. Since the rest of the protein is particulated and a large portion of it may be unaccesible to water, its influence on the structure of water is therefore lower than that of the other cell components.

With regard to the plasma membrane, if we accept that the principal components are lipids and proteins, then lipids usually account for around 40% by weight, the balance being protein because the amount of carbohydrate is usually small. We then assume that the matrix is composed of lipid molecules mainly arranged in a bimolecular leaflet. This last point may not be acceptable to everyone, but I would like to mention additional evidence in support of this structural hypothesis. When utilizing local anaesthetic molecules as NMR probes of the physical properties of lipid bilayers (11), we have observed an identical type of differential broadening of local anesthetics (L.A.) when they are interacting with liposomes and natural membranes (Fig. 2). These same L.A. molecules, however, show a completely different type of alterations in their NMR spectrum when interacting with micellar systems, in spite of the fact that the same kind of polar and non-polar forces are involved (12). In micellar systems the L.A. NMR spectrum shows only very small chemical shifts and practically no differential broadening (Fig. 3). In view of the evidence provided by freeze etching of yeast, we can accept that a large proportion of the membrane protein particles reside in the inner part of the bilayer, and therefore the main components exposed to the external surface of the membrane must be phospholipids. The structured water at the interphase must be due mainly to the polar groups of the lipids and not to the influence of the proteins. Even if we accept that in some cells large portions of the proteins are facing the water, we would find that more water is affected in its mobility by lipids than by proteins because of the higher hydration of the lipids: 40% for lecithin and up to more than 100% for PS. (data obtained by $D_2O$ Resonance (13)), against only 10 to 20% in proteins. Evidence on the importance of the amount of structured water at the cell surface on permeability phenomena, as tested by NMR, fluorescence spectroscopy, etc., as well as on the effect of temperature, pH and ionic strength on lipid mobility has been published (14).

The great importance of the structure of water on polypeptide conformation and stability has also been reported and reviewed by Hagler, Scheraga, and Nemethy (5). These last two properties -conformation and stability- are those in which the specific activities of the cell membrane (recognition, binding, transport, etc. etc.) reside. Therefore, since lipids control the structure of water at the cell membrane and this particular structure of the solvent modulates the conformation and stability of the polypeptides, the very important role of water in membrane structure and function is more than obvious.

An example, of the above,is found in the fact that high ionic strength has a dramatic effect on the binding of insulin to fat-cell membranes (11). Increasing concentrations of NaCl up to 2M cause a dramatic (6-fold) increase in insulin binding. This probably results from the appearence (unmasking) on the membrane of new binding sites for insulin that are kinetically indistinguishable from those normally exposed. The effect of 2M NaCl appears to be qualitatively similar to the effects of digesting the membrane with phospholipases. Both phospholipase treatment and high ionic strength alter (break) the structure of the solvent at the cell membrane-water interphase by reducing the density of negative charge and allow insulin to reach new binding sites.

On the other hand, sodium phosphate preserves or even increases hydration of lipid dispersions, and enhances permeability (leakage of cell components) of Nocardia (1) and yeasts. The glucose transport in Nocardia seems to be of a type that does not require energy but does require phosphate, which can be replaced by sodium arsenate but not by sodium chloride or potassium chloride of identical ionic strength (16). The possible role of these anions (phosphate and arsenate) could be that of facilitating the sugar-transporter interaction at the highly hydrophobic cell surface of Nocardia. The importance of the structure of the solvent is evident in this case since Nocardia cells exposed to 1% glucose in water (hypotonic conditions) are hydrated to a larger extent although they are still incapable of transporting the sugar.

The modification of a large variety of membrane-associated phenomena by changes in pH, temperature, ionic strength or even "specific" ionic effects are perhaps due to local alteration in the structure of the solvent.

Hypertonicity effected by elevation of NaCl concentration (only up to 100 mM) in the growth medium results in a rapid cessation of protein synthesis in Hela cells accompanied by a complete breakdown of polyribosomes. Upon restoration of isotonicity protein synthesis resumes at a normal rate and ribosomes are synchronously reinitiated only at the proper site. Since this can also be accomplished by other salt solutions and even by sucrose (17), another explanation for this phenomenon could be a change in the structure of the intracellular water with marked effects on polyribosomes structure and function.

How can boundary conditions be studied? In other words, how can variations in the structure of the solvent at the membrane surface be measured if the cells or the natural or artificial membranes are suspended in aqueous solutions, the bulk water exchanges at a rapid rate with the structurated water, and, above all, the fraction of modified water represents a minor component of the extracellular medium? It will then be necessary to isolate an interphase with only a small amount of water. The study of inverted micelles in which the polar groups of membrane components can be exposed to controlled amounts of water could be the solution. The effect of water on the membrane components and viceversa can be followed by means of Fourier transform NMR. By utilizing this technique. I have been able to detect the effect of millimolar concentrations of NaCl, KCl, LiCl, etc., even on switterionic lecithin by measuring variations of $T_1$ of the water inside the inverted micelle (data to be reported in detail). It is well known that these salt concentrations are incapable of altering the NMR properties of liquid bulk water. Therefore, their effect is due to their modification of the interaction of water with the hydrophilic region of the lipid.

Is water, then, the directing or the orienting force in membrane structure and function, and in biochemical functionality?

## Acknowledgements

I gratefully acknowledge the aid given to me by Dr. Brian Pethica in the course of many helpfull discussions as well as in providing me with the necessary research facilities at Port Sunlight, Unilever Research Laboratories, England. I also wish to express my profound thanks to the John Simon Guggenheim Memorial Foundation for the fellowship that has enabled me to continue this work.

## References

1. Cerbón, J. Biochim. Biophys. Acta 88 (1964)444.
2. Cerbón, J. Biochim. Biophys. Acta 102(1965)449.
3. Cerbón, J. Biochim. Biophys. Acta 144(1967)1.

4. Cerbón, J. Biochim.Biophys. Acta 211(1970) 389.

5. Hagler, A.T., Scheraga, H.A. and Nemethey, Ann. N.Y. Acad. Sci. 204 (1973) 51.

6. Walter, J.A. and Hope A.B. Progress in Biophysics and Molecular Biology 23 (1971) 3.

7. Outhred, R.K. and George, E.P. Biophys. J. 13 (1973) 83.

8. Glasel, J.A. Nature 218 (1968) 953.

9. Brey, W.S. Jr., Evans, T.E. and Hitzrot, L.J. J. Colloid and Interface, Sci. 26 (1968) 306.

10. Abetsedarskaya, L.A., Miftakhutdinova, F.G., Fedotov, V.D. and Mal'tsev, N.A. J. Mol. Biol. 1(1967) 451.

11. Cerbón, J. Biochem. Biophys. Acta 290(1972)51.

12. Fernández, M.S. and Cerbón, J. Manuscript in preparation.

13. Finner, E.G. and Darbe, A. Personal communication.

14. Cerbón, J. PAABS Symposium 1 (1972) 313.

15. Cuatrecasas, P. J. Biol. Chem. 246(1971) 7265.

16. Cerbón, J. and Ortigoza Ferado, J. J. Bacteriol. 95(1968) 350.

17. Saborio, J.L., Sheng Shung, Pong and G. Koch. J. Mol. Biol. In press.

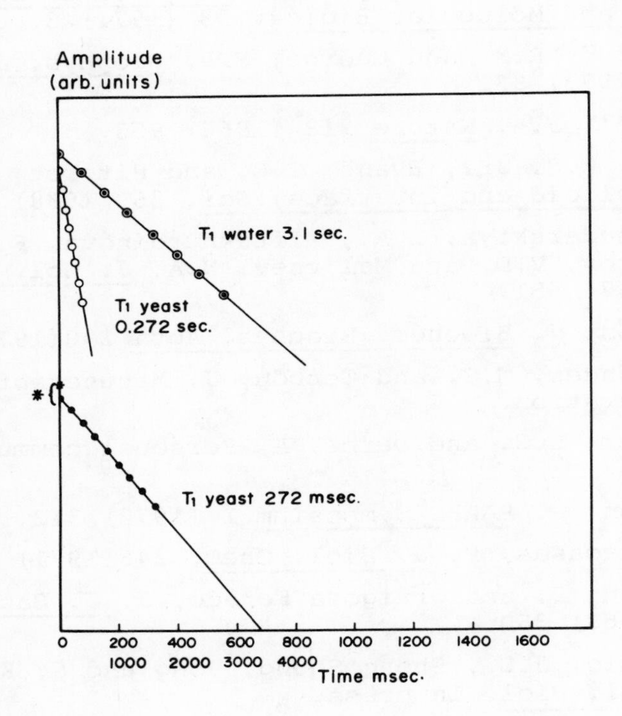


Figure 1. Pulsed NMR measurements of spin-lattice ($T_1$) of yeast cells exposed to 300 mosm. Na Cl.

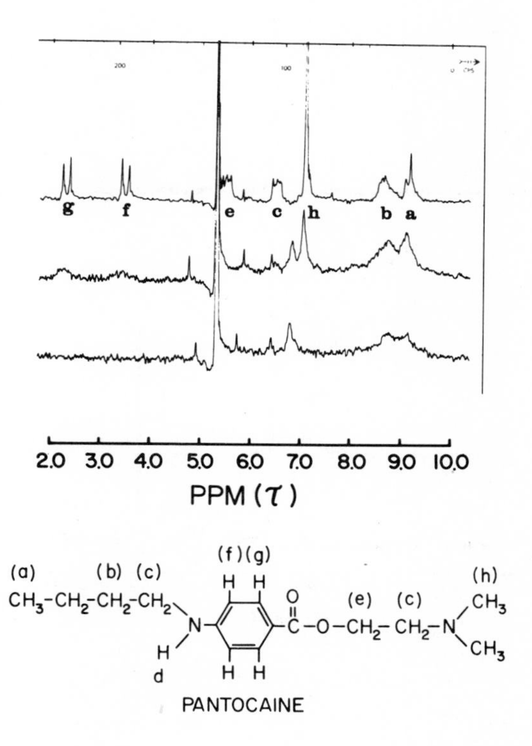


Figure 2. The NMR spectra of Pantocaine-HCl (2% w/v) in $^2H_2O$ upperline, and of pantocaine-HCl (2% w/v) in sonicated egg-phospholipid dispersion (5% w/v) in $^2H_2O$ middleline (Note the broadening of the NMR signals h, f and g of the L.A. molecule), and sonicated egg-phospholipid dispersion (5% w/v) in $^2H_2O$, lower line 33°C.

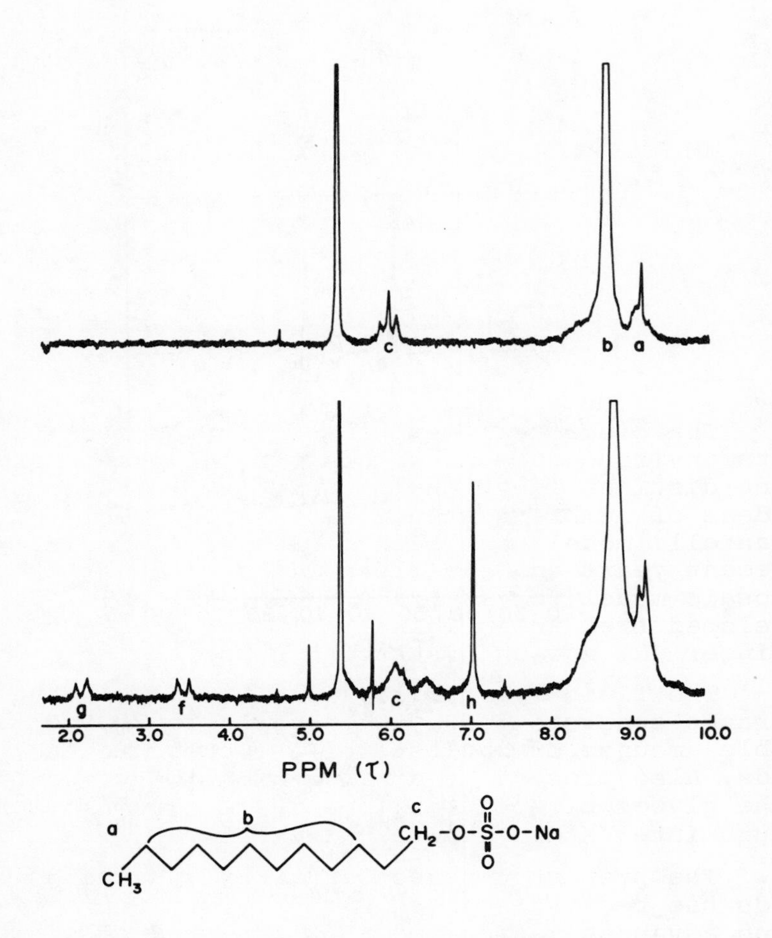


Figure 3. The NMR spectra of sodium dodecylsulphate (SDS) 50 mM in $^2H_2O$ upperline; and pantocaine HCl. (1.5% w/v) plus SDS in $^2H_2O$, lower line. 33°C. Note the absence of broadening in the NMR signals h, f, and g of the L.A. molecule together with the large broadening of signal b of the SDS molecule.

# GLYCOSPHINGOLIPIDS IN BIOLOGICAL MEMBRANES

G.M. Gray

MRC Unit on the Experimental
Pathology of Skin
The Medical School.
The University, Birmingham
England

The biological membrane separates a cell from its environment and, within the cell, separates one distinct functional organelle from another. Ideas of membrane structure based on the Davison-Danielli model have been considerably modified in recent years and the fluid lipid-globular protein mosaic model for membrane organisation (1) developed from the original proposal of Lenard and Singer (2) now enjoys strong support.

The fluid lipid bilayer in most mammalian membranes consists mainly of phospholipids with variable amounts of cholesterol and other neutral lipids. Also present is a minor group of compounds, the glycosphingolipids, and these have aroused much interest in several areas of research.

The present interest in the glycosphingo lipids has been stimulated, firstly, by the outstanding advances made since 1960 in our knowledge of the inherited metabolic defects which cause the series of lipid storage diseases known collectively as the sphingolipidoses (3). Secondly, by the work of Rapport and colleagues who, in 1958, isolated a pure lipid hapten from human epidermoid carcinoma which reacted with rabbit antisera directed against many different types of human tumour (4). The hapten, called Cytolipin H by Rapport, was subsequently identified as lactosylceramide. Their further studies suggested that all the known glycolipids possessed haptenic function

(5) and thus may be immunologically significant. And finally, by the accumulating evidence for a predominantly plasma membrane location of cell glycosphingolipids (6) coupled with the observations that the glycosphingolipid patterns of several cultured cell lines undergo changes following cell transformation by viruses (7).

This information has encouraged unceasing speculation on the possible contribution of the membrane glycosphingolipids to intercellular reactions, to the mechanisms of cell contact and growth control, cell transformation and malignancy. The object of this paper is to consider if it is possible to attribute specific biological functions to glycosphingolipids from the information at present available and to suggest how further relevant information might be obtained.

## The chemical structures of glycosphingolipids

These have been described in detail in several reviews (8-10) and only a brief summary for background information is necessary. The structures are conveniently divided into three classes.

a). The neutral glycosphingolipids. These compounds contain the basic lipid group, N-acylsphingosine (Ceramide) and the sugars, glucose, galactose, N-acetylgalactosamine and, more rarely, N-acetylglucosamine. The simplest compounds from which other glycosphingolipids are formed by the sequential addition of simple carbohydrate units are glucosyl(1→1)-ceramide and galactosyl (1→1) ceramide (Fig. 1).

b). Glycosphingolipids containing fucose. They all possess the basic structure, galactosyl β (1→4)glucosyl(1→1) ceramide to which is attached, by a N-acetylglucosamine β(1→3) link to the galactose, a series of oligosaccharide chains containing fucose, many of which exhibit blood group activity (Fig. 2). There is evidence that glycosphingolipids with blood group A activity can contain from six to sixteen carbohydrate units, though the A determinant grouping is always terminal (11).

c) The gangliosides. The majority of ganglioside structures stem from galactosylβ(1→4) glucosyl (1→1)ceramide (lactosylceramide) by the addition of N-acetyl (or N-glycolyl)neuraminic acid, N-acetylgalactosamine and galactose (Fig. 3).

A number of other glycosphingolipids have been isolated, usually in very minor amounts, which fit the general structural pattern of one of the three groups but differ in one particular sugar. For example a ganglioside with N-acetylglucosamine replacing N-acetylgalactosamine has been found in bovine spleen (12).

## Distribution of glycosphingolipids in mammalian tissues

The mammalian cell appears very selective in its choice of glycosphingolipids. Galactosylceramide, sulphatide (Fig. 1) and the mono- and disialogangliosides (Fig. 3, $GM_1$, $GD_{1a}$) are the major compounds in brain. Brain has a very high concentration of glycosphingolipids and in white matter they account for about 20% of the tissue dry weight. By comparison extraneural tissues have very minor amounts of glycosphingolipids and in most cases they account for less than 0.5% of the dry weight of the tissue. The major glycosphingolipids in extraneural tissues are the neutral compounds (Fig. 1) and the small ganglioside, hematoside (Fig. 3, $Gm_3$). Kidney, spleen and lung tissue are relatively rich in glycosphingolipids compared with liver and heart. Glycosphingolipids show some tissue specificity, the major component in mouse kidney, for example, is sulphatide, in mouse heart, hematoside and in mouse liver, monoglycosyl ceramide. Nevertheless, it is difficult to generalise about glycosphingolipid distribution as there are considerable variations amongst species and these tend to obscure tissue differences. Many extraneural tissues do have traces of the higher gangliosides and may also contain some fucose glycosphingolipids. The latter compounds have so far been identified in erythrocytes, spleen, pancreatic and intestinal cells, and in the plasma lipoproteins (13).

The detailed knowledge of the variety of

chemical structures, the distribution and variety of glycosphingolipid compositions in different tissues and species does not directly suggest particular biological functions. It is possible, however, that different glycosphingolipids can perform similar functions. For example, the glycosphingolipid composition of human granulocytes is almost entirely diglycosylceramide (lactosylceramide) whereas the glycosphingolipids in the granulocytes of the pig (a physiologically similar species) consist mostly of triglycosylceramide (14).

## Location of glycosphingolipids in the mammalian cell.

There is a body of evidence (6, 15) which indicates that the glycosphingolipids are predominantly localised in the plasma membrane of the mammalian cell and are not present in the membranes of some subcellular organelles. In contrast other studies on membrane fractions isolated from cultured cell lines (16, 17) have shown that glycosphingolipids are components of some subcellular membranes as well as plasma membrane and Marcus and Janis showed by an immunofluorescence technique (18) that globoside (Fig. 1) was present in the cytoplasm as well as in the plasma membrane of the epithelial cells in intact adult human kidney. Henning and Stoffel found that rat liver lysosomal membranes prepared from 'tritosomes' contained much higher concentrations of glycosphingolipids than the liver cell plasma membranes (19). The glycosphingolipid composition of the two membrane fractions were also different. However, they were unable to show if the lysosomal glycosphingolipids were derived from the plasma membrane, or originated from subcellular membranes taken up by autophagocytosis or were genuine lysosomal components.

With one or two exceptions the composition of glycosphingolipids in subcellular organelles appears to be similar to that in the plasma membrane and the generalisation that this membrane is the major site for these compounds still holds.

The confident assignment of glycosphingolipids to particular membranes other than the plasma membrane is very much dependent on the techniques of subcellular fractionation and membrane isolation and further improvements in membrane preparation procedures will no doubt modify some of the existing evidence on the location of these compounds.

Along with the acceptance of the plasma membrane site for glycosphingolipids has been the tacit assumption that they are components of the membrane outer surface rather than the inner one. Evidence that this is so for some of the blood group active glycosphingolipids in erythrocytes has been available for some time (10) but only recently it has been shown that at least two of the major glycosphingolipids in the human erythrocyte, triglycosylceramide and globoside (Fig. 1) are exposed at the external surface (20). The erythrocytes were treated with galactose oxidase followed by reduction with tritiated sodium borohydride. This labelled the galactose and galactosamine residues on the cell surface with tritium and the activities of the glycosphingolipids (and glycoprotein) were measured after extraction from the erythrocytes and chromatographic separation. The lactosylceramide was not labelled and therefore could not be proved to be at the external surface. If it is assumed that the compound is at the external surface of the membrane a possible reason for their unavailability to galactose oxidase (MW 75,000) becomes apparent by reference to Fig. 4 which compares the space-filled molecular models of the neutral glycosphingolipids with those of some phospholipids normally present in the fluid lipid layer of the membrane. It is likely that in a lipid layer in which molecules of phosphatidylcholine, sphingomyelin and phosphatidyl ethanolamine predominate the glucose or galactose of a monoglycosylceramide will often be closely associated with, and be within, the surface layer of the charged phosphate and base head groups of the phospholipids. Even the addition of a second molecule, galactose, as in lactosylceramide barely projects the carbohydrate group beyond the charged surface layer of polar groups. Under these conditions it may not normally be possible for enzyme-substrate or antigen-antibody complexes to form. It is worth noting that even after the erythrocyte is treated with trypsin, which exposes the globoside on the surface to antigloboside antibody, the membrane surface is unreactive towards antilactosylceramide antibody.

Recent evidence has indicated that phospholipids in the erythrocyte membrane (21, 22) have an asymmetric distribution, that is, phosphatidyl choline is predominantly on the outside and phosphatidyl-ethanolamine on the inside. It is not known whether glycosphingolipids have a similar asymmetric distribution but this could probably be shown by a comparison of tritium labelled (20) glycosphingolipids from intact erythrocytes and from ghosts.

## Biological functions.

It is difficult to attribute any particular biological function to the glycosphingolipids with any degree of certainty despite the mass of biochemical, chemical and physical data available on them.

Are they structural elements in the organisation of the biological membrane? In general, their minor contribution to the total lipids in a membrane effectively rules out the type of "building block" function often given to phosphatidylcholine. Possible exceptions are the membranes of the myelin sheath which contain high concentrations of galactosylceramide and sulphatide and the membranes of the epithelial brush borders of the small intestine in which glycosphingolipids are reported to account for 50% of the total lipids (23). Even in these tissues a structural role may be secondary to other functions associated with their carbohydrate groups. Nevertheless the Singer and Nicholson membrane model allows a wide range of lipid compositions, all providing a fluid lipid phase under normal physiological conditions, but each suited to the requirements and properties of a particular membrane. There is no difficulty in visualising, therefore, some replacement of phospholipids by glycolipids with a minimal disturbance of the broad molecular organization of the membrane. Support for such interchangeability comes from the work of Minnikin et al (24) which showed that under varying growth conditions phospholipids in Pseudomonas diminuta could be replaced almost completely by glycolipids.

Some phospholipids are required by several membrane-bound enzymes for optimal activity. Are some glycosphingolipids 'activators' for particular enzymes? Karlsson and co-workers (25) have shown an interesting correlation between the activity of the membrane bound $Na^+$, $K^+$-activated ATPase and the level of sulphatide in the salt gland of marine birds and in the kidney medulla of several mammalian species, but evidence is lacking of a specific requirement for sulphatide by this ATPase. Glycosphingolipid: protein interactions may be worthy of special study (see later).

Speculation on the role of gangliosides in brain has continued undiminished for many years. They have been localised in the membranes of nerve endings (26) and it is thought that they may play a role in the transmission of electrical impulses in the synapses. This view is not supported by a recent study (27) on several clonal lines of mouse neuroblastoma cells. The glycosphingolipid profiles of these cells were generally similar but they had widely different membrane electrical activities. It has been suggested (28) that gangliosides may be receptors for serotonin in the synapses since neuraminidase blocked the action of serotonin (a neurotransmitter) on a synaptic preparation and the activity was restored by the addition of gangliosides, the most effective being a disialoganglioside. It has also been suggested that gangliosides may be associated with the membrane receptors for some biogenic amines (29).

Somewhat surprisingly, it is in the extraneural tissue that firm evidence has been obtained of a tissue receptor role for a ganglioside. Holmgren et al (30) have shown that the tissue receptor for cholera exotoxin in the gut mucosa is the monosialoganglioside $GM_1$ (Fig. 3). The elegant work of Cuatrecasas (31) has confirmed and extended this finding. He puts forward the hypothesis that the initial toxin-ganglioside complex at the cell surface is biologically inactive (this would explain the considerable lag period before the action of the toxin) and subsequent changes occur spontaneously by rearrangements within the membrane structure which convert the complex to an active form. The active form would in some way modulate the activity of adenyl cyclase which could cause the observed increase in the intracellular concentrations of cyclic AMP in the epithelial cells of the small intestine following exposure to enterotoxin. This association of toxin and ganglioside may be unique but the gut mucosa is rich in glycosphingolipids as well as surface glycoproteins and the former compounds may represent a variety of receptors able to hold certain molecules at the mucosal surface.

Are glycosphingolipids important in the processes of cell transformation and malignancy? As mentioned earlier, the initial observation that the glycosphingolipid pattern of a cultured cell was changed following viral transformation, the knowledge of the potential immunological activity of glycosphingolipids and their possible importance as antigens in malignancy (32, 33) have stimulated an intensive study of glycosphingolipid metabolism during cell growth and transformation (6). The studies have been carried out on a wide range of cultured cell lines from hamster, mouse, rat, bovine, monkey and human sources and some general conclusions can be drawn. The glycosphingolipid patterns of cultured cell lines can be separated, with some overlap, into two types: those in which the neutral glycosphingolipids and hematoside predominate, with only traces of the higher gangliosides (e.g. monkey kidney cells) and those which have predominantly hematoside and the higher gangliosides and very little of the neutral glycosphingolipids (e.g. some mouse cell lines and adult hamster kidney epitheloid cells).

In vitro transformation of the cells by tumour viruses result in the decrease or loss of one or more of the higher gangliosides or neutral glycosphingolipids present in the normal untransformed cell, often with a corresponding increase in one of the precursor glycosphingolipids. The possibility that the glycosphingolipid pattern changed with, and was simply related to the rapidity of cell growth rather than transformation has been discounted by Hakomori and Siddique who showed that some lines of Morris rat hepatomas that divided more slowly than neonatal rat liver had glycosphingolipid patterns very different from those of both neonatal liver and normal adult liver (34). The patterns of the normal adult and neonatal livers were similar. Recent studies (35, 36) have shown that the loss of the more complex glycosphingolipids is the result of the loss, or inhibition of, a particular glycosyltransferase following the transformation of the cell by virus. Unfortunately this clear-cut biochemical change is probably unrelated to the mechanism of malignancy since the same cell lines can often spontaneously transform to produce highly malignant cells which do not show a decrease in the relevant glycosyltransferase activity or a change in the normal glycosphingolipid pattern (37).

In a number of cultured cell systems the concentrations of some glycosphingolipids are dependent on culture density. For example, the concentrations of tri-, tetra- and pentaglycosylceramides are much higher in NIL 2 (hamster kidney) cells in dense than in sparse culture (38) though the concentrations are not related to actual saturation densities, which can vary over a wide range. Glycosphingolipid metabolism in cells in sparse culture bears some similarity to that in transformed cells in that in NIL 2 cells, for example, the activity of UDP galactose:lactosylceramide-galactosyl transferase is much higher in dense compared with sparse cultures and those glycosphingolipids which are 'density dependent' are specifically depleted in the transformed cells which have very low transferase activity.

The possible importance of cell contact in relation to the density dependent glycosphingolipids has been considered (38) in the light of the hypothesis of Roseman and co-workers (39) which suggests that cellular recognition and adhesion may be mediated by binding of surface glycosyltransferases on one cell to target molecules of substrate on the surface of adjacent cells in a process of transglycosylation. Several observations do not support glycosphingolipid involvement in this mechanism. In cells seeded to sparse from dense culture, the density dependent triglycosylceramide is synthesised for some time even though cell contact is minimal (38) and cells made quiescent in sparse culture by serum depletion synthesise large amounts of triglycosylceramide.

The main results of the work I have discussed are summarised in Fig. 5. Despite the excellent work in this area no common finding has emerged which correlates the glycosphingolipids with a particular membrane property which may be modified during cell growth or transformation. The disparity of many of the findings arising from similar studies on different and even on the same cell lines highlights the limitations of the cultured cell system. It is possible that many of the changes in the metabolism of glycosphingolipids are artifacts of the culture system itself and may not relate to their metabolism in the fully differentiated, adult cell. The fact cannot be ignored that the glycosphingolipid compositions of the baby hamster kidney fibroblast (BHK) cell lines and the adult hamster kidney epithelioid cell lines (HaK) are different from each other and both very different from that of normal kidney tissue irrespective of whether the latter is from a hamster only a few days old or fully adult (40).

Is there a relationship between the glycosphingolipids and glycoproteins of the cell plasma membrane? The plasma membrane of the mammalian cell is covered by a surface coat of glycoprotein. It is not known whether these macromolecules are held by being partially embedded in the fluid lipid phase of the membrane or by their association with other proteins which are an intrinsic part of the membrane organisation. The glycoproteins do have a high turnover rate and appear to be continually shed into the surrounding medium as a process of normal cell function (41). Kapeller et al have suggested that the universally used 'trypsin treatment' of cells probably weakens the surface structure of the cell by its proteolytic action and thus merely enhances the rate of the normal shedding process. Part removal of the glycoprotein on the red cell surface, previously labelled by the lactoperoxidase iodination technique, by a short trypsin treatment exposes a far greater number of available sites for labelling than were present initially on the untreated cell surface (42). This also supports the idea that trypsin 'thins out' a network of glycoprotein chains covering the cell surface.

The glycoproteins which are so easily removed are rich both in fucose and sialic acid residues in terminal linkage. In several established cell lines there are direct correlations between sialic acid levels and cell saturation density (43). In some cell lines transformed by viruses levels of glycoprotein sialic acid (44) and glycosyltransferases (45) are lower than in the normal cells. Thus there are some similarities between the changes that occur to the glycosphingolipids and glycoproteins of cultured cells under different growth conditions and viral transformations. As these changes seem to be caused by the altered activities of one or more glycosyl transferases it is important to study the specificity of the altered transferase towards both glycosphingolipids and glycoproteins.

The carbohydrate chains of particular classes of glycoproteins and glycosphingolipids are considered to be synthesised by the sequential addition of glycosyl units by different multiglycosyltransferase systems (39). In general each glycosyltransferase is specific for a particular acceptor molecule. There are probably exceptions to this rule and there is evidence that the major sialyltransferase in fibroblasts (46) can transfer to glycoprotein and to glycosphingolipid substrates. That erythrocytes probably contain glycoproteins as well as glycosphingolipids with A and B blood group activity suggests that other glycosyltransferases may also have a dual specificity (47).

If we consider that some glycosphingolipids at the cell surface may be receptors for certain biologically active molecules the availability of the receptor site to the 'active' molecule is dependent on two factors: the ability of the approaching molecule to penetrate the glycoprotein coat to reach the glycosphingolipid receptor and the degree of penetration of the receptor into or projection above the glycoprotein coat. The human erythrocyte membrane is a good example. It contains glycosphingolipids with oligosaccharide chains of six to ten or more units which have Lewis blood group activity. Though the amounts of these compounds in the membrane are extremely small (10), the red cell is highly reactive towards the relevant antisera. In contrast the membrane contains large amounts of globoside, a tetraglycosylceramide, but the red cell is unreactive towards antigloboside antibody. However, erythrocytes pretreated with trypsin and fetal erythrocytes are both reactive towards antigloboside antibody presumably because a partial 'thinning' of the glycoprotein coat (or an incomplete coat in fetal cells) has enabled antigen-antibody contact to occur. Since Forssman antigen, a pentaglycosylceramide is also not reactive towards antibody in the intact erythrocyte it suggests that a lipid-bound oligosaccharide chain may require six or more residues to overcome the 'blocking effect' of the glycoprotein on its active terminal group. The available evidence suggests that future studies on the surface properties of plasma membranes involving glycosphingolipids and glycoproteins would benefit if both classes of compound were simultaneously studied rather than either one or the other in isolation.

It is reasonable to assume that compared with the membrane glycosphingolipids the importance of the surface glycoprotein in maintaining surface properties such as surface charge, cell recognition sites, binding sites, antigen pattern, etc., is overwhelming. Nevertheless the glycosphingolipids do provide a 'second line' of carbohydrate chains which are not shed and are firmly bound and in direct contact with the fluid lipid phase of the membrane (Fig. 4).They provide a direct, non-charged, hydrophilic entry into the hydrophobic lipid phase and may, for example, play a special role in transport across tight junctions between cells where, in contrast to desmosomes there is no interposing glycoprotein (48) and the lipid phases of the two plasma membranes are much closer together. As the techniques of membrane fractionation and isolation and surface labelling continue to improve it may soon be possible to determine the composition of tight junction areas of membrane. The recent development of methods to obtain purified antibodies to many of the glycosphingolipids will be valuable in studying the organisation of these compounds in the cell membranes (49).

The Singer membrane model has received much support from the application of a number of physical techniques to the study of membrane organisation. NMR measurements on the lipids of sarcoplasmic reticulum membranes (50) have indicated a possible non-random distribution of lipids in the membrane and the association of some fraction, in this case, some of the phosphatidylcholine, with membrane protein. All the physical measurements have concentrated so far on the major lipid components of the membranes, the phospholipids and cholesterol. These techniques could be applied equally well to the glycosphingolipids in the membranes and may provide significant information on their interaction with proteins and other lipids.

The reader will have concluded that the correlation of any biological function with this fascinating group of lipids is still very much a matter for speculation. They may have significant roles in some of the internal membrane systems of the cell, in the lysosome, for example, where there is some evidence that the gangliosides may act as binding sites for lysosomal enzymes at the internal surface of the membrane (51). But I believe that the main clues towards an understanding of their biological function will emerge from the studies of the factors which effect their metabolism and behaviour in the plasma membrane of the cell. The further lines of study that have been mentioned, of the possible interrelation of glycosphingolipid and glycoprotein metabolism, of glycosphingolipid-protein and even glycosphingolipid-phospholipid interactions though difficult with present techniques will almost certainly soon become practical possibilities if the advances in all aspects of membrane technology continue at their present rate.

## References.

1. Singer, S. J. and Nicholson, G.L. Science 175 (1972) 720.

2. Lenard, J. and Singer, S.J. Proc. Nat. Acad. Sci., U.S.A. 56 (1966) 1828.

3. Brady, R.O. (1972) NIH: Current Topics in Biochemistry. p. 1. ed. Anfinsen, C.B., Goldberger, R.F. and Schechter, A. N. New York Academic.

4. Rapport, M.M., Graf, L., Skipski, V.P. and Alonzo, N.F. Nature 181 (1958) 1803.

5. Rapport, M.M., Graf, L. and Schneider, H. Arch. Biochem. Biophys. 105 (1964) 431.

6. Critchley, D.R. (1972) Membrane Mediated Information. pp. 37. ed. Kent, P.W. London. Medical and Technical Publishers.

7. Hakomori, S. and Murakami, W.T. Proc. Nat. Acad. Sci., U.S.A. 59 (1968) 254.

8. Stoffel, W. Ann. Rev. Biochem. 40 (1971) 57.

9. Martensson, E. (1969) Progress in the Chemistry of Fats and other Lipids. Vol. 10, pp.367 ed. Holman, R.T. London. Pergamon.

10. Hakomori, S. Chem. Phys. Lipids. 5 (1970) 96.

11. Hakomori, S., Stellner, K. and Watanabe, K. Biochem. Biophys. Res. Comm. 49 (1972) 1061.

12. Wiegandt, H. Chem. Phys. Lipids. 5 (1970)198.

13. Marcus, D.M. and Cass, L.E. Science 164 (1969) 553.

14. Levis, G.M. Lipids 4 (1969) 556.

15. Kleining, H. J. Cell Biol. 46 (1970) 396.

16. Weinstein, D.B., Marsh, J.B., Glick, M.C. and Warren, L. J. Biol. Chem. 245 (1970) 3928.

17. Critchley, D.R., Graham, J.M. and Macpherson, I. FEBS Letters 32 (1973) 37.

18. Marcus, D.M. and Janis, R. J. Immunol. 104 (1970) 1530.

19. Henning, R. and Stoffel, W. Hoppe-Seyler's Z. Physiol. Chem. 354 (1973) 760.

20. Gahmberg, C.G. and Hakomori, S. J. Biol. Chem. 248 (1973) 4311.

21. Bretscher, M.S. J. Mol. Biol. 71 (1972) 523.

22. Gordesky, S.E. and Marinetti, G.V. Biochem. Biophys. Res. Comm. 50 (1973) 1027.

23. Forstner, G.G., Tanaka, K. and Isselbacher, K. J. Biochem. J. 109 (1968) 51.

24. Minnikin, D.E., Abdolrahimzadeh, H., Baddiley, J. and Wilkinson, S.G. Biochem. Soc. Trans. 1 (1973).

25. Karlsson, K.-A., Samuelson, B.E. and Steen, G. O. Biochim. Biophys. Acta. 316 (1973) 317.

26. Wiegandt, H. J. Neurochem. 14 (1967) 671.

27. Yogeeswaran, G., Murray, R.K., Pearson, M.L., Sanwal, B.D., McMorris, F.A. and Ruddle, F.H. J. Biol. Chem. 248 (1973) 1231.

28. Woolley, D.W. and Gommi, B.W. Nature 202 (1964) 1074.

29. Lapetina, E.G., Soto, E.F. and De Robertis,E. J. Neurochem. 15 (1968) 435.

30. Holmgren, J., Lonnroth, I. and Svennerholm, L. Infect. Immun. 8 (1973) 208.

31. Cuatrecasas, P. Biochemistry 12 (1973) 3558.

32. Tal, C., Dishon, T. and Gross, J. Brit. J. Cancer 18 (1964) 111.

33. Tal, C. Proc. Nat. Acad. Sci. U.S.A. 54 (1965) 1318.

34. Siddiqui, B. and Hakomori, S. Cancer Res. 30 (1970) 2930.

35. Kijimoto, S., Hakomori, S. Biochim. Biophys. Res. Comm. 44 (1971) 557.

36. Mora, P.T., Fishman, P. H., Bassin, R.H., Brady, R.D. and Mcfarland, V.W. Nature New Biol. 245 (1973) 226.

37. Brady, R.O. and Mora, P.T. Biochim. Biophys. Acta 218 (1970) 308.

38. Critchley, D.R. and Macpherson, I. Biochim. Biophys. Acta. 296 (1973) 145.

39. Roseman, S. Chem. Phys. Lipids 5 (1970) 270.

40. Gray, G.M. Unpublished results.

41. Kapeller, M., Gal-Oz, R., Grover, N.B. and Doljanski, F. Exptl. Cell Res. 79 (1973) 152.

42. Phillips, D.R. and Morrison, M. Nature New Biol. 242 (1973) 213.

43. Grimes, W.J. Biochemistry 9 (1970) 5083.

44. Sakiyama, H., Burge, B.W. Biochemistry 11 (1972) 1366.

45. Den, H., Schultz, A.M., Basu, M., Roseman, S.

J. Biol. Chem. 246 (1971) 2721.

46. Grimes, W.J., Robbins, P.W. (1972) Biochem. Glycosidic Linkage, PAABS Symp. 2:113 New York. Academic.

47. Marcus, D.M. New Engl. J. Med. 280 (1969) 994.

48. Rambourg, A. and Lebloud, C.P. J. Cell Biol. 32 (1967) 27.

49. Laine, R.A. and Hakomori, S. Fed. Proc. 32 (1973) 1468.

50. Robinson, J.D., Birdsall, N.J. M., Lee, A.G. and Metcalfe, J.C. Biochemistry 11 (1972) 2903.

51. Henning, R., Plattner, H. and Stoffel, W. Biochim. Biophys. Acta 330 (1973) 61.

---

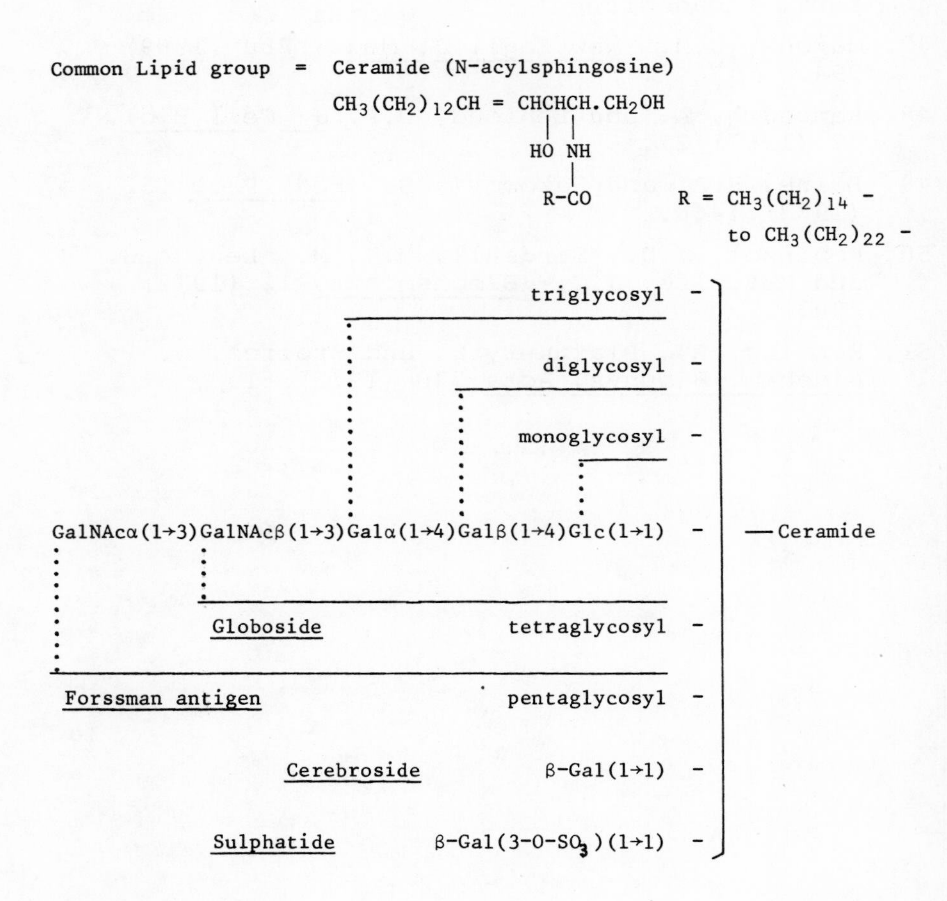


Abbreviations: As for figure 2.

**Figure 1.** Structure of neutral glycosphingolipids and sulphatide.

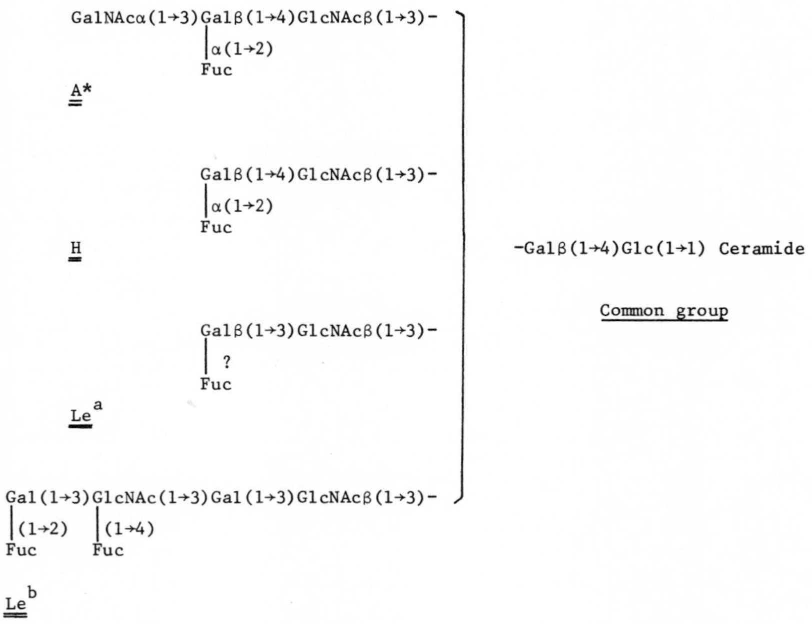

```
GalNAcα(1→3)Galβ(1→4)GlcNAcβ(1→3)-  ┐
                 |α(1→2)
                 Fuc
A*

                 Galβ(1→4)GlcNAcβ(1→3)-
                 |α(1→2)
                 Fuc
H                                                  -Galβ(1→4)Glc(1→1) Ceramide

                                                            Common group
                 Galβ(1→3)GlcNAcβ(1→3)-
                 | ?
                 Fuc
Le^a

Gal(1→3)GlcNAc(1→3)Gal(1→3)GlcNAcβ(1→3)-  ┘
|(1→2)  |(1→4)
Fuc     Fuc

Le^b
```

* Blood group activity

Abbreviations: Glc, Glucose; Gal, Galactose; GlcNAc, Nacetylglucosamine; GalNAc, Nacetylgalactosamine; Fuc, Fucose.

Figure 2. Structures of some glycosphingolipids containing fucose.

Notation*

| | |
|---|---|
| GM3<br>(Hematoside) | NANA(2→3)Galβ(1→4)Glc(1→1) Ceramide |
| GM2 | GalNAcβ(1→4)Galβ(1→4)Glc(1→1) Ceramide<br>|(2→3) [on Galβ(1→4)]<br>NANA † |
| GM1 | Galβ(1→3)GalNAcβ(1→4)Galβ(1→4)Glc(1→1) Ceramide<br>|(2→3) [on Galβ(1→4)]<br>NANA |
| $G_{D1a}$ | Galβ(1→3)GalNAcβ(1→4)Galβ(1→4)Glc(1→1) Ceramide<br>|(2→3) [on Galβ(1→3)] |(2→3) [on Galβ(1→4)]<br>NANA NANA |
| $G_{T1}$ | Galβ(1→3)GalNAcβ(1→4)Galβ(1→4)Glc(1→1) Ceramide<br>|(2→3) [on Galβ(1→3)] |(2→3) [on Galβ(1→4)]<br>NANA NANA(8→2)NANA |

* According to Svennerholm, L., J. Neurochem. 10 (1963) 613.

† NANA = N-acetylneuraminic acid (= Sialic acids; N-acetyl often replaced by N-glycolyl).

Figure 3. The structures of some gangliosides.

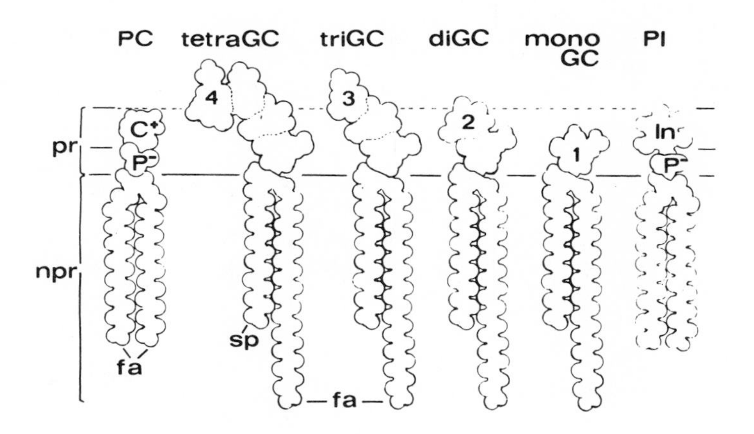


Figure 4. Structural relationships of space-filled molecular models of glycosphingolipids and phospholipids.

pr, polar region; npr, non-polar region; $P^-$, phosphate group; $C^+$, choline; In, inositol; fa, fatty acid; sp, sphingosine; 1, glucose; 2 and 3, galactose; 4, N-acetylga galactosamine; mono GC, monoglycosylceramide; diGC, diglycosylceramide; triGC, triglycosylceramide; tetraGC, tetraglycosylceramide; PC, phosphatidylcholine; PI, phosphatidylinositol.

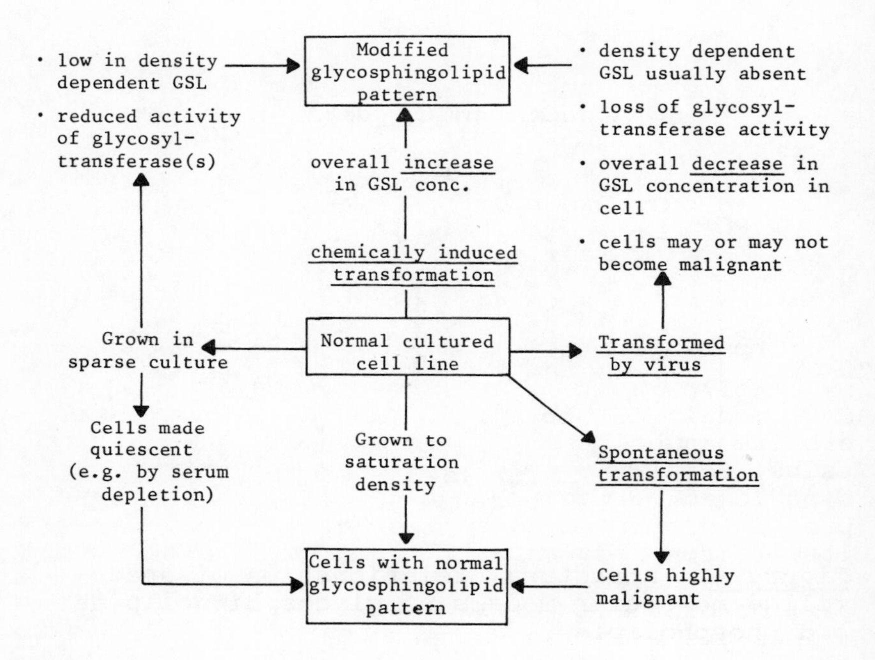


GSL, Glycosphingolipid

Figure 5. Some factors affecting the glycosphingolipid patterns of mammalian cells.

# NMR STUDIES OF LIPID BILAYER SYSTEMS USED AS MODEL MEMBRANES

B.A. Pethica and G.J.T. Tiddy

Unilever Research,
Port Sunlight Laboratory,
Unilever Limited,
Port Sunlight, Wirral, Cheshire,
England. L62 4XN.

Following the early work of Danielli, in most models of biological membranes lipid bilayers are associated in some way with various proteins. Popular models have the lipid bilayers sandwiched between two protein layers, or have protein dispersed as a 'mosaic' in the bilayer itself (see references 1 and 2 for recent surveys). Evidence for these models rests on a comparison of the physical properties of actual membranes and membrane extracts with those of model lipids and proteins (3).

One of the most fundamental membrane properties is the degree of mobility of the different constituents. Magnetic resonance techniques (ESR and NMR) have been used extensively to study (4-7) this property of both membranes and model systems. For the lipid bilayers the model systems chosen have been either the smectic lamellar liquid crystalline phase formed by a wide range of amphiphilic compounds, or the vesicles formed when certain two-phase lipid plus water mixtures are subjected to ultrasonication. The NMR properties of the two model systems are very different, and the same experimental parameter can give different information on the mobility in each system. It is the purpose of the present study to survey the research in this area to date, to point out contradictions in the existing information and to indicate new experiments that resolve these contradictions. While the bulk of the article is concerned with NMR, a section comparing NMR and ESR results is included. We show that the distribution of molecular mobility within any long-chain molecule is very dependent on the structure of the molecule. In a final section we discuss how the data obtained from studies of model lipid lamellar or vesicle systems might be used to interpret the molecular mobility in biomembranes.

## Structure of Model Bilayers

Before discussing the NMR properties, it will be rewarding to recall some basic similarities and differences between the structure of smectic lamellar phases and that of vesicles. Lamellar phases have molecules arranged in bilayers with water between the layers (Figure 1). The thicknesses of the lipid and water layers are constant throughout the sample, and can be measured using low-angle X-ray scattering. The sample is a single thermodynamic phase, and the thickness of the water layer can be varied over a wide range (10-60+Å). Variation of the water content leads to a change in the surface area occupied by the lipid groups (8). A further important fact is that samples with a single orientation can be prepared using heat cycling in a magnetic field or by shearing between two glass slides. Additives dissolved in the lamellar phase will alter the position of the lamellar phase/aqueous phase boundary in the phase diagram. This will result in a change in the surface area occupied by a lipid molecule in the bilayer, and this change can be important.

Vesicles are prepared by sonication of a two-phase lipid and water mixture consisting of a lamellar phase dispersed in an isotropic aqueous phase. The lipid is usually a lecithin and the vesicles consist of a spherical lipid bilayer (radius ∿ 140 Å) around a water interior. The vesicles are suspended in a continuous aqueous phase (9). The system is not in thermodynamic equilibrium although it is stable over a period of days. Clearly, the surface of the bilayer is curved and there are fewer lipid molecules on the inside of the bilayer than on the outside (9). The characterization of the vesicles is not as straightforward as with smectic phases. In particular, the surface packing of the lipid is only known approximately. Given the close relationship between packing and mobility, the lack of definition of molecular packing in vesicles limits their utility as models.

## Nuclear Magnetic Resonance Studies

In NMR studies of bilayer systems the parameters most commonly measured are the spin lattice relaxation time ($T_1$) and the line-width ($\Delta\nu$) or spin-spin relaxation time ($T_2$). All these parameters can give information on molecular mobility, and in the following sections it is proposed to review existing experimental data and to draw conclusions about the similarity of, and differences between, molecular mobility in the different systems.

### *Spin Lattice Relaxation Time ($T_1$)*

The spin lattice relaxation time is determined by molecular motions which couple the nuclear spin energy to the surrounding lattice. The mechanism for this process is usually via the thermally modulated dipolar interactions between different magnetic nuclei in the system. The larger the dipolar interactions are, and the closer the thermal modulation frequency is to the NMR frequency, the faster the relaxation rate ($= 1/T_1$). For two like nuclei on the same molecule undergoing isotropic rotation the relaxation rate is given by equation 1 (10).

$$\frac{1}{T_1} = \frac{3}{10} \frac{\gamma_N{}^4 h^2}{r^6} \left\{ \frac{\tau_c}{1 + \omega_o^2 \tau_c^2} + \frac{4 \tau_c}{1 + 4 \omega_o^2 \tau_c^2} \right\} \quad (1)$$

$\gamma_N$ = Magnetogyric ratio, $\tau_c$ = correlation time for the molecular motion, $2\pi\omega_o$ = NMR frequency, r = distance between nuclei.

If the molecule contains more than one pair of nuclei then the total relaxation rate is the sum of the individual rates given by equation 1. For anisotropic mobility such as that expected for bilayers the general form of equation 1 is still valid, but the expression in the brackets is modified to take account of the two molecular correlation times involved (11).

In the case of two unlike magnetic nuclei (A and B) on the same molecule the relaxation rate for the A nuclei is given by equation 2

$$\frac{1}{T_{1A}} = \frac{\gamma_A^2 \gamma_B^2 h^2}{10r^6}\left\{\frac{\tau_c}{1+(\omega_A-\omega_B)^2\tau_c^2} + \frac{3\tau_c}{1+\omega_A^2\tau_c^2} + \frac{6\tau_c}{1+(\omega_A+\omega_B)^2\tau_c^2}\right\} \quad (2)$$

Again the relaxation rates are additive for more than one magnetic nucleus in the molecule, and the expression in the brackets is modified in the case of anisotropic motion.

Spin lattice relaxation can also occur from dipolar interactions between nuclei in different molecules. These interactions can be modulated by both translational and rotational motions. The theory describing these interactions is complex even for simple molecules (10, 12), and the equations for the translational contribution involve both the self-diffusion coefficient (D), the mean square jump distance and the distance of closest approach of the two nuclei.

Intra-molecular dipolar relaxation is normally the most important mechanism for determining $^{13}C$ relaxation times, while those of protons contain both intra- and intermolecular contributions.

In practice, for proton relaxation it is possible to distinguish between inter- and intramolecular dipolar relaxation by dilution with deuterated molecules. Deuterium has a smaller magnetic moment than hydrogen, and thus gives a smaller intermolecular contribution to the proton relaxation. From a comparison of measurements made in the presence of large amounts of deuterated material with those on the undiluted sample, the intermolecular contribution to $T_1$ can be obtained.

A further complication in the interpretation of $T_1$ measurements in systems with restricted molecular mobility is the phenomenon of spin-diffusion. Where the spin-spin relaxation rate is much faster than the spin lattice relaxation rate the process of inter-nuclear energy exchange (i.e. 'spin diffusion') is much more rapid than exchange of energy with the lattice. In this case a single average $T_1$ is measured for all the different groups in the molecule, although the individual $T_1$ values may be very different (equation 3).

$$\frac{1}{T_1}\,(\text{obs}) = \frac{Pa}{T_{1a}} + \frac{Pb}{T_{1b}} + \ldots\ldots. \qquad (3)$$

Pa, Pb... = fractional population of nuclei with $T_1$ values of $T_{1a}$, $T_{1b}$, etc. Spin diffusion only occurs between like nuclei, and is important for proton relaxation but not for $^{13}C$ relaxation.

Measurements - Smectic Lamellar Phases

Only a single $T_1$ value is observed because the very rapid spin-spin relaxation rate leads to spin diffusion for the alkyl chain protons. If the lipid molecules contain an exchangeable proton in the head group (N<u>H</u> or O<u>H</u>) and the exchange rate with the water is fast enough, then the process of spin diffusion can be extended to include the water protons as well (13). The $T_1$ contribution of water protons can be eliminated by the use of $D_2O$.

The $T_1$ proton values observed for a number of different systems (13-17) are relatively long (0.2 - 1.0s) and depend only very little on the thickness of the water layer. The values increase with increasing temperature and the mechanism originally proposed for the relaxation was an intramolecular process due to rapid $CH_2$ group rotation about the long molecular axis (13, 16, 17), the correlation time for the process being of the order of $<10^{-9}$s. Recently, however, the alternative mechanism of intermolecular relaxation due to translational self-diffusion has been proposed (18, 19, 20). Whichever explanation is correct, the rotation about the long axis of the chain must be very rapid ($\tau_c < 10^{-9}$s). The rotational contribution to $T_1$ is determined by equations of the type (1) and (3) for all the alkyl chain protons. In a molecule where translational diffusion is rapid enough to give a $T_1$ contribution, rotation would also be expected to be rapid. The rotational contribution to $T_1$ must be equal to or less than the observed value, and an upper limit on the rotational $\tau_c$ value can be obtained from this as indicated above. A large self-diffusion contribution to $T_1$ implies a small rotational contribution, and thus a short rotational $\tau_c$ value ($\tau_c < 10^{-10}$s). With $\tau_c$ values as short as this there is little possibility of a large distribution of different rotational rates for different $CH_2$ groups along the chain. The size of the intermolecular relaxation contribution could be determined by experiments with deuterated chains. These are commercially available for the potassium laurate/water system studied by Rigny (15,17) and Roberts (20). In the absence of conclusive proof of either of the proposed mechanisms, we prefer the original explanation of a significant contribution to $T_1$ originating from alkyl group rotations.

We have attempted to investigate the distribution of alkyl group mobility within smectic bilayers by measuring the $T_1$ values of hydrocarbon alkyl chain probes dispersed in a fluorocarbon surfactant/ $D_2O$ lamellar phase (16). We were able to measure changes in $T_1$ due to variation of the head group and alkyl chain length. If a large distribution of molecular mobility along the chain did occur then, since the head group region would be expected to be the most restricted, the $CH_2$ group close to the head group would be expected to have a faster relaxation rate than $CH_2$ groups in the middle of the bilayer. The increment in relaxation rate for the addition of a $CH_2$ group to a short chain would be expected to be larger than that for addition to a larger chain. In practice the relaxation rate was directly proportional to chain length for a series of alcohols (Figure 2). These results were interpreted to indicate that there was little mobility distribution, and that the rate of rotation of the molecules decreased with increasing chain length. As a check we examined the $T_1^{-1}$ value of a diol with OH groups at each end of the chain. If there were a large distribution of rotational correlation times along the chain,then anchoring the chain at two ends should cause the relaxation rate to become twice that of the monohydric derivative. In practice an increase of only ∿ 25% was observed.

If self-diffusion were the cause of the $T_1$ changes it is possible to obtain a rough estimate of the change in D required to produce this effect. The ratio of probe:fluorocarbon was 1:4.31, and the ratio of the intermolecular proton relaxation rates produced by both molecules having the same diffusion coefficients would be 1:2.5 (calculated using standard equations in ref. 10). If it is assumed that the probe does not have a significant effect on the fluorocarbon self-diffusion coefficient (the $^{19}F$ relaxation rates were not altered significantly), then the change in probe self-diffusion required to produce the experimental observations can be calculated, provided an assumption is made about the relative sizes of the probes and fluorocarbon D values. It appears reasonable to assume that D (fluorocarbon) < D (butanol). Thus the $T_1^{-1}$ value for butanol has a contribution of < 28% from butanol self-diffusion. The increase in $T_1^{-1}$ on adding a $CH_2$ group is ∿ 50%, and a change in D of about 3000% is required to produce this. This does not seem likely and in the present case we conclude that $T_1^{-1}$ is dominated by rotational motion.

In a second attempt to investigate the distribution of chain mobilities we measured $^{19}F$ and $^{1}H$ relaxation rates in a lamellar phase of $CF_3CF_2CF_2$ $CH_2)_{10}CO_2Na/D_2O$ over a range of temperatures (21). After making allowances for the different magnetic moments and bond distances of protons and fluorine in $CH_2$ and $CF_2$ groups, and assuming that the relaxation rates were determined by intra-molecular rotation, we found that the difference in average $\tau_c$ values for $CH_2$ and $CF_2$ rotations required to explain the results was only 20%. The value for the $CF_2$ groups was slightly longer than that for the $CH_2$ groups and both had the same activation energy. Lee et al (19) have suggested that this result could be explained by the self-diffusion mechanism rather than by a small distribution of $\tau_c$ values. We do not think that this is correct because of the observed $T_2$ values (see below), but if it were, the low magnitude of the rotational contribution to $T_1$ require would suggest $\tau_c$ values of $<10^{-10}$s for both hydrocarbon and fluorocarbon parts of chain, implying again that $CF_2$ and $CH_2$ groups are rotating at comparable rates.

All the detailed spin lattice relaxation studies have been reported for simple lipid/water systems where the amphiphile has only one chain and a relatively simple head group. The conclusion that only a small distribution of $CH_2$ group rotational correlation times occurs in smectic lamellar phases applies only to this type of compound. The behaviour of molecules with two alkyl chains and complex head groups, such as lecithin, may be considerably different.

## Measurement - Vesicles

$T_1$ values of a range of different nuclei have been reported for lecithin vesicles (5,6, 22-25). Because of the narrow lines observed for each different group of nuclei (see below) it is possible to measure separate $T_1$values for these groups. These demonstrate the existence of a distribution of rotational $\tau_c$ values for different $CH_2$ groups along the alkyl chains, with $\tau_c$ values changing from $\sim10^{-8}$s at the head group to $\sim10^{-10}$s at the terminal alkyl groups. Lee et al

(19) have convincingly demonstrated the existence of a significant intermolecular contribution to the $CH_3$ group proton relaxation rate. From the data given by Lee et al the relative magnitudes of the inter- and intra-molecular contributions for $CH_3$ protons can be calculated to be $0.7s^{-1}$ and $0.45s^{-1}$, with an error of ± 30%. This value of the intra-molecular contribution is larger than that given by Lee et al (19), who made a number of errors in their derivation. Lee et al also suggest that the intermolecular contribution to $T_1$ arises from a self-diffusion mechanism. While this is possible, it seems unlikely that the rotational motions of the neighbouring chains would make no contribution to the intermolecular relaxation.

*Line Width/Spin-Spin Relaxation - Theory*

For Lorentzian NMR lines the spin-spin relaxation time ($T_2$) and line width at half height are related by:

$$\Delta\nu = (\pi T_2)^{-1}$$

Lorentzian lines occur when molecular motion is relatively rapid. The relaxation mechanism is still via dipolar interactions and in the case of isotropic motion $1/T_2$ is given by:

$$1/T_2 = \frac{3}{20}\frac{\gamma^4 h^2}{r^6}\left\{3\tau_c + \frac{5\tau_c}{1+\omega^2\tau_c^2} + \frac{2\tau_c}{1+4\omega^2\tau_c^2}\right\} \qquad (5)$$

For very slow motions the line width is determined by static dipolar coupling, and the line shape is gaussian. In this case it is not possible to define a $T_2$ value, and the line width is dependent on the angle between the normal to the line joining the two nuclei and magnetic field ($\Theta$). For two hydrogen nuclei with slow motion, the dipolar coupling causes a doublet to be observed with the splitting ($\Delta$) being given by equation 6

$$\Delta = \frac{\gamma^2}{r^3}\,(3\cos^2\theta - 1) \qquad (6)$$

In a polycrystalline sample all orientations are present, and a powder spectrum is normally observed. The effect of molecular motion is normally observed. The effect of molecular motion is to 'average out' to zero the dipolar splittings. This happens when the motion is isotropic and has a frequency larger than the splittings. Otherwise the splittings are reduced, but not eliminated.

## Measurements - Smectic Lamellar Phases

The broad lines observed for alkyl chains in lamellar phases arise because of dipolar coupling, but because of the complexity of the theory and the number of possible explanations of the line width it is difficult to extract detailed information on molecular mobility (15,17,26,27). Only in a few cases is it possible to utilise the line width (or $T_2$), unless measurements are made on orientated samples.

In our NMR studies (21) of the partially fluorinated surfactant $DF_3(CF_2)_2(CH_2)_{10}CO_2Na$ we measured $T_2$ values for both $^1H$ and $^{19}F$. After making an allowance for the difference in magnetic moments between hydrogen and fluorine we found that the $^{19}F$ $T_2$ value was slightly shorter than the $^1H$ $T_2$ value. If the $T_2$ values are determined by non-averaged dipolar coupling this indicates that the mobilities of the two parts of the molecule are similar, in agreement with the conclusion from the $T_1$ values.

Strong evidence against the existence of a large distribution of $\tau_c$ values along the alkyl chain is given by the studies of McDonald and Peel (28, 29). Using orientated lamellar phase samples of different chain length monoglycerides they observed a dipolar splitting for the alkyl chains and measured its dependence on water content and temperature. From some chain lengths they observed two separate splittings (29). The magnitude of the splitting for each $CH_2$ group is dependent on the time-averaged orientation of the $CH_2$ group to the magnetic field (equation 6). If $\tau_c$ were doubled we would expect the amount of splitting to be approximately doubled. Thus the observation of one or two discrete splittings is evidence against the existence of a large distribution of $\tau_c$ values in these simple systems. The above conclusions have recently been dramatically confirmed by Charvolin et al in a $^2H$ NMR study of oriented lamellar phase samples containing perdeuterated potassium laurate (30). Seven separate sets of doublets were observed and from the measured quadruple splittings order parameters describing the average orientation of $CH_2$ groups were calculated. Only the terminal $CH_2$ and $CH_3$ groups had order parameters outside the range 0.2 ± 0.1

## Measurements - Vesicles

Lecithin vesicles give rise to narrow NMR lines, while broad lines are observed for the lamellar phase. The mechanism causing the line narrowing has been the subject of much controversy; some groups suggest that it is due to a difference in molecular packing between the two types (31,32,33) and to the microscopic motions of the chains themselves. Finer et al have proposed that the narrowing is due to vesicular rotation, and that the residual line widths are a function of both the molecular motion and the vesicle rotation (34,35). Finer's theory is based on the theoretical approach of Kubo and Tomita (36) and is successful in predicting the effects of medium viscosity on the line width (35). Using a different theoretical approach Seiter and Chan (32) have derived equations relating to line widths in vesicles and in smectic phases. The final equations are similar to those of Finer but these authors conclude that the narrowing is due to a larger amplitude of chain motions in sonicated vesicles than in the smectic phase. Whatever the cause of the narrowing, it is generally agreed that the residual line widths reflect the mobility of the different groups, and again these show an increase in mobility as the distance from the head group increases (22,31-35). Although there is an intermolecular contribution to the line widths (19) part of this would be expected to reflect intermolecular rotations, and the distribution of mobility is in agreement with that deduced from $T_1$ values.

## Discussion

*Alkyl Chain Mobility in Bilayers.* The experimental evidence demonstrates that there is a large distribution of rotational correlation times within vesicular lecithin bilayer systems, and a small distribution in lamellar phases of simple amphiphiles. To understand this it will be useful to review the information available on mobility in simple liquid paraffins. The existence of a distribution of $CH_2$-group rotational rates is shown by both $^{13}C$ relaxation studies on short-chain alkanes (37) as well as $^{13}C$ and $^{2}H$ studies on longer-chain alkanes (38). As with studies on other long-chain alkanes (39,40) the distribution of correlation times was not observed in the $^{1}H$ relaxation behaviour. The relaxation rates for paraffins can be explained by a model in which trans-gauche interconversion about C-C bonds becomes progressively more hindered for C-C bonds towards the middle of the molecule. For very long chains ($C_{36}$) these interconversions become very slow, having a rate of $\sim$ 1 $s^{-1}$. When the isomerisation process is slower than the overall rotation of the molecule, relaxation is determined by the overall rotation process. For example, for the hydrocarbon $C_{36}H_{72}$, the conversions from $C_6$ to $C_{30}$ along the chain make little or no contribution to the relaxation (38). In a bilayer containing a simple amphiphilic molecule one end of the alkyl chain is anchored at the alkyl chain/water interface. One can anticipate that the anchoring will prevent the $CH_2$ groups close to the head group end from undergoing gauche-trans isomerisation at a more rapid rate than rotation of the whole molecule about this axis. Thus, it can be expected that all but a few (<4) $CH_2$ groups will rotate with approximately the same correlation time, in agreement with experimental results. An increase in chain length would be expected to decrease the rate of rotation, also in agreement with experimental results. Where the terminal alkyl groups are modified to raise the barrier to transgauche interconversion, as in the case of $CF_3CF_2CF_2(CH_2)_{10}CO_2Na$, then no distribution of $\tau_c$

values between the terminal group and the rest of the chain would be expected.

The results for a complex amphiphylic molecule such as lecithin would be expected to be quite different. Lecithin contains a large complex head group and two alkyl chains. Thus it has a much larger moment of inertia about the long axis than a single-chain amphiphile, and its overall rate of rotation would be expected to be much slower. Because of this a much larger distribution of alkyl chain $\tau_c$ values can be observed.

The above discussion is open to experimental test in two ways. If oriented samples of lecithin lamellae were prepared, their broad line NMR spectra should indicate a number (>4) of overlapping doublets, with possibly a sharp control line for the $CH_3$ groups which are thought to have very rapid motion. A second test would be the measurement of $T_1$ values (or line widths) for vesicles prepared from single-chain amphiphiles. A requirement for the formation of vesicles is that the amphiphile does not form micelles, and this is met by the monoglyceride/water system. These experiments are at present underway in our laboratory.

*ESR Probes and Alkyl Chain Mobility.* The mobility and orientation of a nitroxide probe can be determined from its ESR spectrum (40-44). For bilayer studies the nitroxide group is usually contained in a 5-membered ring, one carbon of which forms part of the amphiphile alkyl chain. By altering the position of the nitroxide group along the chain it is possible to measure the distribution of probe motion within the bilayer. For lecithin vesicles the ESR results (40-42) show a distribution of mobility similar to that obtained by NMR. However, the ESR results for probes in simple amphiphile lamellar phases also indicate a distribution of motion (43), in contradiction with the NMR results.

The most likely reason for this disparity is the difference in detailed chain motions between the two systems. Where the overall molecular rotation is slow (as in lecithin bilayers), the $CH_2$ group or probe mobility will be determined by the rate of gauche-trans interconversions about C-C bonds. Since the head group has the slowest motion, the NMR relaxation rate for any $CH_2$ group will be determined by rotation about the C-C bond closer to the head group. In the case of the spin probe, rotation about the -C - $CH_2$- bond would

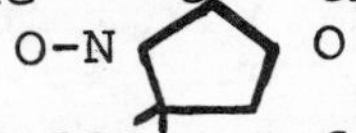

O-N [ring] O

be expected to be very different from that about a -$CH_2$-$CH_2$- bond. But $CH_2$-$CH_2$ bonds β or α to the probe would be far less affected, and would be expected to be similar to the value in the absence of the probe. Thus the ESR probe measurements probably reflect the motion of the β or α C-C bond closer to the head group. In simple amphiphile bilayers molecular rotation about the long axis of the molecule is relatively rapid, and this determines the NMR relaxation rate. Addition of a spin label to the alkyl chain will increase the area of the bilayer occupied by the molecule and will disrupt the motions of neighbouring chains as a consequence of this. This disruption will become easier as the distance of the nitroxide from the head group is increased, giving the observed mobility and orientation distribution. Once again this emphasises the need for caution in using the results of probe experiments to determine details of molecular motions in the absence of the probe.

*Biological Membranes.* In using the data on model lipid lamellar or vesicle systems as a basis for interpreting the molecular mobility in biomembranes, we should note firstly that at temperatures below that for the equilibrium smectic phase, lipid/water mixtures can exist in a gel phase. From calorimetric and NMR measurements we know that the gel phase is characterised by much slower chain motions, and these increase sharply as the temperatures goes through the phase transition. The presence of double bonds or of more than one lipid component usually broadens the phase transition and lowers its temperature (corresponding to a reduction of molecular order) although admixture of cholesterol with phospholipids is a well known exception. In mixtures of pure lipids partial separation of the various species can occur in the gel or smectic phases. A striking demonstration of such separation of near homologues and of cephalins and lecithins has been found recently in phospholipid monolayers at the oil/water interface (45). In biological membranes, the lipids are multicomponent, often unsaturated, and are likely to exist in patches of varying composition. Most important of all, however, is that the membrane properties are strongly dependent on the proteins, which may be expected on genetic grounds to dominate the function of membranes, and therefore of the membrane lipids associated with the proteins. The differences between model bilayers and biomembranes are large. An illuminating example is in the behaviour of the associated water molecules. In the lamellar smectic systems so far studied in detail (13) the associated water is restricted to a couple of molecular layers at the head groups. By contrast, the associated water in the erythrocyte membrane is extensive, and has properties more reminiscent of protein-water systems (4, 46, 47). With these points in mind, we can examine the available data on membranes, ignoring for present purposes all results on sonicated or detergent-modified membranes, since it is clear that these processes can disrupt the natural arrangements. In some cases, the appearance of narrow NMR lines and estimates of the amounts of 'gel' or 'lamellar' type lipid may be misleading, due to membrane breakdown in preparation.

Erythrocyte membranes have been studied in some detail. No thermal transitions are revealed by differential scanning calorimetry (3, 48). Detailed line-width and spin-echo measurements of $T_1$ and $T_2$ on the protons in the membrane lipids and proteins as a function of temperature showed clearly that the constituent molecules are not very mobile (4). The membrane appears as a water-swollen solid matrix in which the lipids are immobilised by the proteins. The quaternary nitrogen head group of some lecithins in the membrane are more mobile, giving narrow NMR lines (49).* A more recent proton NMR study of erythrocytes (50) has confirmed the lack of mobility of the structural constituents of the membrane, at physiological temperatures, and reveals a complex process of increased mobility at temperatures up to 75°C.** The erythrocyte lipids appear to be rather well dispersed into small regions bound to the proteins, to the point where the classical bilayer is virtually non-existent in the erythrocyte membrane. The X-ray evidence (51), which has been quoted as confirming a bilayer structure in the erythrocyte membrane, is not conclusive. The interpretation ignores the protein contribution to the scattering, shown earlier by Segerman to be significant in the high-angle region (52).

Several other membrane preparation, in contrast, have been shown by calorimetric and NMR studies to exhibit much more molecular motion, in the lipid fraction particularly. Degenerate phase transitions and corresponding changes in molecular motion have been found, and there is good evidence for inhomogeneous arrangement and motion of the lipids. With the reservation that the preparation of these membrane fractions is usually less well developed than for erythrocytes, the results may be taken as applying to the natural membranes.

The calorimetric evidence shows broad phase transitions at growth temperatures in membranes of Acholeplasma Laidlawii (53,54) and of Escherichia coli (55). These transitions are confirmed

* This mobility may be of rotation of the symmetrical quaternary group around the C-N bond, rather than of bond flexing and conformation changes in the choline head groups.

**Erythrocyte membrane preparations exhibit high resolution spectra after sonication, showing molecular mobility and clear evidence of membrane disruption as compared with the natural state (48).

by high-angle X-ray studies on the membranes of both these organisms. The transitions are reminiscent of gel-liquid crystal transitions, and both "phases" can co-exist in the membranes. Magnetic resonance experiments with deuterated saturated fatty acids supplementing the lipids of A. Laidlawii show that the chains are more rigid than in the pure smectic liquid crystal phase at similar temperatures, but that motion increases with rising temperature across the transition (56). Similarly, the NMR spectra of palmitic acid labelled with $^{13}C$ at the carboxyl group and incorporated into the membrane showed an increase of disorder through the phase transition (57).

Other membranes studied by NMR are of mitochondria (58), the sarcoplasmic reticulum (59,60), and bovine retinals rods (61). Both proton and $^{13}C$ magnetic resonance measurements on mitochondrial membranes at 18°C show that the protein structures are very immobile, and that the lipids on the membrane are more rigid or more ordered through association with the protein than are the same lipids after extraction (58). A similar picture of the membrane protein is found for bovine retinal rods. The lipids are rich in unsaturated fatty acids and appear to be in a fluid state judged from $^{13}C$ resonance experiments (61). The broad line seen in the proton magnetic resonance at 220MHz could be due to overlapping, but also indicates considerable ordering of the unsaturated lipids (61). Calorimetric studies of invertebrate photoreceptor membranes, which are also rich in unsaturated lipids, show a low-temperature endotherm in ethylene glycol preparations near -30°C, suggesting that the unsaturated lipids are in fluid conformations at physiological temperatures. A further transition near 60°C is irreversible, perhaps due to protein denaturation (62).

Preparations of sarcoplasmic reticulum were studied by Davis and Inesi by proton NMR (59) at several temperatures. A reversible transition was found in the mobility of choline groups. Most of the lipids are in a restricted state of motion, with some 20% (probably the unsaturateds) showing higher motion and good spectral resolution. Sonication increases the fraction of lipid giving higher resolution, and the authors caution that the more "fluid" lipid may not be part of the natural membrane. Certainly, measurements on fractions prepared by modification of the muscle homogenising method would be desirable. Similar considerations apply to later measurements and related speculations on sarcoplasmic vesicles or membranes (63).

In conclusion, it must be admitted that with some exceptions such as the identification of the motion of specific groups such as choline, the application of NMR has not yet given much more insight into the details of molecular motion in membranes than has been afforded by direct thermodynamic experiments. This situation will slowly change as the model systems used to develop the NMR techniques become better understood, as discussed above, and as nuclear labelling and other methods of assigning signals to individual molecules improve.

## References


1. Kreutz, W. Angew. Chem. Internat. 11 (1972) 551.
2. Singer, S.J. and Nicolson, G.L. Science 175 (1972) 720.
3. Pethica, B.A. 1968. In: Membrane Models and the Formation of Biological Membranes. p. 1 ed. Bolis, L. and Pethica, B.A. North-Holland Publishing Company.
4. Clifford, J., Pethica, B.A. and Smith, E.G. Ibid. 19
5. Oldfield, E. and Chapman, D. FEBS Letters. 23 (1972) 285.
6. Lee, A.G., Birdsall, N.J.M. and Metcalfe, J.C. Chem. in Britain, 9 (1973) 116.
7. Keith, A.D., Sharnoff, M. and Cohn, G.E.

Biochim.Biophys. Acta 300 (1973) 379.

8. Winsor, P. A. Chem. Rev. 68 (1968) 1.

9. Finer, E.G., Flook, A.G. and Hauser, H. Biochim. Biophys. Acta. 260 (1972) 49.

10. Abraham, A. 1961. In: The Principles of Nuclear Magnetism. Oxford University Press. Oxford.

11. Woessner, D.W. J. Chem. Phys. 36 (1962) 1 Ibid 36 (1962) 647.

12. Harman, J.F. and Muller, B.H. Phys. Rev. 182 (1969) 400.

13. Clifford, J., Oakes, J. and Tiddy, G.J.T. Spec. Disc. Faraday Soc., 1(1970) 175. Tiddy, G.J.T. J.C.S. Faraday Trans. I. 68(1972) 369.

14. Hanson, J.R. and Lawson, K.D. Nature 225 (1970) 542.

15. Charvolin, J. and Rigny, P. J. Chem. Phys. 58 (1973)3999. Charvolin, J. Theses. Universite de Paris - Sud, 1972.

16. Tiddy, G.J.T. Symp. Faraday Soc. 5 (1971) 150.

17. Charvolin, J. and Rigny, P. Mol. Cryst. and Liq. Cryst. 15 (1971) 211.

18. Chan, S.I., Feigersson, G.W. and Seiter, C.H. A. Nature 231 (1971) 110.

19. Lee, A.G., Birdsall, N.J.M. and Metcalfe, J.C. Biochemistry 12 (1973) 1650.

20. Roberts, R.T. Nature 242 (1973) 348.

21. Tiddy, G.J.T. J.C.S. Faraday Trans. I., 68 (1972) 670.

22. Horwitz, A.F., Horsley, W.J. and Klein, M.P. Proc.Nat.Acad.Sci. USA 69 (1972) 590.

23. Horwitz, A.F. and Klein, M.P. J.Supramol.Str. 1 (1972) 19.

24. Levine, Y.K., Birdsall, N.J.M., Lee, A.G. and Metcalfe, J.C. Biochemistry 11 (1972)1416.

25. Lee, A.G., Birdsall, N.J.M., Levine, Y.K. and Metcalfe, J.C. Biochim.Biophys.Acta 255(1972) 43.

26. Agren, G. J. de Physique. 33 (1972) 887.

27. Wennerström, H. Chem.Phys. Lett. 18(1973)41.

28. Mc.Donald, M.P. and Peel, W.E. Trans. Faraday Soc. 67 (1971) 890.

29. Peel, W.E., Ph. D. Thesis Sheffield. 1972. McDonald, M.P. and Peel, W.E. to be published.

30. Charvolin, J., Manneville, P. and Deloche, B. Chem. Phys. Lett. 23 (1973) 345.

31. Sheetz, M.P. and Chan, S.I. Biochemistry 11 (1972) 4573.

32. Seiter, C.H.A. and Chan S.I. J. Amer. Chem. Soc. 95 (1973) 7541.

33. Horwitz, A.F., Michaels, D. and Klein, M.P. Biochim. Biophys. Acta 298 (1973) 1.

34. Finar, E.G., Flook, A.G., Hauser, H. Biochim. Biophys. Acta 260 (1971) 59.

35. Finar, E.G. J. Magn. Res. In press.

36. Kubo, R. and Tomita, K. J. Phys. Soc. Japan. 9 (1954) 888.

37. Birdsall, N.J.M., Lee, A.G., Levine, Y.K. Metcalfe, J.C., Partington, R. and Roberts J.C.S. Chem. Comm. (1973)757.

38. Tiddy, G.J.T. and White, J. To be published.

39. Cutnall, J.D. and Stejskal, E.O. J. Chem. Phys. 36 (1972) 6219.

40. Cutnall, J.D., Schisia, R.M. and Hammann, W.C. J. Phys. Chem. 77 (1973) 1134.

41. Hubbell, W.L. and McConnell, H.M. J. Amer. Chem. Soc. 93 (1971) 314. Devoux, P. and McConnell, H.M. J.Amer.Chem.Soc. 94(1972)4475.

42. Jost, P., Libertini, L.J., Herbert, V.X. and Griffith, O.H. J. Mol. Biol. 59(1971)77.

43. Sackmann, E. and Trauble, H. J. Amer. Chem. 94 (1972) 4482; Ibid, 4492; Ibid, 4499

44. Seelig, J. J. Amer.Chem.Soc. 92 (1970) 3881; Ibid 93 (1971) 5017; Seelig, J. Limacher, H. and Bader, P. Ibid 94 (1972) 6364.

45. Taylor, J.A.G., Mingins, J., Pethica, B.A. Beatrice, Y.J. Tan and Jackson, C.M. Biochim. Biophys. Acta. 323 (1973) 157.

46. Clifford, J. and Sheard, B. Biophysics, 4 (1966) 1057.

47. Lynch, L.J. and Marsden, X.H. J. Chem. Physics 51 (1969) 5681. Blears, D.J. and Danyluk, S.S. Biochim. Biophys. Acta 159 (1968) 17.

48. Jenkinson, T.J., Kamat, V.B. and Chapman, D. Biochim. Biophys. Acta. 182 (1969) 427.

49. McConnell, H.H. 1970. In: Molecular Properties of Drug Receptors. eds. Poster, R. and O'Connor, M. J. and A. Churchill, London.

50. Sheetz, M.P. and Chan, S.I. Biochemistry 11 (1972) 548.

51. Wilkins, M.H.F., Blaurock, A.E. and Engelman, D.M. Nature 230 (1971) 72.

52. Segerman, E. 1968. In: Membrane Models and the Formation of Biological Membranes. p. 52, eds. Bolis, L. & Pethica, B.A. North Holland Publishing Co., Amsterdam.

53. Steim, J.M., Tourtellote, M.E., Reinert, J.C. McElhaney, R.N. and Rader, R.L. Proc. Nat. Acad. Sci. USA 63 (1969) 104.

54. Melchior, D.L., Morowitz, H.J., Sturtevant, J.M. and Tsong, T.Y., Biochim. Biophys. Acta 219 (1970) 114.

55. Steim, J.M. 1970. In: Liquid Crystals and Ordered Fluids. eds. Porter, R.S. and Johnson J.F. Olenum Press, N.Y.

56. Oldfield, E., Chapman, D. and Derbyshire, W. Chem. Phys. Lipids, 9 (1972) 69.

57. Metcalfe, J.C., Birdsall, N.J.M. and Lee, A.G. FEBS Letters 21 (1972) 335.

58. Keogh, K.M., Oldfield, E. and Chapman, D. Chem. Phys. Lipids. 10 (1973) 37.

59. Davis, D.G. and Inesi, G. Biochim. Biophys. Acta. 241 (1971) 1.

60. Robinson, J.D., Bridsall, N.J.M., Lee, A.G. and Metcalfe, J.C., Biochemistry 11 (1972) 2903.

61. Millett, F., Hargrave, P.A. and Raftery, M.A. Biochemistry 12 (1973) 3591.

62. Mason, W.T. and Abrahamson, E.W. J. Membrane Biol. 15 (1974) 383.

63. Robinson, J.D., Birdsall, N.J.M., Lee, A.G. & Metcalfe, J.C. Biochemistry 11 (1972) 2903

---

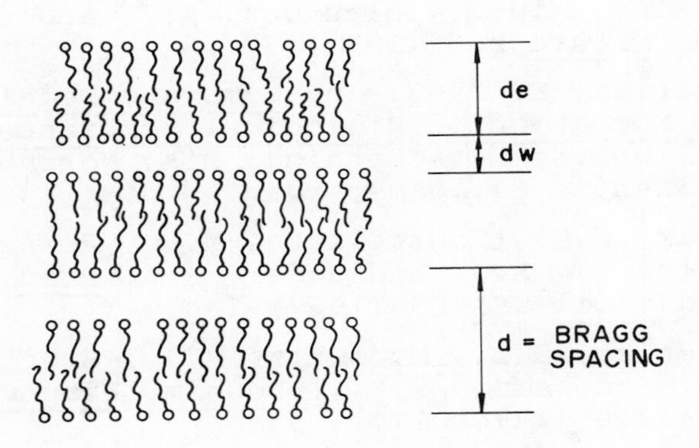


Figure 1. Schematic diagram of smectic lamellar phase; (d = regular repeat distance, $d_{\omega}$ = water thickness $d_{\ell}$ = lipid thickness)

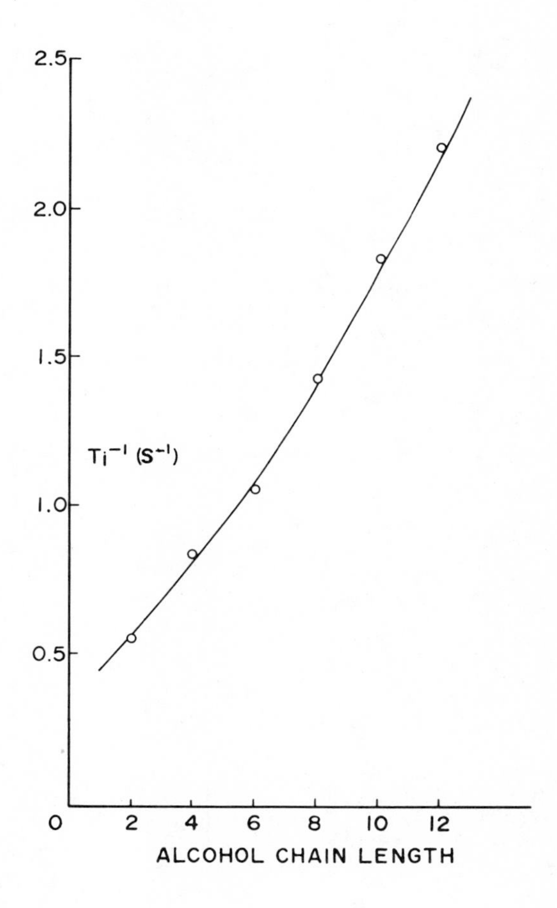


Figure 2. Spin lattice relaxation rate of hydrocarbon chain protons for alcohol dispersed in fluorocarbon surfactant lamellar phase.

# ON THE FLUIDITY AND ASYMMETRY OF BIOLOGICAL MEMBRANES

S. J. Singer

Department of Biology
University of California at San Diego
La Jolla, California 92037

The molecular biology of membranes has made great progress in the last several years. In that short time, new ideas about the molecular organization of membranes have become widely accepted. Even so, our understanding of membrane biology is still rudimentary, and many basic features of membrane structure and function remain to be discovered. Therefore, rather than concentrate in this article on the work of the recent past, we accept the suggestion of the sponsors of this Conference to indulge in "serious speculation" about some few of the many problems that are still to be resolved.

## The fluid mosaic model

The concept that the basic structure of biological membranes is a two-dimensional solution of globular integral proteins dispersed in a matrix of fluid lipid bilayer, as embodied in the fluid mosaic model of membrane structure (1, 2), has received experimental support of three major kinds. First, there is much evidence that the proteins of many m mbranes can move about rapidly in the plane of the membrane. Initially demonstrated for the H-2 antigen in the membranes of mouse and human cell heterokaryons by Frye and Edidin (3), this translational mobility of membrane proteins has since been observed with many surface antigens and receptors in a wide variety of cells (reviewed in (4)), and even with the proteins of myelin (5). Secondly, there is evidence that many proteins are deeply embedded in membranes, as revealed by freeze-fracture methods in electron microscopy (6, 7, 8), and by structural studies of integral membrane proteins such as cytochrome $b_5$ (9, 10), cytochrome $b_5$ reductase (11) and glycophorin (8, 12). And thirdly, the latter three proteins appear to be amphipathic proteins as had been postulated (1, 2, 13, 14) with their hydrophilic ends protruding from the membrane, and their hydrophobic ends embedded in the membrane.

## Constraints on Membrane Fluidity

While fluidity and protein mobility therefore appear to be widespread and characteristic properties of membranes, there are instances where membrane components are clearly constrained from being freely mobile in the plane of a membrane. What mechanisms might be involved in such constraints? Does the existence of these systems contradict the fluid mosaic model?

*Ordered complexes within membranes.* There are specialized structures within certain membranes which show a high degree of internal order, such as synapses (15) and gap junctions (16, 17), or the plaques of halobacteria (18). It is important to realize that these structures do not constitute an entire membrane, but comprise from 2% (gap junctions in liver cells) to 20% (plaques in halobacteria) of the membranes of which they form a contiguous part. The remainder of such membranes do not show any internal order. Single proteins, together with lipids, have recently been shown to constitute gap junctions in liver cell membranes (19) and the plaques of halobacterial membranes (20). This may therefore be the answer to such ordered structures. Specific, oriented protein-protein interactions between identical or closely similar subunits could produce large planar aggregates with pockets of lipid intercalated in interstices between the protein subunits. These ordered aggregates could exist within membranes whose basic structure was a fluid mosaic. In this fluid mosaic portion would presumably be contained all the many other proteins of the membrane that were required for various transport and enzymatic functions. The existence of ordered structures in membranes is therefore neither inconsistent nor irreconcilable with the general validity of the fluid mosaic model.

*Interactions with structures attached to membranes*. In certain cases, the mobility of components in an otherwise fluid membrane may be constrained by interactions with structures (microfilaments, microtubules, extracellular matrices, or juxtaposed membrane surfaces) external to the membrane. The membrane of the adult human intact erythrocyte, remarkably enough, seems to be a case in point. It has been known for many years, long before the fluidity of membranes was widely recognized, that the binding of antibodies to specific antigens on the surfaces of adult human intact erythrocytes did not lead to any clustering of the antigens on the membrane (21, 22), unlike the situation subsequently found with lymphocytes (23, 24) and other cells (25-27). Also, freeze-fracture electron microscopy experiments have shown that the distribution of intramembranous particles of intact erythrocytes is not altered by treatments which produce marked redistributions of these particles in ghosts prepared from these cells (28). These observations suggest that the mobility of components in the membrane of the adult human intact erythrocyte is significantly restricted compared to that of the lysed cell membrane. Of further interest is the indication that in the intact erythrocytes of newborn humans, the membrane components are mobile (22).

We have elsewhere (4) discussed the hypothesis that the protein complex called spectrin (29), known to be peripherally attached over the entire cytoplasmic surface of the erythrocyte membrane (30), is an actomyosin-like complex that acts as a kind of scaffolding which restricts the free diffusion of membrane components in the intact cell. The proposal is that a component of the spectrin complex has a binding site by which it becomes non-covalently bound to a specific integral protein embedded in the membrane. The reversible aggregation of the spectrin complex (presumably an ATP-mediated process) then effectively ties together these integral protein molecules and prevents their independent diffusion in the plane of the membrane. Such a structure may also serve as an effective barrier to the translational diffusion of other integral components in the membrane even though these components are not bound to the spectrin complex.

Some experimental support for this hypothesis has been obtained by Nicolson and Painter (31). They have shown that the attachment of specific affinity-purified anti-spectrin antibodies to the inner surface of human erythrocyte ghosts causes a redistribution of colloidal-iron-hydroxide (CIH) binding sites on the outer surface of the ghosts. The CIH binding sites are clearly sialyl residues on the surface, most likely attached to the glycophorin molecules that are components of the intramembranous particles in the membrane (8). These results suggest, therefore, that some component of the spectrin complex is bound to a region of the intramembranous particle that protrudes from the inner cytoplasmic face of the membrane; and that a redistribution of the spectrin complex results in a redistribution of the intramembranous particles with their bound oligosaccharides protruding from the outer face of the membrane. Other experimental support for the actomyosin-like nature of the spectrin complex are the findings we have recently made (M. Sheetz, R.G. Painter, and S.J. Singer, to be published) that antibodies specific for smooth muscle myosin, isolated from the human uterus, cross-react with one or both of the high molecular weight components (components 1 and 2) of the spectrin complex isolated from human erythrocytes; and that component 5 of human erythrocytes is definitely an actin, by the criterion that when isolated and converted to a fibrous form, it forms characteristic arrowhead complexes with striated muscle heavy meromyosin. Other evidence that implicates the spectrin complex in the mechanochemical properties of the erythrocyte is discussed in reference (4).

Although this proposed mechanism for diffusional restriction may be highly specialized to the adult human erythrocyte, it is conveivable that in other cases other kinds of structures external to the membrane might exist which could (either locally or over large areas) similarly restrict the mobility of integral components of membranes.

*Localized Barriers to Diffusion in Membranes.* Another problem for the fluid mosaic model to explain is the gross compositional difference that is sometimes observed between two regions of an apparently continuous single membrane. If the membrane were fluid, these regions should intermix. Examples of such compartmentalized but apparently continuous membranes are that formed by myelin and oligodendrocyte plasma membranes (32) and the compositionally distinctive elements of the rough and smooth endoplasmic reticulum (33).

Recent freeze-fracture experiments have revealed the presence of remarkable intramembranous structures called "necklaces", which consist of a single circular strand of particles embedded in the membrane, on specific regions of cilia (34) and sperm flagella (35). The nature, origins, maintenance, and functions of these necklaces are still entirely unknown, nor has the existence of any similar structure as yet been demonstrated with other membranes. One may speculate whether similar or related structures may be more commonly found in membranes than is now appreciated, and that their function may be to act as barriers to the intermixing of components in the plane of a membrane from fluid regions on either side.

## The Functions of Fluidity in Membranes

Is fluidity and the mobility of components a casual feature of certain membranes, or is it a critical feature which all cell membranes must exhibit during at least some stage in their development? The fact that bacteria have mechanisms to regulate the degree of unsaturation of the fatty acids they incorporate into their membrane lipids so as to maintain the membrane at least partly fluid at the particular temperature of growth (36) suggests that membrane fluidity is more an essential than a casual feature. Why this might be so is not yet entirely clear. Perhaps the main requirement for fluidity is to permit cell growth and division, by allowing the plasma, intracellular, and organellar membranes to incorporate newly synthesized membrane proteins and lipid precursors and thereby extend themselves. There is evidence, for example, that newly synthesized M protein of *E. coli* membranes can be incorporated into the membranes *in vivo* only when the lipid of the membranes is at least partly fluid (37); and the spontaneous incorporation *in vitro* of purified cytochrome $b_5$ into the membranes of liver cell endoplasmic reticulum occurs readily at 37°C but not at 25°C (38). On the other hand, a cell which is not growing or dividing (either transiently or permanently) may correspondingly *not* require its membrane(s) to be highly fluid. This might be connected with the observations that lectin-receptors on the surfaces of normal mouse fibroblasts in dense monolayers (where cell growth is inhibited) appear to be relatively immobile in the membranes, whereas the same receptors in virus-transformed fibroblasts (dividing cells) are mobile (39); and also with the relative immobility of components in the membranes of adult human erythrocytes mentioned above, since the erythrocyte is a non-dividing cell.

There are many membrane functions besides growth which may also depend upon membrane fluidity. For example, reactions involving an enzyme and a substrate, both of which are membrane bound, may require in some cases that these components diffuse rapidly to each other and the products diffuse rapidly away in the plane of a fluid membrane. The reduction of cytochrome $b_5$ by cytochrome $b_5$ reductase (38) in endoplasmic reticulum membranes is apparently such a case. As another example, the activation of various cells by mitogens, and of antibody precursor cells by antigen, may be triggered by redistributions of components in the plane of the membrane (23), but the detailed mechanism of triggering is not clear.

## Membrane Asymmetry - Proteins.

There is substantial evidence of various kinds, (chemical, physical, and functional) showing that integral proteins of membranes are generally asymmetrically distributed and oriented across the bilayer. Certain proteins protrude primarily from one face of a membrane, others from the opposite face, and still others from both faces but with a certain preferred orientation perpendicular to the plane of the membrane. We may ask, however, what fraction of the molecules of any one protein are oriented in the direction preferred by that protein, and what fraction in the other? Is the asymmetry absolute? (That is unlikely.) Does it vary upon perturbing the membrane or its environment? For no membrane protein is the degree of asymmetry accurately known.

Let us call the ratio of the numbers of molecules of a particular integral membrane protein that are oriented in one direction in the bilayer to those oriented in the other, the asymmetry coefficient $\alpha$. Most methods currently used to demonstrate asymmetry, such as chemical methods comparing right-side-out and inside-out membrane vesicles (40), cannot determine $\alpha$ to better than ± 10%. Therefore, values of $\alpha > 10$ cannot be accurately measured. In one case, however, involving receptor sites for concanavalin A (Con A) on rabbit erythrocyte membranes, as visualized and counted by the use of ferritin-labeled Con A in electron microscopy (41), $\alpha$ is clearly at least $10^2$ (see Fig. 2 in (41)), but might be several orders of magnitude larger. It would be highly desirable to have methods capable of measuring large values of $\underline{\alpha}$ accurately, for reasons that will shortly become clear.

There are two possible explanations of protein asymmetry:

*The distribution is an equilibrium one.* There is a difference in the chemical potential $\mu_p$ for a particular membrane protein oriented facing one way compared to the other; i.e., $\Delta\mu_p = -RT \ln \underline{\alpha}$. The rates of flipping of the membrane protein from one orientation to the other must be sufficiently rapid to maintain this equilibrium.

*The distribution is not an equilibrium one.* The newly synthesized protein is inserted into the membrane in the appropriate orientation. Thereafter, for a time longer than the half-time for the turnover of that protein in the membrane, there is essentially no flipping of the protein to the other orientation; i.e., $\Delta\mu_p^{act}$, the free energy of activation for the flipping process, is very large.

It is a true reflection of our present ignorance about membrane structure that we have no data whatsoever that unambigously favors one or the other possibility, and we therefore do not know for certain the basis of protein asymmetry in membranes. It is an experimentally approachable problem, however, if $\underline{\alpha}$ can be accurately measured. For example, if the distribution is an equilibrium one, then altering the physical or chemical features of one or the other surface of a membrane might be expected to result in predictable changes in $\underline{\alpha}$ for any given integral membrane protein. On the other hand, if the distribution is a non-equilibrium one, $\underline{\alpha}$ should not be changed by any alterations that do not destroy the integrity of the membrane, or as long as $\Delta\mu_p^{act}$ remains very large.

We have suggested on the basis of the fluid mosaic model (2), that the second of the two possible explanations of protein asymmetry is the correct one, and that despite the rapid translational mobility of integral proteins in fluid membranes there is only a negligibly slow rate of trans-membrane rotations of these proteins. In the fluid mosaic model, the integral membrane proteins are postulated to be globular amphipathic molecules oriented with their hydrophilic ends, containing most or all of the ionic residues of the protein, protruding into the aqueous phase. Our argument is that it requires a large expenditure of free energy to transfer the ionic residues of an integral membrane protein from water to the hydrophobic interior of the membrane (1) to get to the other side; hence $\Delta\mu_p^{act} >> 0$, and the rates of flipping are exceedingly slow. Theoretically, for an integral protein molecule bearing 20-30 ionic residues $\Delta\mu_p^{act}$ might well be of the order of magnitude of 100 Kcal/mole (1). Even the most hydrophobic membrane proteins known contain at least that many ionic residues (4).

For these reasons, we have argued that those mechanisms for active transport of hydrophilic small molecules that postulate the rapid rotation of a carrier protein from one surface of a membrane to the other are unlikely to be correct, and that a translocation mechanism that involves a quaternary rearrangement in a subunit protein aggregate which spans the membrane (Fig. 1) is a much more likely alternative (4). Dr. Jack Kyte in our laboratory (42) has prepared specific antibodies to the major polypeptide chain of the $Na^+$, $K^+$-dependent ATPase of canine kidney proximal tubule membranes. This enzyme is the protein involved in the active transport of $Na^+$ and $K^+$ in opposite directions through the membrane. The stoichiometric binding of the antibodies to the enzyme in membrane vesicles has no detectable effect on the $K_M$ or $V_{max}$ of the enzyme. The ATP hydrolysis and the translocation of $Na^+$ and $K^+$ ions by this enzyme are widely thought to be expressions of the same event (42). These antibody results therefore provide evidence that a transmembrane rotation of this transport protein cannot be the mechanism for its translocation of $Na^+$

and $K^+$ since it is inconceivable that the binding of one or more molecules of antibody to the enzyme in the membrane would not have some effect on its transmembrane rotation. On the other hand, a translocation mechanism that involved a quaternary rearrangement in a subunit aggregate in the membrane (as in Fig. 1) might not be seriously affected by the binding of antibody molecules to one or more of the subunits.

Similar studies of the effects of antibodies on other transport systems (e.g., the $Ca^{++}$-ATPase of sarcoplasmic reticulum) are in progress in our laboratory.

If it is indeed correct that the asymmetry of proteins in membranes is the consequence of an initially asymmetric insertion of newly-synthesized protein into the membrane, then it is essential that the principles that govern such insertion processes be understood. What factors determine that the molecules of protein A be inserted into the membrane protruding from one membrane surface, those of protein B protruding from the other, and those of protein C protruding from both surfaces in a preferred orientation? The answer is entirely unknown at the present time.

## Membrane Asymmetry - Lipids

Evidence has recently been obtained (43, 44) that supports Bretscher's suggestion (45) that the phospholipids of the human erythrocyte membrane are asymmetrically distributed in the two halves of the bilayer, with phosphatidylcholine and sphingomyelin predominantly in the exterior layer, and phosphatidylethanolamine and phosphatidylserine predominantly in the interior layer facing the cytoplasm. The same two possible explanations considered in the previous section on protein asymmetry apply here: the asymmetric distribution of lipids is either an equilibrium one, with rapid flipping of the phospholipids from one half of the bilayer to the other; or it is a nonequilibrium one, which is generated by different enzyme-catalyzed lipid synthetic reactions in the two halves of the bilayer, and is maintained because the rates of flipping of the phospholipids is very slow. Although the rates of flipping of spin-labeled phospholipids in synthetic bilayers (46) and in electroplax membranes (47) have been measured by spin-label techniques, their significance is not clear, since these rates were reported to be very slow in the former case and fairly rapid in the latter. More work needs to be done to settle this very important point.

Whatever the causes of the lipid asymmetry in the erythrocyte membrane may be, however, its existence suggests many possible consequences. As one consequence, we propose that the human erythrocyte membrane may behave as a _bilayer couple_, in rough analogy to the behavior of a bimetallic couple. It is suggested that under various perturbations, the two halves of the bilayer may be differentially expanded or contracted, producing corresponding evaginations or invaginations of the membrane (Fig. 2). It is well known that low concentrations of a variety of amphipathic molecules can produce characteristic shape changes in intact human erythrocytes (48), causing the cells to become either crenated (evaginated) or cupped (invaginated). Most of the substances that are crenators are _anionic_ amphipathic molecules (such as dinitrophenol, free fatty acids), while most of the cup-formers are _cationic_ amphipathic molecules (such as chlorpromazine and other tranquillizers, tetracaine and other local anaesthetics). There is good evidence that these compounds are taken up by the membrane and expand its surface area by a few percent (49). We suggest that the compounds that are crenators distribute preferentially in the outer of the two halves of the bilayer of the intact erythrocyte membrane, (Fig. 2, top), whereas those that are cup-formers bind preferentially to the inner of the two halves (Fig. 2, bottom). In the former case, this produces a differential expansion of the outer layer relative to the inner, resulting in evagination or crenation, whereas the opposite is true in the latter case. The reason that cationic amphipathic compounds (with exceptions discussed below) might bind preferentially to the inner half of the bilayer may be because the anionic phospholipid phosphatidylserine appears to be concentrated there (43, 44), whereas anionic amphipathic compounds might similarly be repelled electrostatically from that layer.

P.A.G. Fortes (personal communication) has found that the anionic amphipathic fluorochrome ANS (1-anilinonaphthalene-8-sulfonate) at a concentration of about $10^{-5}$M is an effective crenator of intact human erythrocytes. M. Sheetz and I (unpublished observations) then tested the effect of the cationic amphipathic fluorochrome ethidium bromide, expecting that it might be an effective cup-former. However, ethidium bromide even at concentrations as large as $10^{-2}$M had no significant effect on the shape of human erythrocyte under conditions where $3 \times 10^{-5}$M chlorpromazine immediately cupped the cells. On the other hand, ANS (50) and ethidium bromide (51) are bound to erythrocyte ghosts with about equal affinity. It therefore is possible that ethidium bromide is ineffectual as a cup-former because, added to intact erythrocytes, its rate of flipping from the outer to the inner half of the bilayer is very slow, and no significant expansion of the inner half of the bilayer relative to the outer half occurs. The molecular explanation of such a slow rate of flipping may be that ethidium bromide contains a quaternary nitrogen atom whose positive charge cannot be discharged, whereas chlorpromazine and most other cationic amphipathic cup-formers are amines. The free amine (uncharged) form of these latter compounds may flip relatively rapidly from one side of the membrane to the other, whereas there may be a large free energy of activation for the corresponding flipping motion of the obligatorily-charged ethidium bromide molecule (1). These suggestions are amenable to experimental tests, which we are carrying out at present.

The bilayer couple hypothesis, along with

many refinements that it requires, will be discussed in greater detail elsewhere. If membranes other than that of erythrocytes are also characterized by an asymmetric distribution of their amphipathic lipids as well as proteins, then such membranes may also behave as bilayer couples. This could be related to the mechanisms whereby various surface shape changes are induced in cells, such as occur in microvillus formation and in endocytosis. Compounds that are cup-formers of erythrocytes might then inhibit microvilli-mediated processes with such cells, whereas compounds that are erythrocyte crenators might inhibit endocytotic processes. These considerations are also clearly relevant to the mechanisms of anaesthesia (49).

## References.


1. Singer, S.J., 1971. Structure and Function of Biological Membranes. p. 145, ed. Rothfield, L.I. Academic Press: New York.
2. Singer, S.J. and Nicolson, G.L. Science 175 (1972) 720.
3. Frye, L.D. and Edidin, M. J. Cell Sci. 7 (1970) 319.
4. Singer, S.J. Ann. Rev. Biochem. 43 (1974) In press.
5. Matus, A., de Petris, S., and Raff, M.C. Nature New Biol. 244 (1973) 278.
6. Branton, D. Ann. Rev. Plant Physiol. 20 (1969) 209.
7. Pinto da Silva, P. and Branton, D. J. Cell Biol. 45 (1970) 598.
8. Marchesi, V.T., Tillack, T.W., Jackson, R.L. Segrest, J.P., and Scott, R.E. Proc. Nat. Acad. Sci. U.S.A. 69 (1972) 1445.
9. Ito, A., Sato, R. J. Biol. Chem. 243 (1968) 4922.
10. Spatz, L., and Strittmatter, P. Proc. Nat.

Acad. Sci. U.S.A. 68 (1971) 1042.

11. Spatz, L., and Strittmatter, P. J. Biol. Chem. 248 (1973) 793.

12. Segrest, J.P., Kahane, I., Jackson, R.L., and Marchesi, V.T. Arch. Biochem. Biophys. 155 (1973) 167.

13. Lenard, J., and Singer, S.J. Proc. Nat. Acad. Sci. U.S.A. 56 (1966) 1828.

14. Wallach, D.F.H. and Zahler, P.H. Proc. Nat. Acad. Sci. U.S.A. 56 (1966) 1552.

15. Whittaker, V.P. 1969. The Structure and Function of Nervous Tissue, Vol. III, p. 1, ed. Bourne, G.H. Academic Press: New York.

16. Revel, J.P. and Karnovsky, M.J. J. Cell Biol. 33 (1967)C7.

17. Goodenough, D.A. and Revel, J.P. J. Cell Biol. 45 (1970) 272.

18. Blaurock, A.E., and Stoeckenius, W. Nature New Biol. 233 (1971) 152.

19. Goodenough, D.A., and Stoeckenius, W. J. Cell Biol. 54 (1972) 646.

20. Oesterhelt, D., and Stoeckenius, W. Nature New Biol. 233 (1971) 149.

21. Lee, R.E., and Feldman, J.D. J. Cell Biol. 23 (1964) 396.

22. Blanton, P.L., Martin, J., and Haberman, S. J. Cell Biol. 37 (1968) 716.

23. Taylor, R.B., Duffus, W.P.H., Raff, M.C., and de Petris, S. Nature New Biol. 233 (1971) 225.

24. Loor, F., Forni, L., and Pernis, B. Eur. J. Immunol. 2 (1972) 203.

25. Sundqvist, K.G. Nature New Biol. 239 (1972) 147.

26. Edidin, M., and Weiss, A., Proc. Nat. Acad. Sci. U.S.A. 69 (1972) 2456.

27. Becker, K.E., Ishizaka, T., Metzger, H., Ishizaka, K. and Grimley, P.M. J. Exp. Med. 138 (1973) 394.

28. Pinto da Silva, P. J. Cell Biol. 53 (1972) 777.

29. Marchesi, V.T. and Steers, E., Jr. Science 159 (1968) 203.

30. Nicolson, G.L., Marchesi, V.T. and Singer, S.J. J. Cell Biol. 51 (1971) 265.

31. Nicolson, G.L. and Painter, R.G. J. Cell Biol. 59 (1973) 395.

32. Geren, B.B. Exp. Cell Res. 7 (1954) 558.

33. Meldolesi, J. and Cova, D. J. Cell Biol. 55 (1972) 1.

34. Gilula, N.B. and Satir, P. J. Cell Biol. 53 (1972) 494.

35. Bergström, B.H. and Henley, C. J. Ultra-struct. Res. 42 (1973) 551.

36. Sinensky, M. J. Bacteriol. 106 (1971) 449.

37. Tsukagoshi, N. and Fox, C.F. Biochemistry 12 (1973) 2816.

38. Strittmatter, P., Rogers, M.J. and Spatz, L. J. Biol. Chem. 247 (1972) 7188.

39. Rosenblith, J.Z., Ukena, T.E., Yin, H.H., Berlin, R.D. and Karnovsky, M.J. Proc. Nat. Acad. Sci. U.S.A. 70 (1973) 1625.

40. Steck, T.L. 1972. Membrane Research, p. 71, ed. Fox, C.F. Academic Press: New York.

41. Nicolson, G.L. and Singer, S.J. Proc. Nat. Acad. Sci. U.S.A. 68 (1971) 942.

42. Kyte, J. J. Biol. Chem. (1974) In press.

43. Zwaal, R.F.A., Roelofsen, B. and Colley, C. M. Biochim.Biophys. Acta 300 (1973) 159.

44. Verkleij, A. J., Zwaal, R.F. A., Roelofsen, B., Comfurius, P., Kastelijn, D. and Van Deenan, L.L.M. Biochim. Biophys. Acta. 323 (1973) 178.

45. Bretscher, M.S. J. Mol. Biol.71 (1972) 523.

46. Kornberg, R.D. and McConnell, H.M. Biochemistry. 10 (1971) 1111.

47. McNamee, M.G. and McConnell, H.M. Biochemistry 12(1973)2951.

48. Deuticke, B. Biochim.Biophys.Acta 163(1968)494.

49. Seeman, P. Pharmacol. Rev. 24 (1972) 583.

50. Rubalcava, B., Martínez de Muñoz, D. and Gitler, C. Biochemistry 8 (1969) 2742.

51. Gitler, C., Rubalcava, B. and Caswell, A. Biochim. Biophys. Acta 193 (1969) 479.

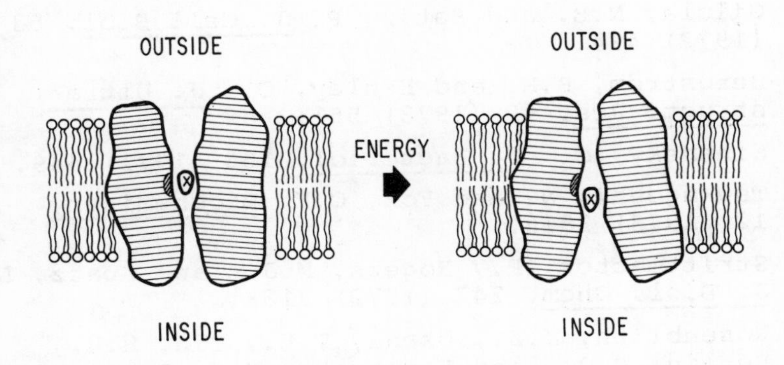


Figure 1. A schematic representation of a proposed mechanism of the translocation event in active transport. A subunit aggregate of integral membrane proteins forms a narrow water-filled central channel, or "pore". On the surface of one or more of the subunits lining the channel is a binding site for the transportable ligand X. Some energy-yielding process (an enzymic phosphorylation, for example) might produce a quaternary rearrangement of the subunits and result in a translocation of X across the membrane. Reversal of the energy-yielding process then restores the aggregate to its initial state.

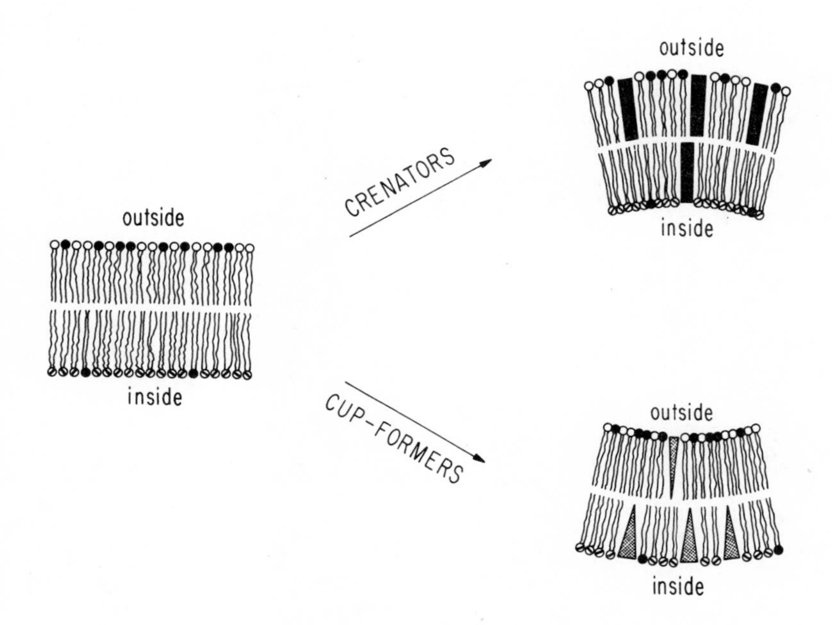


Figure 2. Highly schematic representation of the bilayer-couple hypothesis, and its explanation of the effects of crenating and cup-forming agents on the shapes of intact erythrocytes. It is proposed that these agents bind to (intercalate into) the two halves of the bilayer to different extents, crenators being more concentrated in the outer half, cup-formers being more concentrated in the inner half, thereby inducing the shape changes. The polar head groups of the four major classes of phospholipids are distinguished in the drawing in accordance with the evidence that they are asymmetrically distributed in the two halves of the bilayer. The integral proteins of the membrane are omitted from the drawing for simplicity, but most probably also contribute to the properties of the bilayer couple. See text for further details.

# COUNTERPOINT

Carlos Gitler and Amira Klip

Departamento de Bioquímica
Centro de Investigación y de
Estudios Avanzados del I.P.N.
Apartado Postal 14-740
México 14, D. F.

It is the purpose of this chapter to present additional points to those in the previous chapters of this book in order to give the reader an alternative view pertaining to the interpretation of the available evidence on membrane structure and function. Clearly, it is impossible to do this in but a few areas of those covered in the book. In addition an attempt will be made to single out some procedures and areas of research which in the authors' opinion show a definite perspective towards the resolution of some of the problems we are dealing with in membranology.

The counterpoint is here meant in a musical sense: the technique whereby two or more independent lines are simultaneously combined and made to proceed with emphasis on the melodic aspect.

## On Membrane Structure

*Chemical Labels*. The analysis of the structure of biological membranes obtained by the use of physical methods such as X-ray diffraction and nuclear magnetic resonance has been rather disappointing as indicated by Dr. Luzzati. Not only are the results ambiguous with regards to the location of the membrane proteins but even in the case of the existance of a bilayer, the conclussions depend on the proper solution to the phase problem. Furthermore,the times required to obtain a diffraction pattern were until recently such that only an average immage could be attained (see Gitler (1) for additional discussion). It is surprising in this regard that few if any attempts have been made to use heavy metal-containing amphiphiles and heavy metal-containing nonpenetrating labels of membrane proteins to obtain reference points from which to resolve the current impass.

On the other hand, there has been a sudden increase in the attempt to use chemical labels to define the topological location of some of the membrane elements (See also Bretscher and Simons in this book). Some of the reagents which have been used are: a) SITS, 4-acetamido, 4'-isothiocyanostilbene-2, 2'-disulfonate. This was the reagent introduced by Maddy (2) which pioneered in this area. b) Diazonium salt of sulfanilic acid (3), a highly reactive and thus quite labile reagent. c) Diazonium salt of diiodosulfanilic acid (4), which forms a more stable diazonium salt and can be better controlled. d) 7-diazo-1, 3-naphthalenedisulfonate, a reagent tested by Pardee and Watanabe (5) does not label internal enzymes in Salmonella Typhimurium. e) Trinitrobenzene sulfonate (6, 7, 8) is a stable reagent but disagreement exists about its penetration (6,8) and its reactivity is sensitive to environmental factors (9, 10). f) FMMP, formylmethyonyl methylphosphate is a highly labile reagent and furthermore labels hemoglobin (11). g) Pyridoxal phosphate-sodium borotritium (12) appeared nonpenetrant since pyridoxal requiring bacterial mutants do not grow in its presence. It also labels hemoglobin (see below.)

Several aspects should be emphasized in the interpretation of the results obtained with these compounds. First, the surprising fact emerges that in all cases studied, the so called "non-penetrant" covalent labels are anionic in nature. It is well know in the case of the erythrocyte that this membrane is definitely more permeable to anions than to cations. Furthermore, if exhaustive labeling is attempted (to decrease the kinetic effect discussed below), the use of an anion directed usually to react with free amino-groups would lead to a marked modification or even reversal in the net charge of the labelled membrane elements. This could clearly alter those groups accesible to further reaction. Moreover, in some cases, the label used is very reactive and is added in very small amounts. This imposes kinetic limitations to the interpretation of the specificity of the resulting labelling pattern. That is, only the most reactive moieties will be labelled The reagent will be destroyed in some cases at a rate too high to allow it to interact with poorly reactive groups which might be located towards the outside. There is an additional kinetic consideration to be made. It is well known (Cordes and Gitler (13); Fendler and Fendler (14)) that at interfaces the surface charge can markedly modify the rates of reaction. An anionic surface charge will repell an anionic reagent while it will be strongly attracted by a surface of apposite charge. It is likely that the highly anionic compounds depicted in Table I will not react with amino groups located in regions of high negative charge density. This is specially so since the equilibrium

$$R\text{-}NH_3^+ \rightleftharpoons R\text{-}NH_2^+ + H^+ \qquad (1)$$

will be displaced towards the left in the presence of surrounding anionic-charged groups.

In addition to the membrane proteins labelled from the outside, in many cases where care - full studies have been made, it is also shown that internal soluble proteins such as hemoglobin are also tagged. Our own studies with the use of pyridoxal phosphate-sodium borotritium were quite indicative of this limitation. Labelling of intact erythrocytes under the conditions of Rifkin et al (12) resulted in the tagging of the two external proteins reported by most workers (3,4,11). However we hemolysed the cells and treated the hemolysate with acidified acetone to remove the heme from hemoglobin in order to obtain adequate tritium counts in the globin. About five times more counts were found in the globin than those incorporated into the ghosts. Labelling with pyridoxal phosphate, piridoxal and sodium borotritium resulted in the incorporation of the tritium into nearly all of the membrane proteins. The globin in this experiments contained some ten fold more radioactivity than the ghosts. These results indicate that pyridoxal phosphate can penetrate and label internal soluble proteins. The much higher labelling obtained in the latter case, could be interpreted either that it is due to labelling of internal membrane proteins in addition to those outside the cell or alternatively, that the uncharged pyridoxal, could be reaching sites outside the cell which repell the anionic pyridoxal phosphate. It is probable that any reactive molecule that penetrates the erythrocyte is more likely to interact with hemoglobin -which is present at a great excess- than with internal membrane proteins.

The above consideration indicate that attempts should be made to study cationic covalent labels. They should be more impermeable in most membranes, would retain the original charge of the tagged amino groups and would concentrate at the surface since the potential is negative there. It seems also clear that any considerations made at present about the external location of membrane elements is suggestive but not proven. This seems to be the case with regards to membrane lipids: certain reagents used to label the lipids in the membrane react only sparingly with liposomes, thus indicating that a lack of reaction is due to chemical reasons (mainly charge density) rather than to unavailability of those lipids in the outer surface. That this effect is surface-charge dependant is suggested from the fact that the small labe ing of liposomes is impaired by addition of small amounts of sodium lauryl sulfate (10).

It is also evident that no sweeping conclusion should as yet be made from the use of a single label. A combination of techniques will be required for any definite conclussion to be reached.

*Penetration of membrane proteins into the lipid core.* Most current membrane models (see other chapters in this book) postulate that their integral proteins are attached to the bilayer through an apolar segment which becomes stabilized through hydrophobic interactions with the lipid core. 1) do the proteins actually penetrate into the lipid core? and 2) what are the stabilizing forces which allow this penetration?

Let us first review in part the evidence that membrane proteins penetrate into the lipid core. The initial suggestion (15,16) of an anomalous membrane protein structure based on a 2 nm shift in the circular dichroism spectra has been shown to be a light scattering artefact (14). When some membrane proteins are digested with proteolytic agents they are split into soluble fragments and a segment anchored to the membrane. This latter fragment contains a higher fraction of apolar amino acid side chains than the soluble fractions do. It should be remembered that even soluble proteins contain some 50% of apolar residues because about half of all amino acid side chains are non-polar and even their digestion with proteolytic enzymes leaves a water insoluble core. However, in no case is the fragment left in the membrane found to contain only apolar amino acids.

A segment present in human erythrocyte glycophorin contains a sequence of some 20-25 apolar aminoacids, and it has been proposed (20) that these could form an α-helix (pitch 1.5 Å per residue) which would span the bilayer. However, such a sequence of apolar residues is not unique of membrane proteins (see subtilisin as an example).

Nontheless, some membrane proteins have been found to form proteolipids and to have an overall high proportion of apolar amino acids (18, 19).

The strongest evidence that support the possibility of membrane penetration in the bilayer comes, however, from studies on the freeze fracture of biological membranes. Branton (22) has presented evidence that lipid bilayers split during freeze-fracture in the ω-portion of the fatty acyl chains. When the membranes are thus fractured, a series of globular particles (about 80 Å in diameter) are found imbeded in one or the other of the fracture faces. There is as yet no unequivocal evidence that these globular structures are indeed proteins although they seem to be due to their presence. The evidence discussed by Drs. Martínez Palomo and Pinto da Silva in this book that such structures are associated with some but not all surface determinants is suggestive that they may represent aggregates containing glycoproteins. This is likely to be the case since the single α-helix spanning the bilayer as proposed by Marchesi and coworkers (20) would not be seen in the fracture since its dimentions are below the limit of current resolution (> 20 Å).

Recently we attempted to label those components present within the lipid core including membrane proteins (23). What appeared to be required was a small apolar molecule, insoluble in water or nearly so, which would partition and dissolve within the membrane lipids. Once thus located it required to be capable of conversion into a highly reactive species which could then attach covalently to those inert groups present within the lipid core. Because of the required reactivity we decided on either carbene- or nitrene-generating molecules since their precursors are stable until photoactivated and the reactive species formed are known to insert into carbon-carbon or carbon-hydrogen bonds among others (24,25,26). We have synthesized two initial compounds whose formulae are shown in Fig. 1. These azides were made highly radioactive, 1-azidonaphthalene (AzN) by the use of radical-catalyzed tritium-exchange while 1-azido-4-iodobenzene (AzIB) by the incorporation of $^{125}I$

in the synthesis of the precursor iodobezylamine. Both compounds are quite apolar since their octanol-water partition coefficients are 199 for AzIB and 166 for AzN.

Their addition to sarcoplasmic reticulum membranes shows a rapid incorporation of the compound into the membranes. If kept in the dark, exposure to either a serum albumin solution or a liposome suspension results in the removal of more than 99% of the azides. However, irradiation with light above 300 nm renders 20-30% of the label unremovable under either condition. This suggests that such percentage is covalently bound to the membrane components. Of this label some 60% appears to be bound to membrane proteins while some 40% is attached to the membrane lipids. Exhaustive pronase treatment of the sarcoplasmic reticulum removes 70-80% of the membrane protein but only 15 to 25% of the radioactivity. This would imply that 60 to 75% of the label attached to the proteins is unavailable to the action of the proteolytic enzymes. A more definite estimate will be obtained when the amount of label in a net lipid-free protein preparation can be ascertained. Saponification of the extracted phospholipids indicates that about 85% of the label is attached to the fatty acid residues. This latter result strongly suggests that the azides are present within the lipid core when they are activated. Since their lifetime is $< 10^{-7}$ sec they could diffuse, if at all, along the lipid core.

When a serum albumin solution is irradiated in the presence of either azide, the label is attached covalently to the extent of some 15-30%. However, when equal amounts of trypsin chymotrypsin or ribonuclease are similarly exposed to the azides the insertion of the label is extremely low. These results indicate that covalent insertion can occur in those proteins containing regions designed to bind apolar molecules. We have not tested sufficient enzymes to know whether the azides will bind into the active sites which are believed to have exposed apolar residues. In any case, irradiation of sarcoplasmic reticulum in the presence of the azides does not inactivate significantly the $Ca^{++}$-dependant ATPase.

Disc gel electrophoresis in the presence of sodium dodecyl sulfate of the sarcoplasmic reticulum membranes labelled with the azides are shown in Fig. 2. The radioactivity is found preferentially in the ATPase and in some instances in a band of higher molecular weight. No other proteins seem to be labelled, although we suspect that perhaps a proteolipid which runs in the position of the membrane lipids (27) might also be labelled. This would appear to be the case because delipidation of labelled membranes by several procedures leaves more counts in the protein that can be accounted for in the electrophoretic patterns of Fig. 2.

Thin layer chromatography of the lipids labelled with AzIB $^{125}I$ is shown in Fig. 3. The membranes were applied directly to the plates and these were developed with chloroform:methanol: acetic acid:water (85:20:8:4). The lipids were visualized with ANS, removed from the plates and the $\gamma$-emission counted in a well counter. The recovery was some 70% of that applied probably because of the interferance from the silica gel. In any case it can be observed that not all the lipids appear to be labelled to the same extent. Component 3 which is probably a polar lipid such as phosphatidyl serine or inositide is present at a much lower concentration than phosphatidyl choline and yet is labelled nearly to the same extent. At the origin remain mainly the tagged proteins (no radioactivity is detected in the origin when solvent extracted lipids are similarly applied). It can be seen that some 50-70% of the radioactivity is recovered in this component.

We are aware that much work has as yet to be done with this approach in order to get unequivocal results, however, it is clear (see also below on fluorescence quenching) that these compounds might be a way of establishing the frequency with which a given component is located within the lipid core. The fact that proteins are labelled suggests either that they have regions similar to those of serum albumin or that they have portions of their molecules within the lipid core. In order to establish whether hydrophobic regions of membrane proteins exist facing the aqueous environment, radioactive 7-azidonaphthalene-1-sulfonate is being synthesized. This compound should be capable of insertion-after photoactivation- into such regions. However, the presence of the polar group should preclude its penetration deep within the lipid core. Peptide mapping of the purified membrane proteins should allow some clarification to this regard. We believe that this general approach might allow in addition, the establishment of nearest neighbours for molecules imbeded in a quasi fluid environments such as biological membranes. We are presently synthesizing those compounds shown in Fig. 4 in order to establish what groups are present within biological membranes in the vicinity of fatty acids, detergents and if possible phospholipids. Since carbenes have shorter lifetimes than nitrenes they will give more specific information concerning immediate neighbours.

A further application of this molecules will be to establish which lipids within a liposome or a membrane are fluid at a given temperature. Since the photoactivation is independant of temperature, active nitrenes can be generated within liposomes and membranes at any temperature and they should diffuse and insert only within those pockets which are fluid at the temperature studied.

*Use of encounter fluorescence quenching to determine the location of components in biological membranes.* The quenching of fluorescence emission can often be separated into a static or time independent component and a dynamic, time dependant process first described by Stern and Volmer (28). The first appears to be due to fluorophore molecules being quenched immediately upon excitation due to close proximity to or complex formation with quencher molecules, predating excitation. This quenching is then due to a reduction in the number of free fluorophore molecules available for excitation after complex formation in the ground state. The time-dependant process is associated with diffusion limited encounters between excited fluorophore and quencher molecules. This process leads to a depopulation throughout the lifetime of the excited state.

If the quenching molecule neither absorbs excitation or emission energy nor fluoresces as a result of transfer of energy from the excited fluorophore molecules, a kinetic scheme of the following type can be used to described the various alternatives (After Weller (29); Vaugham and Weber (30); and Spencer (31).

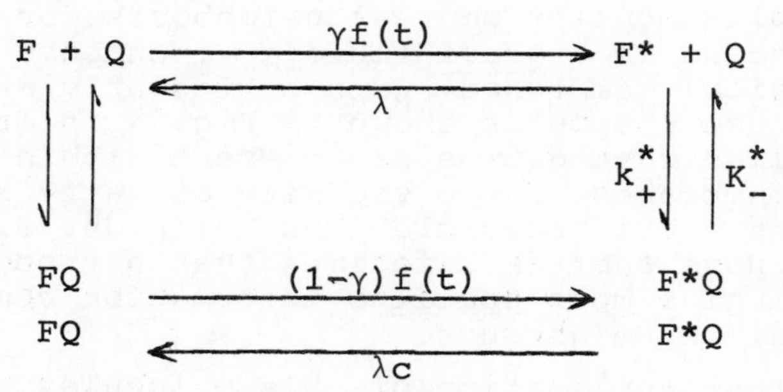


where F is the fluorophore, F* is the excited state and Q is the quencher molecule. The const $\gamma$ is the fraction of absorption transitions by free fluorophore in the ground state relative to the total number of absorptions. Steady state treatment (31) of this kinetic scheme leads to the following equation under conditions where no ground state complex is formed:

$$\phi_0/\phi = 1 + K_+^* \quad \tau_0 \ \{Q\} \qquad (3)$$

where $\phi_0$ and $\phi$ are the quantum yields in the absence and presence of quencher respectively and $\tau_0$ is the fluorescence decay time. This is the Stern-Volmer relation for dynamic quenching of fluorescence. However, under conditions where only static quenching occurs, an essentially similar equation is obtained:

$$\phi_o/\phi = 1 + K\{Q\} \qquad (4)$$

where K is the bimolecular equilibrium constant of association in the ground state. It is clear that under steady state conditions where interaction with Q does not alter the fluorescence spectra, the following relation holds:

$$\phi_o/\phi = I_o/I = 1 + K_Q\{Q\} \qquad (5)$$

where $I_o$ and I are the relative fluorescence intensities in the absence and presence of quencher respectively and $K_Q$ is an overall constant which can either be due to static or dynamic quenching or to both. Thus a plot of $I_o/I$ versus {Q} results in a linear relation which under steady state conditions does not distinguish between the two types of "encounter" quenching. These can be resolved only by determining the lifetimes of the excited states (see Spencer 31). However, for the present purposes, $K_Q$ represents the effectiveness of a given quencher molecule to reach either the ground or excited states of a fluorophore. If the quenching constant, $K_Q$, of a fluorophore with a given quencher is first established in a homogeneous medium such as water or another solvent, the only limitation will be the diffusion rate constants, the K of ground state complex formation or the effectiveness of the collision process. If similar $K_Q$ values are then obtained for the fluorophore attached to a biological membrane, a comparison of the derived $K_Q$ values will be an indication of the accesibility of the fluorophore to the quencher (32,33). If one had available a series of quencher molecules of varying polarities then some strong inferences about the location of the fluorescent moiety within the membrane could be made.

Conti (34) was responsible for drawing our attention to this possibility. Papageorgiou and Argoudelis (32) have shown that chlorophyll a fluorescence from spinach chloroplasts can be quenched by nitroaromatic compounds. Charged nitroaromatics are almost equally effective quenchers of the fluorescence of Chl a *in situ* or when dissolved in methanol; on the other hand, apolar nitroaromatics quench more strongly *in situ* than in methanol. This was taken to suggest that chlorophyll a is located in an apolar environment that also dissolves the apolar quencher. Pownall and Smith (33) have studied quenching of the intrinsic protein fluorescence in apolipoproteins in the presence and absence of the phospholipid by means of the anionic iodide ion and the cationic pyridinium chloride (35). They also studied as a model, the quenching by the same ionic species of the fluorescence of anthracene dissolved in detergent micelles (33). They observed enhanced quenching when the surface charge of the micelle was opposite to that of the quencher molecule. Depressed quenching efficiency occurred when like charges were present.

We have combined these approaches in our own studies (unpublished results). It was previously shown that perylene is almost completely insoluble in water but is readily taken up by detergent micelles, liposomes and biological membranes (36, 37). Studies on the depolarization of the perylene fluorescence emission allowed the calculation of microviscosities in the lipid core of these aggregates. From the fact that the polarization was sensitive to the action of phospholipase c and to the type of fatty acid present in the liposome-phospholipids, the suggestion was made that the fluorescent molecule was present dissolved in the hydrocarbon region of the above mentioned structures. Further evidence to this effect is now derived from quenching studies. We have used as quenchers N-butyl pyridinium bromide, a cationic water soluble quencher, 1,3-dinitrobenzene, 1-azidonophthalene and 1-azido-4-iodobenzene as apolar quenchers. Fig. 5 shows the plots of $I_0/I$ versus (Q) for perylene and quenchers dissolved in methanol while Fig. 6 shows similar plots when perylene was present in phosphatidyl choline liposomes and sarcoplasmic reticulum membranes. Excellent linear relations were obtained in all cases. Table III shows the derived parameters obtained from these plots. It is clear that similar

$K_Q$ values are obtained in a homogeneous solvent (methanol) for all of the quenchers used. On the other hand, butylpyridinium bromide shows a marked decrease in quenching effectiveness in sarcoplasmic reticulum and phospholipid liposomes (a 67 fold decrease). In marked contrast, the $K_Q$ values for the apolar quenchers are significantly enhanced; this implies that these molecules must partition in favour of the liposome and membrane phase. In the case of the 1,3-dinitrobenzene no differences were found for the liposomes and the sarcoplasmic reticulum. However, in the case of the 1-azidonaphthalene and 1-azido-4-iodobenzene, a higher $K_Q$ value resulted with perylene present in the membranes than that in the liposomes. Preliminary studies with the anionic iodide ion show a behaviour essentially equal to that of the butylpyridinium bromide.

Several conclussions can be drawn from the present results: firstly, it has been reported (27) that the polar quenchers are capable of collision encounters with anthracene dissolved in detergent micelles of charge opposite to that of the quencher and ineffective in those of like-charge. However, the present studies show that in liposomes or in the membranes neither the cationic butylpyridinium nor the anionic iodide are capable of reaching the perylene (except at very high concentrations eg., $I_0/I$ equals 2 at 1.25 M butylpyridinium). It is likely that the butylpyridinium because of the apolar contribution of the butyl moiety can penetrate into the region of the anionic phosphate groups; specially in the membranes which probably have a negative surface charge it should be concentrated in the Stern layer and yet it does not quench the perylene fluorescence. Thus, the perylene must have an average residence deeper in the lipid core. Secondly, eventhough in the interior of the liposomes and of the membrane the microviscosity is some 200 fold higher than that in the methanol solution, the $K_Q$ values are enhanced some 26 to 85 fold for the azides. Making a rough calculation based on the volume occupied by the lipids both in the liposomes and in the membranes and assuming purely dynamic quenching, 1-azidonaphthalene must be dissolved in about 1/3 the volume of the lipids while 1-azido-4-iodobenzene in about 2/3 of the volume. These are very rough approximations, yet it is quite obvious that the azides must readily reach the sites, probably the lower portions of the fatty-acyl chains, in which the perylene is located. Thirdly, 1,3-dinitrobenzene must partition in favour of the liposomes and membranes but not as effectively as the azides. This might be due to the partial charge of the nitro-groups keeping it closer to the polar surface groups.

These quenching experiments taken together with the results presented above on the photoactivation and covalent insertion of the azides seem to indicate that these molecules reach the lipid core and are therefore effective agents to detect components present therein.

## Nature of the Lipid-Protein Interactions Within the Lipid Core

It is likely that in addition to apolar interactions between lipids and proteins, polar interaction in regions of low dielectric constant might play a significant role in stabilizing the aggregates. This seems likely because as the dielectric constant of the environment decreases, polar interactions will be favoured. That is, in nonpolar solvents, the solvophobic groups will be those containing polar moieties. A study on the formation of artificial proteolipids was initiated (Gitler and Montal 38, 39) in order to determine whether polar phospholipid-protein interactions are feasible, and, was envisaged that ion-pairs might be formed between acidic phospholipids and the ammonium groups of proteins leading to overall charge neutralization of these groups. However, no phospholipids are available which at neutral pH carry a net positive charge as would be required in order to form ion-pairs with the protein carboxylates. These latter groups could be uncharged under acid pH so that the overall charge of the phospholipid-protein aggregates would tend to zero. Under these conditions and due to the amphipathic nature of the phospholipids it was though that the complex should now become soluble in apolar solvents. This was indeed the case since a proteolipid soluble in decane could be formed between cytochrome c and a mixture of acidic phospholipids -either phosphatidyl serine (PS), phosphatidylinositol (PI) or cardiolipin- and phosphatidyl choline (PC) when extracted at pH 1.9 (38). At neutral pH no cytochrome c could be extracted in the presence of the phospholipids into decane. However, it was later found that on addition of 2.0 mM $Ca^{++}$ nearly quantitative extraction of a PS-PC proteolipid of cytochrome c into decane could be attained (39). Yet, it was not clear whether water was being carried into the decane phase so that microemulsions were actually formed. In order furthermore to separate the different species that could be formed, a technique had to be devised to separate the components present in the decane phase. It has been possible to accomplish this by means of a gradient generated with either decane and chloroform or isooctane and chloroform (40). Four ml of chloroform are placed in a cellulose nitrate tube and on top are added 4.0 ml of a 1:1 mixture of decane or isooctane-chloroform followed by 3.0 ml of decane or isooctane. The tubes are centrifuged for 3 hours in an SW-40 Ti swinging bucket of a Beckman Centrifuge at 4° and 175, 000 x g. The gradient shown in Fig. 7 is thus generated. The density varies from 0.730 at the top to 1.489 at the bottom. When a proteolipid prepared by the procedure of (39) is applied on the top of the tube containing the solvents as indicated above, the patterns of distribution shown in Fig. 8 result after the 3 hours of centrifugation. In decane or isooctane the lipid and $^{45}Ca^{++}$ separate forming a band with approximate density of 1.0. The cytochrome c is found in two bands when extracted and centrifuged in decane (Fig. 8A) while as a single lighter band when extracted and centrifuged in isooctane. The lower band in decane has a density of about 1.35

and contains mainly protein but always a definite amount of phospholipid (the lipid to protein ratio is 11) and of calcium. A drop of water added to one of the tubes appears immediately below the lipid band. Most of the cytochrome c is however found below the water drop. These results indicate that the proteolipids formed are mainly the result of protein-protein and protein-lipid-$Ca^{++}$ interactions and that the water content is such that it does not dominate the sedimentation behaviour. In other words, if a high water containing microemulsion were present it should have banded at the density of water or nearly so. In the absence of lipid or $Ca^{++}$ no proteolipid is formed. Thus, it is clear that ionic interactions must be participating in leading to the overall charge neutralization which allows the aggregate to be soluble in an apolar solvent. The density of the lower cytochrome c band is nearly that of a completely dehydrated protein. It shows that it is mainly the result of protein-protein interactions. It is plausible to envisage that a protein core is formed containing several cytochrome c molecules surrounded by phospholipids, through ion pairs between the amino groups and ternary ionic interactions of (carboxylate$^-$-$Ca^{++}$-phospholipid$^-$). It is of interest that Hart and Leslie (33) have shown that proteolipids soluble in methanol-chloroform containing dansylated lysozyme and cytochrome c show quenching due to energy transfer of the dansyl-fluorescence by the heme of the cytochrome c. These results were interpreted to indicate very close apposition of the two proteins in the proteolipids.

From these findings, the question must be raised whether proteins might not be stabilized within the lipid matrix not only by apolar interactions but also by the formation of ion-pairs or ternary complexes. Ion-pair formation in apolar environments does occur even with soluble proteins such as chymotrypsin. Salt bridges in apolar environments have also been found in the $Ca^{++}$-binding proteins described by Dr. Kretsinger in this volume. Thus, they are thermodynamically feasible with soluble proteins but their role in membrane structure requires further studies. Perhaps the use of water soluble and insoluble carbodiimides (41,42) might lead to covalent cross-linking between the carboxyls of PS and the amino groups of the protein, allowing the study of the frequency of their occurrance.

It is known that proteolipids do exist in many biological membranes (18,19,27). When obtained they are usually associated with acidic phospholipids (18,19). They are best extracted under acidic conditions (18,19).

It seems that 80 Å spherical particles as those observed in freeze fracture are likely to be due to protein aggregates. The apolar regions in the proteins would have to be very large if the formation of the aggregates in the lipid bilayer were only mediated through hydrophobic interactions.

Moreover whenever apolar peptides are found, for example after proteolytic digestion, these always contain an important fraction of polar ionic residues. Thus, it is perhaps likely that the protein aggregates might be neutralized by ion-pair formation to become inmersed in the lipid matrix, as appears to be the case for the cytochrome c proteolipid.

It is also highly suggestive that $Ca^{++}$ might induce marked topological changes of membrane proteins by leading to the formation of protein-phosphate-$Ca^{++}$ complexes or ternary complexes with lipid. Furthermore, a proton gradient as occurs during energy coupling might lead to the neutralization of surface charges, the energy being conserved as a partition coefficient of some surface components. This could lead to either ionic movements or to the synthesis of the anhydride in ATP through the formation of lipid soluble $Mg^{++}$ complexes. This clearly is sheer speculation.

However, the fact that we can solubilize cytochrome c as a proteolipid either with $Mg^{++}$ (39) or with protons (38) implies that such ionic interactions might lead to significant penetration of proteins into the lipid matrix. (see model by Gómez-Puyou).

Several practical aspects have been derived from the techniques developed to form proteolipids with regards to the possibility of incorporating functional membrane proteins into lipid bilayers and liposomes. These are discussed in the chapter by Dr. Montal.

## Some Further General Comments

The reason for the presence of carbohydrate determinants attached both to glycoproteins and glycolipids in biological membranes is not known. Both are capable of functioning as receptors for a series of determinants such as lectins, viruses, toxins (see article by Cuatrecasas and Bennett). It has been suggested (Gitler (1)) that glycolipids might under some circumstances be acquired by a membrane after its formation. This has been shown to be the case *in vitro* (43,44,45) and in tissue culture (46) but there is no evidence that this does occur *in vivo*. The possibility that cells might passively be made to adsorb glycolipids opens up an experimental means of altering surface determinants and studying the resulting functional effects. We have synthesized some simple compounds such as N-hexosiloctylamines, N-pentosyloctylamines and recently the N-dodecyl-derivatives of glucosamine, galactosamine and mannosamine. These could be adsorbed by the cells and the sugar moieties would now be part of the surface determinants (47). Some interesting effects have already been shown for these molecules in normal and leukemic lymphocytes (48). The possibility exists that these molecules might bind to surface enzymes involved in the elongation of sugar chains. If such binding involved the active site, an inhibition in the function of these enzymes could result. More complex sugar-amphiphiles might be used for the above purposes.

One final speculation: lateral diffusion of membrane proteins might allow the formation of distinct aggregates through lateral adhesion of the membrane elements or through the interaction with components present both within and outside the cell (see article by Singer in this book and also Gitler et al (42)). These aggregates can form because there is lateral diffusion. The question exists whether at some stage of the cell cycle such mobility might be restricted, so that the aggregates cannot be formed. On receiving the cell some stimulus, diffusion could be allowed and this would lead to the formation of functional aggregates. The restriction to the mobility could be internal or it could be due to carbohydrate-carbohydrate or carbohydrate-protein interactions. Addition of trypsin or of lectins, for example, might free these latter restrictions and allow for the rearrangements to set in.

## Acknowledgement

The excellent technical help of Lucía García de de la Torre and Eunice Zavala, made these results possible.

## References

1. Gitler, C. Ann.Rev.Biophys.Bioeng. 1(1972)51.
2. Maddy, A.H. Biochim.Biophys.Acta 88(1964)390.
3. Berg, H.C. Biochim. Biophys.Acta 183(1969)65.
4. Sears, D.A., Reed, C.F. and Helkman, R.W. Biochim. Biophys. Acta 233 (1971) 716.
5. Pardee, A.B. and Watanabe, K. J. Bacteriol. 96(1968)1049.
6. Bonsall, R.W. and Hunt, S. Biochim. Biophys. Acta 249 (1971) 281.
7. Arrothi, J.J. and Garvin, J.E. Biochim. Biophys. Acta 355 (1972) 79.
8. Gordesky, S.E. and Marinetti, G.V. Biochem. Biophys. Res. Commun. 50 (1973) 1027.
9. Okuyama, T. and Satake, T. J. Biochem. (Japan) 47 (1960) 454.

10. Gitler, C. (1971) In: Biomembranes. p. 41, Vol. II, ed. Manson, L. Plenum Press, New York.

11. Bretscher, M.S. J. Mol. Biol. 58(1971)775.

12. Rifkin, D.B., Compans, R.W. and Reich, E. J. Biol. Chem. 247 (1972) 6432.

13. Cordes, E. and Gitler, C. (1973) In: Progress in Bioorganic Chemistry. p. 1, vol. II, eds. Kaiser, E.T. and Kézdy, F.J. New York:Wiley.

14. Fendler, E.J. and Fendler, J.H. Advan. Phys. Org. Chem. 8 (1970) 271.

15. Lenard, J. and Singer, S.J. Proc.Nat.Acad. Sci. USA 57(1967) 1043.

16. Wallach, D.F.H. and Zahler, P.H. Proc. Nat. Acad. Sci. USA 56 (1966) 1552.

17. Urey, D.W. and Ji, T.H. Arch.Biochem.Biophys. 128 (1968) 802.

18. Capaldi, R.A. and Vanderkooi, G. Proc. Nat. Acad. Sci. USA 69 (1972) 930.

19. Sandermann, H. and Strominger, J.L. Proc.Nat. Acad. Sci. USA 68 (1971) 2441.

20. Segrest, J.P., Jackson, R.L., Marchesi, V.T. Guyer, R.B. and Terry, W. Biochem. Biophys. Res. Commun. 49 (1972) 964.

21. Enomoto, K.-I.,and Sato, R. Biochem. Biophys. Res. Commun. 51 (1973) 1.

22. Branton, D. Ann.Rev.Plant Physiol. 20(1969)209.

23. Klip, A. and Gitler, C. Biochem. Biophys. Res. Commun. Submitted for publication.

24. Knowles, J.R. Accounts Chem. Res. 5(1972)155.

25. Patai, S. (1971) In: The Chemistry of the Azide Group. London:Interscience.

26. Lwowski, W. (1970) In: Nitrenes. New York: Interscience.

27. MacLennan, D.H., Yip, C.C., Iles, G.H. and Seeman, P. Cold Spring Harbor Symp. Quant. Biol. 37 (1972) 469.

28. Stern, O. and Volmer,M. Physik. Z. 20 (1919) 183.

29. Weller, A. Progress in Reaction Kinetics 1 (1961) 187.

30. Vaugham, W.M and Weber, G. Biochem. 9 (1970) 464.

31. Spencer, R.D. Doctor of Philosophy Thesis, University of Illinois at Urbana-Champain, 1970.

32. Papageorgiou,G. and Argoudelis, C. Arch. Biochem. Biophys. 156 (1973) 134.

33. Pownall, H.J. and Smith, L.C. Biochemistry 13 (1974) 2594.

34. Conti, F. Personal Communication.

35. Pownall, H.J. and Smith, L.C. Biochemistry 13 (1974) 2590

36. Shinitzky, M., Dianoux, A.-C., Gitler, C. and Weber, G. Biochemistry 10 (1971) 2106.

37. Rudy, B. and Gitler, C. Biochim. Biophys. Acta. 288 (1972) 231.

38. Gitler, C. and Montal, M. Biochem. Biophys. Res. Commun. 47 (1972) 1486.

39. Gitler, C. and Montal, M. FEBS Letters 28 (1972) 329.

40. Gitler, C. Submitted for publication.

41. Interiano de Martínez, A.I. Master of Science Thesis, Centro de Investigación y de Estudios Avanzados del I.P.N. México. 1972.

42. Gitler, C., Interiano de Martínez, A.I., Viso, F., Gasca, J.M. and Rudy, B. (1973) In: Membrane Mediated Information. p. 129, vol. II, ed. Kent, P.W. Lancaster:Medical and Technical Publishing, Co.

43. Marcus, D.M. and Cass, L.E. Science 164 (1969) 553.

44. Schroffel, J. Thiele, O.W. and Koch, J. European J. Biochem. 22 (1971) 294.

45. Bara, J. Lallier, R., Brailovsky, C. and Nigam V.M. Eur. J. Biochem 35(1973)489.

46. Franks, D. Biol. Rev. 43(1968)17.
Laine, R.A. and S.-I. Hakomori, Biochem. Biophys. Res. Commun. 54(1973)1039.

47. Gitler, C. (1969) In: Propiedades de las Superficies Biológicas. p. 99, ed. Soberón, G. Ensayos Bioquímicos. México: Prensa Médica Mexicana.

48. Romero-Villaseñor, G. Doctor of Science Thesis. Centro de Investigación y de Estudios Avanzados del I.P.N. México 1967

---

TABLE I

Constants for the Quenching of Perylene in Different Environments

| Quencher | $K_Q$ ($M^{-1}$) Methanol | Liposomes | | Sarcoplasmic Reticulum | |
|---|---|---|---|---|---|
| Butylpyridinium Br | 53.8 | <1.0 | | 0.8 | (-67) |
| 1,3-Dinitrobenzene | 149 | 816 | | 816 | (5.5) |
| 1-Azidonaphthalene | 62 | 2777 | (45) | 5263 | (85) |
| 1-Azido-4-iodobenzene | 88 | 2325 | (26) | 2797 | (32) |

APOLAR AZIDES

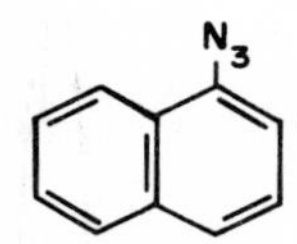


1- AZIDONAPHTHALENE

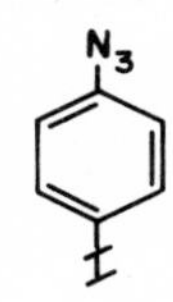


P-IODOPHENYLAZIDE
1-AZIDO-4-IODOBENZENE

Figure 1. Structure of the nitrene-precursors used to label components present within the lipid core of biological membranes.

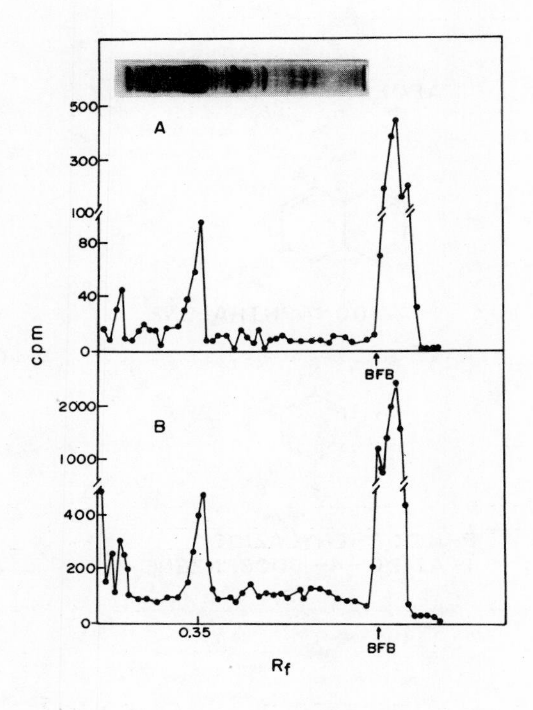


Figure 2. Disc gel electrophoresis of sarcoplasmic reticulum labelled with 1-azidonaphthalene (A) and with 1-azido-4-iodobenzene (B)

| | | cpm | IDENTIFICATION | PERCENT OF RECOVERED | PERCENT OF APPLIED |
|---|---|---|---|---|---|
| FRONT | 11 | 3621* | FREE PRODUCTS NEUTRAL LIPIDS | 10.6 | 7.3 |
| | 10 | 365 | CHOLESTEROL? | 1.1 | 0.76 |
| | 9 | 991* | PE | 2.9 | 2.0 |
| | 8 | 1738 | PC | 5.1 | 3.5 |
| | 7 | 554 | -- | 1.6 | 1.1 |
| | 6 | 279 | -- | 0.8 | .55 |
| | 5 | 168 | -- | 0.5 | .34 |
| | 4 | 168 | -- | 0.5 | .34 |
| | 3 | 1418* | -- | 4.1 | 2.8 |
| | 2 | 160 | -- | 0.5 | .34 |
| ORIGIN | 1 | 24,857* | Protein | 72.4 | 49.8 |

* Main components labelled
Total counts applied 49 867, recovered 34 320 (70%)
Membranes applied directly to the thin layer plate. Developed with ANS.

Figure 3. Thin layer chromatography of sarcoplasmic reticulum labelled with $A_zIB$-$^{125}I$.

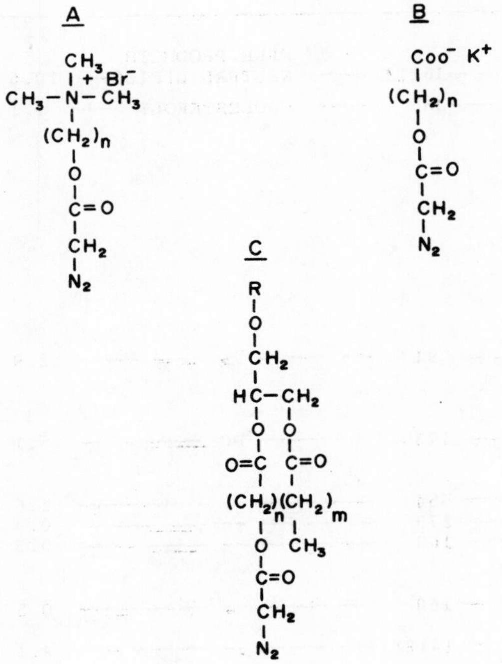

Figure 4. Some carbene-precursors being synthesized to study the components located in the vicinity of these molecules within biological membranes.

A. Diazoacetyl homologues of acetylcholine and of cationic detergents.

B. Diazoacetyl homologue of fatty acid and C of phospholipids.

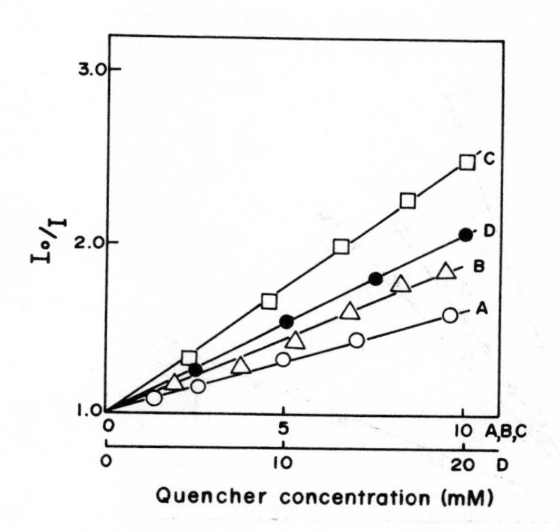


Figure 5. The encounter quenching of the fluorescence of perylene dissolved in methanol as a function of the concentration of 1-azidonaphthalene (A), 1-azido-4-iodobenzene (B), 1,3-dinitrobenzene (C) and N-Butylpyridinium bromide (D). Data plotted according to equation 5 in the text.

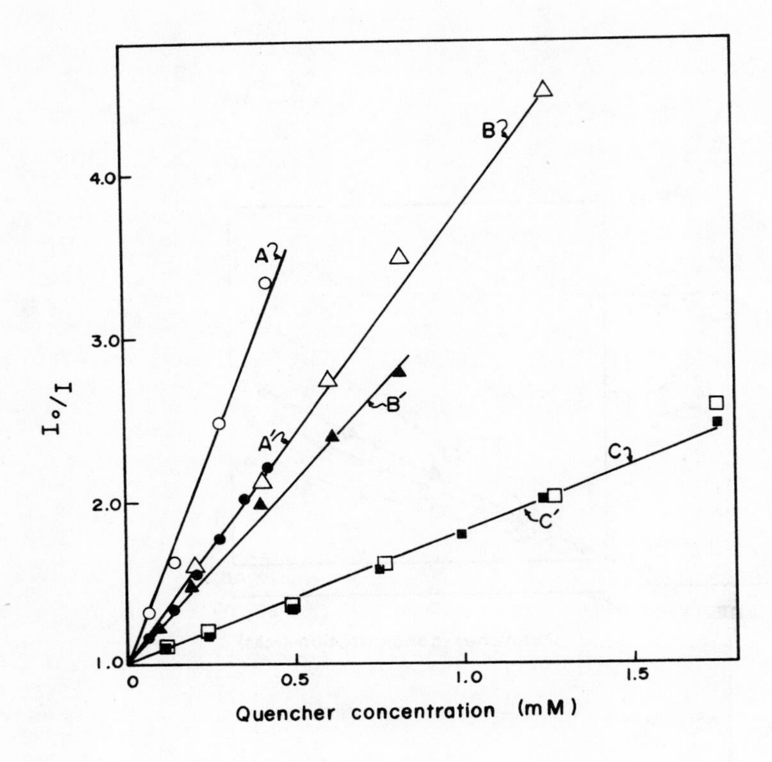


Figure 6. The encounter quenching of the fluorescence of perylene dissolved in phosphatidyl choline liposomes (A', B' and C') and in sarcoplasmic reticulum membranes. Details as in Fig. 5.

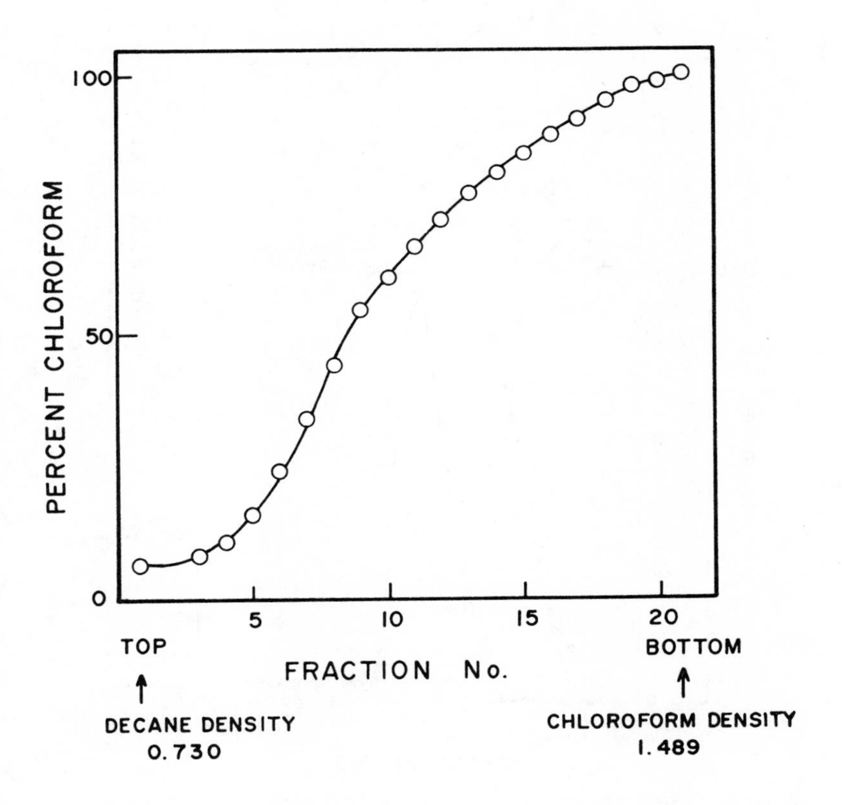


Figure 7. The profile of the nonpolar solvent gradient generated using decane-chloroform. Percent chloroform was determined by the weight loss at 20°.

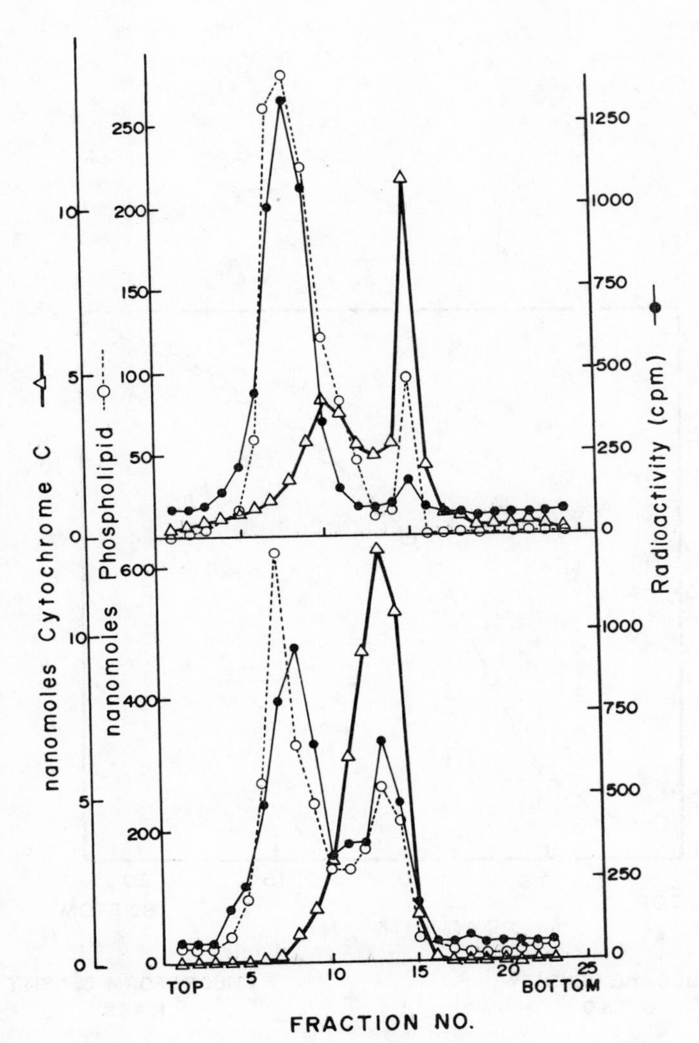


Figure 8. Sedimentation pattern of the cytochrome c-proteolipid. A. Proteolipid extracted into decane sedimented in a decane-chloroform gradient. B. Proteolipid extracted into isooctane and sedimented in an isooctane-chloroform gradient. For extraction procedures see 39 . For the centrifugation see text.

# II

# ION AND METABOLITE DISTRIBUTION

# SODIUM SITES AND ZONULAE OCCLUDENS: LOCALIZATION AT SPECIFIC REGIONS OF EPITHELIAL CELL MEMBRANES

David Erlij and Terry E. Machen*

Departamento de Biología Celular, Centro de Investigación y de Estudios Avanzados del I.P.N., Apartado Postal 14-740, México, D.F. and Department of Physiology and Anatomy, University of California - Berkeley. Berkeley, Calif. 94720

In recent years evidence has been accumulating in support of the notion that the cell membrane is in a fluid state i.e., that molecules incorporated into the lipid bilayer matrix are free to move laterally in the plane of the membrane (1). Such a structure should produce, in the long run, a random distribution of the molecules included in the lipid matrix of the membrane. On the other hand, a number of observations on cell and tissue functions indicate that given membrane molecules have a specific topographical localization on the surface of the cell. There are many striking examples of tissue functions that imply such a precise topographical distribution of membrane sites. In the electric organs of many fishes, acetylcholine sensitivity and the ability to generate action potentials is restricted to the innervated surface of the cells (2). In the neuromuscular junction the acetylcholine receptor is localized only under the

* Dr. Machen's visit to the Centro de Investigación was supported by a Grant from the Consejo Nacional de Ciencia y Tecnología, México.

nerve endings (3,4). In all epithelia that absorb water and solutes, the most useful concept for explaining such transport of substances across the tissue is that of asymmetric membrane properties (5), which holds that transepithelial transport occurs because membranes of the same cell have strikingly different properties at each of the surfaces of the epithelium. In this paper we present data that further emphasize the fact that in a salt-transporting epithelium -the frog skin- the function of specific membrane components is expressed only in specific regions of the cell membrane. Moreover, the appearance of these localized components can be triggered in cells in which they were not previously present.

It has been known since the beginning of this century that the potential difference across the frog skin depends on the presence of either sodium or lithium on the outside surface of the skin (6). This sodium-dependent potential led to the original proposal that at the outside surface of the skin the permeability to sodium is much greater than the permeability to potassium, in contrast with what is observed in most resting cells, where the converse relationship is observed (7). Table I shows the values of the sodium and potassium fluxes across the outer and inner surfaces of the epithelium calculated from determinations made with radioactive isotopes (8,9,10 and 11). The membranes facing the inside surface of the epithelium have a much higher permeability to potassium than to sodium, as is found in most cell types, whereas the outer surface has high sodium permeability and very low potassium permeability.

Table II summarizes the results of experiments carried out to characterize the sodium-selective channel at the outer border of the skin (12). Its features are compared with those of the $Na^+$-selective channel of active nerve which has been also the subject of considerable attention (13). The alkali cation selectivity sequence for both types of channels is similar. The results in both nerve and skin fit the Eisenman sequence indicating that the sites for alkali cation permeation are of high field density (14).

The analogy goes no further. In nerve a number of organic cations can substitute for sodium in the generation of action potentials (15,16). The currents carried by these organic cations are blocked by the specific sodium channel poison tetrodotoxin (16). In the frog skin we have been unable to detect a potential when all the outside sodium is substituted by the organic cations that are effective in nerve. The case of hydroxylamine may be an exception, since it produces a small potential when on the outside of the skin. This potential, however, is not modified by amiloride, a drug that selectively blocks the sodium permeability in frog skin. This finding suggests that the specific sodium channels at the outer border of the frog skin are not involved in the generation of the small hydroxylamine potentials. Finally, Table II shows that the sensitivity of the two channels to blocking agents is markedly different. Keynes and I have observed that squid and frog axons carry normal action potentials in solutions containing $10^{-3}$M amiloride while the sodium conductance of the frog skin is markedly reduced when $10^{-5}$M amiloride is added to the outside solution (17). The converse is true for tetrodotoxin; $10^{-8}$M of the inhibitor blocks action potential generation (18) while concentrations of up to $10^{-4}$M have no effect on the sodium channel of the outside border of the skin.

Concerning the sodium movements across the inside surface of the frog skin, some recent observations of Aceves (personal communication) indicate that they do not occur through an amiloride-sensitive channel. He found that epithelia are loaded with sodium when they are incubated either in potassium-free or in ouabain-containing solutions and that this movement of sodium is not blocked by amiloride.

So far we have collected evidence that strongly supports the notion that at the outer border of the skin there are specific sodium selective sites. We shall now describe experiments that indicate that during the moult cycle of the frog skin the membranes of cells that do not possess specific sodium channels reach the outer border of the epithelium.

We have found that these cells instead behave as if they were potassium-selective. Some of the findings of an experiment in which the sequence of events during the moulting of the skin was followed are illustrated in Figures 1 and 2. In these experiments we have taken advantage of the finding that the addition of aldosterone to isolated skins produces a moult (19,20). We believe that the changes during the aldosterone-induced moult probably also occur during every normal moult cycle of the skin. Figure 1 shows two samples of the skin of the same frog. Figure 1A is the control sample and Figure 1B was fixed two hours after the addition of aldosterone. The hormone had its characteristic moulting effect since the outermost cell layer of the epithelium was shed.

Figure 2 shows the experiment in which the potassium selectivity of the outer border of the same skin was studied. The skin received aldosterone at zero time. To detect potassium permeability we measured the potential difference across the skin when all the sodium in the outside solution had been substituted by potassium. Immediately after this ion substitution the potential across the skin dropped from a control value of 30 mV to values very near zero. Then the potential began to increase gradually until 90 minutes after the addition of aldosterone, when it reached a maximum value of about 30 mV. At this point, the substitution of $K^+$ by $Na^+$ almost completely abolished the potential.

In other experiments we also found that substitution of all outside $K^+$ by tris reduced the transepithelial potential to values near zero.

Figure 2 also shows that Barium ions which selectively inhibit the $K^+$ conductance in several cell types (21,22) also reduced the $K^+$-dependent potential in the skin. On the other hand, the $K^+$-dependent potential was unaffected by the $Na^+$ channel inhibitor amiloride. The results of these ion substitution experiments and the action of the inhibitors suggest that during the aldosterone-induced moult the cell layer containing the sodium-sensitive sites facing the outside solution are shed and that cell surfaces that were facing the inside and which have mainly $K^+$-selective channels are exposed to the outside solutions; later, new $Na^+$-selective sites appear on the outside surface of the epithelium as shown by the recovery in $Na^+$ sensitivity. Since the whole moulting process can be blocked by actinomycin or puromycin (20) it is possible that the appearance of the $Na^+$ conductance may be the result of synthesis of new proteins.

Regardless of whether the reappearance of $Na^+$ selectivity is the result of the synthesis of new protein, it is clear that the increase in $Na^+$ permeability after a moult is restricted to the surface of the cells facing the outside solution, as if a peculiar type of membrane components was added only to this region of the cell membrane.

In the following part of this communication we shall describe another type of cell membrane component that has a specific localization in the epithelium and that also must be renewed during each moult cycle. An important feature of epithelia is that the tissue forms a multicellular barrier separating two compartments that contain solutions of different composition. In order for this barrier to be effective, the extracellular spaces must be obliterated at some point. This obliteration is brought about by a special organization of the cell membrane, the _zonulae occludens_.

The _zonulae occludens_ is characterized by a very close apposition of the cell membranes of adjacent cells resulting in the obliteration of the intercellular space over variable distances. The _zonulae occludens_ have a precise and specified localization on the cell surface of every epithelium. In the particular case of the skin, zonulae occludens are localized at the outer border of the cells in the stratum corneum and between cells of the outer layer of the stratum granulosum (23,24). Several findings show that the close membrane apposition at the region of the zonulae occludens results from the presence in this region of a specific membrane component. First, freeze cleave studies show that in the region of the zonulae occludens there is a branching network of threads or chains of small globular subunits (25,26,27). Second, some observations from our laboratory suggest that the mechanism that maintains close cell membrane apposition is mediated by stable membrane components. For these studies we have judged the ability of the occluding zonules in the frog skin to obliterate the extracellular space of the frog skin epithelium by measuring the reduction in resistance caused by increasing the osmolarity of the solution on the outside of the skin and the recovery in resistance that follows the return to a solution of normal osmolarity. We have shown previously by using $La^{+++}$ as a tracer for electron microscopy that the reduction in resistance and its subsequent recovery respectively correspond to the opening and resealing of the occluding zonules (28).

In our experiments, we have found that the resealing process occurs even when the energy metabolism of the skin has been inhibited by either dinitrophenol ($10^{-4}$M) or antimycin ($3x10^{-6}$M) to such levels that the potential difference and short-circuit current have been blocked almost completely or when the whole experiment is carried out at 4°C (29).

One of these experiments is shown in Figure 3. In this experiment the recovery from the effects of an hypertonic urea solution were first studied, then antimycin A ($3x10^{-6}$M) was added to the inside solution. After the antibiotic had abolished the potential difference and short circuit current we again tested the effects of hypertonic urea solutions. The reduction and recovery of the resistance that follows the removal of the hypertonic solution were similar to those observed during the control period.

The resealing of the occluding zonules after abolishing the energy metabolism or when temperature is reduced is in marked contrast to the available information on another type of junction, the gap junctions, which are responsible for cell-to-cell communication. It has been found that the resistance of the latter junctions is greatly increased when cell metabolism is blocked or when temperature is reduced (30,31).

The lack of effect of low temperature and metabolic inhibitors on the resealing process suggests that the obliteration of the extracellular spaces is brought about by a rather stable membrane component.

Since the one or two outermost layers of the epithelium are shed during moulting and since very soon after a moult we can find zonulae occludens with the electron microscope in layers that previously did not have such structures, we infer that the membrane components responsible for close membrane apposition must have become functional after the moult.

The data summarized in this communication emphasize the essential requirement of specific localization of membrane components for adequate cell function. Furthermore, the data also imply that these components can be exposed in defined positions within a relatively short period. To conciliate the data summarized here with the fluid nature of the lipid membrane matrix it is necessary to postulate elements that will provide the membrane with long-range topographical organization. Although several types of cellular elements could be proposed for maintaining this long-range organization i.e., microtubules, microfilaments, cell surface components and so on, it still remains a task for cell biology to establish the manner in which the function of specific components appear only at precise membrane spots.

References

1. Singer, S.J. and Nicolson, G.L. Science 175 (1972) 720.
2. Grundfest, H. Physiol. Rev. 37 (1957) 337.
3. Kuffler, S.W. J. Neurophysiol. 6(1943)110.
4. Thesleff, S. Physiol. Rev. 40(1960)734.
5. Ussing, H., Erlij, D. and Lassen, U. Ann. Rev. Physiol. 36 (1974)
6. Galeotti, G. Z. Phys. Chem. 49(1904) 542.
7. Koefoed-Johnsen, V. and Ussing, H.H. Acta Physiol. Scand. 42(1958)298.
8. Erlij, D. and Smith, M.W. J. Physiol. 228 (1973) 221.
9. Aceves, J. and Erlij, D. J. Physiol. London 212 (1971) 195.
10. Bibber, T.U.L., Aceves, J. and Mandel, L. Am. J. Physiol. 222 (1972) 1366.
11. Curran, P.F. and Cereijido, M. J. Gen.Physiol. 48 (1965) 1011.
12. Lindley, B.D. and Hoshiko, T. J. Gen. Physiol. 47 (1964) 749.
13. Hille, B. J. Gen Physiol. 59 (1972) 637.
14. Diamond, J. and Wright, E.M. Ann. Rev. Physiol. 31 (1969) 581.
15. Luhgan, H.C. Pflügers Arch. 267(1958)331.
16. Hille, B. J. Gen. Physiol. 58(1971)599.
17. Eigler, J., Kelter, J. and Renner, E. Klin. Wschs. 45 (1967) 737
18. Narahashi, T., Moore, J.W. and Scott, W.R. J. Gen. Physiol. 47 (1964) 965.
19. Nielsen, R. Acta Physiol. Scand. 77(1969)85.
20. Larsen, E.H. Acta Physiol.Scand. 79(1970)453.

21. Sperelakis, N., Schneider, M.F. and Harris, E.J. J. Gen Physiol. 50 (1967) 1565.

22. Schwartz, M., Pacifico, A.D., Mackrell, T.N., Jacobson, A. and Rehm, W.S. Proc. Soc. Exp. Biol. and Med. 127 (1968) 223.

23. Farquhar, M.G. and Palade, G.M. J. Cell Biol. 26 (1965) 263.

24. Martínez-Palomo, A., Erlij, D. and Bracho, H. J. Cell Biol. 50 (1971) 277.

25. Staehelin, L.A., Mukherjee, T.M. and Williams, A.W. Protoplasma 67 (1969) 165.

26. Scott McNutt, N. and Weinstein, R.S. J. Cell Biol. 47 (1970) 666.

27. Claude, P. and Goodenough, D.A. J. Cell Biol. 58 (1973) 390

28. Erlij, D. and Martínez-Palomo, A. J. Membrane Biol. 9 (1972) 220.

29. Erlij, D. and Lázaro, A.R. Fed. Proc. 33 (1974) 215.

30. Loewenstein, W.R. Fed. Proc. 32 (1973) 60.

31. Payton, B.W., Bennet, M.V.L. and Pappas, G.D. Science 165 (1969) 594.

TABLE I

Ion Fluxes Across Inside and Outside Surface of Frog Skin Epithelium

| | Sodium | Potassium |
|---|---|---|
| Outer surface | 1.5 μeq $hr^{-1}$ $cm^{-2}$ | Undetectable |
| Inner surface | 0.050 μeq $hr^{-1}$ $cm^{-2}$ | 1.5 μeq $hr^{-1}$ $cm^{-2}$ |

TABLE II

Characteristics of the Sodium-Selective Sites in Frog Skin and Nerve

| | Nerve | Skin |
|---|---|---|
| Selectivity sequence | $Na^+ \simeq Li^+ > K^+ >> Rb^+ \simeq Cs^+$ | $Na^+ \simeq Li^+ > K^+, Rb^+, Cs^+$ |
| Organic substitution | Sodium≃hydroxylamine≃ hydrazine>ammonium≃ formamidine≃guanidine | Organic compounds do not substitute |
| Tetrodotoxin | Inhibits | Does not inhibit |
| Amiloride | Does not inhibit | Inhibits |

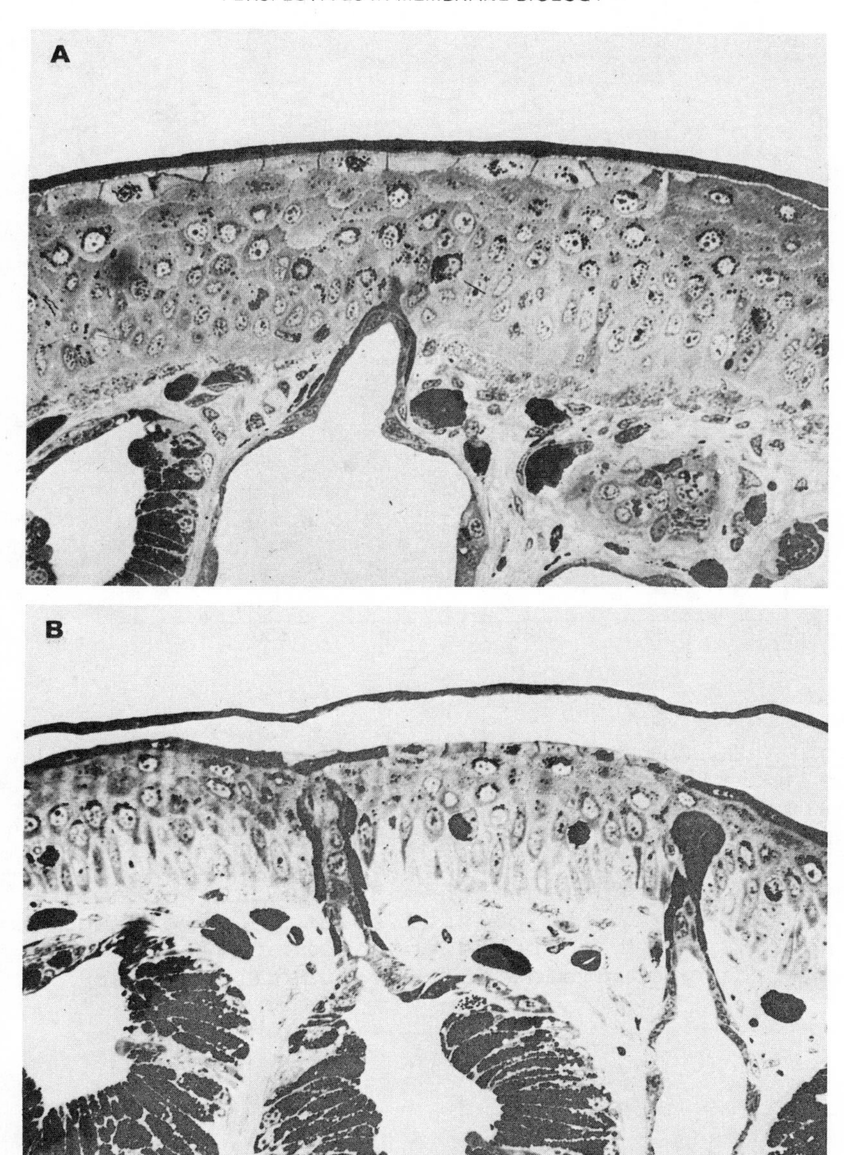


Figure 1. The moulting effect of aldosterone. The sample of skin in A was fixed before the addition of aldosterone. The sample B was fixed two hours after the addition of aldosterone.

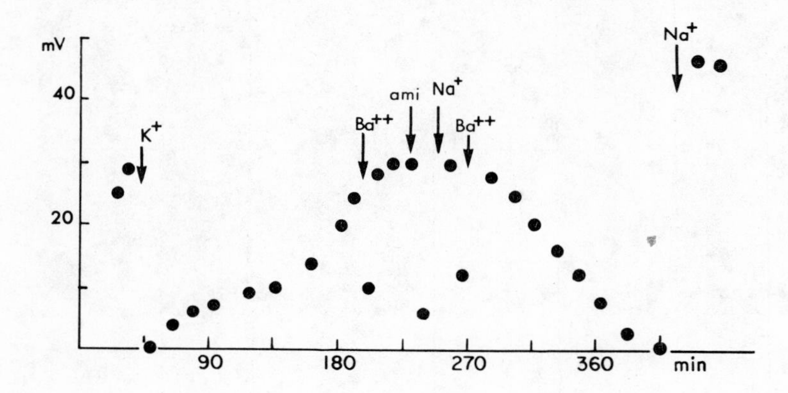


Figure 2. The potential of the frog skin after the addition of aldosterone. First, all the $Na^+$ was substituted by $K^+$. Then when the potential had increased spontaneously in the presence of $K^+$ on the outside the effects of substituting $Na^+$ for $K^+$ and of adding either $Ba^{2+}$ (0.5mM) or amiloride ($10^{-4}$M) to the outside solution were studied. Finally, when the spontaneous potential in $K^+$ ringer dropped again, $Na^+$ was substituted for $K^+$.

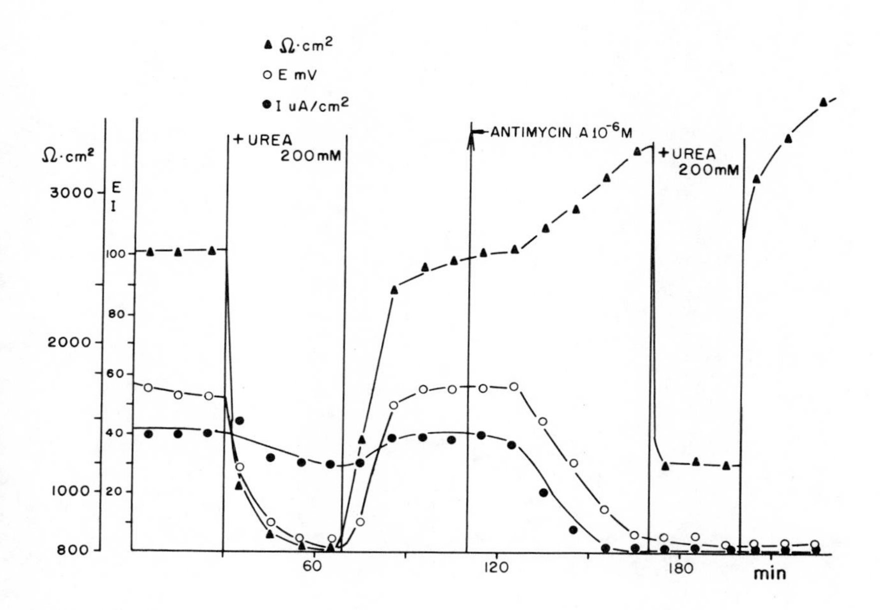


<u>Figure 3</u>. The effects of antimycin A on the recovery from the action of hypertonic urea. First the effects of adding 200 mM urea on the electrical parameters of the skin were tested. Then antimycin A was added and when the inhibitor had abolished the potential difference and short-circuit current, urea was added again.

# THE CONTROL OF METABOLISM BY ION TRANSPORT ACROSS MEMBRANES

Antonio Peña

Departamento de Bioquímica, Facultad de Medicina y Departamento de Biologia Experimental, Instituto de Biología
Universidad Nacional Autónoma de México
Apartado Postal 70-600
México, D. F.

The importance of the interactions of the transport of ions across membranes with metabolic pathways can be visualized by the simple fact that cells or subcellular structures capable of carrying out the transport of ions can greatly increase their metabolic rates as a consequence of the translocation of certain ions.

This phenomenon was to be expected; transport in general represents the means to bring into or out of the cells or subcellular structures the large number of different materials necessary to feed and regulate all the metabolic pathways therein contained or to eliminate their waste products. It is only a point of coincidence that many of these materials are ions. If it is now added that many transport systems require energy to work, and that this energy has to be obtained from metabolism, it is obvious that the cells must direct a good proportion of their metabolism to driving the transport systems.

The effects of ion transport on metabolism can be either direct or indirect. In this presentation, three kinds of effects will be considered:

1. Direct effects of ions on enzyme systems.

2. Effects derived from the fact that some

of the ions which must be transported are enzyme substrates.

3. Changes in the rate of metabolic pathways or enzyme reactions produced as a consequence of the energy requirement of ion transport.

## Direct Effects of Ions on Enzyme Reactions

Numerous cases have been described in which an enzyme action is favoured by the presence of certain ions. On the other hand, there are enzymes which are inhibited by ions. In the first group, one of the best known cases is that of pyruvate kinase (1), which requires $K^+$ and $Mg^{2+}$ to work; furthermore, this requirement cannot be satisfied by any other monovalent or divalent cation; the enzyme in fact shows some sort of ion-selectivity pattern. This case is not rare; in the glycolytic sequence alone, several cases of ionic requirement for the functioning of different enzymes (2-6) have been described. There are also numerous cases of enzymes which are inhibited by the presence of certain ions; pyruvate kinase itself, for instance, is inhibited by relatively low concentrations of $Ca^{2+}$ (7,8) and pyruvate carboxylase is also inhibited by the same cations (9) and by phosphate (10).

These are only a few of the known cases of enzymes whose activity is modified by the presence of certain ions. These effects can be related to ion transport by the simple fact that, in order for actual interaction of ions with intracellular enzymes to occur, the former must be transported to the sites where the latter are located; the effects of $Ca^{2+}$ on glycolysis in Ehrlich cells, for instance, seem to depend on its penetration into the cells (8). Ion-transport systems, furthermore have the capability of changing in one way or another the concentrations of the ions that can modify enzyme rates. Although in many cases the actual or <u>in situ</u> meaning of these activations or inhibitions has not been analyzed, studies have been carried out which have led to the postulation of some regulatory systems.

In general, the ions whose actual effects on metabolism are shown by their in vitro effects on enzymes seem to be those found at rather low concentrations in the medium surrounding the enzymes. Thus, although $K^+$, for instance, has an important effect on the activity of pyruvate kinase, it seems that from the point of view of the actual regulation of the activity of this enzyme within the cells there is a much more important regulatory role of $Ca^{2+}$, whose intracellular concentration is much lower than that of $K^+$ (8). Gevers and Krebs (9) have already pointed out the regulatory effect of $Ca^{2+}$ on the pyruvate kinase and pyruvate decarboxylase activities. On the other hand, no case has been reported in which $K^+$ uptake by cells seemed to modify the activity of this enzyme whithin the cells. In yeast, Peña et al (11) in analyzing the changes that $K^+$ addition to the incubation medium (together with its net uptake) produced on glycolytic intermediates, could not find variations atributable to the activation of this enzyme by $K^+$.

Within this same category, a case of special importance is the effect of the changes in the concentration of $H^+$ on enzyme activity. The system most commonly described as capable of producing changes in $H^+$ concentration in living organisms is precisely proton transport. Both in procaryotic and eucaryotic organisms, $H^+$ transport systems have been described which can actually change the pH of the medium containing different enzymes, and there is no doubt that these pH changes will have effects on at least some steps of metabolic pathways.

In studies carried out in yeast, conditions have been described under which changes in the intracellular pH result from the translocation of $H^+$ (12). When yeast cells are placed in a medium with a pH of 7.5, in the presence of a substrate, a proton expulsion is favoured which produces the alkalinization of the internal medium (12). Owing apparently to the energy requirement of this proton expulsion, there is a decrease of the levels of ATP and an increase of ADP and Pi in the cell, and these changes produce an increased rate of fermentation.

According to the following scheme:

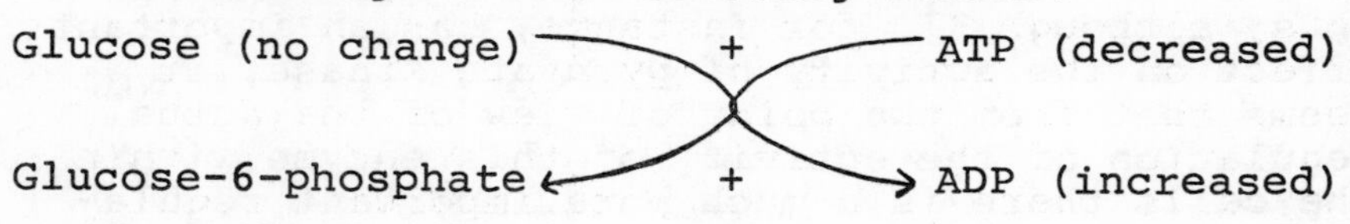


the levels of glucose-6-phosphate at the higher pH should decrease. When the pH of the incubation medium is 7.5, as compared to 3.5, the addition of glucose reduces the glucose-6-phosphate content to lower levels (Fig. 1). However, after some time, the levels at pH 3.5 start to decrease rapidly to reach rather constant levels. At the high pH which on the other hand produces the alkalinization of the inside of the cell, the levels of glucose-6-phosphate, after an initial decrease, show an oscillatory pattern in which, however, they remain always higher than at the low pH of incubation. This effect, in view of the concentration levels of the components of the hexokinase reaction and the increased rate of utilization of glucose-6-phosphate, must be due to a higher rate of phosphorylation of glucose catalyzed by hexokinase which at the increased internal pH is now closer to its optimum pH. This fact seems to have further implications; when the changes of the levels of NADH were recorded, it was possible to observe an oscillatory behavior within certain glucose concentrations at the high pH; however, at the low pH all attempts to obtain this phenomenon failed. Apparently the low pH produces a decrease in the rate of entrance of glucose to the glycolytic pathway which, by lowering the levels of intermediates in the sequence, prevents the functioning of regulatory mechanisms. In this same investigation it was observed that the increased internal pH also produces an increase in the rate of alcohol dehydrogenase (12,13). Similar results were obtained when the increase of the internal pH was produced by the addition of $K^+$ to the incubation medium, (11), which also produces the internal alkalinization (12) by the $K^+/H^+$ exchange which takes place in the presence of the cation. The changes in the concentrations of glucose-6-phosphate are shown in Fig. 2. Under these conditions, by adjusting the concentrations of glucose, it is also possible to reproduce the oscillatory changes of NADH.

Other phenomena were also observed when the interior of the cell was alkalinized either by incubating at a high pH, or at a low pH in the presence of $K^+$. It was found that the maximal respiratory rate, demonstrable with 2,4-dinitrophenol at pH 4.0 with ethanol as substrate, was increased by the incubation of the cells in the presence of $K^+$ (Fig. 3).

In these cells, grown under vigorous aeration, glucose is not a good substrate for respiration, and higher rates could be obtained with ethanol. When $K^+$ was added to the incubation medium, there was a gradual increase in respiration with glucose as substrate. This same kind of increase could be observed when the cells were incubated at pH 7.5 instead of at pH 4.0. This behavior could be due to the low concentrations of glucose-6-phosphate formed at the low pH in the absence of $K^+$ shown in Figs. 1 and 2, or to the lower levels of NADH present under these conditions. However, the measurement of the ATP levels at the low pH in the absence of $K^+$ showed similar figures with ethanol and with glucose as substrates; the differences in the respiratory rates could be due to differences in the transfer of reducing power produced in glycolysis to the mitochondria.

These data show some of the changes in metabolism which can be detected in this experimental subject as a consequence of the change of concentration of hydrogen ions within the cell, changes which, on the other hand, seem to be the result of the transport of $H^+$ across the cell membrane.

This type of experiments brings into consideration another fact also related to the effect of transport on the concentration of certain ions within biological systems. In spite of the fact that $K^+$ has no effect on the activity of hexokinase, it is evident that its presence in the medium results in an increase of its activity, but this is done by facilitating the expulsion of $H^+$ from the cell. Although the effect on hexokinase is due to the pH change, at the low pH of incubation this does not occur unless $K^+$ is added to the medium. There are ions which, when being transported, are linked to or facilitate the transport of other ions which do have influence on some metabolic steps within the cell. There are numerous examples of ions capable of being transported but which, on the other hand, require the balancing movements of other ions to be transported across membranes; for these, although the regulatory mechanism is a consequence of their translocation, it appears as a secondary effect to the transport of other ions.

## Effects Derived From The Fact That Some Ions Which Must Be Transported Are Enzyme Substrates

Ion- transport systems, in some cases, indirectly regulate the function of metabolic pathways, because they carry through membranes organic ions which are the substrates of some enzymes. One of the most apparent cases is that of the substrates of the tricarboxylic acid cycle. All the enzymes are located within the mitochondria, and the substrates, when present outside the mitochondrion, must be transported to the inside of the particle to be utilized. The transport systems have been studied and many of their characteristics elucidated. In addition to the fact that it is beyond any doubt that many substrates have to be transported to be utilized, investigations have shown numerous interesting relations between the several systems involved (14-17). Although mitochondria represents perhaps the clearest example of this type of dependence of metabolism on transport, studies which have been carried out on microorganisms and isolated cells also show a distinct relationship between metabolism and the transport of many substrates to the interior of these structures. Two classes of transport have been identified in this respect; that in which ionic substrates are transferred across the membrane, and that in which the substrates are not ions, but in whose transport ions are in some way involved. In the former case, perhaps the most evident representatives are some members of the tricarboxylic acid cycle: phosphate, the adenine nucleotides and the aminoacids. Some transport systems of non-ionic substrates have been described, however, which seem to be linked to the transport of ions; the transport of glucose, for instance, is linked in many cases to that of sodium (18), and in studies of *Escherichia coli* West and Mitchell (19) have described the coupling of proton transport to that of galactose. In any case, both categories represent instances in which there is a close link between ion transport and the supply of substrates necessary for metabolic activity. There are still other aspects of this kind of interaction between transport and metabolism. First, there are species differences in the distribution of the several transport systems involved which also give differential properties to the functioning of some metabolic pathways of different biological systems because of the possibility of taking up the materials contained in the external medium even when they have the necessary enzymes to carry out their metabolism. Another important point consists in the possibility of metabolic regulation by changes in the levels of the transport systems involved that will permit regulation of the rate of utilization of some substrates by means of the synthesis of the transport systems involved in its uptake by the cellular or subcellular structures. The transport of β-galactosides, for instance, is an inducible system in *Escherichia coli* (20). In these cases, the genetic regulation of metabolism can be mediated by the synthesis of the systems coupling mechanisms in bacterial membrane vesicles. Today, it is well established that ion-transport systems are capable of stimulating metabolism to satisfy their energy requirements, and the main mechanisms involved in this process have been disclosed.

There are, however, important facts related to this energy requirement for ion transport. In many cases the materials transported into or out of membrane structures have obvious utility for the cells, because they are usually required as intermediates of different metabolic pathways or are degradation products which have to be eliminated. In other cases, however, there is a different situation, as occurs with the transport of $K^+$, which is not directly required as a working material. Almost all biological systems invest large amounts of energy in making their potassium-transport systems function. In yeast cells, for instance, at low pH values glycolysis is almost doubled when $K^+$ is present in the incubation medium (11), and ATP seems to be the source of energy for transport (12,13). In spite of the fact that $K^+$ cannot be utilized to the same extent as aminoacids, phosphate and other molecules, it is obvious both from the theoretical point of view and from increasing experimental evidence, that $K^+$ accumulation is very important to living systems.

In mitochondria, bacteria and other systems, it is found that the functioning of many transport mechanisms depends either on $K^+$ transport or on the previous accumulation of the cation. In recent studies of _Streptococcus faecalis_, Asghar et al (30) have shown that $K^+$ accumulation is necessary, and apparently constitutes energy stored in the form of an electrical potential to drive the transport systems of neutral aminoacids. In this same study, it seems clear that it is not even necessary that the cation accumulated be $K^+$, since it can be replaced in this exchange by $Na^+$ with similar results by loading the bacterium with $Na^+$ instead of $K^+$. In addition to this case, there are a number of transport systems whose functioning is known to depend on the accumulation of $K^+$. Among these are: inorganic phosphate in yeast (31), citrate in *Bacillus subtilis* (32), aminoacids in yeast (33), and inorganic phosphate in *Escherichia coli* (34).

Another system in which the accumulation of $K^+$ plays an important role is the generation of bioelectricity (35). In this case it is the double concentration gradient of $K^+$ and $Na^+$ that represents the immediate energy source which the cells utilize to produce bioelectric phenomena, the nerve impulse or the discharge of electricity. Once the gradients are expended in the production of these phenomena they are regenerated by the $Na^+$ pump, which is responsible for the expulsion of $Na^+$ and the uptake of $K^+$. The operation of this pump involves the expenditure of ATP, which must in turn be produced by the usual pathways, mainly oxidative phosphorylation and glycolysis. A new operation of the excitability mechanism uses this gradient, which must then be regenerated, and so on.

In these cases, $K^+$ accumulation represents not only a way of increasing the electrochemical potential of the cell, but also a form of energy storage which can be used *directly* to drive other systems. Although it is evident that the primary translocation systems ($H^+$ or $Na^+$ pumps) are energetically capable of driving the others, it is obvious that the cell finds it advantageous to count on another way of increasing its electrochemical potential by means of the accumulation of this cation which, can also participate directly in, or influence, a series of cell functions or transport mechanisms.

In these cases, then, the accumulation of $K^+$ can be considered as an energy transduction system, which adds a further step to the coupling existent between metabolic pathways and other cell functions. This situation, added to the participation of $K^+$ in the transport of some substrates, as well as the regulatory functions derived as secondary consequences of its transport required for the transport of substrates. It is only to be expected that many cases of cell transformation and differentiation bring about changes in metabolism, not only by changing the enzyme patterns, but also by modifying the synthesis of the transport components of the cell membranes.

## Energy Requirement of Ion Transport

One of the best-studied cases of changes induced in metabolism by ion transport consists in the stimulation produced by transport systems which require energy to function. It has long been known that numerous transport mechanisms do not operate in the absence of substrates from which energy can be obtained, and are inhibited by the so-called metabolic inhibitors. In early times it used to be inferred from these characteristics that transport systems required energy to work; however, it was only recently that the types of coupling between metabolism and transport have been established by the discovery and study of sodium-ATPase, proton pumps, etc. It was perhaps the application of the artificial carriers theory to the study of transport in biological systems that led to more solid investigations into the coupling between metabolic pathways and transport, and it is now known with more or less certainty what kind of linkage exists between these phenomena. Studies by Skou (21) and the postulations of Mitchell (22) added considerably to the comprehension of the mechanisms of energy transformation involved in transport. In the studies of Cockrell et al with valinomycin (23) it was already postulated that the stimulation produced by $K^+$ on mitochondrial metabolism was due to the energy required to transport the cation, and that this energy requirement could be fulfilled either by respiration or by ATP. These findings have been extended by numerous authors to other systems such as *Streptococcus faecalis* (24-27), submitochondrial particles (28), *Micrococcus denitrificans* (29), yeast (12,13), etc. Other authors in this volume will refer to these, and in his manuscript Dr. Kaback will discuss the

and those due to its presence *per se* have contributed to a better understanding of the role of one of the most active transport systems in most living organisms.

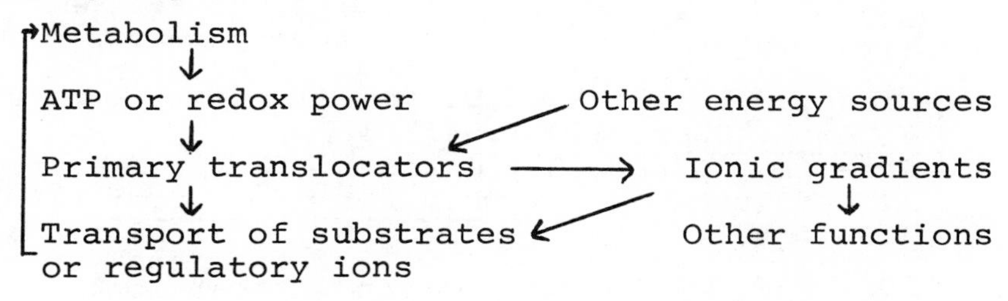


This last diagram attempts to summarize most of the facts emphasized in this presentation. Essentially, it shows two facts: first, that one of the main roles of ion-transport systems is that of feeding and regulating metabolism itself; and second, that ion transport, through the establishment of ionic gradients, can be directly utilized as a form of energy to drive other transport systems, or for the performance of certain vital functions. It must be also pointed out that the interrelations among the various ion-transport mechanisms are in many cases far more complicated, and to establish the linking of some of them to the energy sources requires several intermediate steps.

## References

1. Seitz, I.F., *Biokhimiya*, 14(1949)134. Quoted by Boyer, P.D., *J. Cell. Comp. Physiol*. 42 (1953) 71.

2. Parks, R.E., Ben Hershom, E. and Lardy, H.A. *J. Biol. Chem*. 227 (1957) 231.

3. Atkinson, D.E. and Walton, G.M. *J. Biol. Chem*. 240 (1965) 757.

4. Hanlon, D.P. and Westhead, E.W. *Biochim. Biophys. Acta*. 96 (1965) 537.

5. Black, S. *Arch. Biochem. Biophys*. 34 (1951)86.

6. Von Korff, R.W., *J. Biol. Chem*. 203 (1953)265.

7. Mildvan, A.S. and Cohn, M. J. Biol. Chem. 240 (1965) 238.

8. Bygrave, F.L. Biochem. J. 101 (1966) 488.

9. Gevers, W. and Krebs, H.A. Biochem. J. 98 (1966) 720

10. Green, D.E., Herbert,D. and Subrahmanyan, V. J. Biol. Chem. 138 (1941) 327

11. Peña, A., Cinco, G., García, A., Gómez-Puyou, A. and Tuena, M. Biochim. Biophys. Acta 148 (1967) 673.

12. Peña, A., Cinco, G., Gómez-Puyou, A. and Tuena, M. Arch. Biochem. Biophys. 153 (1972) 413.

13. Peña,A. Cinco, G., Gómez-Puyou, A. and Tuena, M. Biochim. Biophys. Acta 180(1969)1.

14. Papa, S., Lofrumento, N.E., Quagliarello, E., Meijer, A.J. and Tager, J.M. J. Bioenergetics 1(1970)287.

15. Chapell, J. B. Brit. Med. Bull. 24(1968)150.

16. Chapell, J.B. and Crofts, A.R. 1966. In: Regulation of Metabolic Processes in Mitochondria. p. 293, eds. Tager, J.M., Papa, S. Quagliarello, E. and Slater. Elsevier, Amsterdam.

17. Chapell, J.B. and Haarhoff, K.N. 1967. In: Biochemistry of Mitochondria, p. 75, eds. Slater, E.C., Kaniuga, Z. and Wojtczak, L. Academic Press and Polish Scientific Publishers, London-Warsaw.

18. Crane, R.K. Biochem.Biophys.Res.Commun. 17 (1964) 481.

19. West, I.C. and Mitchell, P. Biochem. J. 132 (1973) 587.

20. Nikaido, H., Biochem. Biophys. Res. Commun. 9 (1962) 486.

21. Skou, J.C. Physiol. Revs. 45 (1965) 596.

22. Mitchell, P. Chemiosmotic Coupling and Energy

Transduction. Glynn Research Ltd. Bodmin, Cornwall.

23. Cockrell, R.S., Harris, E.J. and Pressman, B.C. Biochemistry 5 (1966) 2326.

24. Harold, F.M., Baarda, J.R., Baron, C. and Abrams, A. J. Biol. Chem. 244 (1969) 2261

25. Harold, F.M., Baarda, J.R. and Pavlasova, E. J. Bacteriol. 101 (1970) 152.

26. Harold, F.M. and Papineau, D., J. Membrane Biol. 8 (1972) 27.

27. Harold, F.M. and Papineau, D. J. Membrane Biol. 8 (1972) 45.

28. Montal, M.,Chance, B. and Lee, C.P. J. Membrane Biol. 2 (1970) 201.

29. Scholes, P. and Mitchell, P. J. Bioenergetics 1 (1970) 309.

30. Asghar, S.S., Levin, E. and Harold, F.M. J. Biol. Chem. 248 (1973) 5225.

31. Schmidt, G., Hecht, L. and Thanhauser, S.J. J. Biol. Chem. 178 (1949) 733.

32. Willecke, K., Gries, E. M. and Oehr, P. J. Biol. Chem. 248 (1973) 807.

33. Eddy, A.A. and Nowacki, J.A., Biochem. J. 122 (1971) 701.

34. Roberts, R.B. and Roberts, I.Z. J. Cell. Comp. Physiol. 36 (1950) 15.

35. Caldwell, P.C., Hodgkin, A.L., Keynes, R.D. and Shaw, T.I. J. Physiol. 152 (1960) 561.

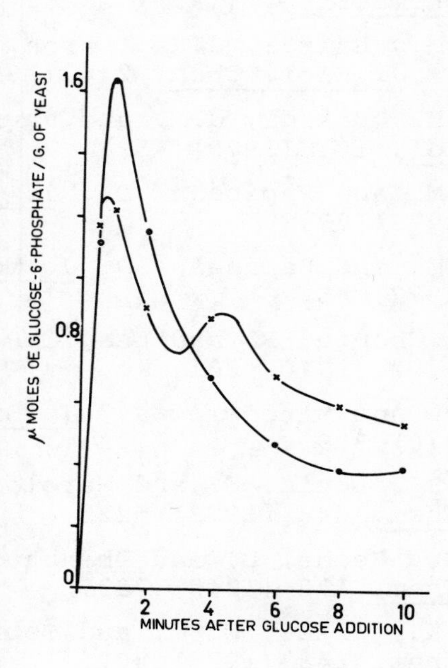


Figure 1. Changes in the concentration of glucose-6-phosphate after the addition of glucose; effect of the pH of the incubation medium. The experimntal conditions have been described elsewhere. o, pH 3.5; x, pH 7.5.

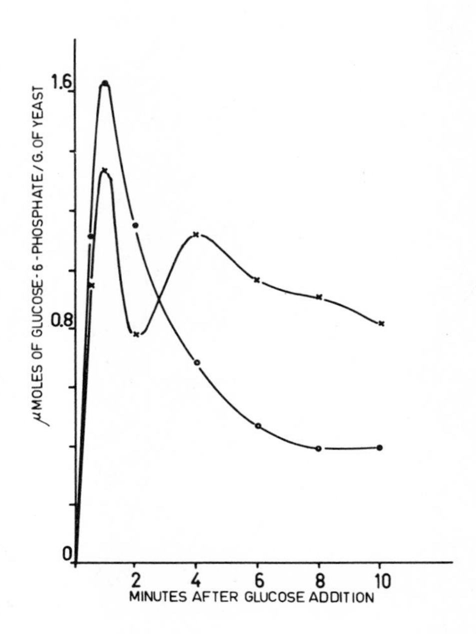


Figure 2. Changes in the concentration of glucose -6-phosphate after the addition of glucose at pH 3.5; effect of the presence of 50 mM $K^+$. The experimental conditions were the same as in Fig. 1: o, no $K^+$ added; x, 50 mM KCl present in the medium.

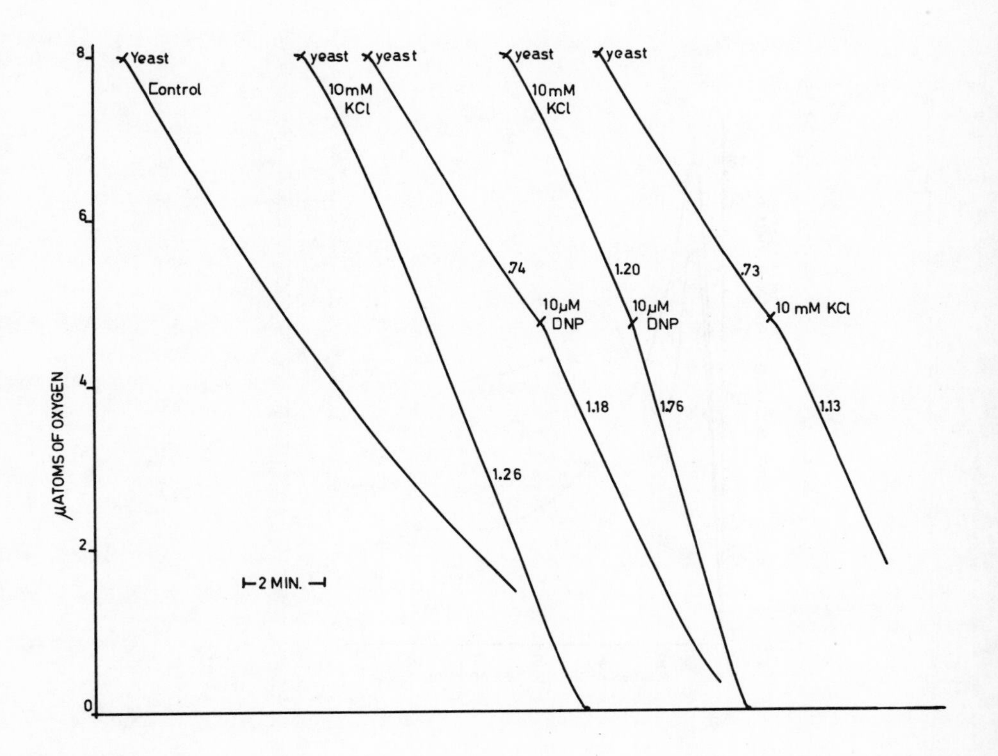


Figure 3. Effect of 2,4-dinitrophenol on the respiratory rate of yeast at pH 4.0 in the presence and absence of $K^+$. Incubation conditions: 10 mM tartrate-triethanolamine buffer, pH 4.0; 87 mM ethanol; 10 mM KCl; yeast cells, 125 mg, wet weight; final volume, 20.0 ml. Dinitrophenol (DNP) was added as a 10 mM solution in dimethylformamide. Figures alongside in the tracings represent respiratory rates as uatoms of oxygen per minute.

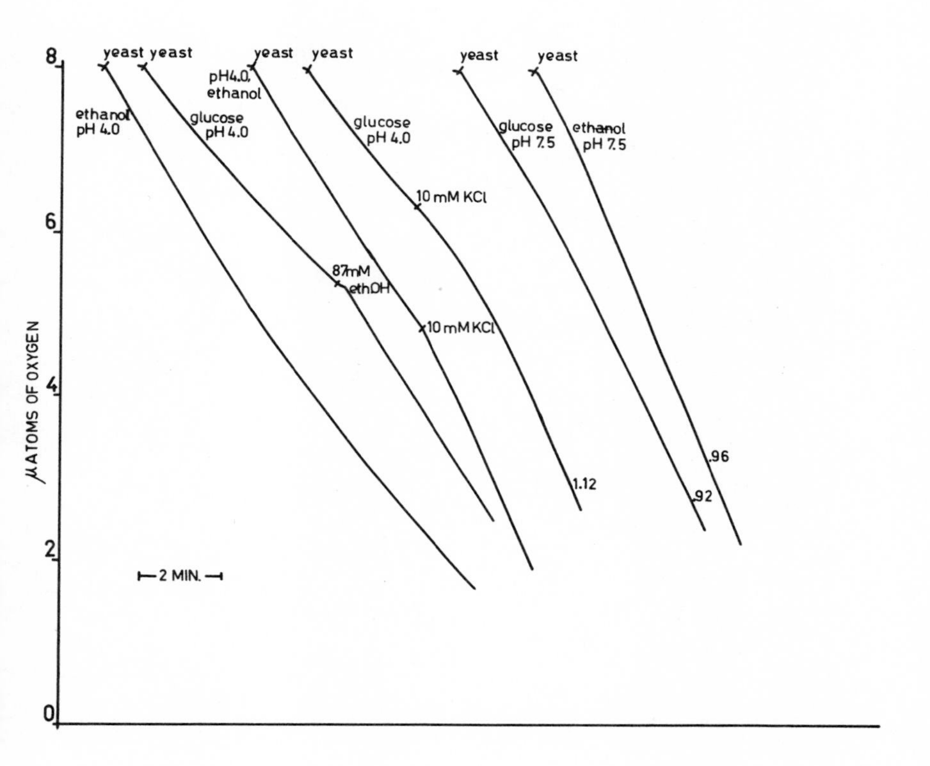


Figure 4. Respiratory rates of yeast, obtained with ethanol or glucose as substrates, and the effect of pH and the presence of $K^+$. The experimental conditions were the same as for Fig. 3, except that at pH 7.5 the buffer employed was triethanolamine-tartrate at the same molarity.

# ACTIVE TRANSPORT IN ISOLATED BACTERIAL MEMBRANE VESICLES

H. Ronald Kaback

The Roche Institute of Molecular Biology
Nutley, New Jersey 07110

Cytoplasmic membrane vesicles isolated from *E. coli*, as well as a number of other organisms, catalyze the active transport of a wide variety of metabolites (1-7) including amino acids, sugars, hexose-P's, hydroxy-, keto-, and dicarboxylic acids, nucleosides, and potassium or rubidium (in the presence of valinomycin) at rates which are comparable to those of the parent whole cells in many cases. In *E. coli* and *S. typhimurium*, most of these transport systems are coupled primarily to the oxidation of D-lactate or reduced phenazine methosulfate (or pyocyanine) *via* a membrane-bound cytochrome chain with oxygen as the terminal electron acceptor, and neither the generation nor utilization of high-energy phosphate intermediates is apparently involved in the mechanism (1-3). The energy-coupling site for transport in *E. coli* vesicles is localized in a segment of the respiratory chain between D-lactate dehydrogenase and cytochrome $b_1$ (1-3), and evidence for an electron-transfer-coupling component mediating the coupling of respiratory energy to active transport has been demonstrated (8).

This discussion is concerned with recent studies pertaining to respiration-linked active transport in the isolated membrane vesicle system. Methods for the preparations of bacterial membrane vesicles, their morphology and other pro perties, and experimental observations related to the role of the P-enolpyruvate-P-transferase system in the vectorial phosphorylation of certain sugars and the regulation of this system will not be discussed, and the reader is referred to other publications (9-15) for information related to these topics.

## Anaerobic Transport

One aspect of the respiration-dependent transport systems that has not been studied is their possible relationship to active transport under anaerobic conditions. Obligate anaerobes or facultative anaerobes growing under anaerobic conditions transport nutrients; moreover, δ-aminolevulinic acid- or heme-requiring mutants of *E. coli* do not manifest transport defects. Although ATP is not involved in active transport under aerobic conditions (1-3, 16-22), evidence has been presented which is consistent with the interpretation that cells can utilize glycolytically-generated ATP to drive transport under anaerobic conditions (17-23). It is also possible, however, that such cells might use the same general type of transport mechanism as that used aerobically, with the exception that an alternative electron acceptor is used rather than the cytochrome chain and oxygen.

Recent experiments (24) carried out in this laboratory in collaboration with Dr. W. N. Konings of the University of Groningen, The Netherlands have shown that anaerobic lactose transport in whole cells and membrane vesicles from *E. coli* is coupled to the oxidation of α-glycerol-P̄ or D-lactate with fumarate as an electron acceptor. Alternatively, anaerobic lactose transport may be coupled to the oxidation of formate utilizing nitrate as electron acceptor. Both anaerobic electron transfer systems are induced by growth of the organism under appropriate conditions. Components of both systems are loosely bound to the membrane, necessitating the use of a modified procedure for vesicle preparation in order to demonstrate anaerobic transport in the vesicle system.

Addition of ATP or an ATP-generating system during preparation of vesicles does not stimulate transport. The results support the conclusion that active transport under anaerobic conditions is coupled primarily to electron flow.

## Dansyl-Galactosides, Fluorescent Probes of Active Transport

Fluorescent compounds that exhibit polarity-dependent fluorescence properties have been used to investigate the structure of biological membranes (25). Two such compounds, 1-anilino-8-naphthalene sulfonate (26) and dansyl phosphatidylethanolamine (27) have been used to study structural changes associated with transport in membrane vesicles of *E. coli*. However, the non-specificity of these probes limits the type of information that can be obtained.

The fluorescent galactosides, 2-(N-dansyl)-aminoethyl-β-D-thiogalactoside ($DG_2$) (28) and 2-(N-dansyl)-aminohexyl-β-D-thiogalactoside ($DG_6$) (Schuldiner, S., Kerwar, G.K., Weil, R., and Kaback, H.R., manuscript in preparation), competitively inhibit lactose transport by membrane vesicles of *E. coli* but are not actively transported. In the presence of membrane vesicles prepared from cells possessing the *lac* transport system, an increase in $DG_2$ and $DG_6$ fluorescence is observed upon addition of D-lactate. The fluorescence increase is not observed in membrane vesicles lacking either the β-galactoside transport system or D-lactate dehydrogenase, and is blocked or rapidly reversed by addition of β-galactosides, sulfhydryl reagents, certain inhibitors of D-lactate oxidation, or uncoupling agents. The fluorescence increase observed with both $DG_2$ and $DG_6$ exhibits an emission maximum at 500 nm and excitation maxima at 345 nm and at 292 nm. The latter excitation maximum is absent unless D-lactate is added, indicating that the bound dansyl-galactoside molecules are excited by energy transfer from the membrane proteins. Titration of vesicles with $DG_2$ and $DG_6$ in the presence of D-lactate dem onstrates that the β-galactoside carrier protein represents about 3-4% of the total membrane protein, a value which is in excellent agreement with that of Jones and Kennedy (29). Moreover, the affinity of the carrier increases as the length of the carbon chain between the galactose and dansyl moieties is increased (i.e., the $K_D$ for $DG_2$ is 30 μM while the $K_D$ for $DG_6$ is 5 μM).

Preliminary nanosecond half-life studies corroborate the results discussed above. Attenuation of the half-life of the excited state of $DG_2$ is obtained in the presence of D-lactate but not in its absence, and no change is observed with membranes prepared from cells devoid of the *lac* transport system.

These data, taken as a whole, indicate that in the absence of D-lactate oxidation, the *lac* carrier protein is unable to bind substrate. There are at least three possible mechanisms by which D-lactate oxidation might lead to the observed increase in dansyl-galactoside fluorescence. D-lactate oxidation could (1) increase the affinity of the *lac* carrier at the external surface of the membrane, (2) make the carriers more accessible to the external medium, or (3) cause translocation of bound dansyl-galactoside into the hydrophobic interior of the membrane. A detailed discussion of these alternatives will not be presented because the data do not allow a clear choice. Furthermore, the three possibilities are not necessarily mutually exclusive. In any case, the data strongly suggest that energy is coupled to one of the initial steps in transport and that facilitated diffusion, therefore, cannot be the rate-limiting step for active transport of β-galactosides.

Recent studies with a photoaffinity-labeled galactoside, 2-azide-4-nitrophenyl β-D-galactoside, have provided independent support for these conclusions (Rudnick, G. and Kaback, H.R., unpublished information). This compound is a competitive inhibitor of lactose transport, and in its presence, lactose transport is inactivated by exposure to light at 400 nm. Under the same conditions, the transport of proline and other amino acids is only mildly inhibited, indicating that the site of inhibition is the _lac_ carrier protein. Finally, the rate of photoinactivation is markedly enhanced by the addition of D-lactate to the reaction mixtures.

## Vinylglycolate, An Internally Generated Sulfhydryl Reagent.

Previous studies (31, 32) demonstrated that an acetylenic hydroxy acid, 2-hydroxy-3-butynoic acid, is an irreversible inactivator of D- and L-lactate dehydrogenases and D-lactate-dependent active transport in isolated cytoplasmic membrane vesicles from _E. coli_. The compound is a substrate for the membrane-bound, flavin-linked D-lactate dehydrogenase which undergoes 15 to 30 turnovers prior to inactivation. The mechanism of inactivation is due to covalent attachment of a reactive intermediate to the flavin coenzyme at the active site.

Inactivation of D- and L-lactate dehydrogenase activity and D-lactate-dependent transport by 2-hydroxy-3-butynoate is highly specific. Moreover, inactivation of D-lactate-dependent transport is blocked by D-lactate, but not by succinate and NADH; and α-glycerol-P-dependent transport in _S. aureus_ vesicles is not inactivated by the acetylenic hydroxy acid.

Both the hydroxy function and the alkyne linkage in 2-hydroxy-3-butynoate are critical for inactivation. 3-Butynoate inhibits neither D-lactate dehydrogenase nor D-lactate-dependent transport; and vinylglycolate (2-hydroxy-3-butenoate) serves as a substrate for D-lactate dehydrogenase, and is thus an effective electron donor for transport.

In view of the high degree of specificity of 2-hydroxy-3-butynoate inactivation, it was surprising when subsequent experiments demonstrated that the compound markedly inhibits vectorial phosphorylation catalyzed by the P-enolpyruvate-P-transferase system in isolated membrane vesicles, and that vinylglycolate, a substrate for D-lactate dehydrogenase, is 50- to 100-times more potent (32, 33).

Vinylglycolate (2-hydroxy-3-butenoic acid) inactivates Enzyme I of the P-enolpyruvate-P-transferase system in *E. coli*. By this means, vinylglycolate is a potent, irreversible inactivator of vectorial phosphorylation in whole cells and membrane vesicles prepared from this organism.

Prior to inactivation of the P-transferase system, the unsaturated hydroxy acid gains access to the intravesicular pool by means of a lactate transport system. Subsequently, the compound is oxidized by membrane-bound, flavin-linked D- and L-lactate dehydrogenases to yield a reactive electrophile (presumably 2-keto-3-butenoate) which reacts with Enzyme I and other sulfhydryl-containing proteins on the inner surface of the vesicle membrane. The following observations support this interpretation:

1) Vectorial phosphorylation catalyzed by whole cells and isolated membrane vesicles undergoes a time-dependent, irreversible inactivation in the presence of vinylglycolate. The rate of inactivation is increased markedly by the addition of ascorbate-phenazine methosulfate, an artificial electron donor system which markedly stimulates transport. Vinylacetate (3-butenoate) has no effect.

2) Membrane vesicles prepared from a D-lactate dehydrogenase mutant undergo inactivation of vectorial phosphorylation and labeling by ($^{14}$C) vinylglycolate at slower rates.

3) Reagents or conditions which block D- and L-lactate dehydrogenase activity or electron transfer diminish the rate of inactivation of the P-transferase system in isolated membrane vesicles. Similarly, the rate of inactivation is diminished by addition of D- or L-lactate, or both, to the reaction mixtures. ($^{14}$C)Vinylglycolate up-

take by the vesicles exhibits similar properties.

4) Dithiothreitol protects against inactivation of P-enolpyruvate-dependent sugar uptake and markedly inhibits ascorbate-phenazine methosulfate-dependent ($^{14}C$)Vinylglycolate labeling in isolated membrane vesicles.

5) Vinylglycolate transport is the rate-limiting step for labeling membrane vesicles. Thus, the rate and extent of ($^{14}C$)vinylglycolate uptake by the vesicles is stimulated 10- to 20-fold by the addition of ascorbate-phenazine methosulfate. Moreover, stimulation by ascorbate-phenazine methosulfate is inhibited by 2,4-dinitrophenol, an uncoupling agent, and by phospholipase treatment, neither of which have any effect on D- or L-lactate oxidation.

D-lactate is a competitive inhibitor of vinylglycolate uptake, exhibiting a $K_i$ which is the same as that of the $K_m$ of the lactate transport system. Vinylglycolate is a competitive, reversible inhibitor of respiration-linked D-lactate transport, but does not inactivate the transport of lactose, proline, or succinate.

6) Treatment of whole cells with vinylglycolate results in inactivation of Enzyme I activity. Enzyme II and HPr activity remain intact, and vesicles treated with vinylglycolate exhibit normal Enzyme II activity.

Vinylglycolate is also a potent inhibitor of growth in E. coli. At 1 μM vinylglycolate, the rate of growth is inhibited by approximately 50%; at 10 μM by over 95%; and at 100 μM, growth is inhibited completely. Unfortunately, however, preliminary studies on the possible efficacy of vinylglycolate as an antibacterial agent in vivo have been negative. This is due perhaps to high circulating levels of lactate in the intact animal.

Since the rate-limiting step for covalent labeling of the membrane vesicles with vinylglycolate is transport, it should be possible to determine the absolute number of membrane vesicles in a given population that are capable of catalyzing active transport. High specific activity ($^{3}$H)vinylglycolate has been prepared by reductive tritation of 2-hydroxy-3-butynoate, and vesicles have been labeled with this compound in the presence and absence of ascorbate-phenazine methosulfate. These preparations are being examined by radioautography in the electron microscope by Dr. Samuel Silverstein of Rockefeller University. Each vesicle capable of active transport in the presence of ascorbate-phenazine methosulfate will be overlaid by grains from the covalently bound ($^{3}$H)vinylglycolate.

As stated above, vinylglycolate must be transported and oxidized before reacting with membrane-bound sulfhydryl groups. This being the case, vinylglycolate should be extremely useful for studying membrane structure, i.e., only sulfhydryl groups exposed on the inside of the membrane should be labeled under the conditions described. On the other hand, generation of the reactive species in the external medium (by addition of lactate dehydrogenase) should label only those sulfhydryl groups exposed on the outside of the membrane.

## Purification and Properties of the Membrane-Bound D-Lactate-Dehydrogenase (D-LDH) from E.coli and Reconstitution of D-Lactate Oxidation and D-Lactate-Dependent Active Transport in dld$^{-}$ Mutants

The membrane-bound D-LDH of *Escherichia coli* ML 308-225 has been solubilized and purified to homogeneity (34,35). The enzyme has a molecular weight of 75,000 ± 7% and contains approximately 1 mole of flavin adenine dinucleotide per mole of enzyme. Its pH optimum before exposure to Triton X-100 is 7.5 to 8.0; after exposure it is 8.0 to 9.5. The purified enzyme has a $K_m$ value of 1.4 x $10^{-3}$ *M* for D-lactate and 3.0 x $10^{-2}$ *M* for L-lactate before exposure to Triton X-100; after exposure its $K_m$ values are 0.5 x $10^{-3}$ *M* and 2.1 x $10^{-2}$ *M* for the same substrates, respectively. The enzyme is not active with D-glycerate, L-glycerate, succinate, malate, D-tartrate, L-tartrate, *meso*-tartrate, 1-propanol, or isopropanol as substrates and oxidized diphosphopyridine nucleotide has no effect on the catalytic conversion of D-lactate to pyruvate. The enzyme is not inhibited by *p*-chloromercuribenzoate, n-ethylmaleimide, or 5,5'-dithiobis (2-nitrobenzoic acid) but is inactivated by treatment with 2-hydroxy-3-butynoic acid.

While this work was in progress, Reeves *et al*. (36) demonstrated that guanidine.HCl extracts from wild type membrane vesicles containing D-LDH activity are able to reconstitute D-lactate-dependent oxygen consumption and active transport in membrane vesicles from *E. coli* and *Salmonella typhimurium* mutants deficient in D-LDH (*dld*⁻). Very recently these studies have been confirmed and extended by Short et al. (37) using the homogeneous preparation of D-LDH described above. Moreover, Futai (38) has independently confirmed many of the observations using *dld*⁻ membrane vesicles and a similar preparation of D-LDH.

Reconstitution of D-lactate oxidation and D-lactate-dependent active transport is carried out by diluting the enzyme preparation (dissolved in 0.1*M* potassium phosphate buffer, pH 6.6, containing 0.1% Triton X-100 and 0.6 *M* guanidine HCl), 25-fold into 0.1 *M* potassium phosphate buffer containing *dld*⁻ membrane vesicles. The vesicles are centrifuged, and resuspended in a small volume of phosphate buffer before assaying transport and dehydrogenase activities. Optimal reconstitution and binding take place in the presence of 0.6 M guanidine HCl, at 25° to 37°, and at pH 6.6, and the vesicles can be washed several times subsequent to reconstitution without loss of activity

Reconstituted *dld*⁻ membranes carry out D-lactate oxidation and take up a number of transport substrates when supplied with D-lactate. D-Lactate is not oxidized, and will not support transport of any of these substances in unreconstituted *dld*⁻ vesicles. Addition of enzyme to wild type membranes produces an increase in D-lactate oxidation but has little or no effect on the ability of the membranes to catalyze active transport. Addition of increasing quantities of D-LDH to dld⁻ membranes produces a corresponding increase in D-lactate oxidation, and transport approaches an upper limit which is similar to the specific transport activity of wild type membrane vesicles. However, the quantity of enzyme required to achieve maximum initial rates of transport varies with different transport systems.

As discussed above, binding of 2-(N-dansyl) aminoethyl-β-D-galactoside ($DG_2$) to membrane vesicles containing the lac transport system is dependent upon D-lactate oxidation, and this fluorescent probe can be utilized to quantitate the number of lac carrier proteins in the membrane vesicles. When dld⁻ membrane vesicles are reconstituted with increasing amounts of D-LDH, there is a corresponding increase in the binding of $DG_2$. Assuming that each lac carrier protein molecule binds one molecule of $DG_2$m it can be estimated that there is at least a 7- to 8-fold excess of lac carrier protein relative to functional D-LDH in reconstituted dld⁻ vesicles. A similar determination can be made for wild type vesicles. These vesicles contain approximately 0.07 nmoles of D-LDH per mg membrane protein (based on the specific activity of the homogeneous enzyme preparation) and about 1.1 nmoles of lac carrier protein per mg membrane protein, yielding a ratio of about 15 for lac carrier protein relative to D-LDH. It should be emphasized that these are minimal values as other carriers which are also energized by D-lactate oxidation are not measured by this technique.

Although the rate and extent of transport of a number of substrates increases dramatically with reconstitution, the rate and extent of labeling of dld⁻ vesicles with ($^{14}C$) vinylglycolate remains constant. As discussed previously, this compound is transported via the lactate transport system, and oxidized to a reactive product by D- and L-lactate dehydrogenases on the inner surface of the vesicle membrane. The reactive product (presumably 2-keto-3-butenoic acid) then reacts with sulfhydryl groups in the membrane. The observation that reconstituted dld⁻ membranes do not exhibit enhanced labeling by vinylglycolate indicates that bound D-lactate dehydrogenase is present on the outer surface of the vesicles. In this case, the reactive products released from D-LDH would be diluted into the external medium, whereas if the enzyme were on the inner surface of the vesicle membrane, the rate of labeling would be expected to increase with reconstitution, since the reactive product should accumulate within the vesicles to higher effective concentrations.

The flavin moiety of the holoenzyme appears to be critically involved in binding to the membrane. Treatment of D-LDH with $(1-^{14}C)$hydroxybutynoate (see above discussion) leads to inactivation of D-LDH by covalent binding of hydroxybutynoate to the flavin adenine dinucleotide coenzyme of the enzyme. Enzyme labeled in this manner does not bind to dld⁻ membranes. Although preliminary, these findings suggest that the flavin coenzyme itself may mediate binding or alternatively, that covalent inactivation of the flavin may result in a conformational change that does not favor binding. Efforts to reversibly dissociate the flavin from the native and inactivated enzyme are in progress so that appropriate binding studies can be carried out.

## Mechanism of Active Transport

Despite various lines of evidence which indicate that redox-coupled conformational changes are intimately involved in the molecular mechanism of active transport (1-3), chemiosmotic phenomena (36, 37) may also play a very important role in active transport. By the latter means, it is postulated that a membrane potential (positive outside) generated by the asymmetric release of protons during substrate oxidation is the immediate driving force for active transport. Support for the chemiosmotic hypothesis has been presented recently (38, 39) and some of the findings have been corroborated and extended in this laboratory (Lombardi, F. J., Schuldiner, S. and Kaback, H.R., unpublished observations). Thus, it has been demonstrated that during D-lactate oxidation, lipophylic cations such as safranine and dimethyldibenzylamine (in the presence of tetraphenylboron) are accumulated, indicating that a membrane potential is established. Moreover, when membrane vesicles treated with valinomycin (a potassium-specific ionophore) and loaded with potassium are rapidly diluted into media lacking potassium, lactose is taken up (potassium-lactose antiport). These findings suggest that a membrane potential generated by the passive diffusion of potassium can drive lactose uptake.

Despite the attractiveness of the chemiosmotic hypothesis and the experiments discussed above which support this concept, a number of inconsistencies remain (40). Exchange of dimethyldibenzylamine accumulated in the intravesicular pool is inhibited by sulfhydryl reagents. Moreover, neither the rate nor extent of lactose uptake generated by a potassium diffusion gradient approaches that obtained with D-lactate or reduced phenazine methosulfate. Finally potassium-lactose antiport is inhibited by a variety of respiratory inhibitors such as cyanide, HOQNO, and amytal, while potassium-safranine antiport (i.e., the membrane potential) is not inhibited by the same compounds. In addition to these observations, a number of other experimental observations (1-3, 40, 41) are inconsistent with the chemiosmotic hypothesis in its stated form.

Recent experiments demonstrate that potassium-lactose antiport is markedly stimulated by D-lactate oxidation suggesting that a membrane potential acting in concert with redox-coupled reactions is necessary for active transport. (Schuldiner, S. and Kaback, H.R., unpublished observations.)

## References.


1. Kaback, H. R. Biochim. Biophys. Acta 265 (1972) 365.
2. Kaback, H. R. and Hong, J.-s. 1973 In: CRC Critical Reviews in Microbiology, Vol. 2, p. 333. Eds. Laskin, Allen I. and Lechevalier, Hubert. CRC Press, Ohio.
3. Kaback, H.R. 1973. In: Bacterial Membranes and Walls, Vol. I, p. 241. Ed. Leive, L. Marcel Dekker, New York, N.Y.
4. MacLoed, R.A., Thurman, P. and Rogers, H.J. J. Bacteriol. 113 (1973) 329.
5. Stinnett, J.D., Guymon, L.F. and Eagon, R.G. Biochem. Biophys. Res. Commun. 52 (1973) 284.
6. Komatsu, Y. and Tanaka, K. Biochim. Biophys. Acta 311 (1973) 496.
7. Murakawa, S., Izaki, K. and Takahashi, H. Agr. Biol. Chem. 37 (1973) 1905.
8. Hong, J.-s. and Kaback, H. R. Proc. Nat. Acad. Sci. U.S.A. 69 (1972) 3336.
9. Kaback, H.R. Ann. Rev. Biochem. 39 (1970) 561.
10. Kaback, H. R. 1970. In: Current Topics in Mem branes and Transport, p. 35. Eds. Bronner, F. and Kleinzeller, A. Academic Press, New York, N. Y.
11. Kaback, H. R. 1971. In: Methods in Enzymology Vol. XXII, p. 99. Ed. Jakoby, W.B. Academic Press, New York, N.Y.
12. Roseman, S. J. Gen. Physiol. 54 (1969) 1385.
13. Lin, E.C.C. Ann. Rev. Genetics 4 (1970) 225.
14. Oxender, D.L. Ann. Rev. Biochem. 41 (1972)777.
15. Boos, W. Ann. Rev. Biochem. 42 (1972) In press.
16. Prezioso, G., Hong, J.-s., Kerwar, G.K. and Kaback, H. R. Arch. Biochem. Biophys. 154 (1973) 575.

17. Schairer, H. U. and Haddock, B.A. Biochem. Biophys. Res. Commun. 48 (1972) 544.

18. Klein, W.L. and Boyer, P.D. J. Biol. Chem. 247 (1972) 7257.

19. Berger, E.A. Proc. Nat. Acad. Sci. U.S.A. 70 (1973) 1514.

20. Butlin, J.D. 1973 Ph.D. Thesis, Aust.Nat. Univ., Canberra City, Aust.

21. Parnes, J. R. and Boos, W. J. Biol. Chem. 248 (1973) 4429.

22. Or, A., Kauner, B. I. and Gutnick, D.L. FEBS Letters 35 (1973) 217.

23. Abrams, A. and Smith, J.B. Biochem. Biophys. Res. Commun. 44 (1971) 1488.

24. Konings, W.N. and Kaback, H. R. Proc. Nat. Acad. Sci. U.S.A. 70 (1973) 3376.

25. Radda, G.K. 1971. In: Current Topics in Bioenergetics p. 81. Ed. Sanadi, D. R. Academic Press. New York.

26. Reeves, J. P., Lombardi, F.J. and Kaback, H. R. J. Biol. Chem. 247 (1972) 6204.

27. Shechter, E., Gulik-Krzywicki, and Kaback, H. R. Biochim. Biophys. Acta 274 (1972) 466.

28. Reeves, J.P., Shechter, E., Weil, R., and Kaback, H.R. Proc. Nat. Acad. Sci. U.S.A. 70 (1973) 2722.

29. Jones, T. H. D. and Kennedy, E.P. J. Biol. Chem. 244 (1969) 5981.

30. Rudnick, G., Weil, R. and Kaback, H.R. Unpublished observations.

31. Walsh, C.T., Abeles, R.H. and Kaback, H.R. J. Biol. Chem. 247 (1972) 7858.

32. Walsh, C.T. and Kaback, H.R., Annals N. Y. Acad. Sci., In press.

33. Walsh, C.T., and Kaback, H.R. J. Biol. Chem. 248 (1973) 5456.

34. Kohn, L. D. and Kaback, H.R. J. Biol. Chem. 248 (1973) 7012.

35. Futai, M. Biochemistry (1973)

36. Harold, F.M. Bacteriol. Rev. 36 (1972) 172.

37. Mitchell, P. Bioenergetics 4 (1972) 265

38. Hirata, H., Altendorf, K., and Harold, F.M. Proc. Nat. Acad. Sci. U.S.A. 70 (1973) 1804.

39. Kashket, E.R. and Wilson, T.H. Proc. Nat. Acad. Sci. U.S.A. 70 (1973) 2866.

40. Lombardi, F.J., Reeves, J.P., Short, S.A. and Kaback, H.R. Annals N.Y. Acad. Sci. 1974. In press.

41. Lombardi, F.J., Reeves, J.P. and Kaback, H. R. J. Biol. Chem. 248 (1973) 3551.

# CALCIUM BINDING PROTEINS AND NATURAL MEMBRANES

R.H. Kretsinger

Department of Biology
University of Virginia
Charlottesville, Va. 22901

## Introduction

Drs. Estrada-O. and Gitler asked us to present "serious speculation" on perspectives in membrane biology. I welcome the opportunity to explore the concept of evolution as a guide to understanding membrane function. As can be seen from the outline, I will review our work on the structure of fish muscle calcium binding parvalbumin (FMCBP) as well as its evolutionary relationship to the calcium binding component of troponin (TN-C). These experimental results are seen in better perspective by reviewing general patterns of calcium coordination and protein evolution. A consideration of those systems controlled by or controlling calcium suggests certain general groupings. I predict that the calcium mediating protein(s) in several of these systems is homologous to FMCMP and to TN-C. Further it seems quite a reasonable speculation that this homology, seen both in terms of protein structure and a general pattern of action, may be found in membrane systems.

## Fish Muscle Calcium Binding Parvalbumin

*Background.* Contradictory as it may at first seem, I will base some rather far reaching predictions about calcium control systems on the structure of a protein, whose function remains unknown

FMCBP is found in the white muscle (but not red muscle) of fish and amphibia but not in higher vertebrates. Usually two to five closely homologous components are found in one fish in total amount about 3g/kg wet tissue (1,2). The same fish white muscle also has troponin (3).

FMCBP is acidic with pK values ranging from 3.95 (carp) to 5.4 (coelacanth). It is an albumin, that is soluble in salt free water, and is easily washed out of ground muscle. It is small and monomeric. The carp (component pK 4.25) MCBP, whose three-dimensional structure we solved, has 108 amino acids. Pike (4) and coelacanth (5) have one additional C-terminal alanine and coelacanth has two additional N-terminal residues. The N-terminus is always acetylated excepting for coelacanth (component pK 5.4). A source of initial interest was the 9% content of phenylalanine and frequent absence, depending on the species, of tyrosine, histidine, proline, cysteine, methionine and tryptophan. No enzymatic function has been found nor does examination of the structure reveal any grooves or pits, the sorts of sites to date found in all enzymes.

Most importantly, FMCBP binds two calcium ions with association constant, $K_a = 2 \cdot 10^7\ M^{-1}$ (6).

*Structure.* The arguments supporting the proposed mechanism of cooperative binding and release of calcium by the FMCBP monomer depend on a rather detailed knowledge of the structure. Since our recent description of the structure and its determination (1), we have completed seven cycles of difference Fourier refinement with a residual, $R = \Sigma(F_{obs}-F_{calc})/\Sigma F_{obs}$ for all observed reflections of 0.27 (7). Bond distances and angles are constrained to near canonical values by use of the Diamond model building program (8). The average error in atomic coordinates is less than 0.25 Å except for the seven residues at the N-terminus which are definitely disordered.

Before describing the course of the main chain, I will emphasize four features of the structure, two seemingly destabilizing and two energetically favorable.

Arg-75 and Glu-81 are invariant in the seven known sequences (figure 4). Their side chains are joined together in a polar hydrogen bond, or salt bridge. This strong dipole is shielded from the solvent by amino acids 19 through 27, that is the AB loop.

In carp MCBP, all side chain protons, bonded to either oxygen or nitrogen, are in hydrogen bonds to the protein or accessible to solvent. However it was surprising to find that some ten peptide protons are not involved in hydrogen bonds. Although there is some question as to reasonable bond lengths and angles for a hydrogen bond, many of the definitely nonbonded peptide protons occur in critical regions of helix D and the antiparallel β-pleated sheet connecting the two calcium binding loops.

All of the side chain atoms with significant dipoles are at the surface of the molecule excepting those associated with the Arg75, Glu81 salt bridge, and those associated with calcium coordination. This general principle of "hydrophilics out" applies to most proteins. However, the other half of the maxim "hydrophobics in" must be qualified. Some 40% of the (non-hydrogen) atoms of the side chains of valine, leucine, isoleucine and phenylalanine are exposed to solvent. Nonetheless, the remaining 60% form a compact hydrocarbon core which accounts for 15% of the protein volume. The main chain passes beside, not through this core. In contrast to the internal dipole and lost hydrogen bonds, this core is quite definitely a stabilizing influence.

The coordination of the two calcium ions is best visualized in terms of the vertices of two octahedra, figure 3. The first ligand in sequence defines the +X vertex, and the second +Y. The following four ligands are assigned the appropriate vertex designation. The two octahedra are related by an approximate intramolecular two fold axis and the sequence of ligands follows the same patterns (+X, Asp-59 and Asp-90; +Y, Asp-53 and Asp-92; +Z, Ser-55 and Asp-94; -Y, Phe-57 and Lys-96; -X, Glu-59 and Gly-98($H_2O$); -Z, Glu-62 and Glu-101). The -Y ligands (Phe-57 and Lys-96) coordinate with carbonyl oxygen atoms; all others with side chains. The calcium ion in the CD loop is six coordinate, corresponding to the six vertices. The EF calcium is formally eight coordinate because the carboxylate groups of Asp-92 and Glu-101 coordinate with both oxygen atoms. Most important, residue 98 is glycine; the -X ligand of the EF calcium is $H_2O$. In contrast, the CD calcium ion is completely surrounded by protein ligands and is not accessible to water without altering the protein configuration. The hydrophobic core and calcium ions stabilize the molecule.

The main chain forms six helices denoted as A through F. Helix C is related to helix E by the intramolecular approximate two fold axis which relates the two calcium octahedra. Helix D is likewise related to helix F. As illustrated in figure 1 helices C and D as well as the CD calcium binding loop are related to the EF region by the two fold axis.

This relationship is symbolically represented in figure 2 by two right hands. The extended forefingers represent helices C and E running from finger tip to base. The calcium binding loops CD and EF enclose the two octahedra with the middle fingers running from base to tip. Note that the middle parts of the two fingers pass one another in approximate antiparallel β-pleated sheet conformation.

Finally helices D and F are represented by thumbs running from base to tip. The D thumb is bent since helix D is quite distorted. Four hydrogen bonds are "lost" in forming this "tensed" conformation.

The hands are viewed from the inside of the molecule. The palmar surfaces of the four helices face inward and the hydrophobic side groups point in. The side chains of the thumbs interact with both forefingers. There is also thumb-thumb contact but none between forefingers. Helices A and B can also be visualized as a right hand with the thumb (helix B) drawn in so that the two helices run in an almost antiparallel sense. This third hand covers the hydrophobic region exposed to the viewer in figures 1 and 2. Again the palmar surface of the AB hand is hydrophobic and faces inward.

The loop joining helices D and E contains the Arg-75, Glu-81 salt bridge. It is the AB loop, residues 19 thru 27, which covers the salt bridge and makes it internal.

*Gene Triplication.* The near identity of the CD and EF region plus the same sequence of ligands about the two calcium octahedra is interpreted as having evolved from a gene duplication, splicing and subsequent divergent evolution (9). Here arises a quite intriguing and unprecedented situation. Eventhough the CD and EF regions are nearly identical in structure, their amino acid sequences have evolved so far that their relationship is just at the edge of statistical significance.

In fact the AB region shows as great an amino acid sequence similarity to EF as does CD. Even though the AB loop does not bind calcium it is reasonable to suggest that the AB region is homologous to CD and EF. Although this suggested third homologous region is not proven (10, 11) nor is it really necessary for the subsequent discussion of protein evolution, this postulate provided the original idea underlying the model of function (1, 9).

*Model of Function.* The model of the cooperative release and binding of calcium (1, 12) was originally suggested by: (1) the implication that FMCBP is not an enzyme but a mediator of calcium ion concentration. (2) the structural characteristics previously outlined, (3) the subjective feeling that there must be a functional reason for the gene triplication. Subsequently the results of three experiments have been consistent with the model; nonetheless its details remain unproven.

1. The EF calcium ion rapidly exchanges with the solvent in the -X direction where it is coordinated by water. (see figure 3).

2. The coordinating carboxylate groups of residues 90(+X), 92(+Y), 94(+Z) and 101 (-Z) as well as the nearby 100 will repel one another and swing out causing, or allowing, a slight conformational change in the EF loop (middle finger in Figure 2).

3. The antiparallel β-pleated sheet made of the EF and the CD loop, can regain three "lost" hydrogen bonds: 99→56, 97→58 and 60→95.

4. The CD loop is drawn over to meet the EF loop in making the four hydrogen bond sheet, thereby distorting the CD coordination sphere.

5. The affinity for the CD calcium ion is thereby reduced and it can diffuse away. This half of the model predicts cooperative calcium binding and release and "explains" the gene duplication and two fold axis. Instead of a dimer, as often seen in cooperative systems, the two halves were spliced together.

6. The four coordinating carboxylate groups: 51, 53, 59 and 62, as well as nearby 60 and 61 will be mutually repulsive without the divalent cation and hence swing out.

7. Helix D (the bent thumb of figure 2) can now "untense" and form a regular α-helix thereby regaining its four "lost" hydrogen bonds.

8. The Arg-75, Glu-81 salt bridge breaks. Arg-75 is at the end of the D helix and is pointed inside the molecule by the "tensing" of helix D. When Arg-75 is drawn out to the solvent, the negative charge of Glu-81 remains buried inside under the AB loop.

9. The AB loop shifts to expose the carboxylate group of Glu-81 to the solvent.

10. The general AB loop, DE loop region will now have an altered affinity for another cellular component, be it part of the muscle complex or membrane. This "altered affinity" is the result of calcium ion mediation. The latter half of the model suggests a functional reason for the third homologous region, AB. Instead of binding a calcium ion in the AB loop, the invariant and obviously critical Arg-75, Glu-81 salt bridge takes the place of calcium.

Three experiments support the model. FMCBP does bind calcium cooperatively; Benzonana et al. (6). It should be emphasized that it has not yet been shown that FMCBP reversibly binds one, let alone two calcium ions, in its physiological role.

Second, difference Fourier crystallographic studies have shown (13) that lanthanide ions can quite easily replace the EF calcium ion. The CD calcium ion cannot be removed without first exposing FMCBP to strong calcium chelators such as EGTA.

The most exciting experiment was performed by Gerday et al. (14) who found that reaction of Arg-75 with cyclohexanedione abolished the ability of FMCBP to bind calcium. The salt bridge is 20Å from the calcium binding sites.

Although some details of this model may require modification, the general outlines are certainly correct.

*Evolution of FMCBP.* The taxonomic relationship between the seven available sequences of FMCBP is shown in figure 6a. Pechere et al. (15) have calculated slightly different branch lengths. The important point is that this simple protein taxonomy is inconsistent with well established evolutionary and taxonomic relationships for the species. For instance, pike is closer to carp than it is to frog. This apparent contradiction is often encountered in such studies and is resolved by postulating two groups, say $\alpha$ and $\beta$, of homologous proteins. The two groups diverged from a common precursor, previous to divergence of the species. By way of illustration, human and whale myoglobins are more closely related than are human myoglobin and human hemoglobin. This is because myoglobin and hemoglobin diverged from a common precursor previous to divergence of the mammals.

However, the modified $\alpha,\beta$ FMCBP evolution tree in figure 6b presents a more revealing contradiction. Most proteins, for example cytochrome c or the globins (11) show a constant rate of evolution, i.e. accepted point mutations (P.A.M.) per elapsed time. This somewhat unexpected result was interpreted as implying a constant evolutionary pressure, or a constant molecular environment, in the various species which provided the proteins. Yet for FMCBP we see that the $\alpha$ component accumulate some 27 P.A.M.'s per $3.5.10^8$ years since pike and coelacanth diverged (16). While there are only 18 P.A.M.'s from the common ancestor of frog, carp and hake, $4.0.10^8$ years ago. These rates, 7.7 and 4.5 P.A.M/$10^8$ years, are significantly different.

Since the function of FMCBP remains unknown it is futile to speculate as to why different fish have different components. The note worthy point is that FMCBP is subject to different evolutionary pressure in different species. This variation may be expected to apply to its homologues as well.

## Troponin Component C

*Function*. Calcium controls the contraction of muscle. Myosin, the principal constituent of thick filaments, has Mg-ATP'ase activity as does purified actomyosin. If however the fibrous protein tropomyosin is present on the actin, as it naturally is in thin filaments, the Mg-ATP'ase activity is abolished (17).

Troponin is a trimer consisting of a tropomyosin binding component (TN-T), an inhibitory component (TN-I) and a calcium binding component (TN-T). When cytosol free calcium ion concentration rises from $3 \cdot 10^{-8}$ to $3 \cdot 10^{-6}$M, TN-C binds two more calcium ions (18), in addition to the two which appear to be bound in the resting state, and thereby weakens the binding of troponin to the thin filament, allowing contraction to occur. Calcium has no direct effect on myosin Mg-ATP'ase or on the entire actomyosin complex (19). Troponin is the mediator of calcium activation.

*Gene Tetraplication.* The prediction (9) that TN-C is a replicated form of the basic EF region of FMCBP has been confirmed by the amino acid sequence determination of Collins et al. (20). The alignment of the four homologous regions of TN-C with the nodal sequence of FMCBP is shown in figure 4. Knowing the key structural features of the "EF hand" one can quickly recognize four EF hands in TN-C. The essential characteristics are shown at the bottom of figure 4. There should be calcium coordinating side groups at positions +X, +Y, +Z, -X and -Z. (Or residues 90, 92, 94, 98 and 101 in EF hand numbering). Since residue 96 (-Y) coordinates with a carbonyl oxygen atom, it can be any residue. Helix E as well as helix F will show hydrophobic insides at defined intervals. This pattern is unmistakably present four times in TN-C.

Once the four EF hands are recognized in TN-C, their characteristics can be used to extend the definition of the EF hand when looking for this pattern in other proteins. Hydrophobic side chains (L* in figure 4) are expected at positions 82/85, 86/89 in the E helix and at 102/105, 106/109 in the F helix. Aspartic acid has so far always occurred at +X(90) and glutamic acid at -Z (101); however, there may be other oxygen containing ligands at +Y(92), +Z(94) and -X(96). Glycine usually occurs in the loop at position 95; however, in FMCBP residue 95 is aspartic acid in hake. The isoleucine at position 97 joins the loop to the hydrophobic core of the molecule. In hake valine is at position 58 (EF97).

*Structure Prediction.* I assume the four EF hands of TN-C to be isostructural with the EF hand of FMCBP. Helix E begins with residue 79 and helix F runs through 109.

Second I assume the hands to occur in pairs with the same palmar, hydrophobic interactions as occur in CD, EF. That is EF and EF' of TN-C are related by a two fold axis as in FMCBP.

The third assumption is critical. The bands pair 1-2 and 3-4 (not 1-3, 2-4 or 2-3). Subjectively this is appealing because, as discussed later, it necessitates fewer duplication steps in the evolution of TN-C. The 1-3, 2-4 pairing is awkward, but not disproven, by topological criteria; it is difficult to join. The 1-4, 2-3 pairing is possible, but I find the 1-2, 3-4 pairing attractive for two structural reasons.

In TN-C glycine occurs at the end of helix $F_1$ (position 47) and at the end of helix $F_3$ (122) or in EF hand numbering, both at 110. In a pair of EF hands there are just enough residues to connect 109 of $EF_1$ (or of $EF_3$) with 79 or $EF_2$ (or of $EF_4$). Because glycine can assume a broad range of $\phi,\Psi$ values and has no side chain it is particularly suited to make the turn preceeding the $EF_1$-$EF_2$ connection.

We now have two pairs of hands, a two fold axis relating the two hands within each pair. Each pair resembles a half orange with hydrophilic groups on the surface, calcium loops near the stem and hydrophobic groups at the cut surface. (In FMCBP the AB region covers the hydrophobic surface with its palmar surface). The second attraction of the 1-2, 3-4 pairing is that the two cut surfaces, both hydrophobic, can be opposed to one another by aligning the two, local two fold axes. Hence at low resolution at least, the structure prediction consists of determining two numbers, the distance between the two pairs and their relative rotation. The length of peptide chain available (83 to 89 TN-C numbering) limits the range of these parameters.

Several other predictions regarding TN-C structure and function can be made from this model. Just as one of the two calcium ions in TN-C is accessible to solvent (at -X, Gly-98) so too will calcium$_1$ of TN-C be coordinated by water at +Z, Gly-31. This particular calcium should readily exchange with lanthanides. It now seems well established that only two of the four calcium ions can be removed under physiological conditions (18). Arg-75 is not present in TN-C; however, there is a lysine at the homologous position 87 (TN-C numbering) between hands 3 and 4. Further Glu-81 is preserved at the beginning of all four helices E. One suspects that the Glu's-81 are involved in the conformational changes (21) which accompany calcium binding. The model has approximate two fold symmetry. TN-I and TN-T may be expected to bind at symmetry related sites on TN-C and hence to have some regions of TN-I, TN-T similarity. Consistent with this suggestion, Perry (22) has observed 2:1 TN-I:TN-C binding.

Several specific experiments have been suggested by the FMCBP model of function and the recognition of TN-C homology. I will argue that similar benefits will come from extending these ideas to other calcium control systems.

## Survey of Calcium Coordinating Structures

In order to place the preceeding description of FMCBP and of TN-C in perspective, I will summarize the results of my survey (23) of eighteen organo-calcium crystal structures. In general these results are similar to those obtained from inorganic structures. Table I compares the calcium coordination of FMCBP (and of TN-C) with those observed in the three other calcium binding proteins of known structure.

*Organic Ligands of Calcium.* Oxygen is the co-ordinating atom preferred by calcium. Nitrogen co-ordinates in only two of the six structures having potential nitrogen ligands available. Calcium chelators, as EGTA or oxalate, have high oxygen contents. In all eight structures having chloride or bromide as the counteranion, the halide is not in the primary coordination sphere.

The coordination number is usually 8 but in

five of the eighteen examples it is 7. Interestingly in one example, $(CH_2)_6N_4 \cdot 10H_2O$, calcium is 6 coordinate, all to water. The primary coordination sphere is well defined with calcium-oxygen distances ranging from 2.29Å to 2.65Å. Other potential ligands are over 3.0Å away from the calcium ion.

Calcium-oxygen bond distances correlate with coordination number: eight coordinate 2.45Å, seven coordinate 2.40Å and in the hexahydrate 2.33Å. The formal charge of the primary coordination sphere ranges from -2 to 0. As noted the halide anion is not observed in the primary coordination sphere. The coordination number shows no discernible relationship to the nature or electronegativity of the oxygen ligands. The calcium ion is usually hydrated, however there is no water in the glucoisosaccharate crystal.

Oxygen-oxygen distances as close as 2.54Å are seen in the 7 coordinate blepharismin. Several authors have suggested that coordinating oxygen atoms are hydrogen bonded to one another although the actual position of the hydrogen atom is uncertain. Even in eight coordinate complexes it is not the oxygen-oxygen contact which limits the calcium-oxygen bond distance.

Usually the multidentate ligands form five member rings, however some six member rings are formed in conjunction with the formation of five member rings.

The calcium ion certainly induces changes in the structure of the ligand. For instance Cook and Bugg (24) have observed that "calcium binding to adjacent hydroxyl groups results in a decrease of about 0.2Å in the intermolecular spacing between the hydroxyl oxygen atoms".

Carboxylate groups coordinate oxygen with either oxygen or with both oxygen atoms. Of particular relevance to thermolysin with a double calcium site and to concanavalin A with a calcium-magnesium site, four of the structures have calcium-calcium distances ranging from 3.624Å to

4.792Å. In each of these double calcium structures a single oxygen atom of either a carboxylate or phosphate group coordinates to both calcium ions.

Calcium itself links together the coordinating ligands threby building up chains, sheets or three dimensional networks of alternate ligand, calcium,ligand, etc.

In terms of calcium-macromolecular interactions these structures are best regarded as providing a series of precedents of possible and probable configurations. Unfortunately these structures are not usually related to thermodynamic data.

*Protein-Calcium Structures*. The calcium coordinations of the four proteins of known structure, as well as the postulated coordination by TN-C, are summarized in table I. Two of the carboxylate groups in concanavalin-A bridge calcium and magnesium ion separated by 4.5Å. The site one and site two calcium ions of thermolysine are 3.8Å apart and are bridged by three carboxylate groups. In none of these bridges is it interpreted that a single oxygen atom coordinates both calcium ions.

The charge in the primary coordination sphere ranges from -5 at one of the TN-C sites to -1 at the weak thermolysin site four. As seen in the small molecule complexes the coordination charge can be as low as 0, but it does seem probable that the -4 and the -5 coordinations bind the calcium ion more strongly.

It is rather suggestive that in all of the sites, excepting thermolysin site four, one carbonyl oxygen coordinates calcium.

In each case I have defined the coordinate system by assigning +X to the first amino acid in sequence and +Y to the second. The trace of the main chains of nuclease, concanavalin and thermolysin are all different from that observed in FMCBP. No redefinition of axes, and in particular common assignment of the carbonyl oxygen atom to -Y, improves the polypeptide main chain fit.

*Summary of Calcium Coordination.* In context of the present discussion, the most important points are: (1) Coordination numbers in the proteins tend to be lower than in small molecule complexes. (2) The proteins tend to use approximate octahedral geometry. (3) The proteins employ more carboxylate groups. (4) The protein main chain is in all of the known structures involved in calcium coordination. (5) None of the coordination geometries in the protein complexes is really unprecedented in terms of the smaller complexes.

The one point I wish to emphasize is that eventhough there is a basic similarity in the coordination of the various protein complexes, there are distinctly different protein conformations which provide calcium ligands. None of these other proteins are homologous to FMCBP and TN-C. They have arrived at their general octahedral coordination patterns by quite different evolutionary routes.

## Patterns of Protein Evolution.

There are many examples of protein families, such as the cytochromes c or the globulins, whose members are related by point mutations or short deletions and of families containing segments derived from gene duplications, as ferrodoxin, or higher order replications--immunoglobulin light chain (x2), γ heavy chain (x4), and μ heavy chain (x5) (see Dayhoff (11) for documentation and further examples). Members of these families have been identified by their amino acid sequences. Subsequent crystallographic studies have shown that the homologous regions have structures very similar to one another.

As proteins diverge further apart their similarity is recognized only by comparing their structures. Consider the sex (families of) proteins illustrated and described in figure 5 (see Branden et al. (25) for complete references). The basic structure of the molecule (for the dehydrogenases one half of the molecule) consists of a β-parallel pleated sheet whose strands are connected by similar patterns of helical and nonhelical elements. A constant pattern of nucleoside binding sites is formed on the two sides of the sheet.

A wide variety of proteins fit this general pattern. Most surprising of all is subtilisin which has a sixth and a seventh parallel strand added relative to adenylate kinase. The binding site for the aromatic side chain of the polypeptide substrate again corresponds to the nicotinamide binding site of the dehydrogenases.

Although it is not yet proven, it now seems most reasonable to assume that they all have a common evolutionary origin. It is of course possible that this general configuration represents a basic thermodynamic stability, as does the α-helix, arrived at by convergent evolution. Nevertheless it is very exciting to think that all of these proteins are evolutionarily related. Such extensive homologies may be expected to exist in other protein families.

Further it is reasonable to suspect that some proteins have evolved by splicing together genes of different homologue families. In fact mutants of Salmonella typhimurium have already been selected (26) in which a pair of frameshift mutations eliminate the termination signal thereby fusing two adjacent genes to produce a new protein with both functions and with structures derived from two different protein families.

## Systems Controlled By Or Controlling Calcium.

I have now described in some detail the structure of FMCBP and its homologue, TN-C, as well as the characteristics of calcium coordination. Since I want to explore the idea of protein evolution, it seemed appropriate to present an example, the dehydrogenases, of a complex scheme. Now I present an overview of calcium control systems and will conclude by suggesting that the EF hand homology will extend to these systems.

Most of the systems controlled by calcium involve movement-either the contraction of muscles, or of actin-like microfilaments or possibly microtubules. Calcium apparently modulates the activities of some enzymes, but I will emphasize that not all calcium binding proteins are involved in calcium control processes.

Finally the relationship between calcium and membranes is particularly intriguing. On the one hand the membrane controls the passage of the calcium ion to its physiological goal. Yet on the other, the passage and/or storage of the calcium ion may significantly effect the membrane.

*Movement.* Not only is movement usually the primary physiological response of most calcium control but (nearly) all movement in Biology is controlled by calcium.

In vertebrate and arthropod muscle, calcium control is exerted via troponin as discussed in section III.

Lehman et al. (27) have recently found that in mollusca and brachiopods, calcium control is exerted through the thick filament. This myosin has an "extra" EDTA extractable light chain, 18,000 molecular weight, which confers the calcium sensitivity. The universality of the actomyosin system is demonstrated by the fact that desensitized molluscan myosin can be activated and controlled by vertebrate thin filaments. The isolated EDTA light chain itself does not bind calcium but is responsible for increasing the binding affinity of molluscan myosin for calcium from 1.5 to 2.5 ions per molecule.

Various annelids and insects contain, apparently within the same cell, both the thick filament and the thin filament control systems.

Lehman et al. (27) have speculated, one that "animals from phyla which evolved early have a myosin-linked regulation", two that troponins of vertebrates and arthropods "evolved by convergent pathways" because they have slightly different sizes and binding affinities" and three that "the myosin-linked and the troponin-linked regulatory systems are apparently unrelated" evolutionarily. As will be elaborated later, I propose that all TN-C components will prove to be closely homologous and also that the molluscan type EDTA light chain will contain one or several copies of the EF hand.

In contrast to muscle, microfilaments and/or microtubules are responsible for the movement of individual cells, cytoplasmic streaming and the relative movement of cell organelles. Microfilaments are composed of an actin homologue and usually tropomyosin. Although the amino acid sequence of rabbit actin is the only one determined to date (28), the cyanogen bromide peptides have a very characteristic composition including the rare amino acid 3-methylhistidine. The microfilaments can be identified in electron micrographs by their being "decorated" by added heavy meromyosin. The myosin attaches to form easily recognized "arraowhead" formations (17). Processes, which involve microfilaments are indicated by their susceptibility to cytochalasin B.

I suggest that the dependence on external calcium observed in slime mold aggregation and in blood platelet contraction (29) will apply to all microfilament systems. Thrombosthenin A consists of myosin and actin like proteins. Platelets also contain a tropomyosin (30) and have calcium regulation associated with their light chains. Recently Cohen et al. (31) have observed in acrylamide gel electrophoresis patterns a band corresponding to a protein of 18,000 molecular weight. It seems very probable that this is TN-C.

The suggested (32) involvement of microfilaments in vesicle secretion may prove to be their most significant role. As is discussed in the reviews of Rasmussen (33) and of Sutherland (34) a remarkably wide variety of systems respond to either nervous or hormonal stimuli by release of the second messengers, adenosin 3', 5'-monophosphate (cAMP) and calcium ion. In the membrane section I will outline the proposed mechanisms of cAMP, calcium release and action. The important point now, in the discussion of microfilaments, i is that frequently the response of the cell following the nervous or hormonal stimulus is to secrete a vesicle, which contains enzyme, hormone or neurotransmitter. In at least two of these systems the vesicle secretion is inhibited by cytochalasin B( 35, 49). The obvious suggestion to be made at this point is that in many (or all) of the systems which secrete a vesicular product in response to a calcium second messenger, the vesicle does not merely diffuse to the cell membrane and then fuse with it. The vesicle, myosin complex would be drawn along an actin thin filament thread whose calcium control is mediated by a troponin homologue.

The other group of organelles involved in cellular movement uses microtubules as the basic structure. In addition to the well known example of ciliar and flagellar beating, microtubules are involved in chromosome spindle shortening (36). In other systems they are found together with microfilaments, possibly functioning as antagonists, as in body contraction of the ciliate (37) or in axonal streaming. The microtubule is recognized in the electron microscope as a 200 to 250Å hollow tube usually of considerable length. Its involvement in a process such as transcellular movement of procollagen (38) is inferred from its sensitivity to colchicine. Several processes involving secretion of vesicles,for instance of insulin (39), are blocked by colchicine. The *in vitro* assembly of microtubules from tubulin is inhibited by $3.10^{-4}$M calcium (40).

There is no direct evidence that calcium controls microtubule assembly or shortening by either a TN-C like molecule or by activating a phosphorylating enzyme (to be considered later). Nonetheless it is tempting to wonder whether microtubule systems might be under the same sort of calcium control as are muscle and microfilament systems.

*Calcium Binding Proteins* Given the amazingly broad spectrum of physiological processes (possibly) controlled by calcium as well as membrane systems to be described in the next section, one naturally wants to characterize the calcium binding proteins which are involved. Unfortunately only a few have been isolated and even fewer assigned a function.

First it is important to realize that in many proteins, calcium is permanently bound "merely" to stabilize the protein or to serve at its active site. It is not reversibly bound and released as part of an information transfer or control process. A few examples are concanavalin A, nuclease and thermolysin already discussed in section IV, as well as serum albumin and trypsinogen. The calcium ion bound by actin (28) probably plays no regulatory function. I see no reason to suspect that any of these proteins contain an EF hand.

The classification of several other proteins is ambiguous: α-amylase (42), prothrombin (43), smooth muscle alkaline phosphatase (44) nicotinamide nucleotide transhydrogenase (45) and 25-hydroxycholecalciferol-1-hydroxylase (41). Aequorin (46) binds calcium very specifically but the value of its light emission to the jellyfish remains unknown.

There are a few enzymes whose catalytic activity is regulated directly by calcium, not via a mediator. Muscle fructose diphosphatase is inhibited by calcium release during contraction in the rabbit (47) and in bumblebee (48) thereby controlling fructose diphosphate cycling in coordination with muscle contraction. Gluconeogenesis and glycolysis are also calcium controlled (33). Calcium inhibits cAMP phosphodiesterase (33, 49) apparently in a regulatory role.

Finally there are three calcium binding proteins isolated from nervous tissue which are candidates for calcium mediators. The S-100 protein (soluble in 100% saturated ammonium sulfate) is found in the nervous systems of many animals, binds two calcium ions specifically and reversible with accompanying conformation changes and has molecular weight 24,000 (50). In squid axoplasm one protein accounts for 5% of the total protein. It has molecular weight about 14,000, isoelectric point 4.2 and binds two to three calcium ions at $K_d = 2.5 \cdot 10^{-5}$M (51). The third example is the acid protein unique to the adrenal medulla (and other adrenergic tissue). It has a molecular weight 12,000, is monophosphorylated, and binds one calcium ion at $K_d = 1.7 \cdot 10^{-5}$M. Brooks and Siegel (52) have suggested that it is associated with the release of vesicles of noradrenaline. In a similar vein it is natural to suspect that the axon protein is associated with axonal streaming and S-100 with neurotransmitter release.

*Calcium Membrane Interactions.* Membranes are the primary source and regulatory agent of informational calcium. Before discussing the cell membrane which is of course the original receptor of the nervous and hormonal stimuli already mentioned, I will first consider the sarcoplasmic reticular vesicles and the mitochondria. These systems are both simpler and better defined. Also they actually account for most of the calcium transported.

A nervous impulse causes the release of packets of acetylcholine (or noradrenaline from postganglionic sympathetic nerves). The neurotransmitter diffuses some 500 Å across the endplate gap and initiates a depolarization, which is propagated down the transverse tubule system and somehow transmitted to the sacroplasmic reticulum, which releases its stored calcium. Homogenized muscle yields sarcoplasmic reticulum which actively sequesters two ions of calcium with the coupled hydrolysis of one molecule of ATP (53). MacLennan et al. (54) have characterized the protein constituents of isolated vesicles.

| Protein | Location | Composition | M. W. x1000 | $Ca^{2+}$ n, | $K_d$ |
|---|---|---|---|---|---|
| ATP'ase | interior | 90% | 102, | ? | $10^{-6}$ |
| acid protein | in. surf. | 1% | 54, | 25 | $10^{-3}$ |
| calsequestrin | in. surf. | 5% | 44, | 50 | $10^{-3}$ |
| 3 acid prot's | out.surf. | 1% | 20,→32, | 3 | $10^{-6}$ |
| proteolipid | interior | 2% | 6,→12, | ∿ | ∿ |

The Mg-ATP'ase is calcium specific and maximally active in the $10^{-7}$ to $10^{-6}$M range. The calcium accumulating speed and capacity of isolated vesicles is quite adequate to account for skeletal muscle relaxation. Calsequestrin and the 54,000 molecular weight acid protein are not highly specific for calcium and may serve simply as calcium binding resins.

Mitocondria are found in high concentration in cells which have high rates of oxidative phosphorylation, and this is usually considered to be their primary function. Borle (55) has emphasized that in a "typical" mammalian cell the inner mitochondrial membrane can maintain a calcium concentration gradient over 1000-fold with a half-time of calcium uptake about 8 sec. These values correspond to those for the cell membrane. However, since the total mitochondrial surface is some thirty times greater than that of the cell surface, mitochondria might account for some 97% of the cell's calcium buffering capacity. Batra (56) also has pointed out that calcium uptake by mitochondria can account fully for the relaxation of myometrium. Although calcium transport can be coupled to either respiration or ATP hydrolysis, no specific ATP'ase has yet been characterized. In contrast to the sarcoplasmic reticulum the mitochondrion apparently does not have a calcium storage protein like calsequestrin but may instead store calcium in micropackets of $Ca_3(PO_4)_2$ crystals. Hence the maximal concentration of free calcium ion is inversely proportional to the phosphate ion concentration.

It is, though, the outer cell membrane which receives the initial external signal. (In the case of the nerve terminus the "external" signal passed along the axon of the same cell.) Assume for a moment the same idealized response for all cells. Within a second there is a marked increase in the cytosol concentration of calcium ion and cAMP, with the subsequent cellular response of contraction. Several important questions are left unanswered:

1) What is the actual source of the calcium? Most cells, not being muscle, have no sarcoplasmic reticulum. The microsomal fraction usually does not accumulate nor store calcium (55, 57). The mitochondria which do have stores of calcium are not coupled to the cell membrane.

Although removing or replacing the calcium in the external medium will usually abolish the response to an external stimulus, it is by no means proven that the stimulus induced increase in free calcium ion concentration in the cytosol was the direct result of an inward flux of calcium ions. It is possible that the external calcium functions to maintain a certain conformation of a receptor in the cell surface. The stimulus causes a conformational change of the receptor, which communicates with the inner aspect of the membrane thereby releasing bound calcium. The external calcium ion need not have crossed the membrane during a single stimulus.

Regardless of the source of calcium it has definitely been demonstrated by Miledi (58) that intracellular injection of calcium causes release of acetylcholine from squid axon nerve termini, even with magnesium or manganese replacing external calcium.

2) What is the relationship of $Ca^{2+}$ and cAMP? Apparently calcium regulates the membrane bound adenylate cyclase (33) and subsequently the cAMP activates a protein kinase with an ultimate result being the phosphorylation of some still unidentified proteins. Yet calcium is required subsequent to control of adenylate cyclase, possibly in activating other enzymes or as I would like to suggest by binding to a troponin homologue.

There are (in addition to these idealized systems of a stimulus causing the membrane release of calcium) several other well studied systems involving membrane calcium interaction. The intestinal mucosa responds to vitamin D stimulus by synthesizing a calcium binding protein, which is involved in calcium uptake. The form from chicken and cows has molecular weight 28,000 with four calcium binding sites, $K_a=10^6M^{-1}$ (59). Most mammals have a 13,000 molecular weight form, $K_a= 4.10^6M^{-1}$ (60). It is reasonable to suspect that these are related by gene duplication. One should check the amino acid sequence, when available, for an EF hand.

It is promising to find at least one system, temperature sensitive behavioral mutants of Paramecium (61), in which calcium membrane activation can be studied genetically.

Obviously we still have many more questions than answers concerning membrane-calcium interaction. The ATP'ase(s) associated with calcium transport in sarcoplasmic reticulum, mitochondria and cell membrane is an intrinsic protein by Singer's definition (62). I suspect that the targets of the calcium, be they enzymes or mediators may be extrinsic. As suggested by the functional model for FMCBP, and the observations for TN-C, the affinity of these extrinsic proteins for the membrane might be significantly reduced following calcium coordination. Invoking a teleological argument, why have them floating about withoug calcium; they should be stored where needed. In contrast such mediators should not be "waiting" on the inner aspects of the sarcoplasmic reticular or mitochondrial membranes.

## Postulate of General Homology.

As I illustrated in section IV, Calcium Coordination, there are many protein conformations which can be involved in calcium coordination even if the ligands conform to general octahedral geometry. The EF hand, or a pair of EF hands, does seem a reasonable structure for a calcium ion mediator. Any alteration of the loop can be effectively radiated out the finger and thumb as lever arms or conversely alteration of the helices can be transmitted back to the loop. We have seen how the pair of hands can interact through their antiparallel middle fingers. In all of the other families of homologous proteins one sees that not only do the homologues differ by single amino acid replacements but frequently by short deletions or insertions. In the comparison of EF hands, whether within or between molecules, there are no deletions within the hand region (save for the AB loop which has "lost" calcium). The deletions are all between the hands. The hand is quite finely designed; it can tolerate no changes of register.

Reasonable as these arguments may seem, I have really no idea why Nature initially "chose" this hand. Yet once it was first chosen for some sort of calcium mediating function, I suggest that it was used time and again for new calcium mediation roles. As schematized in figure 6c, I postulate that all thin filament control is mediated by the four-hand TN-C. This same pattern should apply to control of microfilaments, although there need not necessarily be four hands.

Before extending the homology, let us return to the precedent of the dehydrogenases, figure 5. Eventhough the basic pleated sheet pattern is maintained, perhaps because of its thermodynamic stability, various pieces are spliced on and off outside of the sheet. As I noted, different non-homologous genes can be spliced together. Let me then speculate that calcium controlled enzymes and calcium transporting ATP'ases have one or two EF hands spliced onto a catalytic function derived from a different homologue family.

Once one has determined the amino acid sequence of a calcium control protein, one should be able to recognize, even at great evolutionary distance, the EF hand pattern:

L LL LD D DG ID EL LL L.

This identification would not be so certain in an evolutionary unit which more readily allowed insertions and deletions.

Having made this identification one should immediately have some insight into the structure and mode of function of the protein. Given the limited success we have had in crystallizing membrane proteins, I think that this argument by homology may prove to be one of our best sources of information about the structure and action of membrane proteins.

## References.

1. Kretsinger, R.H. & Nockolds, C.E. J. Biol. Chem. 248 (1973) 3313.
2. Piront, A. & Gerday, C. Comp. Biochem. Physiol. 468 (1973) 349.
3. Pechere, J. F. Personal communication.
4. Frankenne, F., Joassin, L. & Gerday, C. FEBS Letters 35 (1973) 145.
5. Demaille, J., Dutruge, E., Capony, J.P. & Pechere, J.F. In: Calcium-binding Proteins, W. Drabikowski, ed., Elsevier, Amsterdam, In press.
6. Benzonana, G., Capony, J.P. & Pechere, J.F. Biochim. Biophys. Acta 278 (1972) 110.
7. Moews, P.C. & Kretsinger, R.H. Manuscript in preparation.
8. Diamond, R. Acta Cryst. B21 (1966) 253.
9. Kretsinger, R.H. Nature New Biol. 240 (1972) 85.
10. MacLachlan, A.D. Nature New Biol. 240 (1972)
11. Dayhoff, M.O. Atlas of Protein Sequence.
12. Kretsinger, R.H., Moews, P.C., Coffee, C.J. & Bradshaw, R.A. In: Calcium-binding Proteins W. Drabikowski, ed., Elsevier, Amsterdam, In press.
13. Moews, P.C. & Kretsinger, R.H. Submitted to

Biochemistry.

14. Gosselin-Rey, C., Bernard, N. & Gerday, C. Biochim. Biophys. Acta 303 (1973) 90.

15. Pechere, J.F., Capony, J.P. & Demaille, J. J. Syst. Zool. In press.

16. Young, J.Z. 1952. The Life of the Vertebrates. p. 255, Clarendon Press.

17. Huxley, H.E. Nature 243 (1973) 445.

18. Hitchcock, S.E. Biochemistry 12 (1973) 2509.

19. Bremel, R.D. & Weber, A. Nature New Biol. 238 (1972) 97.

20. Collins, J.H., Potter, J.D., Horn, M.J., Wilshire, G. & Jackman, N. FEBS Letters. 36 (1973) 268.

21. Murray, A.C. & Kay, C.M. Biochemistry 11 (1972) 2622.

22. Head, J.F. & Perry, S.V. Biochem. J. 135 (1973). (and personal communication).

23. Kretsinger, R.H. Manuscript in preparation.

24. Cook, W.J. & Bugg, C.E. Acta Cryst. B29 (1973) 2404.

25. Branden, C.I., Liljas, A. & Rossmann, M.G. 1975. In: The Enzymes. Vol. X. P. D. Boyer ed.

26. Yourno, J., Kohno, T. & Roth, J.R. Nature 228 (1970) 820.

27. Lehman, W., Kendrick-Jones, J. & Szent-Gyorgyi, A.G. Cold Spring Harbor Symy. Quant. Biol. 37 (1972) 319.

28. Elzinga, M., Collins, J.H., Kuehl, W.M. & Adelstein, R.S. Proc. Nat. Acad. Sci. 70 (1973) 2687.

29. Hanson, J.P. Repke, D.I., Katz, A.M. & Aledort, L.M. Biochim. Biophys. Acta 314 (1973) 382.

30. Cohen, I. & DeVries, A. Nature 246 (1973) 36

31. Cohen, I., Kaminski, E. & DeVries, A. FEBS Letters 34 (1973) 315. and Thorens, S., Schaub, T.S. & Luescher, E.F. Experientia 179 (1973) 441.

32. Berl, S., Puszkin, S. & Nicklas, W. Science 179 (1973) 441.

33. Rasmussen, H. Science 170 (1970) 404.

34. Sutherland, E.W. Science 177 (1972) 401.

35. Wessells, N.K., Spooner, B.S., Ash, J.F., Bradley, M.O., Luduena, M.A., Taylor, E.L. Wrenn, J.T. & Yamada, K.M. Science 171 (1971) 135.

36. Schroeder, T.E. Biol. Bull. 137 (1969) 413.

37. Huang, B. & Pitelka, D.R. J. Cell. Biol. 57 (1973) 704.

38. Ehrlich, H.P. & Bornstein, P. Nature 238 (1972) 257.

39. Williams, J.A. & Wolff, J. Proc. Nat. Acad. Sci. 67 (1970) 1901.

40. Olmsted, J.B. & Borisy, G.G. Biochemistry 12 (1973) 4282.

41. Suda, T., Horiuchi, N., Sasaki, S., Ogata, E., Ezawa, I., Nagata, N. & Kumura, S. Biochem. Biophys. Res. Com. 54 (1973) 512.

42. Levitzki, A. & Reuben, J. Biochemistry 12 (1973) 41.

43. Benson, B.J., Kisiel, W. & Hanahan, D.J. Biochim. Biophys. Acta 329 (1973) 81.

44. Limas, C.J. & Cohn, J.N. Nature New Biol. 245 (1973) 53.

45. Rydstrom, J., Hoek, J.B. & Hojeberg, B. Biochim. Biophys. Res. Com.52 (1973) 421.

46. Shimomura, O. & Johnson, F.H. Biochim. Biophys. Res. Com. 53 (1973) 490.

47. Van Tol, A., Black, W.J. & Horecker, B.L. Arch. Biochem. Biophys. 151 (1972) 591.

48. Clark, M.G., Bloxham, D.P., Holland, P.C. & Lardy, H.A. Biochem. J. 134 (1973) 589.

49. Whitfield, J.F., Rixon, R.H., MacManus, J.P. & Balk, S.D. In Vitro 8 (1973) 257.

50. Calissano, P., Moore, B.W. & Friesen, A. Biochemistry 8 (1969) 4318.

51. Alema, S., Calissano, P., Rusca, G. & Giuditta, A. J. Neurochem. 20 (1973) 681.

52. Brooks, J.C. & Siegel, F.L. J. Biol. Chem. 248 (1973) 4189.

53. Makinose, M. Cold Spring Harbor Symp Quant. Biol. 37 (1972) 681.

54. MacLennan, D.H., Yip, C.C. Iles, G.H. & Seeman, P. Cold Spring Harbor Symp Quant. Biol. 37 (1972) 469.

55. Borle, A.B. Fed. Proc. 32 (1973) 1944.

56. Batra, S. Biochim.Biophys.Acta 305(1973)428.

57. Inesi, G. Ann.Rev. Biophys.Bioeng. 1(1972)191.

58. Miledi, R. Proc. Roy.Soc.Lond. 183(1973)421.

59. Fullmer, C.S. and Wasserman, R.H. Biochim. Biophys. Acta 317 (1973) 172.

60. Hitchman, A.J.W. and Harrison, J.E. Can. J. Biochem. 50 (1972) 758.

61. Chang, S.Y. and Kung, C. Science 180(1973)1197.

62. Singer, S.J. and Nicolson, G.L. Science 175 (1972) 720.

63. Cotton, F.A., Bier, C.J., Day, V.W., Hazen, E.E. and Larson, S. Cold. Spring Harbor Symp. Quant. Biol. 36 (1971) 243.

64. Edelman, G.M., Cunningham, B.A., Reeke, G.N., Becker, J.W., Waxdal, M.J. and Wang, J.L. Proc. Nat. Acad. Sci. 69 (1972) 2580.

65. Hardman, K.D. and Ainsworth, C.F. Biochemistry. 11 (1972) 4910.

66. Matthews, B.W., Colman, P.M., Jansonius, J.N., Titani, K., Walsh, K.A., Neurath, H. Nature 238 (1972) 41.

TABLE I

Protein-Calcium Complexes

| Protein | | Octahedral Vertices | | | | | |
|---|---|---|---|---|---|---|---|
| | X | Y | Z | -Y | -X | -Z | charge |
| FMCBP CD | D-51 | D-53 | S-55 | F-57 C=O | E-59 | E-62 | 4 |
| EF | D-90 | D-92 (2) | D-94 | K-96 C=O | $H_2O$ | E-101 (2) | 4 |
| TN-C 1 | D-27 | D-29 | $H_2O$ | D-33 C=O | S-35 | E-38 | 3 |
| 2 | D-63 | D-65 | S-67 | T-69 C=O | D-71 | E-74 | 4 |
| 3 | D-102 | N-104 | D-106 | Y-108 C=O | D-110 | E-113 | 4 |
| 4 | D-138 | D-140 | D-142 | R-144 C=O | D-146 | E-149 | 5 |
| Nuclease (63) | D-19 | D-21 | ? | E-43 | D-40 | T-41 C=O | 4 |
| Con-A (64,65) | D-10* | Y-12 C=O | D-19* | ? | N-14 | $H_2O$ | 1 |
| Thermo lysin (66) 1 | D-138 | E-177* | $H_2O$ | G-189 C=O | D-185* | E-190* | 2.5 |
| 2 | E-177* | N-183 C=O | D-185* | E-190* | $H_2O$ | $H_2O$ | 1.5 |
| 3 | D-57 | D-59 | Q-61 C=O | $H_2O$ | $H_2O$ | $H_2O$ | 2 |
| 4 | Y-193 C=O | T-194 | T-194 C=O | I-197 C=O | $H_2O$ | D-200 | 1 |

The coordination of calcium can be described in terms of the vertices of an octahedron, except for the fourth site of thermolysin where water provides a seventh ligand in a direction between the +X and the -Z vertices. See legend to figure 4 for the one letter amino acid code. "C-O" denotes coordination by a main chain carbonyl oxygen atom. "(2)" means that both oxygen atoms of the carboxylate group coordinate calcium. An asterisk means that the carboxylate group bridges two metal ions, one oxygen atom bonding to each. The number of coordinating carboxylate groups is listed in the column headed "charge"; bridging carboxylate groups count 0.5 to each site.

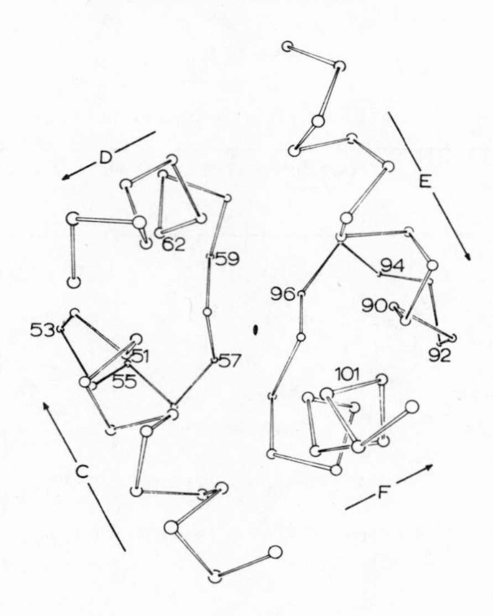


Figure 1. The α-carbon trace of the CD and the EF regions of FMCBP.

---

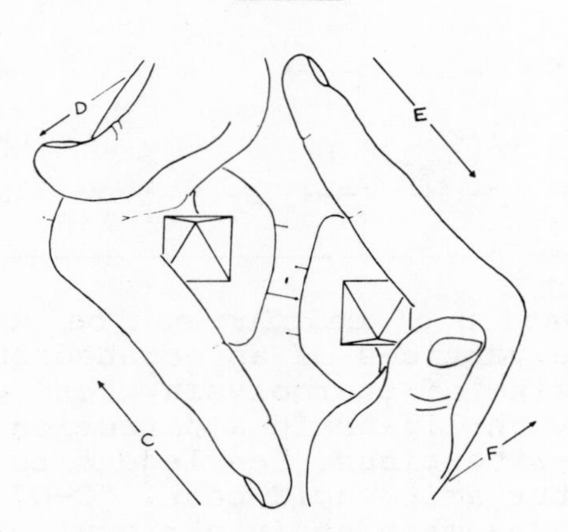


Figure 2. Right hand representation of the CD and the EF regions of FMCBP: The CD and EF loops are represented by middle fingers and enclose the calcium coordination octahedra shown in figure 3. The loops are in β-antiparallel sheet conformation and are connected by one hydrogen bond 58→97.

Three potential hydrogen bonds (99→56, 97→58 and 60→95) are not formed in the calcium bound form of the protein. The hands are viewed from the inside of the molecule.

---

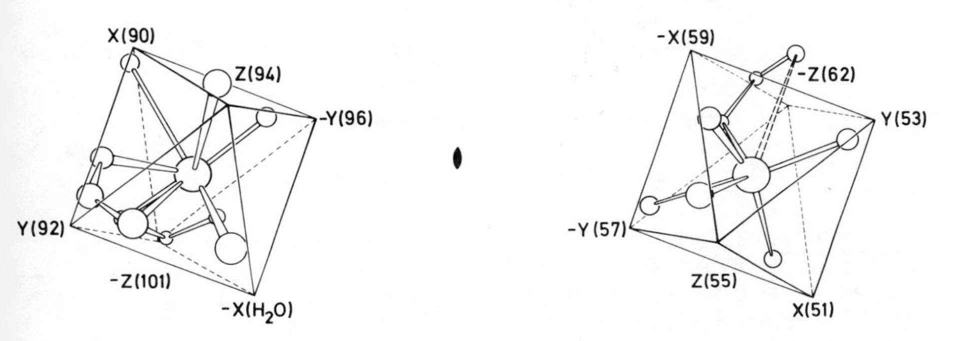


Figure 3. CD and EF calcium coordination octahedra: The CD calcium ion is coordinated by oxygen atoms from six amino acids (Asp-51, Asp-53, Ser-55, Phe-57, Glu-59 and Glu-62) and EF by five amino acids (Asp-90, Asp-92, Asp-94, Lys-96 and Glu-101) plus water. Residues 92 and 101 coordinate with both oxygen atoms of their carboxylate groups; the second oxygen of Glu-62 is not close enough to coordinate and is indicated by a dashed line. The first and second residues in sequence define the +X and +Y vertices of the two octahedra. The octahedra are related by the intramolecular approximate two fold axis.

---

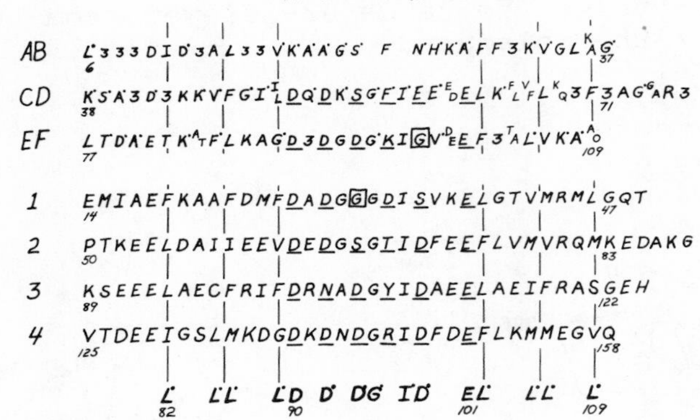


Figure 4. From the amino acid sequences and evolutionary tree described in figure 6, I have derived the nodal sequence of the precursor of both α and β groups of FMCBP (see Dayhoff '72 for procedure). The one letter code is: A Ala, C Cys, D Asp, E Glu, F Phe, G Gly, H His, I Ilu, K Lys, L Leu, M Met, N Asn, P Pro, Q Gln, R Arg, S Ser, T Thr, V Val, Y Tyr and O deletion. A single letter with no asterisk indicated that the residue was invariant in the seven sequences. An asterisk indicates that even though the principle of minimizing the number of mutations clearly indicates one residue, it is not invariant. Only 37 of 109 positions are invariant. The listing of two residues indicates a two fold ambiguity. Except at positions 36, 50 and 84 the ambiguity reflects the α (upper amino acid) and β (lower) division. Those positions having a three (or higher) fold ambiguity are indicated by "3". Except at residue 43 (Val, Leu, Ala or Ilu) all of the three fold ambiguities involve hydrophilic groups. At position "95" Gly is always found excepting coelacanth Asp (95) and hake Asp (56). Likewise "97" is Ilu excepting hake Val (58). Gly (98) of FMCBP is enclosed in a box; here water coordinates calcium at direction -X. Similarly Gly (31) of TN-C is enclosed. The homologous regions (1,2,3 and 4) of TN-C are aligned with FMCBP. The distinguishing

characteristics of the basic "EF hand" are the hydrophobic residues (L* = Leu, Val, Ilu, Phe or Met) on the inner aspects of helices E and F. The calcium ligand at the +X vertex is always Asp, and at -Z, Glu. The -Y, +Z and -X have D* that is either Asp, Glu, Ser or Asn. (Gln or Thr might also be expected to provide oxygen). Since a carbonyl oxygen atom is at -Y, there is no predictable choice of amino acid.

---

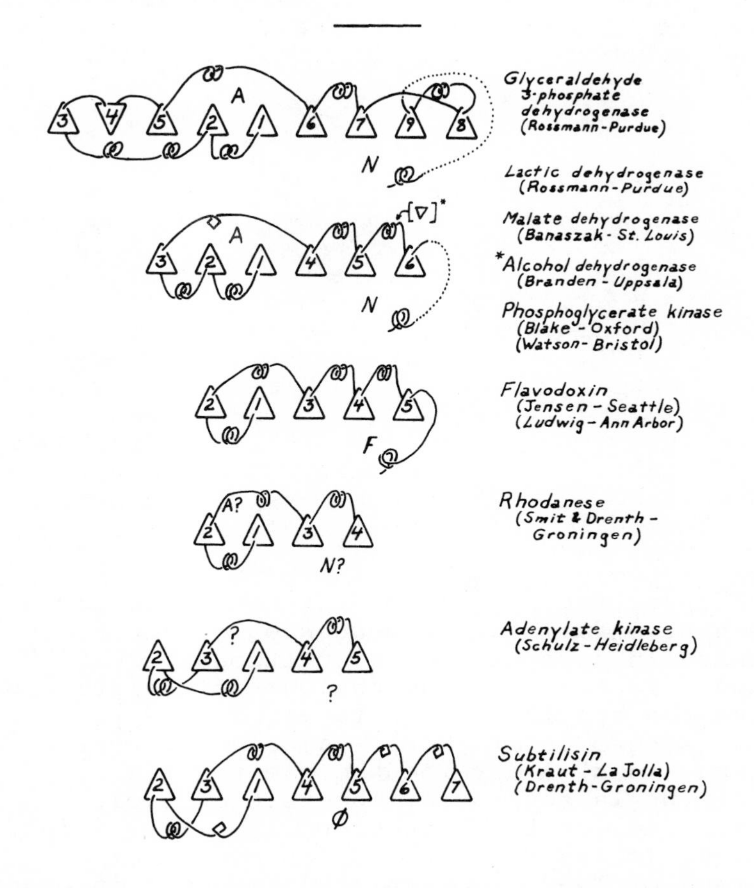


Figure 5. The triangles represent strands, pointing up from the paper, of a β-parallel pleated sheet. They are numbered in order from the N-terminus. In GPD the fourth strand runs down in antiparallel conformation and is represented by

an inverted triangle. The strands are connected by extended chains, represented by straight lines, or by α-helices represented by coils, by non α-helical structural elements represented by a square. In the dehydrogenases the second half of the molecule is omitted and is represented by a dotted line, which leads to a C-terminal helix, which forms part of the nicotinamide binding site, "N" of NAD. Similarly "A" represents the adenine binding site; "F" flavodoxin. In adenylate kinase "?" refers to supposed ADP sites. "Ø" is a pocket for hydrophobic sidechains of the substrate.

---

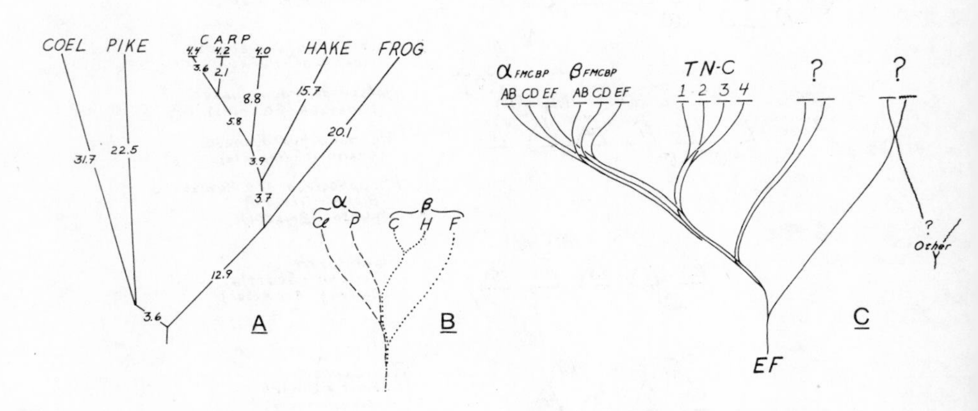


Figure 6. Evolution of calcium binding proteins: In the upper figure (a) the taxonomic distances between seven amino acid sequences of FMCBP are expressed in accepted point mutations per 100 residues. In figure b the phylogenetic distances between the species are expressed in $10^8$ years. The two subgroups of FMCBP --α(dashed lines) and β(dotted lines)-- diverged before the fish and amphibia phyla separated. The rate of evolution (P.A.M./time) of the α subgroup is twice as fast as for β. In figure c the CD, EF precursor of FMCBP and of TN-C is postulated to occur in other calcium mediating proteins. Further it is speculated that the basic EF hand may be found spliced to different basic protein functions (-·-·-·) to generate calcium control enzymes.

# THE ($Na^{+}$+$K^{+}$)-ACTIVATED ENZYME SYSTEM

J. C. Skou

Institute of Physiology
University of Aarhus
8000 Aarhus C
Denmark

## Introduction

There is good evidence that it is the membrane-bound ($Na^{+}$+$K^{+}$)-activated enzyme system which is responsible for conversion of the chemical energy from ATP into a vectorial movement of sodium out and potassium into the cell against electrochemical gradients (see 1). A detailed knowledge about the system and the way it functions may therefore lead to a better understanding of the transport process and more generally to an understanding of how chemical energy from ATP in a biological system is converted into work.

To understand the transport process it is necessary at least to know the answer to the following questions:

1) What is the molecular structure of the system?

2) What is the relationship between the effect of sodium, potassium, magnesium and ATP on the system and the sequence of the steps in the reaction which leads to the hydrolysis of ATP?

3) What happens at the molecular level when the system reacts with sodium, potassium, magnesium and ATP?

These questions cannot be answered at present. But in the following some of the problems related to the questions will be discussed.

## Main Characteristics of the System

The system is located in the cell membrane. It contains an enzyme which can catalyze the hydrolysis of ATP to ADP and Pi in the presence of sodium and potassium (2). Potassium is needed on a site which in the intact cell is facing the external solution, the o-site, and sodium on a site which is facing the internal solution, the i-site (3,4,5). There is a competition between sodium and potassium, both for the i- and the o-site. The number of cations necessary for activation is unknown, but if the activator sites are identical with the carrier sites for the transport of the cations, at least three sodium ions must be necessary on the i-site and two potassium ions on the o-site (see 6).

With the isolated system in the test tube it is not possible to establish an asymmetric situation with a sodium medium in contact with the i-site and a potassium medium in contact with the o-site. The ratio between the concentrations of sodium and potassium necessary to give half maximum saturation of the i-site differs, however, so much from the ratio necessary to give half maximum saturation of the o-site that it is possible with certain concentrations of sodium and potassium in the test tube to have a situation where a major part of the system is on the active $K_m^o/Na_n^i$ form, fig. 1 (o for outside, i for inside, m and n are numbers). The left ascending part of the curve in fig. 1 shows, how sodium activates in the presence of a potassium:sodium concentration ratio which is high enough to saturate the o-site with potassium, i.e. it reflects the effect of sodium on the sodium-site, the i-site of the system. With non-limiting concentrations of ATP, the concentration of sodium necessary to give half maximum activation is 37 mM in the presence of 113 mM potassium, i.e. the sodium:potassium affinity ratio is about 3-4:1. The right part of the curve shows how potassium activates in the presence of a sodium:potassium concentration ratio high enough to saturate the sodium-site with sodium, i.e. the curve reflects the effect of potassium on the potassium-site, the o-site of the system. The concentration of potassium necessary to give half maximum activation is about 1.5 mM in the presence of 148.5 mM sodium, i.e. the sodium:potassium affinity ratio for the o-site is about 1:100. The asymmetry of the curve reflects how much the apparent sodium:potassium affinity ratio for the i-site differs from that of the o-site.

From the sodium:potassium affinity ratios for the two sites it can be estimated that the activity of the system, with the concentrations of sodium and potassium which give maximum activity 130 and 20 mM, respectively, is about 80-85% of the activity which could be obtained with the i-sites in contact with a sodium solution and the o-sites with a potassium solution.

It is furthermore a characteristic, and for the identification of the system very important property, that it is inhibited by cardiac glycosides (see 1), as is the transport system in the intact cell (7).

## Molecular Structure of the Enzyme System

There are two main problems related to the purification of the enzyme system. One is to find means to separate the enzyme system in an active form from the other membrane components. The other is to define a pure system, i.e. decide which components are integrated and necessary parts of the system.

Rich sources of the enzyme system are the electric organ of the electric fish (8), the outer medulla of kidney (9) and the rectal gland of dog fish (10). Detergents have been used in most of the purification procedures. They open the vesicles formed by part of the membranes during homogenisation and by this uncover the latent activity which apparently is due to lack of access of the ligands to both sides of the vesicular membranes (11,12,13). They furthermore dissolve protein from the membranes.

Detergents have been used in low concentrations which uncover the latent activity and dissolve inactive protein but not the enzyme system from the membranes (14). Followed by density gradient and rate centrifugation this has given membrane preparations from outer medulla of rabbit kidney, and with deoxycholate as detergent with a specific activity of about 1500 μmoles Pi per mg protein per hour, and with a molecular activity of about 12,000 Pi per minute (15); and with sodium dodecylsulfate as detergent a preparation with a specific activity of about 2200 and a molecular activity of about 9000 (16). The isolated membranes prepared in this way contain practically no detergent (11).

Detergents have also been used in higher concentrations which "dissolve" the membranes (17-22); this is followed by procedures like density gradient centrifugation, gel filtration and ammonium sulfate precipitation to separate the enzyme system from inactive protein and detergent. The final precipitated enzyme is insoluble in water and it contains still a certain amount of detergent. This has given preparations with a specific activity of about 1500 μmoles Pi per mg protein per hour and a molecular activity of about 6500 Pi per minute from rectal gland of dog fish with lubrol WX as detergent (10) and from outer medulla of canine kidney with deoxycholate (21). From brain it has given a very active but very labile preparation with a specific activity which varies from 700-7000 μmoles Pi per mg protein per hour and with a molecular activity of about 12,000 Pi per minute (22).

The number of enzyme molecules per mg protein, which is necessary to know to calculate the molecular activity, has been determined from the number of g-strophanthin (11,16,21,23-27) and ATP (27-29) binding sites and from the incorporation of $P^{32}$ from $ATP^{32}$ (10,16,18,19,22,23,30). There seems to be a 1:1 relationship between the g-strophanthin and ATP binding sites (16,27) and the determination of the number of enzyme molecules per mg protein is based on the assumption that only one g-strophanthin or ATP molecule is bound per enzyme molecule. The $P^{32}$ incorporation varies from about 1:1 (27,28) to about 2:1 (16, 23) per g-strophanthin and ATP binding site for preparations from different tissues.

The molecular activity for different preparations varies from about 3000 to 15000 Pi per minute (30). In some of the preparative procedures the molecular activity stays constant during the purification (11,16,22) while in others it decreases with the increase in the specific activity (19,20).

There are methodological problems involved in the determination of the number of sites, which may explain some of the differences. Others may be explained as due to species differences. But there is yet another possibility, namely that differences and especially a decrease in molecular activity during purification is due to an inactivation of enzyme molecules without a change in ability to bind the ligands used for site determination.

The system requires lipids besides proteins for activity (for references see 1). It is unknown whether the lipids are specific activators or the lipids form a "solvent" for the proteins. It is also unknown whether specific lipids are necessary for activation or the apparent specificity has to do with the ability of these lipids to reenter the system.

A removal of lipids by detergents or an incorporation of detergent molecules in between the lipid molecules during purification may therefore inactivate without interfering with the site binding capacity of the proteins and give a too low value for the molecular activity. Another possibility is that detergents may remove proteins, for example low molecular weight proteins, which may be necessary for activity but not for binding of the ligands used for site determination. Specific activities can therefore not be used as a basis for a comparison of the composition of different preparations without taking the molecular activity into account.

A problem is that the "true" molecular activity of the "unspoiled" system is not known and it is not known if the "true" molecular activity varies for preparations from different tissues and species.

The above mentioned preparations with specific activities up to 2200 show two main protein bands on sodium dodecyl sulfate polyacrylamide gel electrophoresis (10,16,18-21). One sharp band with a molecular wight of about 95,000 daltons and another more diffuse band with a MW of about 55,000 daltons. $P^{32}$ from $ATP^{32}$ is incorporated in the 95,000 dalton polypeptide under conditions which are specific for the enzyme system (18,20) showing that a polypeptide of this MW is part of the system. The 55,000 dalton polypeptide seems to be a glycoprotein (10,18,21,31).

The highly purified preparation with a specific activity of up to 7000 μmoles Pi per mg protein per hour and a molecular activity of about 12,000 Pi per minute shows only one main protein band with a molecular weight of about 100,000 daltons, while the 55,000 dalton protein band apparently has disappeared (22).

A specific activity of 7,000 and a molecular activity of 12,000 give a molecular weight of about 100,000. It suggests that only one 100,000 MW polypeptide is part of the system (22) and that this preparation may be practically pure. The lipid composition of this preparation is unknown, and it is not possible to exclude that polypeptides with a low molecular weight which are difficult to see on the gel, may also be part of the system besides the high molecular weight polypeptide.

That only one of the $P^{32}$ binding proteins is part of the system disagrees, however, with observations on the less active preparations, which suggest that two of the $P^{32}$ polypeptides are part of the system and that the minimum MW of the system based on protein is 200,000 daltons (10,16).

It is at present not possible to solve the problem, but it must be emphasized that there are problems in the determination of the activity and the protein in the presence of the detergent concentrations used for preparation of the labile, highly active system (see 22).

There is thus still a number of problems to solve to obtain and define a pure preparation as a first and necessary step to obtain information about the molecular structure of the system and of each of the components in the system.

## The Sequence of the Steps in the Reaction

*Effect of ATP on the sodium:potassium affinity of the i-site.* There is a potassium-sodium competition for the sodium-site, the site which in the intact membrane is facing the internal solution, the i-site (see section II). The fraction of the enzyme molecules with the i-site on the sodium form must therefore be a function of the sodium:potassium concentration ratio. It is, however, also a function of the ATP concentration.

Without ATP the apparent affinity of the i-site is about 2.5 times higher for potassium than for sodium. ATP increases the apparent affinity for sodium relative to potassium, and with non-limiting concentrations of ATP the apparent affinity for sodium is about 3 times higher than for potassium, i.e. ATP increases the apparent affinity for sodium relative to potassium 7-8 times (33).

The effect of ATP is seen both without and with magnesium. With magnesium the effect of ATP is given by the total concentration of ATP; at a given total ATP concentration it is independent of a variation in the ratio between free ATP and MgATP (33). It shows that the effect is independent of the catalytic activity, that free ATP and MgATP must have the same effect, and that the site on which they exert their effect must have the same affinity for the two components, i.e. free ATP and MgATP must compete for a common site.

Potassium decreases the affinity of the system for ATP and vice versa, while sodium, if it has an effect, gives a slight increase in affinity (28,29,32). It is potassium on the sodium-site, the i-site of the system, which decreases the affinity for ATP (32). As examplified in the following scheme this can explain that ATP increases the apparent affinity for sodium relative to potassium on the i-site.

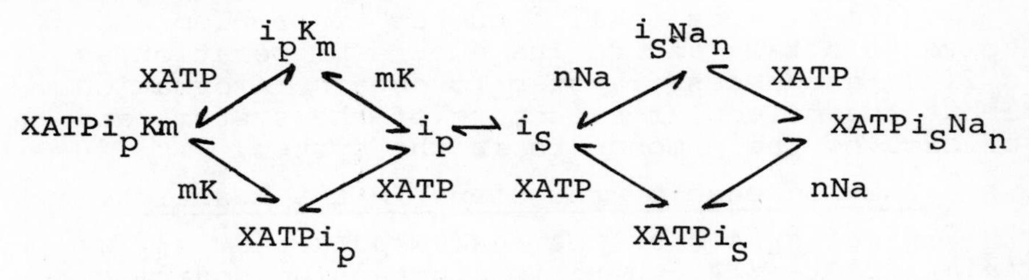


$i_p$ and $i_S$ are the potassium and sodium forms, respectively, of the i-site, m and n are numbers. XATP can either be free ATP or MgATP, and XATP has a lower affinity for the system with the i-sites on the potassium than on the sodium form. Due to this, XATP shifts the equilibrium towards the sodium form at a given sodium:potassium ratio.

The effect on the apparent sodium:potassium affinity ratio is relatively specific for ATP. ITP has practically no effect, 3 mM CTP and GTP has about the same effect as 10 μM ATP; ADP, on the other hand, has an effect like ATP in concentrations which are about two times higher while AMP has no effect (32).

The effect is due to ATP as such, free or complexed with magnesium, and not due to a phosphorylation of the system from ATP or to a hydrolysis of ATP (33).

According to the above given scheme it is the total concentration of ATP independent of magnesium which determines the $K_m i_p:Na_n i_S$ ratio at a given potassium:sodium concentration ratio. The ratio between the $Na_n i_S$ forms which have reacted with free ATP, and with MgATP, is given by the ratio between free ATP and MgATP in the solution. It is, however, only the $Na_n i_S$ forms which have reacted with the one of the two components, free ATP or MgATP, which is the substrate, which can hydrolyze ATP (and transport sodium). It is, however, not possible at present to tell if it is MgATP or free ATP which is the substrate, or with which combinations of $Mg^{2+}$, MgATP and free ATP the system has to react in order to hydrolyze ATP (33).

*The hydrolysis of ATP.* In the presence of sodium but no potassium the reaction of the system with ATP and magnesium leads to a high rate of phosphorylation of the system, followed by a slow rate of dephosphorylation; potassium added to the prephosphorylated enzyme increases the rate of dephosphorylation (34,35). In agreement with this, the system can accomplish a magnesium-sodium-dependent ADP-ATP exchange reaction. The requirement for magnesium for the sodium-dependent exchange reaction is, however, lower than for the overall hydrolysis in the presence of sodium plus potassium, Fahn et al. (36). This has led to the suggestion that two magnesium molecules and two phosphoenzymes are involved in the reaction. One magnesium molecule for which the system has a high affinity and which is necessary for the formation of the phosphorylated intermediate which takes part in the exchange reaction, and which can be dephosphorylated by ADP, but not by potassium, $E_1{\sim}P$ (I and II in the following scheme (36)). And another magnesium for which the affinity is lower and which is necessary for the transformation of $E_1{\sim}P$ into $E_2\text{-}P$, in which the phosphate is in a low energy bond, which is not dephosphorylated by ADP, but by potassium (II, III and IV in the scheme).

$$E + Mg^{2+} \rightleftharpoons Mg\text{-}E \qquad \text{I}$$

$$Mg\text{-}E + ATP \xrightleftharpoons{Na^+} Mg\text{-}E_1{\sim}P + ADP \qquad \text{II}$$

$$Mg\text{-}E_1{\sim}P + Mg^{2+} \rightleftharpoons Mg\text{-}E_1{\sim}P\text{-}Mg \qquad \text{III}$$

$$Mg\text{-}E_1{\sim}P\text{-}Mg \xrightleftharpoons{(Na^+?)} Mg\text{-}E_2\text{-}P\text{-}Mg \qquad \text{IV}$$

$$Mg\text{-}E_2\text{-}P\text{-}Mg + H_2O \xrightarrow{K^+} Mg\text{-}E + Pi + Mg^{2+} \qquad \text{V}$$

This proposal was apparently supported by the observation by Post et al. (37). In experiments at 0°C they found that the phosphoenzyme formed in the presence of a low concentration of magnesium (1 mM magnesium and 3 mM EDTA) was dephosphorylated by ADP while the dephosphorylating effect of potassium was low suggesting that it was $E_1{\sim}P$; when formed with a high magnesium concentration (1mM magnesium without EDTA) it was insensitive to ADP, but sensitive to potassium suggesting that it was $E_2$-P. They also found that NEM-treated enzyme formed $E_1{\sim}P$, but not $E_2$-P.

In experiments at 15°C and with a rapid mixing technique, Fukushima and Tonomura (38) have, however, shown that with the NEM-treated enzyme the formation of the ADP sensitive phosphoenzyme, $E_1{\sim}P$, is preceded by the formation of the ADP insensitive phosphoenzyme, $E_2$-P, in contrast to what was suggested from the scheme by Fahn et al. (36) and Post et al. (37). As pointed out by Fukushima and Tonomura, it means that the ADP insensitive phosphoenzyme, $E_2$-P in the scheme, must also be a high energy form.

Fukushima and Tonomura were not able to confirm the observations by Post et al. that the phosphoenzyme formed in the presence of a low magnesium concentration was ADP sensitive. On the contrary, they found that the phosphoenzyme formed with a low magnesium concentration, 10 μM, was ADP insensitive as the phosphoenzyme formed with a high magnesium concentration. The explanation of the different results in the two sets of investigations seems to be that the formation of the ADP sensitive phosphoenzyme, $E_1{\sim}P$, found in the experiments by Post et al., is due to an effect of the EDTA used to decrease the magnesium concentration and not to the low concentration of magnesium, Klodos and Skou (39). Fukushima and Tonomura concluded from their experiments that the ADP sensitive phosphoenzyme is not an intermediate in the normal reaction. Only one phosphoenzyme is formed in the presence of sodium and magnesium in high or low concentrations, and this phosphoenzyme is insensitive to ADP, but sensitive to potassium, and it is a high energy form. Due to the insensitivity towards ADP they suggested that it is an $E^{\cdot\cdot ADP}{\sim}P$ form.

*Effect of potassium.* Potassium added to a prephosphorylated enzyme system not only increases the rate of hydrolysis of the phosphoenzyme, but also gives rise to formation of ATP, suggesting that potassium influences an

$$EATP \longleftrightarrow E\sim P^{\cdot\cdot ADP} \text{ equilibrium (38).}$$

Potassium in the medium at the start of the reaction together with sodium and magnesium decreases the steady state level of phosphoenzyme which can be obtained, and it becomes practically zero with the concentration of potassium which is optimum for hydrolysis. In experiments where $ATP^{32}$ was used as substrate and with sodium, magnesium and with a suboptimal concentration of potassium which gives a certain level of $E\sim P^{32}$, the addition of cold ATP to stop the formation of $E\sim P^{32}$ leads to a liberation of an amount of $Pi^{32}$ which is larger than the decrease in $E\sim P^{32}$, Kanazawa et al. (40). To explain this, they suggested that there was an irreversible $E_1ATP \longrightarrow E_2ATP$ step involved in the reaction, and that the extra Pi was due to the hydrolysis of ATP bound to $E_2$:

$$E{+}ATP \rightleftharpoons E_1ATP \longrightarrow E_2ATP \rightleftharpoons E\sim P^{\cdot\cdot ADP} \longrightarrow E{+}Pi{+}ADP \qquad VI$$

In this scheme the pathway for the reaction in the presence of sodium plus potassium is the same as in the presence of sodium alone, but potassium influences the different steps in the reaction.

It seems, however, possible to explain the above mentioned results without assuming an irreversible $E_1ATP \longleftrightarrow E_2ATP$ step. The hydrolysis of ATP by the system on the potassium-sodium form may follow a pathway which is different from the pathway by the system on the sodium form, and it may precede without formation of a phosphoenzyme (35). There are some experiments which suggest that the i- and the o-site on the system exist simultaneously (41,42,43). This would mean that the system in the presence of sodium alone is on the $Na^o_m/Na^i_n$ form, while with potassium plus sodium the enzyme molecules will consist of a mixture of the $K^o_m/Na^i_n$, $K^o_m/K^i_n$, $Na^o_m/Na^i_n$ and $Na^o_m/K^i_n$ forms (assuming that no hybrid forms exist of the i- and the o-sites) (1,44). The ratio between the different forms is determined by the sodium:potassium affinity ratio for the o- and the i-sites, which for the i-site depends on the ATP concentration, and on the sodium:potassium concentration ratio in the medium.

As discussed in a previous section, the apparent potassium:sodium affinity ratio is about 100:1 for the o-site and 0.3:1 for the i-site in the presence of non-limiting concentrations of XATP (free ATP or MgATP). With a high sodium concentration, a suboptimal potassium concentration and a saturating concentration of ATP and magnesium (e.g. 150 mM sodium, 1.5 mM potassium, 3 mM magnesium and 3 mM ATP) the main part of the enzyme molecules then consists of a mixture of the $K^o_m/Na^i_n$ and the $Na^o_m/Na^i_n$ forms.

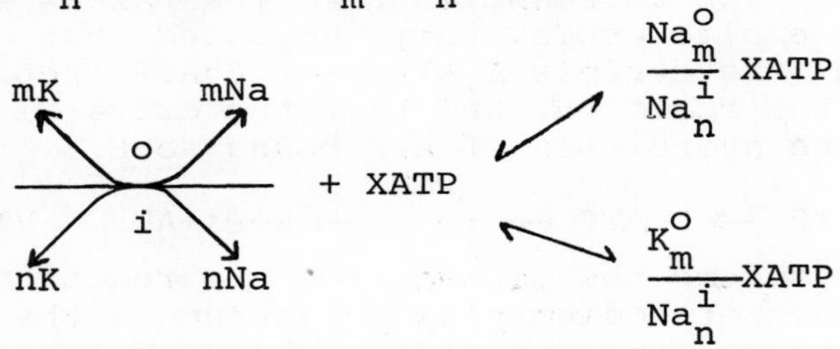


Both forms hydrolyze ATP and according to the above discussed, the $Na^o_m/Na^i_n$ form hydrolyzes ATP via formation of an ADP insensitive phosphoenzyme, $Na^o_m/Na^i_n \cdot\cdot ADP \sim P$.

With both forms in the test tube, the amount of $Pi^{32}$ released is larger than, the decrease in the phosphoenzyme when $P^{32}$ labelling from $ATP^{32}$ is stopped by adding cold ATP (see above). In a two-pathway reaction this can be explained without implying an irreversible $E_1ATP \longrightarrow E_2ATP$ step by assuming that the hydrolysis along the pathway due to the $K^o_m/Na^i_n$ form proceeds without formation of a phosphoenzyme. But via formation of a transient enzyme-ATP complex in which the γ-phosphate is bound in a bond which is intermediary between an electrostatic and a covalent bond. Formation of the complex leads to and it only exists while potassium from outside is exchanged for sodium from inside and the exchange of the cations is intimately connected to the formation of the complex and the hydrolysis of ATP.

$$\frac{Na^o_m}{Na^i_n} + ATP \leftrightarrows \frac{Na^o_m}{Na^i_n}ATP \leftrightarrows \frac{Na^o_m \cdot ADP}{Na^i_n \sim P} \rightarrow \frac{Na^o_m}{Na^i_n} + Pi + ADP \qquad IX$$

$$\frac{K^o_m}{Na^i_n} + ATP \leftrightarrows \frac{K^o_m}{Na^i_n}ATP \leftrightarrows \frac{K^o_m \cdot ADP}{Na^i_n \cdot P} \xrightarrow{Na^+ \; K^+} \frac{Na^o_n}{K^i_m} + Pi + ADP \qquad X$$

Magnesium is necessary for the reaction, but it has been omitted for two reasons. One is that it is unknown which combination of $Mg^{2+}$, MgATP and free ATP it is that the system reacts with, and in which sequence. Another is that it is unknown whether the system on the $Na^o_m/Na^i_n$ form reacts with the same combination of the three components and in the same sequence as on the $K^o_m/Na^i_n$ form.

At present it is not possible to decide between the reaction schemes.

## The Molecular Events Related to the Reaction

Due to lack of a pure system and therefore lack of information about the molecular structure of the system it is not possible to relate the steps in the reaction to molecular events. It is neither possible to explain how the system can discriminate between sodium and potassium nor to explain how the cations are translocated.

## Conclusion

The localization of the transport process to a certain enzyme system in the cell membrane has opened the possibilities of obtaining information about the transport process. But there is still a long way to go before it will be possible to answer the question raised in the introduction and thereby to understand the transport process.

## References


1. Skou, J.C. J. Bioenergetics 4 (1973) 203
2. Skou, J.C. Biochim. Biophys. Acta 23(1957)394.
3. Glynn, I.M. J. Physiol. (Lond.) 160(1962)18P.
4. Laris, P.C. and Letchworth, P.E. J. Cell. Comp. Physiol. 60 (1962) 229.
5. Whittam, R. Biochem. J. 84 (1962) 110.
6. Glynn, I.M. Br. Med. Bull. 24 (1968) 165.
7. Schatzmann, H.J. Helv. Physiol. Acta. 11 (1953) 346.
8. Albers, R.W., Fahn, S.L. and Koval, G.J. Proc. Nat. Acad. Sci USA 50(1963) 474
9. Jørgensen, P.L. and Skou, J.C. Biochem. Biophys. Res. Commun. 37 (1969) 39
10. Hokin, L.E., Dahl, J.L., Deupree, J.D., Dixon, J.F., Hackney, J.F. and Perdue, F. J. Biol. Chem. 248 (1973) 2593.
11. Jørgensen, P.L. and Skou, J.C. Biochim. Biophys. Acta. 233 (1971) 366.
12. Møller, O.J. Exp. Cell Res. 68 (1971) 347.
13. Rostgaard, J. and Møller, O.J Exp. Cell. Res. 68 (1971) 356.
14. Skou, J.C. Biochim. Biophys. Acta 58 (1962) 314.
15. Jørgensen, P.L., Skou, J.C. and Solomonson, L.P. Biochim. Biophys. Acta 233 (1971) 381
16. Jørgensen, P.L Annals N.Y. Acad. Sci. In press.
17. Towle, D.W. and Copenhaver, J.H.Jr. Biochim. Biophys. Acta 150 (1968) 41.
18. Uesugi, S., Dulak, N.C., Dixon, J.F., Hexum, T.D., Dahl, J.L., Perdue, J.F. and Hokin, L.E. J. Biol. Chem. 246 (1971) 531.

19. Kyte, J. J. Biol. Chem. 246 (1971) 4157.

20. Kyte, J. Biochem. Biophys. Res. Commun. 43 (1971) 1259.

21. Lane, L.K., Copenhaver, J. H., Lindenmayer, G.E. and Schwartz, A. J. Biol. Chem. 248 (1973) 7197.

22. Nakao, T., Nakao, N., Nagai, F., Kawai, K., Fujihira, Y., Hara, Y. and Fujita, M.J. Biochem. (Tokyo) 73 (1973) 781.

23. Albers, R.W., Koval, G.J. and Siegal, G.J J. Mol. Pharmacol. 4 (1968) 324

24. Elleroy, J.C. and Keynes, R.D. Nature 221 (1969) 776.

25. Baker, P.F. and Willis, J.S. Nature 226 (1970) 521

26. Hansen, O. Biochim.Biophys.Acta 233(1971)122.

27. Hansen, O., Jensen, J. and Nørby, J.G. Nature, New Biol. 234 (1971) 122.

28. Hegyvary, C. and Post, R.L. J. Biol. Chem. 246 (1971) 5234.

29. Nørby, J.G. and Jensen, J. Biochim. Biophys. Acta 233 (1971) 104

30. Bader, H., Post, R.L. and Bond, G.H. Biochim. Biophys. Acta 150 (1968) 41.

31. Kyte, J. J. Biol. Chem. 247(1972) 7642.

32. Skou, J.C. Biochim.Biophys.Acta. In press.

33. Skou, J.C. Biochim.Biophys.Acta. In press.

34. Post, R.L., Sen, A.K. and Rosenthal, A.S. J. Biol. Chem. 240 (1965) 1437.

35. Skou, J.C. Physiol. Rev., 45 (1965) 596.

36. Fahn, S., Koval, G.J. and Albers, R.W. J. Biol. Chem., 241 (1966) 1882.

37. Post, R.L., Kume, S., Tobin, T., Orcutt, B. and Sen, A.K. J. Gen. Physiol. 54(1969)306s.

38. Fikushima, Y. and Tonomura, Y. J. Biochem. (Tokyo) 74 (1973) 135.

39. Klodos, I. and Skou, J.C., Unpublished.

40. Kanazawa, T., Saito, M. and Tonomura, Y.J. Biochem. (Tokyo), 67 (1970) 693.

41. Hoffman, P.G. and Tosteson, D.C. J. Gen. Physiol. 58 (1971) 438.

42. Garay, R.P. and Garrahan, P.J. J. Physiol. (Lond.) 231 (1973) 297.

43. Skou, J.C. Biochim. Biophys. Acta. In press.

44. Skou, J.C. 1971. In: Current Topics in Bioenergetics. Vol. 4. p. 357. ed. D.R. Sanadi Academic Press. N.Y.

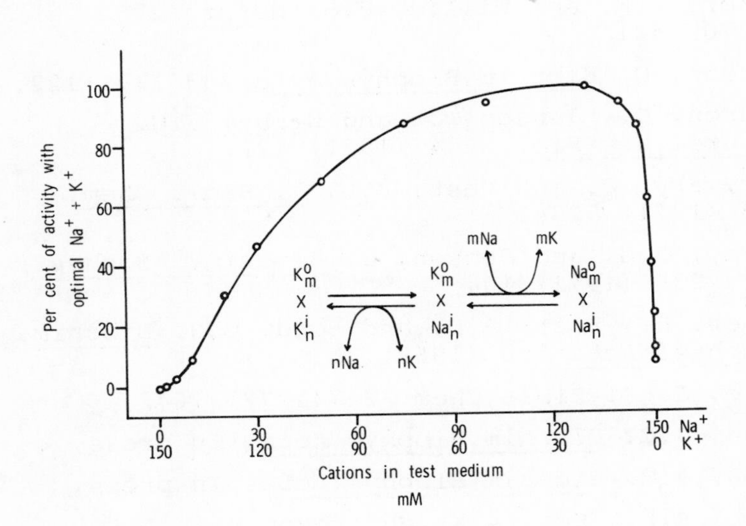


Figure 1. Activity of $(Na^+ + K^+)$ activated enzyme system as function of sodium plus potassium. The enzyme was prepared from ox brain; a specific activity was measured with 3 mM ATP, 3 mM magnesium, pH 7.4, 37°C with the concentrations of sodium and potassium shown on the abcissa.

# III

# THE CONSERVATION OF ENERGY IN MEMBRANES

# THE MECHANISM OF ALKALI METAL CATION TRANSLOCATION IN MITOCHONDRIAL MEMBRANES

Sergio Estrada-O.

Departamento de Bioquimica
Centro de Investigación y de
Estudios Avanzados del I.P.N.
Apartado Postal 14-740
México, D.F.

> No one believes an hypothesis excepts its originator but everyone believes an experiment except the experimenter....
>
> *W. Beveridge. The Art of Scientific Investigation.*

The transport of ions is an event intimately associated with the process of energy conservation occurring in the majority of living organisms of the biosphere. Although the nature of the primary event in energy conservation still remains to be elucidated, it is apparent that components of the electron transport chain and ATP-ase systems from mitochondria, chloroplasts or photosynthetic bacteria directly participate in mediating proton translocation across such lipid membranes (1,2,3,4). Proton transport can be coupled to promote ATP synthesis or electron flow against redox potential gradient in energy transducing membranes (5,6,7). Thus, despite the fact that none of the above proton translocating systems can be adequately described in molecular terms (i.e. channels or mobile translocators) it seems easy to envisage the possible intracellular functions of proton transport (1,2,5).

A somewhat different picture emerges on trying to define the role of alkali metal cation translocation in energy conserving membranes. Its molecular mechanism still remains obscure. Moreover, apart from its possible relationship to energy conservation, the overall physiological significance of intracellular cation movements is by no means obvious.

*The fiction of an energy-dependent translocation of $Na^+$ or $K^+$ occurring across mitochondrial membranes.* Twenty years ago, Spector (8), Stanbury and Mudge (9) and McFarlane & Spencer (10) were the first to show that isolated mitochondria are able to retain alkali metal cations against a concentration gradient. These observations as well as further work by Davies (11), Judah et al (12), Share (13) and Gamble (14) indicated the need of oxidizable substrates or ATP for the organelles to maintain their internal $K^+$ content constant. These data (cf. 15) as well as more recent work carried out in intact mitochondria with neutral $K^+$ ionophores of the valinomycin group (16,17) were considered as evidence for the existence of a natural $K^+$ pump in the mitochondrial membranes (17). However, it has been clearly documented in recent years (for a review see ref. 2), that the transport of ions stimulated by ionophorous antibiotics in mitochondria, chloroplasts or photosynthetic bacteria does not validate the existence of an energized $K^+$ or $Na^+$ pump in energy-conserving membranes. If the chemiosmotic hypothesis (1,2) is correct, the effects of valinomycin-like ionophores on the process of coupled phosphorylation of energy-conserving membranes can easily be interpreted on the basis of their well known molecular properties. In fact, "ion pump" mechanisms for $Na^+$ or $K^+$, (16,17) require in mitochondrial and chloroplast membranes a number of assumptions that seem to lack justification. Nonetheless, it is fair to mention that the data obtained with some antibiotic translocators in model systems (18) do not completely rule out the above-mentioned possibility.

*Alkali ion translocation is mediated by a cation/proton exchange system in liver mitochondria.* Mitchell and Moyle (19) were the first to indicate the existence of an electrically silent alkali metal cation/$H^+$ exchange system naturally operating in the inner mitochondrial membrane. As a result of their examination of the kinetics of proton movements (20) and the passive swelling of liver mitochondria treated with respiratory inhibitors and incubated in isotonic salts of $Na^+$ or $K^+$ (21), these authors proposed the existence of a cation/$H^+$ antiport translocator with a large preference for $Na^+$ over $K^+$. Data by Cockrell and Racker (22) and by Cockrell (23), indicating the natural occurrence of a $K^+$ uptake process in beef heart submitochondrial sonic vesicles (A-particles), further strenghtened Mitchell's suggestions. Also in support of the $Na^+$>$K^+$ selectivity of such cation/$H^+$ translocator was the observation (24) that $Na^+$>$K^+$ inhibits $H^+$ uptake in submitochondrial sonic vesicles ($Mg^{2+}$-ATP particles). Moreover, direct cation flux measurements carried out in rat liver submitochondrial sonic particles by Douglas and Cockrell (25), have recently shown that mitochondrial membranes do indeed possess an $Na^+$/$H^+$ exchanger. Nevertheless, the studies so far oriented to disclose in intact mitochondria the aforementioned antiport translocator have a number of limitations. Most important in this respect is the fact that the measurement of either ion fluxes or swelling variations in intact mitochondria (20,21,25) have required either long term incubation or anaerobic conditions, circumstances which may obviously alter the functional properties of the mitochondrial membranes (25).

Work in our laboratory (26) using the antibiotic monazomycin (27) provided in 1971 a clue to the existence of the $Na^+$>$K^+$/$H^+$ antiport translocator in intact mitochondria actively metabolizing oxidizable substrates. Monazomycin is a channel-forming ion translocator (18) which contains 16 OH-groups, 1 galactose, 1 dissociable amino group and no aminoacids (28). Its selectivity for transporting ions across model membranes is similar for $Na^+$ and $K^+$ (18). If ATP and acetate or phosphate are present in the medium the translocator stimulates either $Li^+$, $Na^+$, $K^+$, $Rb^+$, $Cs^+$ or $H^+$ uptake into mitochondria with only minor variations in the cationic selectivity of the transport process (26). However, when energy is provided by succinate oxidation in media free of added phosphate or acetate, all alkali ions except $Na^+$ inhibited the efflux of $H^+$ and the oxidation of substrate stimulated by the antibiotic. The uncoupling of respiration caused by $Na^+$ and monazomycin in the absence of added penetrant anions suggested to us the parallel functioning of the antibiotic, and a $Na^+$ more active than $K^+/H^+$ antiport translocator present in the mitochondrial membrane (26).

A further strong support for disclosing the existence of a membrane-located $Na^+>K^+/H^+$ ion exchanger came through the examination of the possible mechanism by which the antibiotic beauvericin mediates alkali ion movements across mitochondrial membranes (29,30). Beauvericin is a cyclic hexadepsipeptide ion translocator containing three D-α-hydroxyisovaleryl and three N-methyl-L-phenylalanine residues (31). Its ionic selectivity measured by conductometric techniques (32) is: $Rb^+>Cs^+>K^+>Na^+>Li^+$; in bulk partition measurements the antibiotic extracts $Na^+$, $K^+$ and $Rb^+$ with similar effectiveness while $Li^+$ about one-third as effectively as $Na^+$ (33). Nonetheless, in contrast to the above mentioned $Na^+$, $K^+>Li^+$ selectivity, Dorschner and Lardy (29) found that beauvericin apparently induced the accumulation of $Li^+$ and $K^+$ more than of $Na^+$ in intact mitochondria-oxidizing glutamate.

Experiments carried out in collaboration with Gómez-Lojero and Montal (30) provided an explanation for the apparent variability in the ionic discrimination manifested by beauvericin in liver mitochondria. It was observed that by measuring the oxygen-uptake rates linked to alkali ion movements the antibiotic showed distinct transitions in the $Na^+$-$K^+$ selectivity as a function of the type of carboxylic substrate oxidized and the presence of inorganic phosphate in the medium. It was observed that in the absence of added phosphate the substrates oxidized in membrane-bound enzymes, such as succinate, (34) allowed a $Na^+ > K^+$ selectivity whereas those oxidized in the mitochondrial matrix space, such as glutamate, (34) allowed a $K^+ > Na^+$ discrimination. Moreover, when the media are supplemented with phosphate, the $Na^+$-$K^+$ discrimination of beauvericin is considerably modified by respiratory substrates, giving a $K^+ > Na^+$ selectivity with succinate and $Na^+ > K^+$ with glutamate plus malate. As will be noted below, it is likely that the substrate-dependent transitions in $Na^+$-$K^+$ selectivity observed may be related to one of the proposed cellular functions of the mitochondrial cation/$H^+$ exchanger (2,25), namely the control of oxidizable substrate anion translocation between mitochondria and the cytosol.

*A natural $Na^+ > K^+/H^+$ antiport translocator is responsible for the transitions in $Na^+$-$K^+$ selectivity observed for monazomycin and beauvericin in mitochondria.* Figure 1 provides a scheme where an explanation of the $Na^+$-$K^+$ selectivity transitions of monazomycin and beauvericin in mitochondria emerges. In accordance with the chemiosmotic mechanism (1,2), upon initiation of electron transfer reactions, $H^+$ move out from the mitochondria. This charge migration builds up a gradient associated to the electrochemical activity of $H^+$ across the membrane. This transmembrane proton-motive force consists of two components: a concentration term due to the differential distribution of $H^+$ across the membrane (a pH gradient) and an electrical term due to charge transfer across the membrane (an electrical gradient). Both beauvericin and monazomycin are able to translocate into mitochondria down the electrochemical gradient with similar effectiveness either $Na^+$ or $K^+$. When either of the antibiotics is joined in parallel with a natural $Na^+/H^+$ antiporter more active than a $K^+/H^+$ translocator (2,19,25, 26,30), a marked increase in the turnover of $Na^+$ over $K^+$ occurs. Under these conditions, the antiport catalyzes the efflux of $Na^+$. The concerted action of the antibiotics and the $Na^+$ selective antiport system provide a situation in which the membrane is "short-circuited", $Na^+$ and $H^+$ being continuously moved in and out, with the consequent dissipation of the protonmotive force generated during respiration. Thus the rate-limiting step for the uncoupling is the translocation of a monovalent cation/$H^+$. This mechanism is very similar to that promoted by the combination of valinomycin plus nigericin both in submitochondrial particles (22,23) or in intact liver mitochondria supplemented with succinate in the absence of phosphate (16). When $Na^+$ is replaced by $K^+$, uncoupling is obtained only in the presence of phosphate, indicating that if a $K^+/H^+$ antiporter exists, it is less active than the $Na^+/H^+$ exchanger; therefore, uncoupling requires the translocation of phosphoric acid in order to collapse the transmembrane pH gradient (30). With glutamate or glutamate plus malate the additional constraint of limited substrate anion translocation imposes a requirement for phosphate (35,36,37) to show the beauvericin + $K^+$-dependent uncoupling effect.

*The rate of alkali ion transport mediated by the cation/proton exchanger is controlled by the ΔpH existent across the mitochondrial membrane.* The rate of $Na^+$ versus $K^+$ transport is affected in mitochondria not only by the relative concentration of $Na^+$ and $K^+$ (25) but also by the type of respiratory substrate oxidized (30) and by the differential distribution of protons across the membrane. As shown in Fig. 2, beauvericin mediates a large pH-dependent efflux of $H^+$ in media containing $K^+$ and succinate, whereas it does not cause a significant $H^+$ outflux in $Na^+$-containing media. This observation is entirely compatible not only with previous findings by Mitchell and Moyle (20,21) but also with the scheme displayed in Fig. 1, since the presence of a highly selective $Na^+/H^+$ antiport system in the mitochondrial membrane tends to facilitate $Na^+$ efflux in exchange for protons that would be accumulated into mitochondria.However, it is of great interest that the apparent $Na^+$-$K^+$ discrimination of the antibiotic plus antiport transport system is also affected by $H^+$ concentration differences existent across the mitochondrial membrane. Table I indicates that the $Na^+>K^+$ selectivity of the oxidation of succinate stimulated by beauvericin is reversed to a $K^+>Na^+$ discrimination by the independent addition of either one of two electrically silent cation/proton exchangers such as nigericin and monensin A. Nigericin (38) is a highly selective $K^+/H^+$ exchanger (38,17) whereas monensin A is more selective for mediating $Na^+/H^+$ exchange than $K^+/H^+$ translocation (17,39). Yet despite their different cationic selectivity both carboxylic ionophores promoted a $K^+>Na^+$ sequence for succinate oxidation in the presence of beauvericin. It is likely that if the rate-limiting step in the uncoupling is the exchange of monovalent cation/$H^+$ the addition of exchanger would release this constraint. Since their cationic selectivity differs, however, the only property that obth carboxylic antibiotics have in common is their ability to modify the $\Delta pH$ change without causing net charge transfer. Thus, a reasonable assumption is to consider that the $\Delta pH$ change caused by the carboxylic ionophores may be primarily involved in the observed transition of $Na^+$:$K^+$ selectivity. In fact, Table II strongly indicates that such is the case. It is apparent that at near neutral pH, beauvericin mediates a $Na^+>K^+$ selectivity with succinate while allowing a $K^+>Na^+$ discrimination with glutamate plus malate. However, when the pH of the medium is rapidly lowered to 6.2 the selectivity found for succinate reverts to a $K^+>Na^+$ sequence. The glutamate plus malate pair maintains its $K^+>Na^+$ discrimination at acid pH. These findings are entirely consistent with previous data presented by Mitchell and Moyle (20,21) which indicate that the mitochondrial membrane significantly increases its permeability to $K^+$ at acid pH values.

*Alkylguanidine derivatives selectively inhibit the concerted ion movements mediated by antibiotics of wide cation selectivity and the natural $Na^+/H^+$ antiport translocator.* Antibiotics which like monazomycin and beauvericin (18,26,30) exhibit a poor discrimination between $K^+$ and $Na^+$ in lipid membranes, show drastic modifications in their apparent mitochondrial selectivity patterns dependent on pH and the nature of the anionic species present. This is in contrast to the negligible effects of anions and protons on the ionic selectivity of ionophores highly discriminative for $K^+$ such as valinomycin, and the nonactin homologs (17). Clearly, this difference implies that the former but not the latter group of antibiotics are able to mediate ion fluxes in parallel with the $Na^+/H^+$ translocator present in the inner mitochondrial membrane.

Aditional evidence to support the above statement is provided by the use of alkylguanidine derivatives as inhibitors of the ion movements mediated by the assembly of monazomycin or beauvericin in parallel with the natural antiporter existent in mitochondria.

As shown in Fig. 3, 300 μM octylguanidine completely inhibits the uptake of $K^+$ as well as the extrusion of H+ (caused by the efflux of $H^+$ and the hydrolysis of ATP) promoted by both monazomycin or beauvericin. Octylguanidine does not affect ATP hydrolysis or $K^+$ movements mediated by valinomycin or nonactin in the presence of ATP (Fig. 3). A quantitative estimate of the inhibition by octylguanidine of the $K^+$ influx and the ATP-dependent hydrogen ion concentration changes promoted by beauvericin is shown in Fig. 4. It is clear that lower concentrations of octylguanidine are required to inhibit $K^+$ uptake with respect to those needed to block ATPase activity.

Since the inhibitory effect of octylguanidine is exerted with ATP or succinate (not shown) as energy source for transport, it is apparent that the guanidine derivative is not preventing $K^+$ uptake by selectively blocking electron or energy transfer at the NADH-cytochrome b segment of the respiratory chain (40,41).

It is very likely that the effects of alkylguanidines on mitochondrial ion transport may be related to non-specific actions associated with their relative hydrophobicity and their general cationic properties. However, if such is the case, it is not immediately apparent why octylguanidine shows a striking pH-dependent ability to block the respiration linked to the movements of $K^+$ but not those of $Na^+$ promoted by beauvericin in mitochondria (Fig. 5). Moreover, in view of the results obtained with aliphatic substituted ammonium ions for the inhibition of the $K^+$ channel of nerve (42), it is tempting to suggest that alkylguanidines may be useful tools for helping to elucidate the mode of action of natural or model ion translocators in biological membranes. Thus, it is reasonable to consider that the rapid efflux of $K^+$ and the inhibition of respiration caused in *Escherichia coli* K-12 by guanidine-containing antibiotics such as streptomycin and neomycin (43, 44) as well as the efflux of $K^+$ and the inhibition of respiration caused by octylguanidine in yeast cells of *Saccharomyces cerevisiae* (45) could eventually be interpreted in the context of the latter proposal. In fact, the use of a substituted ammonium ion such as the alkyl guanidine compounds may be advantageous to elucidating the ionic selectivity fingerprints of natural or model ion translocators. As Hille and Eisenman et al have shown, different substituted ammonium ions not only replace $Na^+$ for the process of transport in the $Na^+$ channel of nerve, (46) but also form stable complexes with some ionophorous antibiotics (47).

*Selective inhibition by $Mg^{2+}$ of the mitochondrial $Na^+ > K^+/H^+$ antiport translocator.* As shown in Fig. 6A, concentrations of $MgCl_2$ below 5 mM inhibit the increase in the turnover of $Na^+$ but not that of $K^+$, promoted by the concerted action of beauvericin and the mitochondrial $Na^+/H^+$ antiport translocator. Thus, it is apparent from our work that $Mg^{2+}$ inhibits with an striking $Na^+ > K^+$ selectivity the alkali ion/$H^+$ exchange which takes place in intact mitochondrial membranes. It is very unlikely that the clearcut $Na^+/K^+$ selectivity changes observed could be mediated by $Mg^{2+}$ through a primary inhibition of the beauvericin mode of action. The known lack of discrimination of the antibiotic for complexing $Na^+$ or $K^+$ (32, 33) would be difficult to reconcile with such a possibility; nonetheless, the transitions of ionic selectivity observed (Fig. 6 A-B) are almost identical to those found in intact mitochondria for the operation of the alkali ion/$H^+$ antiport translocator. Thus, it is apparent that $Mg^{2+}$ primarily acts upon the microenvironment of the endogenous $Na^+/H^+$ antiport translocator rather than on the mode of action of beauvericin to transport ions.

As shown by Douglas and Cockrell (25) $Mg^{2+}$ does not inhibit but markedly stimulates $Na^+ > K^+$ movements in submitochondrial sonic vesicles (A particles) where the directionality of cation and $H^+$ fluxes is opposite to that observed in the intact membranes (22,23). It is clear that the divalent ion inhibits $Na^+ > K^+$ uptake in intact mitochondria, stimulating $Na^+ > K^+$ influx in submitochondrial sonic particles. Therefore, it seems logical to propose that the above described effects indicate an important physiological role of $Mg^{2+}$ in the cellular control of $Na^+$, $K^+$ and $H^+$ fluxes across energy-conserving membranes.

*The importance of the mitochondrial $Na^+ > K^+/H^+$ antiport translocator in metabolic regulation.* The distribution and flow of anionic metabolites between mitochondrial and cytoplasmic compartments are important functions in metabolic regulation. Studies of isolated mitochondria have indicated that gradients of various carboxylic acid substrates may be determined either by cation (48) or proton fluxes (49). Despite several suggestive reports which indicate the requirement of $K^+$ for the uptake of some oxidizable substrates in mitochondria (48), there is as yet no substantial evidence to indicate that cation movements per se represent the primary driving force which facilitates such anion translocation. On the other hand, anion fluxes readily respond to variations in the ΔpH existent across the mitochondrial membrane (49). Clearly, the main action of an active cation/proton antiport translocator would be to generate transmembrane pH gradients elicited as a function of its relative selectivity for $Na^+$ or $K^+$. Thus, the mitochondrial cation/$H^+$ exchanger emerges as the main controlling factor determining the rate and extent of distribution of a great variety of anionic substrates between mitochondria and the cytosol. The electrically silent properties of the mitochondrial cation/$H^+$ exchange process further helps to prevent not only drastic changes in the mitochondrial volume but also the potential uncoupling of phosphorylating respiration due to the uptake of anionic substrates.

The above proposal is also consistent with recent data by Dubinsky and Cockrell (50) which indicate that $Na^+$ but not $K^+$ significantly decreases the oxidation of a variety of substrates in isolated liver cells. As suggested by these authors, lower uncoupled respiratory rates in $Na^+$ -treated cells may reflect the competition of a mitochondrial $Na^+/H^+$ exchanger with substrate uptake for the mitochondrial transmembrane pH gradient.

*Some perspectives for the study of ion translocators in biological membranes.* Critical to the understanding of the mechanism of ion transport in energy conserving membranes is not only the isolation and characterization of the mitochondrial cation/$H^+$ antiport translocator but also its complete reconstitution in natural and model membranes. In this respect, claims have been raised (51) for the isolation of a valinomycin-like neutral $Na^+/K^+$, ionophore from beef heart mitochondrial membranes. Nonetheless, the results described here as well as those of Douglas & Crockrell (25) are incompatible with the existence of a uniport-type of $Na^+$ or $K^+$ pump translocator in liver mitochondria. It is possible to predict that the alkali metal cation translocator from mitochondrial membranes would be an anionic low molecular weight ionophore (< 2000 m.w.) similar in its mode of action to some carboxylic antibiotics of the nigericin family (38, cf. 2). The translocator could exist in the membrane in free form (as a mobile carrier) or else anchored (in channel form) to native membrane elements such as membrane proteins or lipids. The anchored form of the translocator would have a more favourable partition into the lipid membrane, since mobile translocators tend to have a discrete partition into water when solubilized into lipid-water interphases.

The reconstitution of the antiport translocator in model or natural membranes is one of the main goals to pursue in mitochondrial membrane research in the near future. In this respect, it seems appropriate to postulate, for its eventual isolation and characterization, that the antiport translocator would have many of the physical and chemical properties of some channel-forming or ionophorous antibiotics. Moreover, it is tempting to suggest that the transport of ions may be catalyzed by low molecular weight antibiotic-like translocators not only in energy conserving membranes (mitochondria, chloroplast and photosynthetic bacteria) but also in the membranes from large populations of microorganisms and yeast cells. For the case of ATPases specialized in the translocation of $Na^+$-$K^+$, $H^+$ or $Ca^{2+}$, it is likely that a polypeptide (channel-like) moiety of low molecular weight, different from that of the catalytic site and the prosthetic group of the enzyme, could be the molecular segment responsible for the translocation of ions. That this latter proposal may be true is strongly suggested by the isolation and reconstitution in lipid bilayers of a $Na^+$-specific ionophore (molecular weight of 2000) extracted by Shamoo (52) from $Na^+$-$K^+$ ATPase preparations.

## Acknowledgements

The author kindly acknowledges the expert collaboration of Miss Guadalupe Gallo and Mrs. Esthela Calderón in the above referred experiments. Also the criticism of Drs. Carlos Gómez-Lojero and Mauricio Montal.

## References

1. Mitchell, P. 1966. In: Chemiosmotic Coupling in Oxidative and Photosynthetic Phosphorylation. p. 26. Bodmin, Cornwall, England.
2. Mitchell, P. 1968. In: Chemiosmotic Coupling and Energy Transduction. p. 9. Bodmin, Cornwall, England.
3. Hinkle, P.C., Kim, J.J. and Racker, E. J. Biol. Chem. 247 (1972) 1338.
4. Thayer, W.S. and Hinkle, P.C. Fed. Proc. Abstracts (1973) 2568.
5. Skulachev, P. 1971. In: Current Topics in Bioenergetics. Vol. 4, p. 127. ed. Sanadi, R. D. Academic Press, New York.
6. Racker, E. 1974. In: Perspectives in Membrane Biology. eds. Estrada-O., S. and Gitler, C. This volume.
7. Crofts, A.R. 1974. In: Perspectives in Membrane Biology. eds. Estrada-O., S. and Gitler, C. This volume.
8. Spector, W.G. Proc. Roy. Soc. B141(1953)268.
9. Stanbury, S.W. and Mudge, G.H. Proc. Soc. Exp. Biol. Med. 82 (1953) 675.
10. McFarlane, M.G. and Spencer, A.G. Biochem. J. 54 (1953) 569.
11. Bartley, W. and Davies, R.E. Biochem. J. 96 (1965) 1c.
12. Christie, G.S., Ahmed, K., McLean, A.E.M. and Judah, J.D. Biochim. Biophys. Acta 94 (1965) 432.
13. Share, L. Am. J. Physiol. 194 (1958) 47.

14. Gamble, J.L. Jr. J. Biol.Chem. 228(1957)955.

15. Harris, E.J., Judah, J.D. and Ahmed, K. 1966 In: Current Topics in Bioenergetics. vol. 1, p. 255. ed. Sanadi,R.D. Academic Press. N. Y.

16. Pressman, B.C., Harris, E.J., Jagger, W.S. and Johnson, J.H. Proc. Nat. Acad. Sci. U.S.A. 58 (1967) 1949.

17. Pressman, B.C. Fed. Proc. 27 (1968) 1283.

18. Mueller, P. and Rudin, D.O. 1966. In: Current Topics in Bioenergetics. vol. 3, p. 157, ed. Sanadi, R.D. Academic Press, New York.

19. Mitchell, P. Biol. Rev. 41 (1966) 445.

20. Mitchell, P. and Moyle, J. Biochem. J. 105 (1967) 1147.

21. Mitchell, P. and Moyle, J. Eur. J. Biochem. 9 (1969) 149.

22. Cockrell, R. and Racker, E. Biochem. Biophys. Res. Commun. 35 (1969) 414.

23. Cockrell, R. J. Biol. Chem. 248 (1973)6828.

24. Papa, S., Guerrieri, F., Simone, S., Lorusso, M. and Larossa, D. Biochim. Biophys. Acta. 292 (1973) 20.

25. Douglas, M.G. and Cockrell, R. J. Biol. Chem. (1974) In press.

26. Estrada-O., S. and Gómez-Lojero, C. Biochemistry Wash. 10 (1971) 1548.

27. Akasaki, K., Karasawa, K., Watanabe, M., Yonehara, H. and Umezawa, H. J. Antibiot. Ser. A 16(1963) 127.

28. Mitscher, L.A., Shay, A.J. and Bohonos, N. Appl. Microbiol. 15 (1967) 1002.

29. Dorschner, E. and Lardy, H.A. 1968. In: Antimicrobiol. Agents and Chemotherapy. p. 11, ed. Hobby, G.L. Ann. Arbor, Mich.

30. Estrada-O., S., Gómez-Lojero, C. and Montal, M. Bioenergetics 3 (1972) 417.

31. Hammil, R.L., Higgins, H., Boaz, E. and Gorman, M. FEBS Letters 49 (1969) 4255.

32. Ovchinnikov, Y.A., Ivanov, V.T. and Mikhaleva, I.I. FEBS Letters 2 (1971) 159.

33. Roeske, R.W., Issac, S., Steinrauf, L.K. and King, J. Fed. Proc. 20 part II (1971) 1340.

34. Sottocasa, G.L., Kulyenztierna, B., Ernster, L. and Bergstrand, A. J. Cell. Biol. 32 (1967) 415.

35. Henderson, P.F.J., McGivan, J.D. and Chappell J. B. Biochem. J. 111 (1969) 521.

36. Estrada-O., S. and Calderón, E. Biochemistry (Wash.) 9 (1970) 2092.

37. Ferguson, S.M.F., Estrada-O., S. and Lardy, H.A. J. Biol. Chem. 246 (1971) 5645.

38. Graven, S.N., Estrada-O., S. and Lardy, H.A. Proc. Nat.Acad.Sci. U.S.A. 56 (1966) 654.

39. Estrada-O., S., Rightmire, B. and Lardy, H.A. 1967. In: Antimicrobial Agents and Chemotherapy. p. 279, ed. Hobby, G.L. Ann. Arbor. Mich.

40. Chance, B. and Hollunger, G. J. Biol. Chem. 236 (1961) 1577.

41. Pressman, B.C. 1963. In; Energy Linked Functions of Mitochondria., p. 181. ed. Chance, B. Academic Press, New York.

42. Armstrong, C.M. and Hille, B. J. Gen. Physiol. 59 (1972) 388.

43. Modollel, J. and Davies, B.D. Proc. Nat. Acad. Sci. U.S.A. 67(1970) 1148.

44. Kanner, B.I. and Gutnik, D.L. J. Bacteriol. 3 (1972) 287.

45. Peña, A. FEBS Letters 34 (1973) 117.

46. Hille, B. J. Gen. Physiol. 58 (1971) 599.

47. Krasne, S. and Eisenman, G. Abstr. Meeting Biophys. Soc. Columbus, Ohio (1973).

48. Harris, E.J., van Dam, K. and Pressman, B.C.

Nature 213(1967)1126.

49. Palmieri, F. and Quagliariello, E. Eur. J. Biochem. 8 (1969) 473.

50. Dubinsky, W.P. and Cockrell, R. Biochem. Biophys. Res. Commun. 56 (1974) 415.

51. Blondin, G.A., de Castro, A.F. and Senior, A. E. Biochem. Biophys. Res. Commun. 43(1971)28.

52. Shamoo, A. Abstr. Proc. New York Academy of Sciences Symp. on $Na^+$-$K^+$ ATPases. New York (1974).

TABLE I

Effect of the Carboxylic Ionophores Nigericin ($K^+$-Selective) and Monensin ($Na^+$-Selective) on the $K^+$ -$Na^+$ Selectivity of Succinate Oxidation Stimulated by Beauvericin in Liver Mitochondria

| Antibiotic Addition | $K^+$ | $Na^+$ |
|---|---|---|
| | *nAtoms $O_2$/min./mg. protein* | |
| None | 155 | 169 |
| Beauvericin | 276 | 352 |
| Nigericin | 296 | 213 |
| Nigericin + beauvericin | 775 | 479 |
| Monensin A | 257 | 338 |
| Monensin A + beauvericin | 747 | 568 |

Except for the fact that L-histidine was omitted and the pH of the medium was 7.4, the reaction mixture was similar to that of Fig. The concentrations of beauvericin, nigericin and monensin A used were $3.8 \times 10^{-6}$M, $1.2 \times 10^{-6}$M and $5.7 \times 10^{-6}$M respectively.

TABLE II

Effect of the pH Change on the $Na^+$-$K^+$ Selectivity of Succinate or Glutamate + Malate Oxidation Stimulated by Beauvericin in Liver Mitochondria

| Oxidizable Substrate | pH 6.2 $K^+$ | pH 6.2 $Na^+$ | pH 7.4 $K^+$ | pH 7.4 $Na^+$ |
|---|---|---|---|---|
| | *Respiratory Coefficient* | | | |
| Succinate | 1.9 | 1.0 | 1.1 | 1.8 |
| Glutamate + Malate | 2.0 | 1.0 | 2.3 | 1.3 |

Except for the indicated pH values (adjunted with HCl) conditions were similar to those of Fig. 1. The concentration of glutamate and L-malate was 5mM respectively.

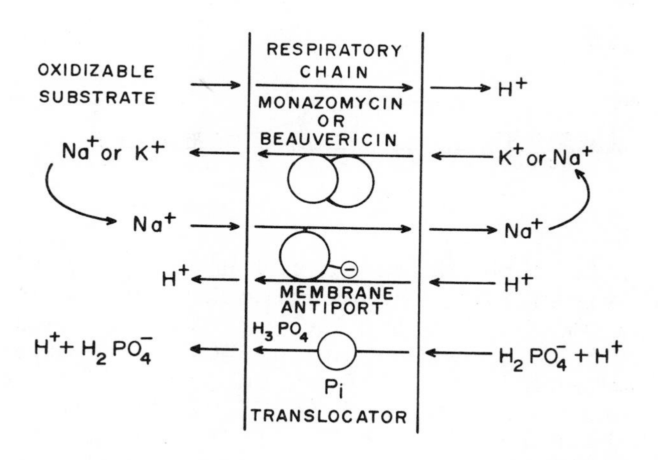


Figure 1. Synchronous correlation of the alkali metal cation and proton movements catalyzed by the function in parallel of the $Na^+ > K^+/H^+$ antiport translocator, the respiratory chain, monazomycin or beauvericin and the phosphate translocator in intact mitochondrial membranes.

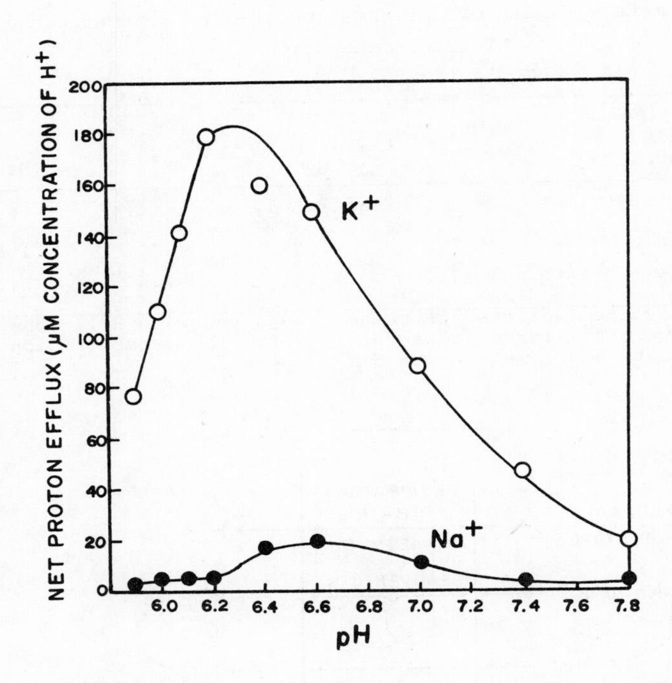


Figure 2. pH dependance of the $H^+$ efflux mediated by beauvericin from intact mitochondria suspended in medium which contains $Na^+$ or $K^+$. The medium contains: 5 mM triethanolamine-HCl, 8 mM histidine, 5 mM succinate, 150 mM sucrose, 15 mM of the chloride salts of $Na^+$ or $K^+$, 2.0 mg mitochondrial protein per ml and $3.8 \times 10^{-6}$M beauvericin at the indicated pH values.

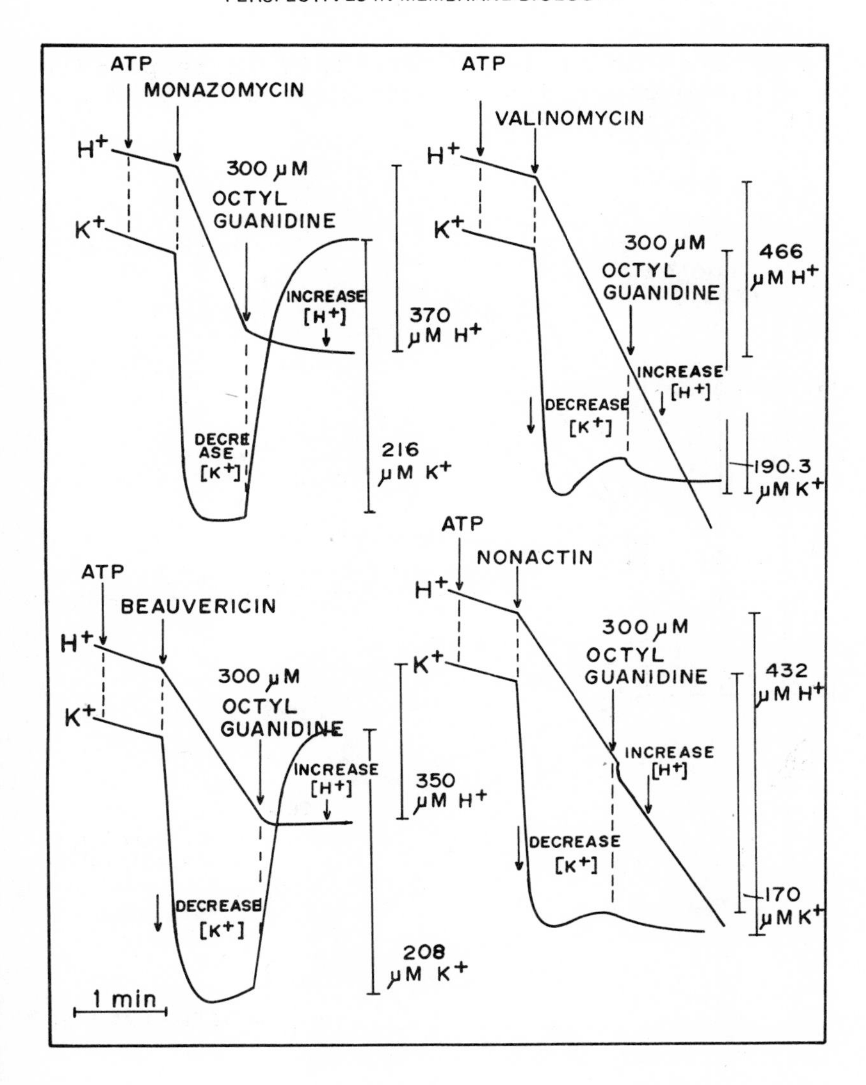


Figure 3. Effect of octylguanidine on the ATP-dependent $K^+$-movements and hydrogen ion-concentration changes mediated by monazomycin, beauvericin, valinomycin and nonactin in intact mitochondria. The medium contains: 3mM triethanolamine-HCl pH 7.4, 5 mM acetate-triethanolamine, 1.5 mM ATP-tris, 8mM KCl, 180 mM sucrose and 2 mg mitochondrial protein/ml. Valinomycin and nonactin were added at

a concentration of $6x10^{-7}$ and $1x10^{-6}$M respectively; monazomycin and beauvericin at $3.5x10^{-6}$M and $13x10^{-6}$M respectively.

---

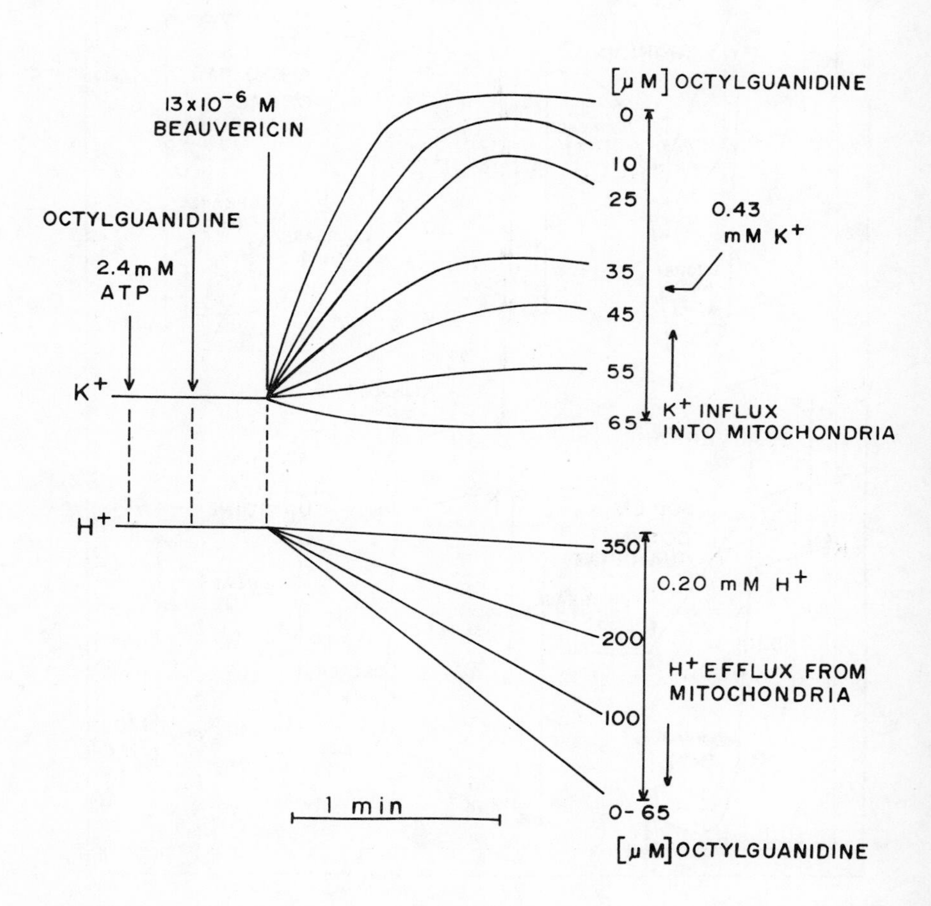


Figure 4. Effect of increasing concentrations of octylguanidine on the ATP-dependent $K^+$-movements and hydrogen ion-concentration changes mediated by beauvericin in intact mitochondria. Except where indicated, conditions are similar to those of Figure 3.

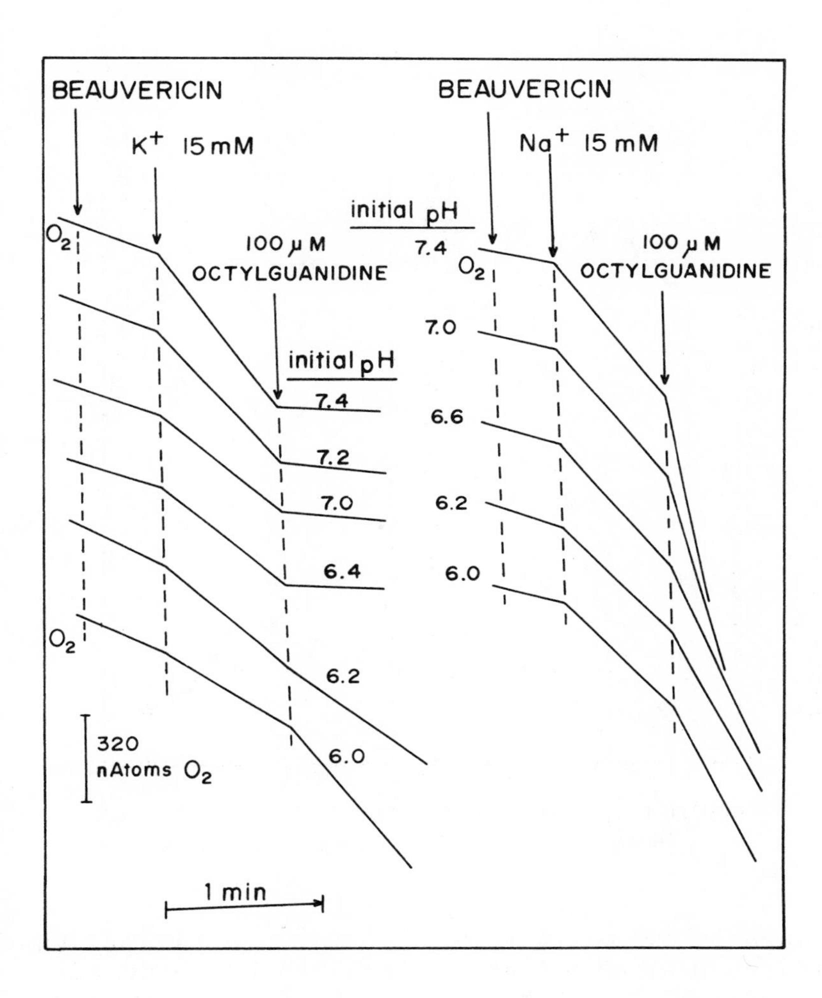


Figure 5. Effect of the pH change on the ability of octylguanidine to inhibit the oxidation of succinate stimulated by beauvericin in medium which contains $Na^+$ or $K^+$. Except for the indicated additions of octylguanidine, the conditions of these experiments were similar to those indicated for Figure 1.

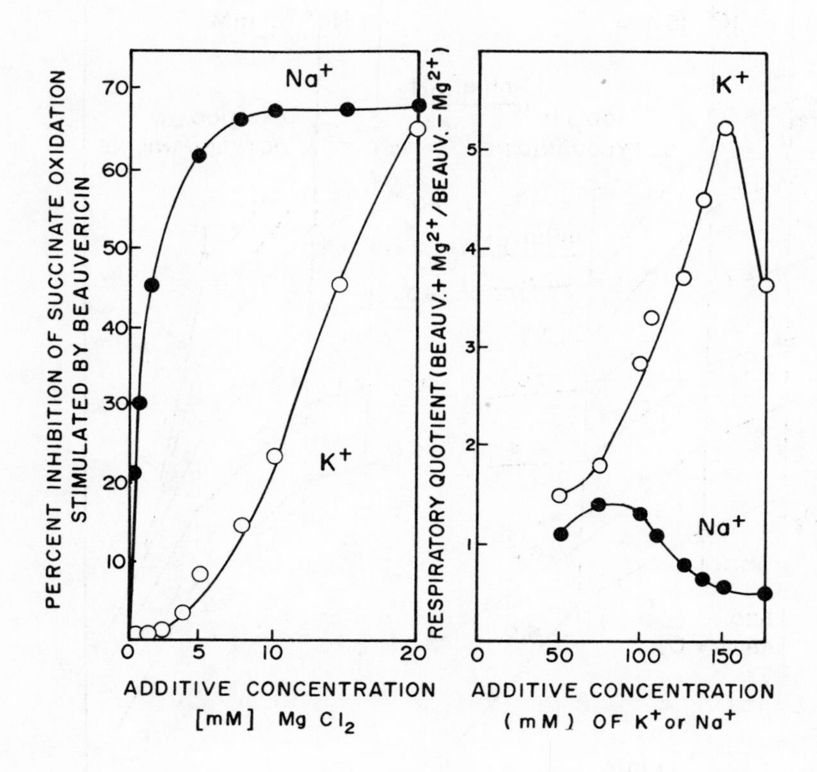


Figure 6. The selectivity of $Mg^{2+}$ to inhibit the oxidation of succinate stimulated by beauvericin in media which contains $Na^+$ or $K^+$. Except for the indicated variations in the concentration of either Mg $Cl_2$ or alkali metal cation other conditions, maintaining constant osmolarity at the expense of sucrose, were similar to those of Figure 1.

# SITE AND MECHANISM OF ACTION OF CATIONS IN ENERGY CONSERVATION

A. Gómez-Puyou and M. Tuena de Gómez-Puyou

Departamento de Biologia Experimental
Instituto de Biologia y
Departamento de Bioquímica
Facultad de Medicina
Universidad Nacional Autónoma de México
Apartado Postal 70-600
México, D.F.

At the present time it is not known exactly how electron transfer coupled to ATP formation occurs in the mitochondrion. Moreover, we are not completely certain as to how electrons are transfered from one electron carrier to another. Both electron transfer and phosphorylation are membrane phenomena, and thus it is our impression that until we know more about the structure of the membrane and the mechanism of reactions that occur within the mitochondrial membrane we will be unable to solve some of the current problems of oxidative phosphorylation.

The first part of this manuscript will summarize some experiments which indicate that cations affect oxidative phosphorylation. It is our hope that the experiments described in the second and third part of the manuscript will yield some insight into the site of action of cations in the membrane, as well as, into the mechanism through which cations affect electron transfer at Site I and the way in which this might relate to oxidative phosphorylation.

## Effect of $Na^+$ and $K^+$ on Oxidative Phosphorylation in Intact Mitochondria

Pressman and Lardy (1,2) reported some 20 years ago that $K^+$ was required for oxidative phosphorylation. Nevertheless some later reports proved to be contradictory; some researchers described a beneficial action of $K^+$ on oxidative phosphorylation (3,4) while others found an uncoupling action of $K^+$ (5); and still others failed to detect any effect of $K^+$ (6). In view of these controversial reports, we decided to explore the possible role of $K^+$ in oxidative phosphorylation.

During the course of these studies it was found that $Na^+$, in the presence of EDTA and phosphate, behaved as an uncoupling agent through a process that was prevented and partially reversed by $K^+$ (7,8). More recently Weiner and Lardy (9) made almost identical observations in mitochondria from kidney and also found that $Na^+$ altered the pattern of oxidation and reduction of pyridine nucleotides.

These findings obviously indicate that cations may affect the formation of ATP during coupled electron transport. However, although in the presence of phosphate and EDTA the effect that $Na^+$ has on mitochondria can be satisfactorily studied, the role that $K^+$ may have on oxidative phosphorylation is difficult to assay. This is due to the fact that mitochondria possess more than 100 nmoles of $K^+$ per mg of protein. In other words, when oxidative phosphorylation is measured in intact mitochondria, a possible action of $K^+$ cannot easily be appreciated due to the high endogenous $K^+$ content of the particles. Therefore we searched for a mitochondrial preparation with a low content of $K^+$ in which oxidative phosphorylation could be properly studied with respect to its possible $K^+$ requirements.

Curiously, it was found that the incubation of mitochondria in the same conditions in which uncoupling by $Na^+$ was observed (7,8) induced the release of most of the intramitochondrial $K^+$ (10). That is, the incubation of mitochondria with $Na^+$,

EDTA, phosphate and oxidizable substrate for less than 3 minutes reduces the $K^+$ content of the mitochondria to a value of approximately 10 nmoles per mg of protein (Table I); $Na^+$, EDTA, phosphate and substrate being essential for maximal release of $K^+$. It was also observed that EDTA could not be replaced by EGTA.

One of the advantages of this mitochondria is that the mitochondria once depleted of $K^+$ can be easily separated from the incubation mixture and washed with sucrose. In the resulting mitochondria all the reactions related to oxidative phosphorylation that have been studied (electron transport, phosphorylation, $^{32}P$-ATP exchange reaction and ATPase activity) are well preserved, but present some peculiarities (11-13). Some of the main characteristics of these $K^+$-depleted mitochondria will be described below.

The oxidative phosphorylation process of $K^+$-depleted mitochondria measured in the absence of added $K^+$ presents some interesting features. The State 3/State 4 ratios with NAD-dependent substrates are low and in most of the experiments they are below 2.0(10). With these substrates the P:O are also low and in the range of 0.6(12). The question was obviously raised as to whether the observed alterations were the result of damage to mitochondria and not the consequence of $K^+$-depletion. This possibility was ruled out, however, when it was found that the addition of $K^+$ to the incubation media significantly enhanced the magnitude of the P:O ratios and the respiratory control of $K^+$-depleted mitochondria (Table II). Moreover, it was also found that valinomycin at limiting concentrations induced a farther enhancement of the value of the P:O ratios.

One of the most important characteristics of oxidative phosphorylation in $K^+$-depleted mitochondria is that the phosphorylation of ADP that accompanies the aerobic oxidation of succinate is not affected by $K^+$-depletion; furthermore, the oxidation of succinate in the absence of $K^+$ also takes place at normal rates (8,10,12). That is, the effect of $K^+$-depletion can only be observed with NAD-dependent substrates.

The preceding experiments indicate that cations influence oxidative phosphorylation in intact mitochondria. However, with respect to the action of cations in oxidative phosphorylation, several points should be emphasized. First, in order to observe an effect of cations, in particular the uncoupling action of $Na^+$, the incubation of mitochondria must be carried out in a mixture that contains EDTA and in which $Mg^{2+}$ is omitted (7,9). Although an effect of EDTA on ion movements in mitochondria has been observed by other workers (14,15), the mechanism through which EDTA exerts its action is not known. Experiments should be carried out to ascertain the role of EDTA in cation fluxes across the mitochondrial membrane.

Second, the requirements of oxidative phosphorylation for $K^+$ can only be satisfactorily studied in $K^+$-depleted mitochondria. This is perhaps the reason why an effect of $K^+$ on oxidative phosphorylation has not been consistently observed (see references 1-7 and 9).

Third, valinomycin, which facilitates the movement of $K^+$ across the mitochondrial membrane (16), enhances the beneficial action of externally added $K^+$ in $K^+$-depleted mitochondria. This suggests that either the influx of $K^+$ or the internal concentration of $K^+$ is the factor that determines the efficiency and the rate of oxidative phosphorylation. In view of experiments which show that there is relation between intramitochondiral $K^+$ and State 3/State 4 ratios (8,10), we believe that the internal concentration of $K^+$ controls oxidative phosphorylation.

Fourth, the detrimental effect of $K^+$ depletion on oxidative phosphorylation could only be observed with NAD-dependent substrates. With succinate no effect of $K^+$ or $K^+$-depletion has been detected. This suggests that the three sites of oxidative phosphorylation may not have similar requirements.

Apparently only the first site of oxidative phosphorylation depends on the intramitochondrial concentration of $K^+$.

Finally, one of the most remarkable properties of oxidative phosphorylation in $K^+$-depleted mitochondria is that not only $K^+$ but also other cations exert a beneficial action on oxidative phosphorylation (Table III). Indeed, $Li^+$ is more effective than $K^+$. The lack of specificity of the mitochondria to cations is of particular importance in view of the experiments that will be described below.

## Localization of the Site of Action of Cations in the Inner Membrane of the Mitochondria

The study of the action of $K^+$ in oxidative phosphorylation in intact mitochondria is complicated by the fact that $K^+$ influences the uptake of substrates (17,18); moreover, a possible direct effect of cations on a membrane component may be difficult to evaluate due to permeability problems. Therefore, to overcome some of these difficulties, studies were carried out in inner membrane particles of the mitochondria, i.e. the EDTA-submitochondrial particles.

These particles show a peculiar behavior with respect to cations. Racker et al (5) found that monovalent cations induced an uncoupling action in submitochondrial particles. However, it is interesting that the data of Racker et al. show that the uncoupling occurs at the expense of an increase in the rate of respiration rather than a decrease in the phosphorylation rate; this clearly separates the mechanism of action of monovalent cations from that of classical uncouplers of oxidative phosphorylation.

This phenomena was explored further, and it was found that by incubating submitochondrial particles with NADH in a low salt media, $K^+$ induced a very significant increase in the rate of electron transport; the rate of respiration increased more than twofold by the mere addition of

$K^+$ (19). Most probably this is related to the coupling effect of $K^+$ in whole mitochondria, since no such effect of $K^+$ was observed when succinate was the oxidizable substrate (Fig. 1).

Moreover, just as cations affected oxidative phosphorylation of intact mitochondria, other cations besides $K^+$ enhanced the rate of respiration of submitochondrial particles oxidizing NADH. That is, both intact mitochondria and submitochondrial particles respond unspecifically to a wide variety of cations (Table IV). It is interstting to note that the latter particles also respond to divalent cations; indeed, $Ca^{+2}$ is the most effective in the enhancement of the respiratory rate.

Pressman (21) reported that octylguanidine interferes with oxidative phosphorylation, but only with NAD-dependent substrates. As octylguanidine possesses a positive charge at pH 7, it was thought that perhaps octylguanidine was acting on the site through which cations affected the rate of electron transport. When the effect of octylguanidine on the respiration of submitochondrial particles was assayed the results shown in figure 2 were encountered. Octylguanidine inhibited the oxidation of NADH, but not the aerobic oxidation of succinate. Moreover, it was also found that octylguanidine behaved as a competitive inhibitor of the stimulatory action of monovalent cations (Fig. 3).

This is an intriguing finding, mainly because octylguanidine with its positive charge seemed to act on the same site of the water soluble cations and yet it inhibited oxygen uptake. One of the most obvious differences between cations that stimulate oxygen uptake and octylguanidine is their degree of lipophilicity. Thus it was thought that perhaps the qualitative and quantitative effects of a given cation on the respiration of submitochondrial particles could be controlled by varying the lipophilicity of the cation. As noted in a previous paper, it was found that the inhibiting action of guanidines depended on the length of the alkyl chain (20); the larger the degree of lipophilicity of the guanidine, the more effectively it inhibited the rate of respiration.

A similar experiment on the effect of various amines with alkyl chains of variable length showed that ammonium and ethylammonium stimulated oxygen uptake; butylammonium at low concentrations stimulated oxygen uptake and at higher concentrations inhibited it. On the other hand, hexylammonium and octylammonium inhibited respiration; octylammonium being more effective than hexylammonium. These experiments in submitochondrial particles suggested that the inhibiting action of lipophilic cations resulted from hydrophobic interaction between the cation and the membrane. On the other hand, in the absence of these interactions, as in the case of the hydrophilic cations, an enhancement of the respiratory rate is observed.

The hydrophobic interaction between lipophilic cations and hydrophobic structures of the membrane was more clearly visualized when the effect of a limiting concentration of octylguanidine was assayed on the respiratory rate of submitochondrial particles oxidizing NADH at various temperatures (20). An arrhenius plot of this experiment is shown in Figure 5.

In the absence of octylguanidine, regardless of the presence of $K^+$, a sharp break in the logarithm of the velocity occurred when the temperature was raised from 35° to 40°. In the presence of octylguanidine, the logarithm of the velocity increased linearly as the temperature was increased. The octylguanidine-induced change in the Arrhenius plot must be due to the interaction of the 8 carbon chain with the hydrophobic structures of the inner membrane, since $K^+$ did not alter the "break" in the Arrhenius plot of submitochondrial particles oxidizing NADH.

The findings that octylguanidine exerts a competitive action with monovalent cations and that octylguanidine interacts with the hydrophobic region of the membrane strongly suggest that at Site 1 of oxidative phosphorylation there is a negative charge region or group which is in close contact with a hydrophobic structure of the inner membrane of the mitochondria. Also the results indicate that the interaction of this negative group with a cation exerts a strong influence on the rate of electron transport.

Although it was considered that this negative region or group lay on the surface of the membrane, the experiment shown in figure 6 made this possibility rather unlikely. Triethylamine exerts a strong activating effect on electron transport, but triethylamine attached to Sephadex fails to stimulate oxygen uptake. Rather, the effect induced by DEAE-sephadex is an inhibiting one. Clearly, the site of action of cations is not at the surface of the membrane, otherwise DEAE-sephadex would have stimulated the respiratory rates.

On the other hand, when a similar experiment was carried out with a guanidine attached to Sepharose through a 10 carbon chain, the results shown in Table V were encountered. Sepharose-bound decamethylene diguanidine exerted a strong inhibiting effect through a process that was fully reversed by $K^+$.

Since submitochondrial particles are considered to be "inside-out" (24), the results obtained with DEAE-Sephadex and Sepharose-bound decamethylene diguanidine indicate that the site of action of cations is not on the inner surface of the internal mitochondrial membrane, but rather at a place that is buried within the membrane.

It is also valid to conclude that the site of action of cations must be at a distance of not more than 15 Å from the inner surface of the membrane, which roughly corresponds to the length of a 10-carbon chain. Moreover, according to the rationale mentioned above, this site should be in close contact with the hydrophobic structures of the membrane.

## Mechanism of Action of Cations

In order to gain insight into the mechanism of action of cations on electron transfer at site I, it should be considered that hydrophobic cations inhibit respiration while hydrophilic cations stimulate it. This suggests that the extent of the degree of interaction of the cation with the membrane modifies the response of the electron transport chain to cations. Certainly if the site of action of cations is not at the surface of the membrane, it would be expected that hydrophobic cations would show a stronger interaction that might result in respiratory inhibition.

According to this reasoning, the experimental results depicted in figure 7 are of interest. Amines are much more effective than guanidines of equal polarity in enhancing the respiratory rate, while guanidines are better inhibitors of electron transport (see figures 2 and 4). According to the results of Table IV, the size of the cation must not be very important and other factors must be involved.

As Lewin reported (25) that guanidines bind more strongly than amines to negatively-charged carboxylates on phosphates, it is possible that the dissociation constant of the cation from its site is also involved in the effect of a cation on the respiratory rate of submitochondrial particles. In other words, it is possible that it is not only the binding of a cation to a negative-charged group that is responsible for its effect on respiration, but that most probably it is the binding and release of the cation at its site that controls the rate of electron transport.

The data of figure 7 show that this is a very likely possibility, since guanidines which bind strongly to carboxylates and phosphates inhibit the stimulatory action of $K^+$. On the contrary, amines which show less important H-bonding have an additive effect with $K^+$ on the respiration of submitochondrial particles. Apparently guanidines interfere in the action of $K^+$ through their stronger binding to the negatively-charged group.

Therefore it would seem that the rate of electron transport is controlled by the rate of binding and release of the cation at a negative charged group. If the cation is strongly bound to this hypothetical site, either by hydrophobic interactions or H-bonding, inhibition of respiration is observed; otherwise stimulation of oxygen uptake is induced.

The validity of this hypothesis was explored with the use of the thallous ion (Fig. 8). The dimensions of this ion are almost identical to those of $K^+$ (26), yet thallous ion binds some 10 times more strongly to proteins (27). Therefore, if the hypothesis that the release of cations from their site controls the respiratory rate is valid, it would be expected that $Tl^+$ would be less effective than $K^+$ in stimulating oxygen uptake. The experiments shown in figure 8 indicate that the Km for $Tl^+$ is larger than the Km for $K^+$. Furthermore, $Tl^+$ is an inhibitor of the action of $K^+$.

## Conclusions

The results described in the present manuscript indicate that cations are required for the coupling of oxidative phosphorylation, but only with NAD-dependent substrates. With succinate we have failed to detect an alteration of oxidative phosphorylation in $K^+$-depleted mitochondria or upon the addition of $K^+$ to these mitochondria. In submitochondrial particles it has also been found that cations influence the oxidation of NADH, but not the oxidation of succinate. Therefore it is likely that the two actions observed are part of the same process.

With respect to one of the main current controversies on oxidative phosphorylation, it should be mentioned that we have not been able to explain our data according to the chemiosmotic hypothesis (28). First, it is difficult to explain according to this hypothesis why, in both intact and submitochondrial particles, the effect of a cation is limited to Site I. Second, it is also difficult to explain why the quantitative and qualitative effect of cations depend on their lipophilicity. Rogers and Higgins (29) have found a similar problem with their studies on the action of lipophilic cations on oxidative phosphorylation.

However, it is not our intention to criticize a hypothesis merely because of the impossibility of explaining some of the data. Rather, we prefer to mention some of the possibilities that could account for the results obtained. These are illustrated in Figure 9.

According to the possibility shown in A, the stimulatory action of a cation on the respiratory rate is visualized as a consequence of the movement of the cation through the membrane; this might occur through a pore or through any other mechanism. The attachment of a lipophilic cation at this structure through hydrophobic interactions would result in respiratory inhibition. This possibility, however, would seem unlikely mainly because it would be expected that valinomycin would lower the Km for $K^+$ in submitochondrial particles and this does not occur (unpublished data). Moreover Montal and Chance (30) have shown that $K^+$ movements mediated by valinomycin and nigericin do stimulate the respiratory rate, but with either NADH or succinate as substrates. Our experimental results have only been observed during the oxidation of NADH.

The second possibility shown in Figure 9 results from the experiments designed to explore the position of the site of action of cations in the membrane. This site is probably not on the surface of the membrane, but buried within the membrane and in close contact with hydrophobic structures. Stimulation of respiration is thought to result from the interaction and release of a cation from its site. Again, the irreversible attachment of the cation either through hydrophobic interactions and/or H-bonding would result in respiratory inhibition. Although this mechanism would fit the experimental results, it is difficult to accept that a polar group lay in the midst of hydrophobic structures. Indeed, Singer (31), on thermodynamic grounds, considers this possibility as very unlikely.

Finally, the mechanism shown in C (Fig. 9) arose from the findings of Gitler and Montal (32). These investigators found that the interaction of water-soluble proteins with cations in the presence of acidic phospholipids drastically changed the solubility properties of the protein in such a way that the protein became soluble in decane. According to our data, it is visualized that a negatively charged group (5) in an electron carrier protein diminishes its degree of hydration upon interaction with a cation. Due to a modification in the solubility properties of the protein, the protein would undergo a change in its position within the membrane. This may result in the facilitation of the transfer of electrons to the next electron carrier. Upon transfer of the electron to the next carrier and also because of the instability of the cation in the hydrophobic environment, the cation will tend to migrate to a more favorable thermodynamic environment which would induce a return of the electron carrier protein to its original position and accept a new electron.

The model shown in Figure 9C would also explain why hydrophobic cations inhibit respiration; the electron carrier with a hydrophobic cation attached to its negative site would become stabilized by hydrophobic interactions within the membrane (second drawing of Fig. 9C). The electron carrier in this latter position would not be able to accept another electron and respiratory inhibition would result.

Although the mechanism detailed in Figure 9C may indeed be operating at Site I, it must be emphasized that we do not have a clear idea of how it might relate to the phosphorylation of ATP. Certainly at Site I the interrelation between lipids and proteins seems to be more critical than at the other phosphorylating sites. The important studies of Ragan and Racker (33) indicate that the reconstitution of Site I is highly dependent on the type and ratio of the phospholipids employed. Also Luzikov et al (34) showed that Site I is more sensitive to phospholipases than the other sites, and in agreement with these data Rogers and Higgins (29) have concluded that Site I possesses a higher degree of lipophilicity.

On the other hand, we are certain that cations are required for oxidative phosphorylation at Site I (see Table II). Therefore, future studies on the structural and mechanistic relation between lipids and proteins and the mechanism through which ions affect these interrelations should clarify the intimate mechanism of oxidative phosphorylation at Site I.

## References

1. Pressman, B.C. and Lardy, H. A. J. Biol. Chem. 197 (1952) 547.

2. Pressman, B.C. and Lardy, H.A. Biochim. Biophys. Acta 18 (1955) 482.

3. Krall, A.R., Wagner, M.C. and Gozanski, D.M. Biochem. Biophys. Res. Commun. 16(1964)77.

4. Blond, D.M. and Whittam, R. Biochem. J. 92 (1965) 158.

5. Christiansen, R.O., Loyter, A., and Racker, E. Biochim. Biophys. Acta 180(1969)207.

6. Opit, L.J. and Charnock, J.S. Biochim. Biophys. Acta 110(1965)9.

7. Gómez-Puyou, A., Sandoval, F., Peña, A., Chávez, E. and Tuena, M. J. Biol.Chem. 244(1969) 5339.

8. Gómez-Puyou, A., Sandoval, F., Peña, A., Chávez, E. and Tuena, M. Biochem. Biophys. Res. Commun. 36 (1969) 316.

9. Weiner, M.W. and Lardy, H.A. J. Biol. Chem. 248 (1973) 7682.

10. Gómez-Puyou, A., Sandoval, F., Chávez, E. and Tuena, M. J. Biol. Chem. 245(1970)5239.

11. Sandoval, F., Gómez-Puyou, A., Tuena, M., Chávez, E. and Peña, A. Biochemistry 9(1970) 684.

12. Gómez-Puyou, A., Sandoval, F., Tuena de Gómez-Puyou, M., Peña, A., and Chávez, E. Biochemistry 11 (1972) 97.

13. Gómez-Puyou, A., Sandoval, F., Chávez, E., Freites, D. and Tuena de Gómez-Puyou, M. Arch. Biochem. Biophys. 153(1972)215.

14. Packer, L. J. Biol. Chem. 235 (1960)242.

15. Azzi, A. and Azzone, G.F. Biochim. Biophys. Acta 105 (1965) 265.

16. Pressman, B.C. Proc. Nat. Acad. Sci. U.S.A. 53 (1965) 1076.

17. Graven, S.N., Estrada-O., S. and Lardy, H.A. Proc. Nat. Acad. Sci. U.S.A. 56 (1966) 654.

18. Harris, E.J., Bangham, J.A. and Wimhurst, J. M. Arch. Biochem. Biophys. 158 (1973) 236.

19. Pinto, E., Gómez-Puyou, A., Sandoval, F., Chávez, E. and Tuena, M. Biochim. Biophys. Acta 223 (1970) 436.

20. Lotina, B., Tuena de Gómez-Puyou, M. and Gómez-Puyou, A. Arch. Biochem. Biophys. 159 (1973) 520.

21. Pressman, B.C. J. Biol. Chem. 238(1963)401.

22. Gómez-Puyou, A., Sandoval, F., Tuena de Gómez-Puyou, M. and Pinto, E. Bioenergetics 3(1972) 221.

23. Lotina, B., Gómez-Puyou, A., Tuena de Gómez-Puyou, M. and Chávez, E. Arch. Biochem. Biophys. 159 (1973) 517.

24. Christiansen, R.O., Loyter, A., Steensland, H., Saltzgaber, J. and Racker, E. J. Biol. Chem. 244(1969) 4428.

25. Lewin, S. J. Theoret. Biol. 23(1969)279.

26. Handbook of Chemistry. p. 177, ed. West, R.C. Ohio, The Chemical Rubber Co.

27. Manner, J.P., Morallee, K. G. and Williams, R.J.P. Chem. Commun. (1970) 965.

28. Mitchell, P., Biol. Rev. 41(1965) 445.

29. Rogers, K.S. and Higgins, E.S. J.Biol.Chem. 248(1973)7142.

30. Montal, M., Chance, B. and Lee, C.P. J. Membrane Biol. 2 (1970) 201.

31. Singer, E.J. 1971 Structure and Function of Biological Membranes p. 146. New York, Academic Press.

32. Gitler, C. and Montal, M. Biochem. Biophys. Res. Commun. 47 (1972) 1486.

33. Ragan, C.I. and Racker, E. J. Biol. Chem. 248 (1973) 6876.

34. Luzikov, V.N., Kupriyanov, V.V. and Maklis, T.A. Bioenergetics 4 (1973) 521.

TABLE I

Release of $K^+$ from the Mitochondria

| Experiment | Incubation Media | Release of $K^+$ nmoles of $K^+$ $mg^{-1}$ |
|---|---|---|
| | Complete | 92 |
| A | -NaCl | 41 |
| | -Phosphate | 16 |
| | -Glutamate | 39 |
| B | Complete | 85 |
| | -EDTA + EGTA (1 mM) | 13 |

Mitochondria were incubated for 3 minutes in the conditions described. The complete medium contained 100 mM NaCl, 10 mM phosphate, 10 mM glutamate, 1 mM EDTA, 80 mM Sucrose and 20 mM Tris-HCl pH 7.3

TABLE II

Effect of KCl and Valinomycin on the P:O Ratios and Respiration of $K^+$-Depleted Mitochondria

| | -Valinomycin | | +Valinomycin | |
|---|---|---|---|---|
| KCl (mM) | P:O | natoms of $O_2$ consumed/$min^{-1}$/$mg^{-1}$ | P:O | natoms of $O_2$ consumed/$min^{-1}$/$mg^{-1}$ |
| - | 0.7 | 12 | 0.7 | 12 |
| 4 | 0.9 | 16 | 1.4 | 35 |
| 10 | 1.2 | 16 | 1.6 | 35 |
| 20 | 1.3 | 20 | 1.7 | 47 |
| 40 | 1.0 | 22 | 1.7 | 41 |

The incubating conditions were 2 mM phosphate, 1 mM EDTA, 4 mM Tris-HCl (pH 7.3), 5 mM glutamate, 200 mM Sucrose, 2.6 ngrams of valinomycin per mg of protein and 1.6 mM ADP. (Reproduced from Biochemistry 11 (1972) 97).

---

TABLE III

Effect of Monovalent Cations on the Oxygen Uptake and P:O Ratios of $K^+$-Depleted Mitochondria

| Salt added 20 mM | natoms of $O_2$ consumed/$min^{-1}$/$mg^{-1}$ | P:O |
|---|---|---|
| - | 17 | 0.6 |
| LiCl | 83 | 3.0 |
| K Cl | 21 | 2.0 |
| RBCl | 18 | 2.0 |
| CsCl | 19 | 1.7 |
| TMACl | 17 | 1.8 |

The incubating conditions were as in Table II with 20 mM of the indicated salts and in the absence of valinomycin. TMACl = tetramethylammonium Chloride. (Reproduced from - Biochemistry 11, (1972) 97).

TABLE IV

Effect of Various Salts on the Respiration of Submitochondrial Particles

| Cation | Rc/Rk |
|---|---|
| $Li^+$ | 0.6 |
| $Cs^+$ | 0.9 |
| $Na^+$ | 1-4 |
| $Tris^+$ | 1-3 |
| Triethanolamine | 1-1 |
| $Ca^{2+}$ | 2.6 |
| Ammonium | 1.8 |
| Protamine (Sulfate) | 0 |

The incubating conditions were as in Figure 1 with NADH as substrate, except that the indicated chloride salts (8 mM) were added at 80% oxygen saturation. The enhancement of the rate of respiration attained after the addition of the salt is taken as Rc, while the enhancement attained by 8 mM KCl is Rk (Reproduced from Archives of Biochemistry and Biophysics 159 (1973) 520).

TABLE V

Effect of Sepharose-Bound Decamethylene Diguanidine on the Oxygen Uptake of Submitochondrial Particles

| Additions | Respiratory Rate (natoms $O_2$/min$^{-1}$/mg$^{-1}$) |
|---|---|
| – | 248 |
| Sepharose 0.4 ml | 323 |
| Sepharose-bound DMD (118 nmoles of DMD) | 78 |
| 11 = 33 mM KCl | 232 |

The incubating conditions were as in Figure 1. Sepharose-bound decamethylene diguanidine DMD was synthesized as described elsewhere (23). Sepharose 4B and Sepharose-bound decamethylene diguanidine were added as suspensions that had the same packed volume. (Reproduced from Archives of Biochemistry and Biophysics 159 (1973) 517).

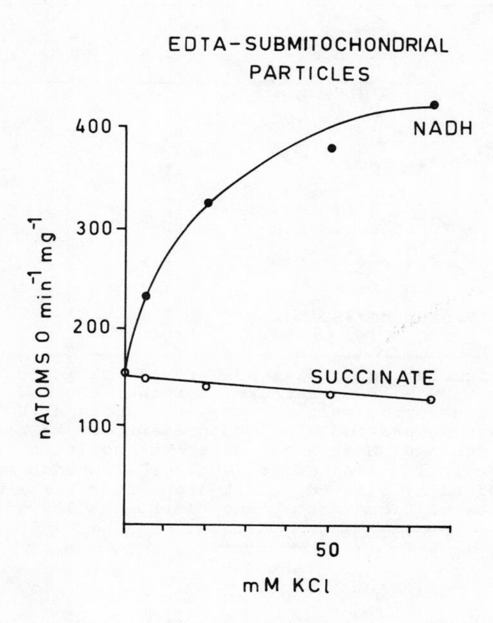


Figure 1. Effect of $K^+$ on the respiration of submitochondrial particles oxidizing NADH and succinate. EDTA-submitochondrial particles were incubated in 0.25 M sucrose, 1 mM EDTA, 4 mM tris-HCl pH 7.3, 0.5 mM NADH or 4 mM succinate and the indicated concentrations of KCl.

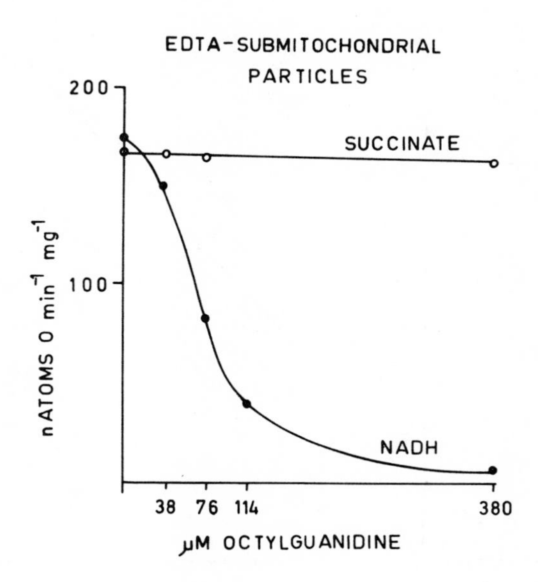


Figure 2. Effect of octylguanidine on the respiration of submitochondrial particles oxidizing NADH and succinate. The incubating conditions were as in Figure 1.

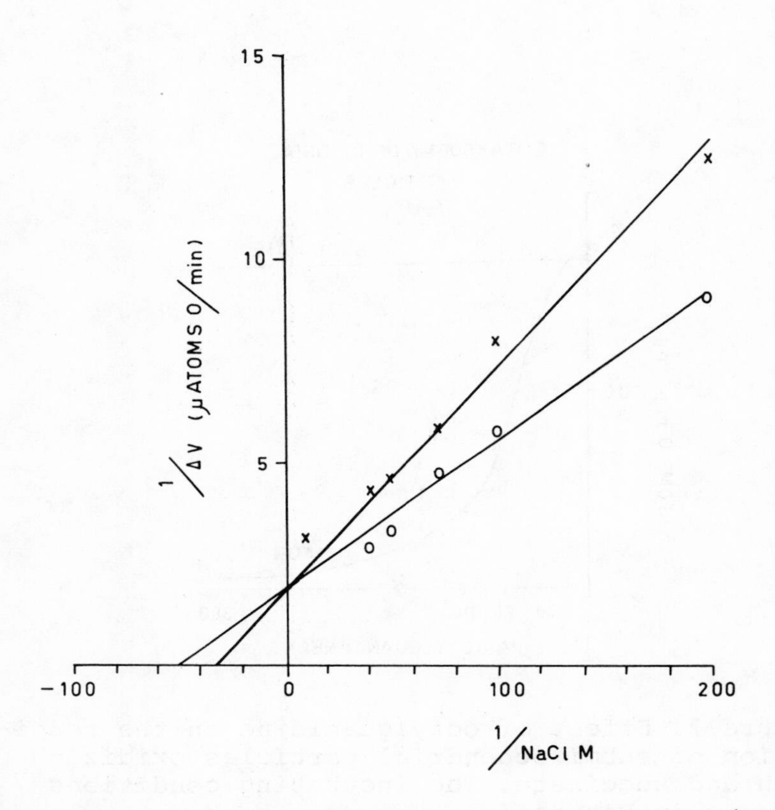


Figure 3. Effect of octylguanidine on the $Na^+$ stimulated oxygen uptake of submitochondrial particles. The experimental conditions were as in Figure 1 (Reproduced from Bioenergetics 3 (1972) 221).

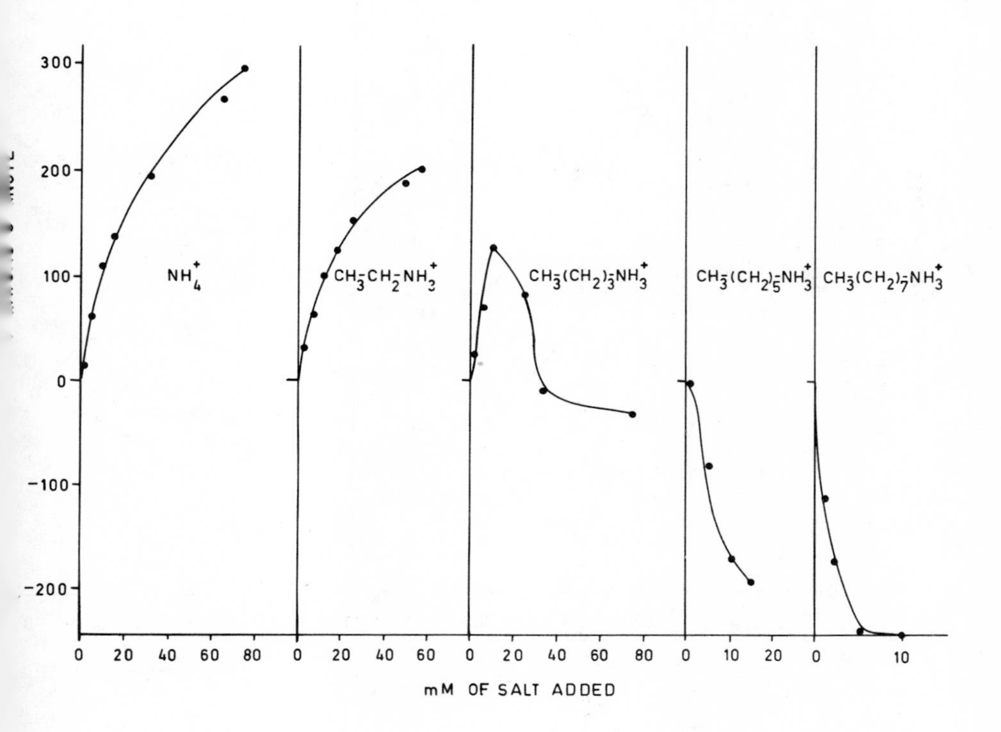


Figure 4. Effect of amines with different alkyl substituent on the respiratory rate of submitochondrial particles. The conditions were as in Figure 1.

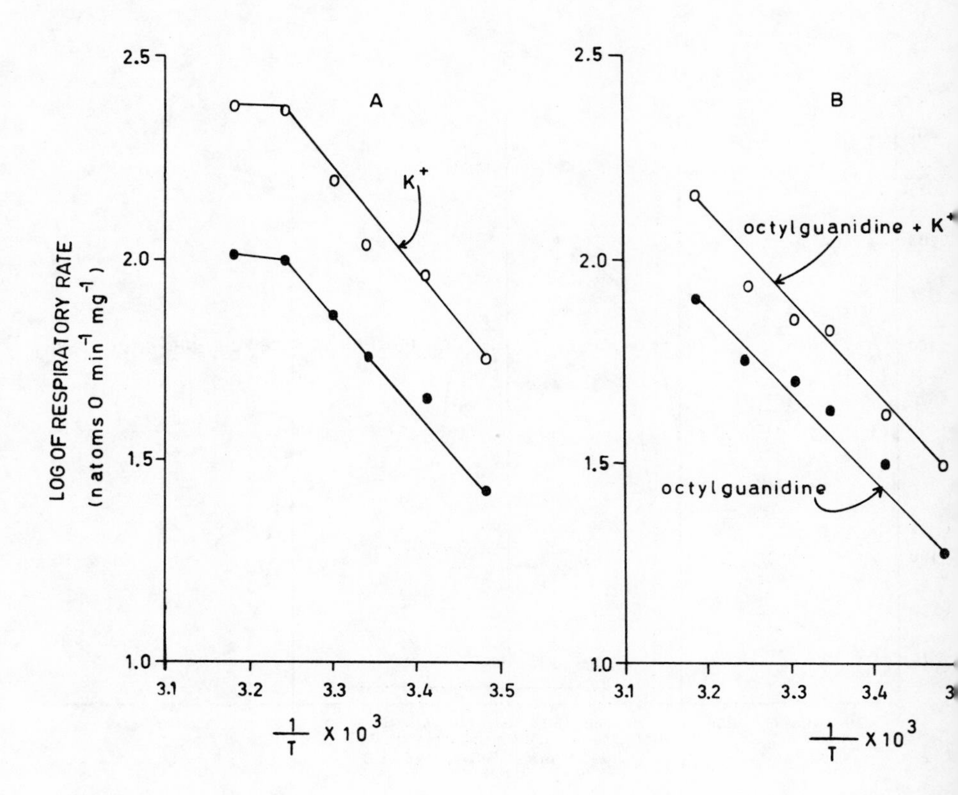


Figure 5. Arrhenius plot of the respiration of submitochondrial particles incubated with octylguanidine and $K^+$. Oxygen uptake was measured as in Figure 1 at the temperatures indicated (Reproduced from Arch. Biochem. Biophys. 159(1973) 520).

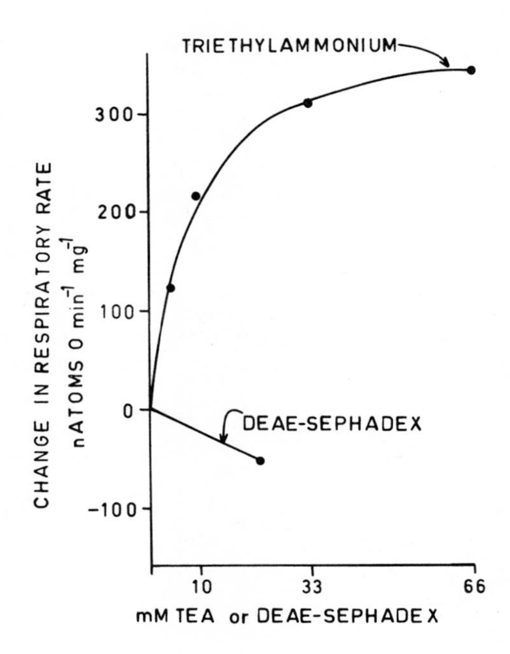


Figure 6. Effect of triethylamine and diethylethylamine-sephadex on the respiration of submitochondrial particles. The conditions were as in Figure 1. Diethylethylamine-sephadex also inhibited by an almost equal percentage the oxidation of succinate.

---

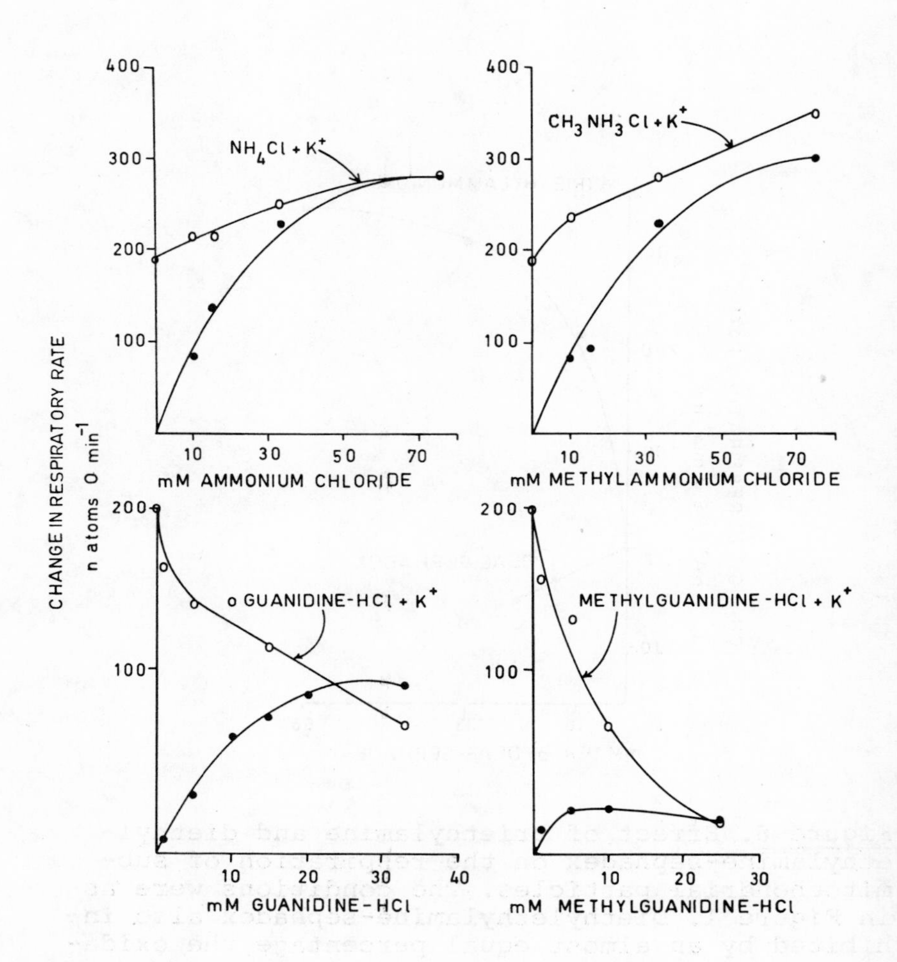


Figure 7. Comparative effects of guanidines and amines on respiration and on the action of $K^+$. The conditions were as in Figure 1; KCl (33 mM) was added at 20-40% oxygen saturation.

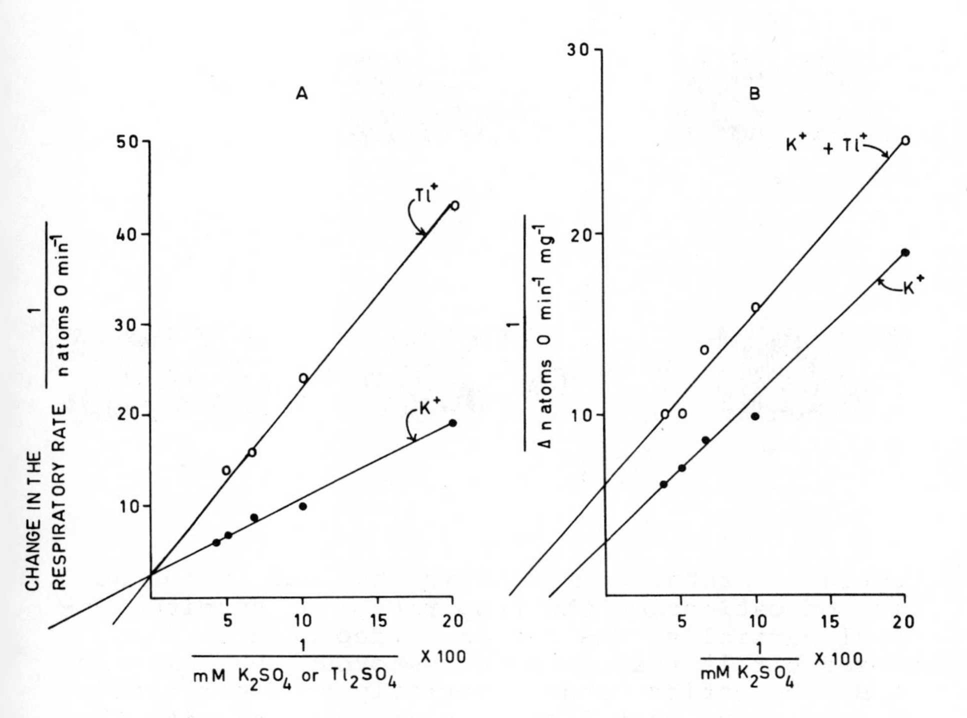


Figure 8. Comparative effect of $K_2SO_4$ and $Tl_2SO_4$ on the respiration of submitochondrial particles The experimental conditions were as in Figure 1. In B the effect of 2.5 mM $Tl_2SO_4$ was assayed on the action of $K_2SO_4$.

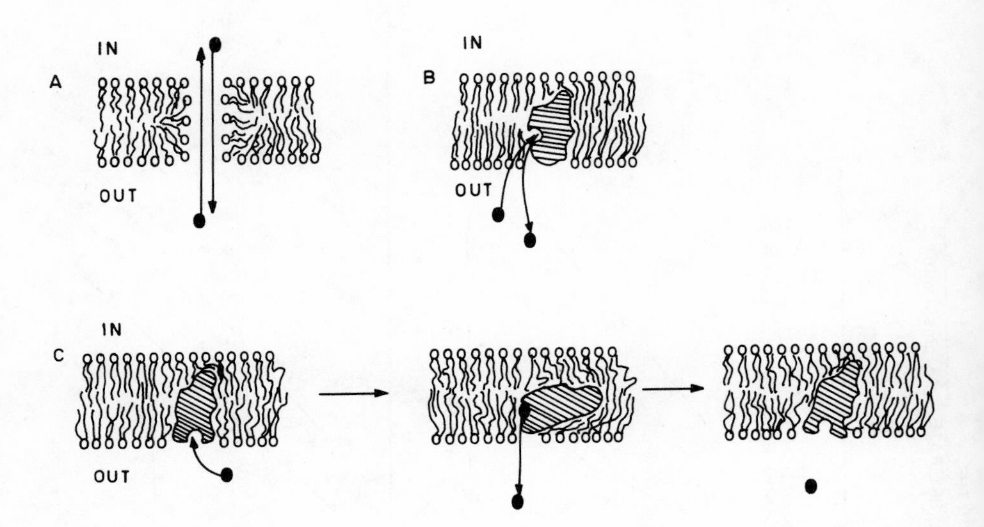


Figure 9. Probable mechanisms involved in the action of cations on the respiration of submitochondrial particles. The cation is depicted as a closed circle that in A moves across the membrane. In B, the cation binds reversibly to a negative-charged group that lies in close contact with the hydrophobic structures of the membrane; thermodinamically this would be an unlikely situation. In C the binding of the cation to a protein changes the solubility properties of the protein and thereby induces a change in the position of the protein within the membrane. This is a reversible change that facilitates the transfer of electrons from one carrier to another. For details see text.

# LOCALIZED AND DELOCALIZED POTENTIALS IN BIOLOGICAL MEMBRANES

Britton Chance, Margareta Baltscheffksy[1], Jane Vanderkooi and Wen Cheng

Johnson Research Foundation, School of Medicine, University of Pennsylvania, Philadelphia, Pa. 19174, U.S.A. and [1]Department of Biochemistry, Arrhenius Laboratory, University of Stockholm, S-104-05 Stockholm, Sweden

One approach to the relationship between membrane structure and energy coupling is to develop probes that occupy regions of the membrane where the energy coupling reaction is occurring. Such probes may be sensitive to the localized or delocalized fields in the membrane as well, and since they may be optimally "designed" for specific purposes their use may afford a more logical, less phenomenological approach than do the intrinsic carotenoid probes. A class of merocyanines called "Brooker" dyes (1,2) exhibit large responses to changes of the dielectric constant or "Z-value" of solvents, termed a "solvatochromicity"; they are also termed "electrochromic" by Platt (3). Drs. G. Strichartz and L. Cohen called our attention to the merocyanines by elegant studies of the fluorescence response of MC-I as an "electrochromic" indicator in squid axons (4-6). (In their most recent paper on this subject, MC-I and MC-II are referred to as "dye I" and "dye 158" respectively.

MC-I is a dicarbocyanine (cf Figure 1) with a sulfonic acid group, like anilinonaphthalene sulfonic acid or ANS (7). An aliphatic chain allows the molecule to reside in the membrane so that its butyl substituents may be in the hydrophobic region of the membrane. This probe may be similar to anthracene stearic acid (AS) or anthracene palmitic acid (AP) (8), since all three are collisionally quenched by the oxidized form of membrane-bound ubiquinone.

Eastman Kodak has provided us with another merocyanine, MC-II (6) which lacks the butyl groups but still occupies the hydrocarbon region of the membrane, as indicated by fluorescence quenching by oxidized ubiquinone. The two aromatic groups of MC-I and MC-II are linked by conjugated double bonds and charge separation between the two is possible (2,3).

While the location of the ANS and AS probes in model membranes has been studied by a variety of approaches (8) including fluorescence intensity, emission spectra, X-ray diffraction, and nuclear magnetic resonance (cf Figure 2), the location of MC-I suggested here is based only on the fluorescence quenching, on energy transfer data described below, and on transition moment data (9).

## Properties of the Probes

The rationale for using these probes is that they occupy the lipid phase of the membrane as well as the aqueous interface, and that substantial portions of the localized or delocalized charge separations of electron transport or energy coupling reactions will appear across the dye, leading to electrochromic responses. Such fields can cause isomerization of the probe to a dipolar form with corresponding rearrangements in the conjugated double bond system, as shown in Figure 3 (3). The absorption bands of the probe, in addition to being affected by the number of conjugated double bonds, are also dependent upon the environment of the two ends ($\underline{b}_L$ and $\underline{b}_R$) of the molecule, described by Platt's (3) equation:

$$E^2 = E_R^{\ 2} + (\underline{b}_R - \underline{b}_L - V_{R-L})^2 \qquad (1)$$

or:

$$h\Delta\nu = \frac{\underline{b}_R - \underline{b}_L}{E} V_{R-L} + \frac{1}{E}(V_{R-L})^2 \qquad (2)$$

where E is the intensity of the absorption band,

$E_R$ is the spectral energy of the isoenergetic wavelength of the first absorption band of a given chain length, $V_{R-L}$ is the potential difference between the end-groups, and *b* is the basicity, which is generally related to the local dielectric constant at R and L. The wavelength of the absorption band will vary with both *b* and V, and in the latter case, with $V_{R-L}$ in both a linear and a quadratic fashion (10).

H. Bücher et al (10) and Schmidt and Reich (11-14) have explored this relationship further and favor equation 3 below, derived as well by W. Cheng in our laboratory (15).

$$H\Delta\vec{\nu} = \Delta\vec{\mu} \cdot \vec{F} - \frac{1}{2} \Delta\vec{\alpha}F^2 \qquad (3)$$

where $\Delta\vec{\mu}$ is the difference of the permanent dipole moment and $\Delta\vec{\alpha}$ is the difference of the induced dipole moment between the ground and excited states, $\Delta\vec{\nu}$ is the frequency shift and $\vec{F}$ is the electric field. The shift of the absorption band is dependent upon the potential difference in both a linear and a quadratic fashion.

It is important to note that whereas a simplistic interpretation of Platt's basic formula (3) suggests that the quadratic response is accompanied by a blue shift of the probe with increasing potential, the equation employed by Bücher (10) and Cheng (15) suggests a red shift, in accordance with our experimental observations. Furthermore, neither equation applies to the fluorescence change, and any relation in this case is empirical.

In order to obtain an electrochromic response from a merocyanine probe according to equation 2, the following considerations must apply:

1) If a localized field is to be measured, the probe should be located near the charge separation and be of comparable dimensions.

2) The probe must occupy the membrane at sites sufficiently near the local charge separation.

3) the probe must be appropriately oriented

that permanent and induced dipole moments, $\vec{\mu}$ and $\Delta\vec{F}$, are perpendicular or at a significant angle with respect to the local field, <u>or</u> perpendicular to the plane of the membrane in the case of a transmembrane potential.

The field will then be detected by

a) a change in the permanent dipole moment, $\vec{\mu}$, unless $\vec{\mu}_{excited} = \mu_{ground}$, so that $\Delta\vec{\mu} = 0$. If $\vec{\mu}_{ex}$ and $\vec{\mu}_{grd}$ are randomly oriented, or $\Delta\vec{\mu}$ is perpendicular to $\vec{F}$, and $\Delta\vec{\mu}\cdot\vec{F} = 0$. Note that this linear effect will reverse with the direction of the field.

b) a change in the induced dipole moment, $\alpha\vec{F}$, unless $\alpha\vec{F}$ is perpendicular to $\vec{F}$; not that this quadratic effect will not reverse with the direction of the field.

By combining the probe in a fatty acid multilayer, evaporating aluminum electrodes across it, and employing alternating potentials (see Figure 4) Bücher et al (10) find that the light transmitted by the multilayer varies at both the same frequency and at twice the frequency as the field. By measuring the wavelength dependence not only at the same frequency (linear) but also at twice the frequency (quadratic) they obtained a blue shift and a red shift, respectively, at both room and low temperatures. It is of considerable interest that this cyanine dye had a large permanent dipole moment and showed both linear and quadratic responses, of roughly the same amplitude. The relationships between the potential and the magnitude of the red shift are not determined in detail, particularly because the probe is nearly perpendicular to the applied field in Bücher's study; in spite of this, the spectra are clearly delineated at the two temperatures. No experimental studies are available on the electrochromicity of the fluorescence oscillator, but our preliminary data suggest that the oscillators responsible for the red shift and for the fluorescence emission are different.

## Experimental Methods

*Chemicals*. As stated throughout the text, the probes themselves were obtained directly from Dr. Alan Waggoner and their properties are described in detail in Reference 6. One probe, however, identified as a component in the preparation of Brooker's M-84 and contained in his fraction R-6 (1) together with M-6 itself obtained from Dr. Sol Harrison, has been resynthesized by Messrs. Greg Stern and Paul Russ in this laboratory. These probes ran as a single component in a thin-layer chromatogram (J. Vanderkooi, personal communication). Spectroscopically, the probes appeared to be homogeneous. Their stability properties are as yet unknown.

*Preparations*. Submitochondrial particles were prepared according to standard methods developed by Dr. C.P. Lee (16) and pigeon heart mitochondria were prepared by the method of Chance and Hagihara (17). Chromatophores were prepared from R. rubrum as described by Baltscheffsky (18). Other aspects of the preparation, reaction media, and sources of antibiotics are provided in Reference 19.

*Optical Methods*. A variety of optical methods were used and these have been specifically described in relation to their application to merocyanine probes by Chance and Baltscheffsky (19). Routine measurements of fluorescence emission were made in a Hitachi MP-4F fluorometer. The lifetimes of the AS probe were determined in the ORTEC photon-counting fluorescence lifetime device. The lifetimes of the merocyanine probes were determined with a mode-locked laser system by Drs. James Callis and Jane Vanderkooi through the kindness of Dr. Robin Hochstrasser of the Materials Science Center of the University of Pennsylvania.

## Applications of Electrochromic Probes

The application of the merocyanine probes to the problem of measuring energized states of various energy-transducing membranes are briefly summarized in terms of binding sites, quantum yields, lifetimes, energy transfer, and membrane localization.

## The Responses of MC-1 in Intact Mitochondrial Membranes

*Anthracene Stearic Acid* (AS) *and MC-I*. In view of the relatively precise location of AS in the hydrocarbon phase of the membrane (8) energy transfer between the two probes can be of considerable significance. The experiments indicated in Figures 5 and 6 illustrate energy transfer from the probe AS bound to pigeon heart mitochondrial membranes; fluorescence excited at 366 nm and emitted at 460 nm is gradually quenched with increasing concentrations of MC-I (0 to 57 μM) while at the same time, the emission of MC-I fluorescence at 585 nm is indicated to increase with each successive addition of MC-I.

A verification of this reaction is seen in the progressive decrease of the lifetime of AS fluorescence as measured with the ORTEC monophoton technique as the concentration of merocyanine added to the membrane is increased from 0 to 90 μM as seen in Figure 6.

## The Response of MC-I and MC-II in Submitochondrial Particles

The response of the electrochromic probes in submitochondrial vesicles has been discussed in part elsewhere (19). Merocyanines I and II readily bind submitochondrial vesicles, showing a small solvatochromism causing a slight red shift that extends over several minutes, illustrated in Figure 7.(Traces 1-4 of Figure 7A and Traces 1-3 of 7B).The interdigitation of the probe into the lipid chains may require fluctuations in the lipid structure to facilitate its entry. After ten minutes, the binding is complete; tests with bovine serum albumin show no free probe initially or during the transition from the de-energized to the energized state.

The responses of MC-I and MC-II due to membrane energization by succinate oxidation in

OESMP are shown in Figures 7A and 7B. On the top, the response of MC-I is seen to consist of the binding of the probe to the membrane and an increase in the intensity of the 572 nm band. Addition of succinate causes, within the first 44 sec, a red shift of about 5 nm which is stable during traces 5-11. The enhancement of the peak at 578 nm is followed by a slower reaction (Traces 5-11) in which the peak becomes less intense and a trough appears at 556 nm. Thus, slow and fast phases of membrane energization are associated with an initial rapid red shift as a primary event, and with a decrease of extinction coefficient of the main band of the probe as a second event.

With MC-II, which absorbs at a longer wavelength region beyond the cytochrome absorption bands, the corresponding traces for the binding of the probe to the membrane are #1 - #3, which show a small red shift and very little peak shift during this interval. Following trace 3, succinate is added, and the reaction is half-complete at 44 sec, as observed by the progressive red shift. Traces 7 through 14 show no further increase of the red shift. Analogous to the change of MC-I, energization causes a slow decrease of the absorbancy peak at 630-640 nm so that by the 13th trance, a considerable fraction of the absorption of the main band has disappeared, the red shift remaining unaffected.

Evidence of the selective changes of MC-I in the presence of a proton gradient and a diffusion potential is shown in Figure 8 in which difference spectra are displayed. The submitocondrial particles have already been energized by the oxidation of succinate and the baseline Trace 1 represents the computer output which has normalized the absorption of the energized probe and the submitocondrial particles to a straight line. The initial addition of KCl and nigericin causes, in Traces 2 - 7, the recovery of absorption at 571 nm which is very nearly complete by the seventh scan, at 44 sec per scan; no shift of absorption is evident. Addition of valinomycin then leads immediately to Trace 8, which shows superimposed on Trace 7 a large blue shift of the MC-I absorption, denoted by a decrease of absorption at 587 nm and an increase of absorption at 559 nm. An explanation for these results is that the nigericin-induced exchange of $K^+$ for $H^+$ causes the reintensification of the absorption band at 571 nm (cf Figure 4) while the diffusion potential induced by the addition of valinomycin leads to the loss of the red shift. Thus, the merocyanine probe appears to be responsive at different parts of its spectrum to the proton gradient (571 nm) and to the diffusion potential (587 and 559 nm).

## Energy-dependent Responses of the Probe MC-II

The following figures present in more detail the complex energy-dependent responses of the probe MC-II in submitochondrial particles caused by energization and de-energization phenomena. The results are presented as absolute and difference spectra. In the simplest case, the energization is recorded in the presence of $K^+$ and nigericin (20,21) (Figure 9A). Trace 1 represents the spectrum of the membrane-bound probe with the absorption and light-scattering of the submitochondrial particles removed by the baseline memorization procedure. Addition of succinate then gives a slow progression of the red shift until a plateau is reached by Trace 6. The interval between scans is 44 sec. It should be noted that the response is principally a red shift, with little change occurring in the region of 631 nm.

In Figure 9B, the experiment is repeated except that $K^+$ and nigericin are omitted. Trace 1 indicates the spectrum of the dye bound to the membrane. Trace 2 represents a spectrum immediately after adding the probe (77 sec per scan) and corresponds closely to Trace 6 of Figure 9A. Thereafter, the two recordings differ and the band at 631 nm sinks in intensity until by Trace 12 a rather flat spectrum is observed.

Thus, two phases of the energization process

are observed; one depends upon whether a $H^+$ gradient can be established, a phenomenon also seen in the response of MC-I but even more extensively affecting the spectrum of MC-II.

Figure 10 illustrates the initial phases of membrane energization and de-energization using MC-II as a probe. In Figure 10A, the traces are recorded at 44 sec per scan (as contrasted to 77 sec per scan in Figure 9B) and $K^+$ and nigericin are omitted, in contrast to Figure 9A; the reaction medium is 5mM Tris-sulfate in contrast to Figure 9B. It is seen that the energization is approximately two-thirds complete by the time of the first scan (44 sec) and becomes complete by the third or fourth scan. The red shift increases to 23 nm under these particular conditions and the drop of intensity at 627 nm is less than under the conditions of Figure 9B.

Following energization, the spectrum of the probe-labelled preparation with added succinate is redrawn (Figure 10B) and step-wise de-energization of the membrane fragments occurs following addition of 5 mM KCNS (Traces b-d). Further de-energization occurs following the addition of 20 μM pentachlorophenol. The red shift at the time of the last trace has diminished to 7 nm and progresses slowly towards the original trace of Figure 10A. It is apparent from these traces that the degree of red shift on energization depends upon the experimental conditions, and that the de-energization can be carried out by thiocyanate or by pentachlorophenol, which eventually restores the membrane to the de-energized condition.

The energization-de-energization process can also be represented by difference spectra where, in Figure 11 the time course of the acquisition of the red shift (increased absorption at 685 nm) is compared with the time course of the decrease of absorption at 632 nm. It is clearly seen that half of the maximal red shift is obtained at the time when only 10% of the decrease of absorption at 632 nm can be observed.

The de-energization process is indicated in

Figure 11B, where a trace corresponding to the completion of the reaction of Figure 11A, Trace 1, is shifted towards the initial phases, the first caused by the addition of KCl and nigericin which, in the first eleven traces, diminishes the peak at 635 nm by about 60% and causes scarcely any change at 680 nm. Thereafter, in Traces 12-18, valinomycin is added and all the absorption change at 680 nm disappears, together with the remainder of that at 635 nm. Thus, we can show by absolute and difference spectrocopy, two components of the response of the probe corresponding to its differential sensitivity to $H^+$ gradients and to diffusion potentials.

## Uncharged Probes

The probe illustrated in Figure 12 represents one of a class of tricarbocyanines that differ from MC-I and MC-II by an isoxazolone end-group and by an additional conjugated double bond (1,2). The latter causes a shift of the absorption and fluorescence maxima further to the red region (2). This type of probe is thus especially suited to chromatophore studies, since changes of bacteriochlorophyll absorption at 605 nm will not interfere with probe absorption changes at 630 nm or 660 nm, or with fluorescence emission at the latter wavelength.

Other features of this MC-V probe differ from those of MC-I and MC-II. MC-V is uncharged and thus would be less likely to move electrophoretically in the membrane. A useful feature of this type of probe is the wide spectrum of membranes with which it interacts; not only does it bind submitochondrial particles and mitochondria, as do MC-I and MC-II, but also the plasma membranes of various types of cells such as ascites tumor cells, and of tissues such as heart and brain as well. In addition and most important, MC-V readily occupies the membranes of chromatophores from various types of bacteria. L. Cohen (22) indicates that it will also bind the squid axon, showing a voltage-sensitive response that is one-fiftieth that of MC-I (L. Cohen, B.N.

Salzberg, and W.N. Ross, personal communication).

The structure shown in Figure 12 is that of R-6, a starting material for Brooker's M-84 (1). A contaminant of this fraction is more effective as a probe, although its structure has not been completely determined. Mr. Greg Stern and Mr. Paul Russ, with the cooperation of Dr. Barry Cooperman, have purified the R-6 fraction extensively, and the purified fractions have been shown to exhibit similar properties in the chromatophore membranes as does the structure shown in Figure 12.

*Fluorescence lifetime measurements.* In collaboration with Dr. James Callis, preliminary measurements of the lifetime of MC-V have been made with a mode-locked laser system (courtesy of Dr. Robin Hochstrasser and his colleagues at the Materials Science Center). Table I give values of the fluorescence lifetime as a function of time after the addition of succinate to submitochondrial particles. The data yield the sequence of lifetimes which clearly show a decrease, either by one- or two-exponential deconvolution, in relation to the passage of time after succinate addition. This identifies the decrease of fluorescence emission occurring on energization of mitochondrial membranes with a decrease of quantum yield.

When added to artificial liposomes, the probe MC-V shows a considerable enhancement of fluorescence, as indicated in Figure 13. In this case the probe fluorescence is excited at 605 nm and the emission is measured from 620 to 700 nm. The emission peak in water is at approximately 630 nm and is shifted to 655 nm on addition of the phospholipid vesicles obtained from sonicated dipalmitoyl lecithin. Two additions of $K_2SO_4$ cause no shift of the maximum and a slight increase of emission intensity. Thus the probe readily occupies artificial lipid membranes.

The titration of MC-V with egg lecithin vesicles is illustrated in Figure 14 where the excitation is at 577 nm and the emission is measured in a broad band near 650 nm. The fluorescence increase with increasing egg lecithin concentration is shown for two concentrations of the merocyanine (2 μg per ml and 5 μg per ml). It is seen that the fluorescence increases with the egg lecithin concentration, reaching a plateau at approximately 100 μM egg lecithin with the lower concentration of MC-V. It should be noted, however, that at the higher concentration of MC-V the high ratios of probe to lipid give a sufficiently high occupancy of the membrane by the probe that concentration quenching is observed, and thus the trace starts with approximately zero initial slope (open circles). This phenomenon probably also occurs at the lower concentration of MC-V and at lower concentrations of egg lecithin.

The titrations of the red shift, as measured by the dual wavelength technique at 630-720 nm, follow rectangular hyperbolae very closely since concentration quenching does not occur in these absorbancy measurements.

*Red shifts associated with membrane energization: ATP effects.* Figure 15 presents three examples of the red shift observed on energization of various vesicular preparations by ATP and pyrophosphate. This figure shows three traces; running from top to bottom, they indicate the red shifts corresponding to succinate energization of OESMP (23), ATP energization of magnesium-ATP particles (24) and ATP energization of a reconstituted, membrane-bound ATPase composed of phospholipid vesicles and a hydrophobic protein (25). It is seen that the extent of the red shift is greatest for the electron transport activated system, next largest for the reconstituted system, and least for the magnesium-ATP particles even though 2.5 as much protein is present. Basically, these responses are very similar, but the local charge distribution may differ, particularly in the magnesium particles. It is unlikely that the lipid composition of the membrane is a factor in the probe response, in view of the very large difference between the soybean phospholipid vesicles and the natural mitochondrial membrane obtained from beef heart.

Energization of the chromatophores is indicated in Figure 16 by the addition of ATP or pyrophosphate to R. rubrum chromatophores in the dark. In Figure 16A, pyrophosphate is the energy donor. Trace I represents the corrected baseline in which the absorption of carotenoid, probe and chlorophyll have been cancelled by the baseline correction computer. The succesive scans #2 - #8 show the progression of the red shift both in intensity and wavelength, as a function of time after the addition of 330 μM pyrophosphate (26). The spectral shift persists as long as the pyrophosphate is present.

In Figure 16B, Trace #1 represents the baseline in which the majority of the absorption of carotenoid, of the added probe, and of the bacteriochlorophyll has been cancelled out by the normalization process. In Trace #2, 330 μM ATP is added, and since the scanning region covers only a portion of the merocyanine band, the prominent feature is the loss of absorption near 630 nm, corresponding to the preceding figure where phospholipid vesicles were used. The energization is complete within a single scan and persists as long as the ATP is present. The wavelength shift is identical to that caused by pyrophosphate and of very nearly the same amplitude of absorbancy change.

*Effect of a valinomycin-induced diffusion potential upon MC-V fluorescence and absorption.* Figure 17 illustrates the effect of increasing $K^+$ concentrations upon the red shift and fluorescence change of MC-V occupying the membrane of R. rubrum chromatophores in the presence of valinomycin. It is seen that there is an approximately linear relationship between the external $K^+$ concentration and the change of fluorescence. The red shift, approaches a linear relationship at high KCl concentrations but deviates significantly at lower KCl concentrations, suggesting a combination of linear and quadratic (or higher power) responses. The purpose of the graph is to identify that MC-V is responsive to the diffusion potential caused by a valinomycin-stimulated $K^+$ movement. The exact nature of the relationship is difficult to determine, particularly in view of the hypothesis proposed by Witt and his co-workers that fixed charges could displace the operating point of the electrochromic probes to a linear region of a quadratic characteristic (14).

*Membrane energization by illumination of chromatophores.* In addition to the methods described above for energization of the chromatophore membrane in the dark by addition of ATP or pyrophosphate, and the induction of the red shift by a valinomycin-induced $K^+$ gradient, it is possible to cause similar effects by illumination. These are shown as an absolute and a difference spectrum in Figure 18. In Figure 18A (left) the initial baseline of Trace 1 corresponds to the dark chromatophores containing 11 μM bacteriochlorophyll. Illumination at 860 nm causes small carotenoid - like shifts in the region at 530 nm but no significant shifts in the region where probe absorbancy changes would be expected (Trace 2). 16 μM MC-V is then added, and its absorption band is shown clearly in Trace 3. Illumination at 860 nm (which causes no detectable interference with the spectral recording) then causes a clear red shift of approximately 10 nm, as shown in Trace 4. The response is reversible on cessation of illumination.

If now the experiment is repeated and the computer memory is used to cancel out the absorption of both the probe and the chromatophores, then the flat baseline "A" is obtained in the dark chromatophores (Figure 18B). Illumination at low intensity give Trace B with a trough at 630 nm and a peak at approximately 650 nm. Increasing the light intensity causes significant increases in the amplitude of the absorbancy changes but most important, causes the isosbestic point to shift approximately 2 nm to the red, signifying the shift of the absorption band rather than the formation of a new band. Further intensity changes cause further shifts. In fact, a plot of the intensity changes or, indeed, of the position of the isosbestic point as a function of the light intensity gives a non-linear response due to an alteration of the polarizibility of the excited state of the probe (see equation 3).

In summary, four methods have been employed to induce the red shift of the MC-V probe: ATP, pyrophosphate, a KCl-induced diffusion gradient, and finally, illumination of the chromatophores.

*Kinetics of the MC-II response.* A further approach to the identification of the nature of the probe response is afforded by the kinetics of the red shift. These should correspond to the establishment of the localized or delocalized potential to which the probe is responding. The simplest method of initiating energization kinetcis is by flash illumination of a chromatophore suspension. Delivery of oxygen pulses to anaerobic, de-energized suspensions of submitochondrial particles has been used as well, but this process is relatively slow, as is shown in the time-resolved spectra. Thus, chromatophores are employed since they can most readily be activated and furthermore show the most rapid response yet observed.

The kinetics of the red shift in response to a 2 msec light flash at 860 nm for both the response of wild type and the blue-green mutant of R. rubrum are shown, respectively, in the lower and upper traces of Figure 19, The half-time for the response of the wild type is 55 msec, and of the mutant, about 50 msec. Less of the mutant membranes are used than in the case of the wild type, and the response is smaller. However, the kinetics of the response are similar in the presence or absence of carotenoid in the chromatophore membranes. Both show an energy-dependent red shift with a characteristic half-time of about 50 msec at 23°. The following results, characterizes the kinetic response; 1) a half-time of about 50 msec; 2) 15% of the maximal steady state response is obtained in single saturated flashes (19); 3) the half-time for the fluorescence decrease is of the same order and possibly longer than that for the red shift (19); 4) the responses of both the red shift and the fluorescence change are completely eliminated by $K^+$ and valinomycin. Similar responses are obtained from chromatophores derived from wild and mutant R. rubrum and from Rps. spheroides.

## Discussion

In a previous summary of work on deep and shallow probes of natural and artificial membranes (8) it was emphasized that the energy-dependent responses of fluorescent probes were observed in the region of the aqueous interface; as the probe chromophore became deeper embedded in the membrane, so was there less response to membrane energization, and more response to collisional quenching by ubiquinone component of the hydrocarbon phase. The merocyanine probes are unique to the point that they are collisionally quenched by ubiquinone. identifying that they enter the hydrocarbon phase of the membrane, and yet they exhibit energy-dependent responses. Furthermore, the merocyanine probes demonstrate a fin fine structure of energy-dependent response, a dist distinction between nigericin and valinomycin that affects separate regions of the spectrum of the probe. Thus, separate portions of the probe response might be assigned to different localizations of charge separations in the membrane.

Considering first the question of how the merocyanines respond simultaneously to ubiquinone collisional quenching and to membrane energization, the extended nature of the probe bridges the aqueous interface and the hydrocarbon region of the membrane. Thus, one end of the probe may be at or near the interface and responds to the environmental changes observed in the case of ANS and similar probes localized in that region (8). In addition, the merocyanine probe is long enough to span the region from the aqueous interface to the hydrocarbon phase of the membrane, and thus is responsive to collisional quenching by ubiquinone. Supplementary evidence that the merocyanine, MC-V, lies with its transition moment at a significant angle with respect to the plane of the membrane is already determined by preliminary optical studies (9). The response of the probe to localized or delocalized charge separations depends upon thedipolar nature of the probe itself; thus, charge separation across membrane proteins located in the aqueous interface would be sensed by the probe, together with that portion of a delocalized or transmembrane charge.

The question of whether or not the probe responds to both local and transmembrane charge separations appears to be clearly indicated by these experiments, namely, the diffusion potential induced by a $K^+$-gradient causes similar changes of the red shift and the fluorescence intensity of MC-V, according to the data of Figure 17 above. At the same time, the fluorescence intensity and the absorption of the probe is altered by the replacement of $H^+$ by $K^+$, (effect of nigericin), and by a variety of anions; to mention but a few, $CNS^-$, $SO_4^=$, and $Cl^-$, dilute acetate being one of the few anions which is without effect upon the probe responses to membrane energization.

On this basis we conclude that the probe responds to both a localized and delocalized charge separation. It becomes of the greatest interest to compare its response with that of the carotenoid, which has been stressed by the groups at Bristol (27) and Berlin (11-14) as a "membrane potential indicator". The responses of these two probes have been compared from a number of aspects in a previous communication (19), and those findings may be summarized briefly here.

While localization studies of the merocyanine probes are incomplete, energy transfer, collisional quenching, and transition moment studies suggest an extension of the probe between the aqueous interface and the hydrocarbon phase. Similar studies on the location of carotenoid are not available, largely because carotenoid is not fluorescent, and is not highly solvatochromic, as are the merocyanine probes. However, it is a reasonable assumption, based just on solubility data, that carotenoid occupies the hydrocarbon phase of the membrane, and based upon extraction studies, is intimately associated with the reaction center.

The electrochromic possibilities for carotenoid and merocyanine seem to differ distinctly according to the work of Schmidt and Reich and Witt, who found that the carotenoid, lutein, can show only a quadratic electrochromic response. Assuming lutein to be a representative example of the carotenoids in the chloroplasts and chromatophores, (neurosporene appears to be the principal component involved in the light-induced responses of Rps. spheroides chromatophores (27)), it can be concluded that the electrochromic responses of carotenoid must be quadratic in nature. Experimental observations, however, particularly those of Jackson and Crofts (28, 29) show linear rather than quadratic electrochromism, as determined by the red shift of the carotenoid to KCl-induced diffusion gradients in the presence of valinomycin. Merocyanines, on the contrary, exhibit both linear and quadratic electrochromic responses under conditions of KCl-induced diffusion potential in a chromatophore membrane. It appears that the fluorescence response is approximately linear, while the red shift is distinctly non-linear with respect to the diffusion potential, suggesting that both linear and quadratic electrochromism can be indicated by these two optical parameters of the merocyanine probe.

The explanation offered by Witt (14) for reconciling the discrepancy between the quadratic electrochromism of lutein observed in vitro, and the linear electrochromism of all carotenoids observed in situ is that the carotenoid is located near a very large fixed charge in the membrane, roughly ten-fold greater than the possible membrane potential. This fixed charge shifts the electrochromism of the carotenoid to a region of the quadratic response over which the comparatively insignificant variation of the membrane potential would result in linear electrochromism. This extreme hypothesis leads, of course, to a number of experimentally testable points, particularly, the expectation of a large shift of carotenoid absorption on extraction from the membrane, the expectation of a small change of the "fixed charge" as being more important in altering the probe spectrum than is the transmembrane potential, and finally, the expectation of a very small shift of wavelength due to a very small red shift of the probe due to the membrane potential (Crofts most recently (27) reports a 10 nm shift, a very large shift -approximately 10 nm- for neurosporene in *Rps. spheroides* mutant). Finally, and most important, is the experiment on linear and quadratic electrochromism in which a reversal of the membrane potential would cause a reversal from a red shift to a blue shift in the case of linear electrochromism and a continued red shift in the case of quadratic electrochromism. In fact, according to Witt's explanation, a reversal of the membrane potential would *not* overcome the fixed potential and thus would not reverse the potential across the probe. However, Crofts (30) reports that the reversal of the membrane potential in *Rps. spheroides* chromatophores causes a reversal of the electrochromic response, supporting the concept that the carotenoid responds to diffusion potentials *in situ* that cannot be explained in model membrane experiments with lutein.

It seems impossible at the present time to reconcile the conflicting theoretical and experimental observations on carotenoids as electrochromic probes, except to suggest that an extensión of the experiments on KCl-induced diffusion potentials over the widest possible ranges-seems essential to detect nonlinearity in the response. Also, the possibility that lutein is not representative of carotenoids in general in its failure to exhibit a linear electrochromism suggests further in vitro experimentation with other types of carotenoids, particularly neurosporene.

A further point in comparing merocyanine and carotenoid probes is a kinetic one. Characteristically, carotenoids show a very rapid red shift, cited to be in the nanosecond region by many observers, but most recently found to be in the picosecond region(31). This most recent result suggests that the carotenoid response occurs before the transmembrane potential can be established. Similarly, and on a much slower time scale, is a further slow phase that follows the fast phase (32); i.e., the "charge separation" is not complete during a short flash.

A second point is that the full "potential" is not established in a single saturating flash. On the basis of the hypothesis that the maximal charge separation across a single reaction center is caused by a single flash and is separated across the membrane as well, we find the fast and slow phases of the light-induced carotenoid red shift as yet uninterpretable as membrane potentials. However, a more realistic interpretation is that the carotenoid response is mainly to a localized charge separation, and that only a portion of this appears as a membrane potential, which in turn depends upon the number of units that have been activated by steady state illumination.

This complicated response may be contrasted with the apparently simple response of the merocyanines in chromatophores; it is completely energy-dependent, no nanosecond fast phase is seen; instead, the response is monotonic, with a halftime of 50 msec in R. rubrum and Rps. spheroides chromatophores.

One interpretation of these observations is that the initial phase of the carotenoid response is primarily to local charge redistributions across the reaction center. This explains why the initial phase is not abolished by uncoupling agents and ionophores, and why there is no fast phase of the merocyanine response. The question of why the merocyanines do not respond to the local charge separation across the reaction center is an important one, and is not yet completely resolved. Two points are relevant: the observations of L. B. Cohen (22) indicate that MC-V responds to the transmembrane potential in the squid axon, and thus the probe is capable of responding to a rapid event. The second is that the location of the MC-V may be relatively distant from the reaction center, as compared with carotenoid.

The possibility of a correlation between the slow phase of the carotenoid response and the total merocyanine response is not eliminated by these studies, although preliminary data on a membrane occupied simultaneously by merocyanine and carotenoid suggest that the merocyanine reaches its maximal red shift during the decay phase of the carotenoid response, rather than during the rising phase.

This result is complicated by the observed acceleration of the carotenoid response by added merocyanine together with a diminution of the carotenoid response under these conditions. These results were obtained in spite of the fact that no uncoupling of chromatophore energy conservation is caused by concentrations of the merocyanine probe employed. At the present time, however, there is no clear correlation between the carotenoid and the merocyanine responses to a single flash in *R. rubrum* and *Rps. spheroides* chromatophores. Thus, on a conservative basis, it is concluded that both probes respond principally to local charge redistributions in the single flash response of chromatophores. Yet both respond to the diffusion potential in the valinomycin supplemented membrane.

On this basis, the sequence of events that occurs on illumination of chromatophores is first the abovementioned "instantaneous" charge (or structure change charge) separation or structure change in the reaction center. This is followed by secondary changes in the carotenoid response related to localized charge redistributions in electron carriers, which can be detected by both probes. In the case of the merocyanine, the most likely response observed is a charge redistribution across the ATP-ase itself.

Summary

Voltage-sensitive probes offer a new insight into processes of membrane energization and deenergization. Four conditions are studied here: substrate and oxygen activation, ATP activation, pyrophosphate activation, and light activation, in mitochondria, submitochondrial particles, and chromatophores. Tentative evaluations of the probe location as mapped by fluorescence characteristics, by energy transfer, by ubiquinone quenching, and by transition moment determinations suggests that the merocyanines bridge the gap between the deep and shallow probes (for example, anthroyl stearic acid, a deep probe, and ANS, a shallow probe) in that they respond to membrane energization on the one hand (a shallow response) and ubiquinone collisional quenching on the other hand (a deep probe response). The merocyanine probes respond to membrane energization with the red shift of 10 to 20 nm depending upon the system, a decrease of the absorbance maximum together with a fluorescence decrease. The time course and the ion responsiveness of the two phenomena differ, the red shift is relatively rapid (down to 50 milliseconds in light activated chromatophores) while the absorption intensity and fluorescence intensity are relatively slow in submitochondrial particles. Fluorescence intensity and absorption intensity changes are eliminated by nigericin plus potassium while the red shift is eliminated only by valinomycin plus potassium or other uncouplers, the total responses being entirely energy dependent.

A comparison of electrochromic responses of the intrinsic probe carotenoid and the extrinsic probe merocyanine suggests that the two are in different locations responding to different localized events. While both probes can be demonstrated to respond to delocalized potentials as evidenced by their response to valinomycin stimulated potassium diffusion potentials. They may respond to either localized or delocalized potentials in membrane energization and de-energization. This distinction is further complicated by the fact that the potassium valinomycin responses are essentially linear with diffusion potential while the basic response of the carotenoid probe is quadratic while that of the merocyanine is both linear and quadratic.

---

This research has been supported mainly by NIGMS-12202 and in part by NINDS, NIMH, and NCHD.

## References

1. Brooker, L.G.S., Craig, A.C., Heseltine, D.W. Jenkins, P.W. and Lincoln, L.L. J. Am. Chem. Soc. 87 (1965) 2443.

2. Platt, J.R. J. Chem. Phys. 25 (1956) 80.

3. Platt, J.R. J. Chem. Phys. 34 (1961) 862.

4. Strichartz, G. Personal communication.

5. Cohen, L.B. Nature New Biol. 241 (1973) 159

6. Cohen, L.B., Salzberg, B.M., Davila, H.V., Ross, W.N., Landowne, D., Waggoner, A.S. and Wang, C.H. J. Memb. Biol. In press.

7. Azzi, A., Chance, B., Radda, G.K., and Lee, C.P. Proc. Nat. Acad. Sci. USA 62 (1969) 612.

8. Chance, B. Proc. 4th. Intern. Biophysics Congr. (Symposia) (Moscow, 1972; L. Kayushin, Ed.) In press.

9. Blasie, J.K. and Harrison, S. Personal Communication.

10. Bücher, H., Wiegand, J., Snavely, B.B., Beck, K. H., and Kuhn, H. Chem. Phys. Lett. 3 (1969) 508.

11. Reich, R. and Schmidt, S. Ber. Bunsenges. Physik. Chem. 76 (1972) 589.

12. Schmidt, S. and Reich, R. Ber. Bunsenges. Physik. Chem. 76 (1972) 599.

13. Schmidt, S. and Reich, R. Ber. Bunsenges. Physik. Chem. 76 (1972) 1202.

14. Schmidt, S., Reich, R. and Witt, H.J. Paper delivered at the Second Intern. Congress on Photosynthesis, Stresa, 1971.

15. Cheng, W. A. Katchalsky Memorial Volume, J. Memb. Biol. In press.

16. Chance, B. and Hagihara, B. Proc. Vth Intern. Congr. Biochem. V (1963) 3.

17. Lee, C.P. and Ernster, L. Methods in Enzymol. X (1967) 543.

18. Baltscheffsky, M. Nature 216 (1967) 241.

19. Chance, B. and Baltscheffsky, M. A. Katchalsky Memorial Volume, J. Memb. Biol. In press.

20. Montal, M., Chance, B., Lee, C.P. and Azzi,A. Biochem. Biophys. Res. Commun. 34 (1969) 104.

21. Montal, M., Chance, B. and Lee, C.P. J. Memb. Biol. 2(1970)201.

22. Cohen, L.B., Salzberg, B.N. and Ross, W.N. Personal communication.

23. Lee, C.P. and Ernster, L. 1966. In: Regulation of Metabolic Processes in Mitochondria. Vol. 7, p. 218, eds. Tager, J.M., Papa, S. Quagliariello, E. and Slater, E.C. BBA Library, Elsevier, Amsterdam.

24. Löw, H. and Vallin, I. Biochim. Biophys. Acta 69 (1963) 261.

25. Kagawa, Y., Kandrach, A. and Racker, E. J. Biol. Chem. 248 (1973) 676.

26. Baltscheffsky, M. Biochem. Biophys. Res. Commun. 28(1967)270.

27. Jackson, J.B. and Crofts, A.R. Eur. J. Bio-Chem. 18 (1971) 120.

28. Jackson, J.B. and Crofts, A.R. FEBS Letters 4 (1969) 185.

29. Crofts, A.R. This volume.

30. Crofts, A.R. Personal Communication.

31. Nedzel, T.L., Rentzepis, P.M. and Leigh, J.S. Science 182 (1973) 238.

32. Baltscheffsky. Personal communication.

---

A

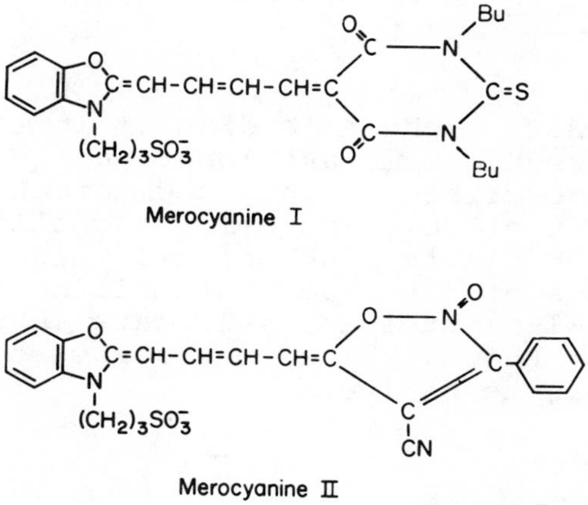

Merocyanine I

B

Merocyanine II

LBC-3

Figure 1. Chemical structures for two of the merocyanine dyes used in these studies. (LBC-3).

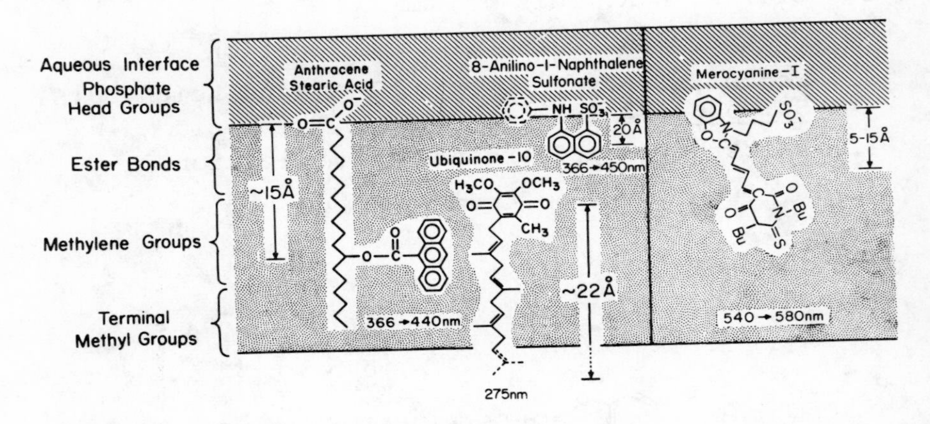


Figure 2. A schematic diagram indicating the structure and membrane location of three of the probes described in this communication, together with that of ubiquinone-10, a collisional quencher of the fluorescence of all three probes. The locations of anthracene stearic acid (AS) and 8-anilino-1-naphthalene sulfonate (ANS) are accurately known, while that for merocyanine-I (MC-I) is tentative. (CPL-452)

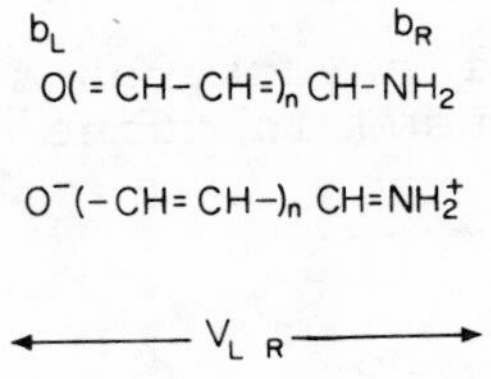

---

$b_L$ $b_R$

$$O(=CH-CH=)_n\,CH-NH_2$$

$$O^-(-CH=CH-)_n\,CH=NH_2^+$$

$\longleftarrow V_{L\ R} \longrightarrow$

Figure 3. Illustrating dipolar ion formation in merocyanine dyes. (ME-229)

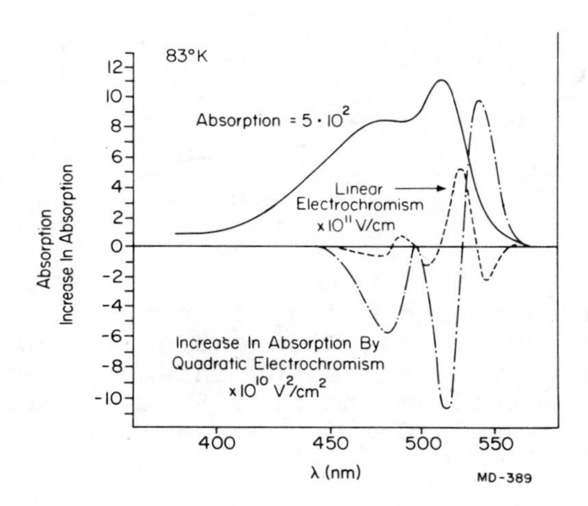


Figure 4. Illustrating linear and quadratic electrochromic shifts. The solid trace represents the absorption spectrum; the dashed trace, linear electrochromism the dash-dot trace, quadratic electrochromism. The scale at the left applies with the appropriate scale factors indicated on the diagram. The merocyanine dye is 1-stearyl-2-(3'-nitro-3'-formyl-allyliden)-1-2-dihydrochinolin. The spectra were recorded at 83° K and 6° K. (Redrawn from Ref. 10, Bücher et al.)

(MD-389)

---

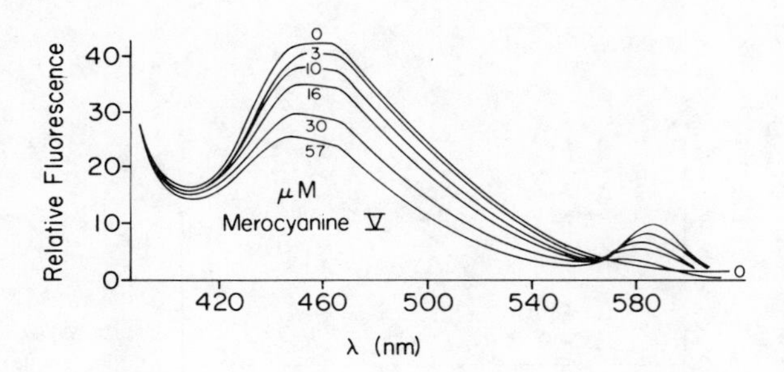


Figure 5. Illustrating energy transfer from the AS probe and added MC-I. This energy transfer is identified with the increasing intensity of the fluorescence emission band at 585 nm as the concentration of added MC-I is increased from 0 to 57 μM. (JMV-67)

---

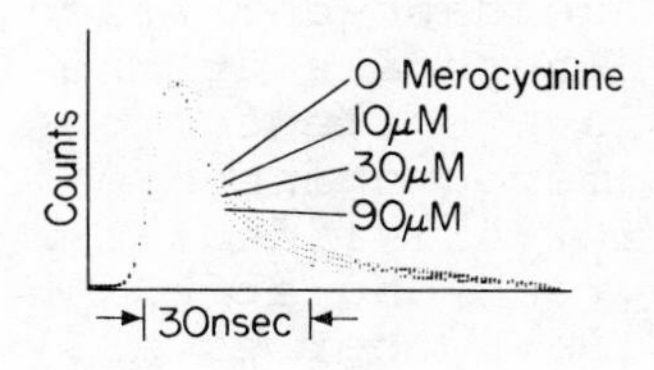


Figure 6. Decrease of lifetime of the AS probe with increasing concentration of MC-I, as recorded from corrected traces of the ORTEC photon-counting lifetime machine (experiments in collaboration with Dr. Jame Callis). (JMV-67).

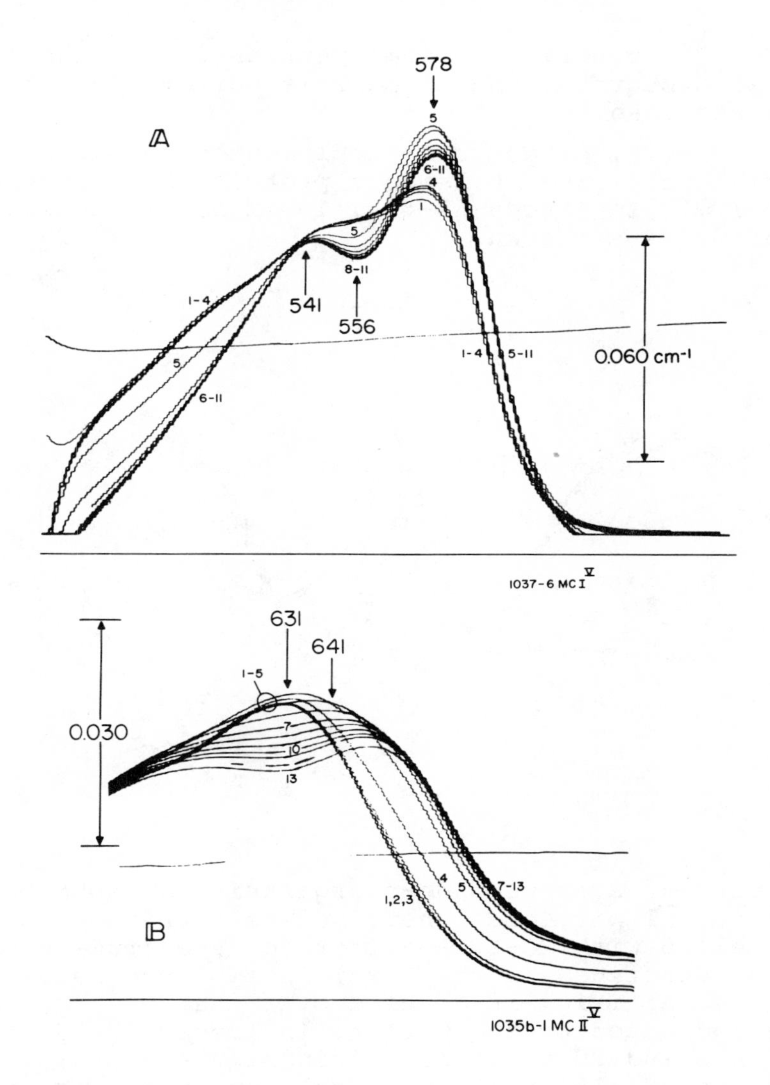


Figure 7. Illustrating the energy-dependent responses of MC-I (A) and MC-II (B). In 7A, the submitochondrial particles, prepared by the method of Lee and Ernster (17), 0.2 mg of protein/ml are supplemented with 1 μg of oligomycin/ml and 7 μg of MC-II. In Traces #5-#11, 10 mM succinate is ad-

ded. The reaction medium contains 0.2 M mannitol, 0.25 M sucrose, and 10 mM Tris buffer, pH = 7.4. 44 sec/scan.

In 7B, oligomycin-supplemented submitochondrial particles, 0.2 mg of protein/ml, 3.5 μg of MC-I/ml. In Traces #3-#13, 10 mM succinate is added. 44 sec/scan.

(1037-6; 1035b-1)

---

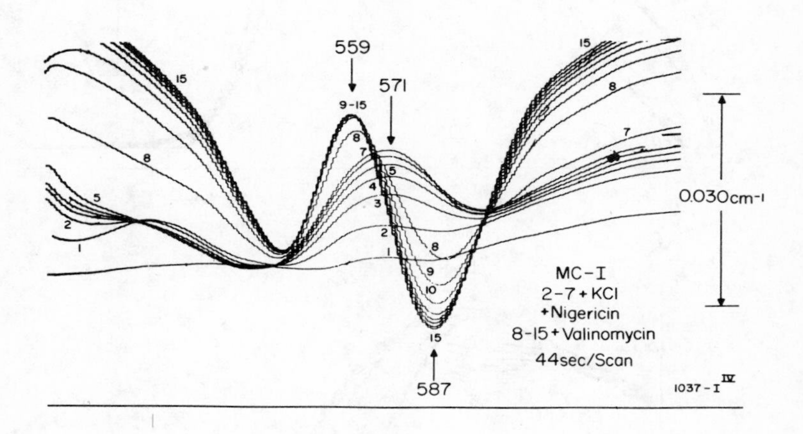


Figure 8. Step-wise de-energization of submitochondrial particles energized as in Figure 7A and labelled with MC-I. A difference spectrum recording (described in the text) is employed here, and the sensitivity is doubled over that of Fig. 7A, as indicated by the marker. In Traces #2-#7, 50 mM KCl and 20 μg of nigericin/ml are added; in Traces #8-#15, 0.13 μg of valinomycin/ml is present.

(1037)

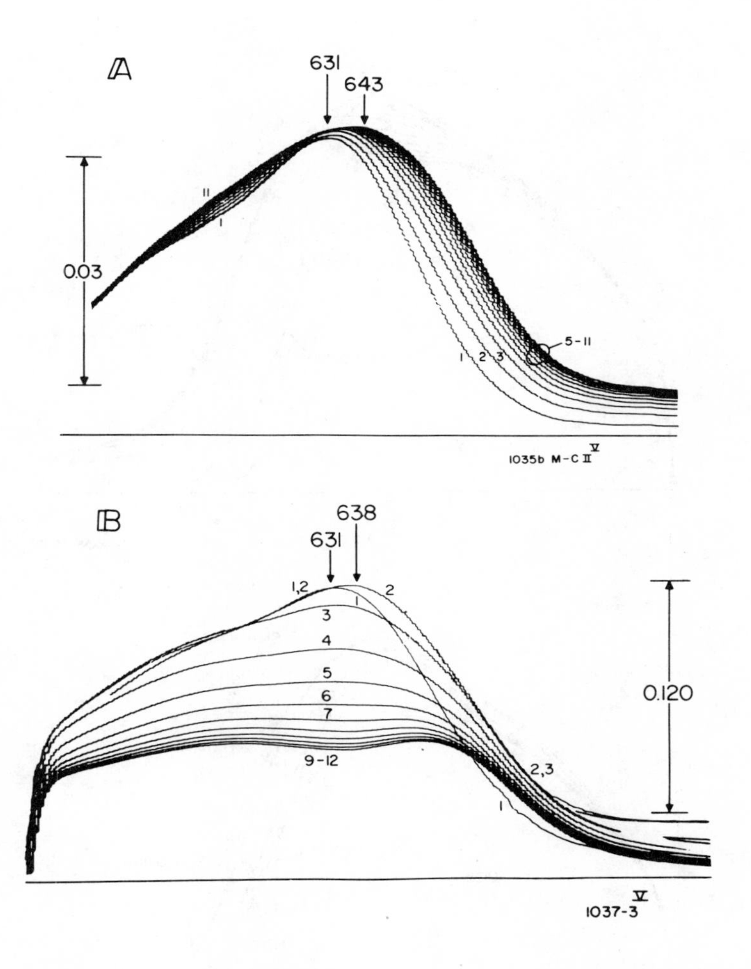


Figure 9. Energization of submitochondrial membrane fragments by succinate oxidation in the presence and absence of KCl and nigericin, using MC-II as a probe. Submitochondrial particles, 0.2 mg of protein/ml in 0.25 M mannitol-sucrose-Tris-Cl buffer, pH = 7.4; 6.7 μg of MC-II/ml, 10 mM succinate. In A, 50 mM KCl and 1 μg of nigericin/ml are present.

(1035b; 1037-3).

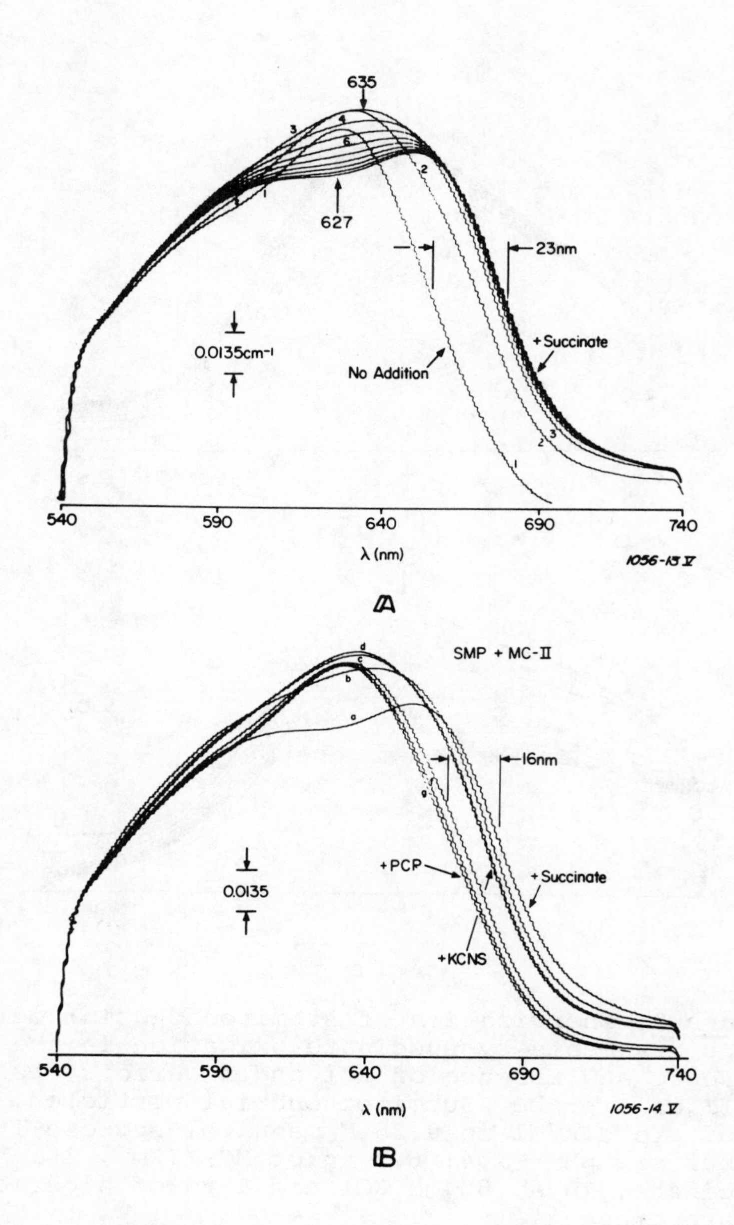


Figure 10. (See legend on the next page).

Figure 10. Energization/de-energization responses of submitochondrial particles probe -labelled with MC-II. Submitochondrial particles, 0.3 mg of protein /ml, supplemented with 1 μg of oligomycin/ml and 3 μg of MC-II/ml.

A. Energization. Trace #1 represents probe only; in Traces #2 onwards, the preparation is supplemented with 5 mM succinate.

B. De-energization by addition of 5 mM KCNS (Traces b - d) followed by 20 μM pentachlorophenol (Traces f - g).

(1056-13; 1056-14)

---

Figure 11. Energization/de-energization responses of submitochondrial particles supplemented with MC-II. Oligomycin supplemented submitochondrial particles, 0.2 mg of protein/ml, labelled with 6.7 μg of MC-II per ml in mannitol-sucrose-Tris, pH = 7.4. The absorption of the probe is cancelled out by the computer memorization.

A. Energization. In Traces #2 - #10, 20 mM succinate is added.

B. De-energization. In Traces #2 - #11, 20 mM KCl and 20 μg of nigericin/ml are added; in Traces from #12 onwards, 0.13 μg of valinomycin/ml is present.

(1039-5,6)

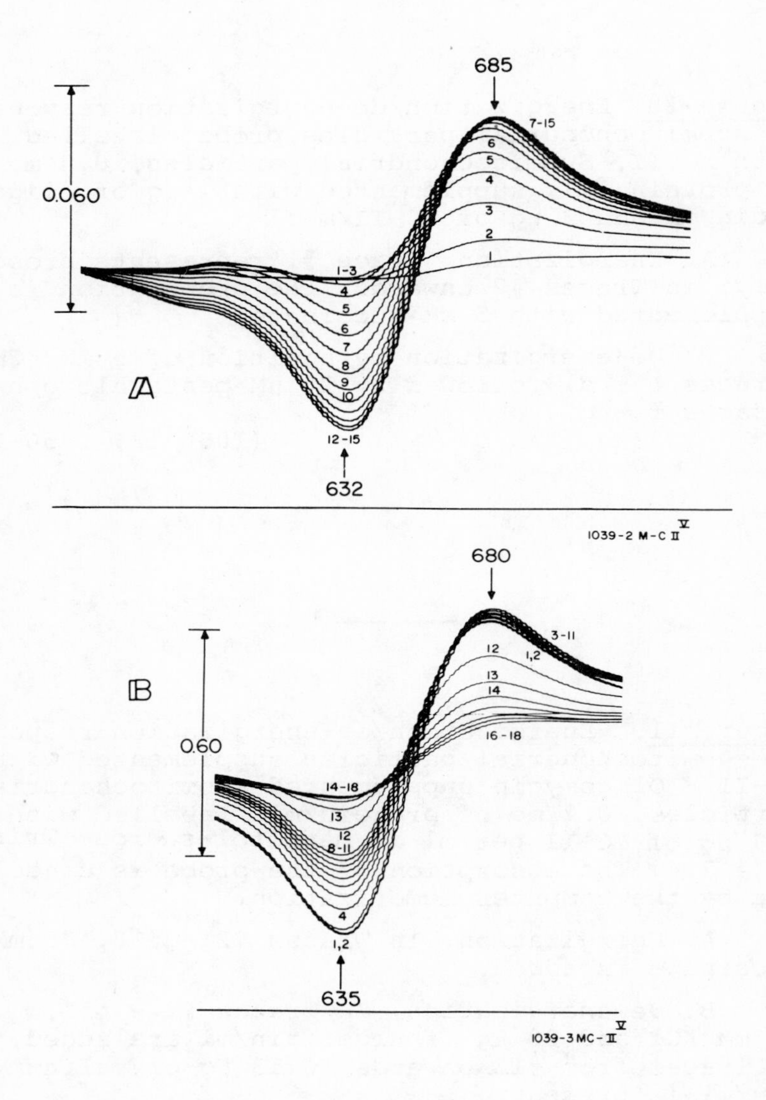


Figure 11. (See legend on the previous page).

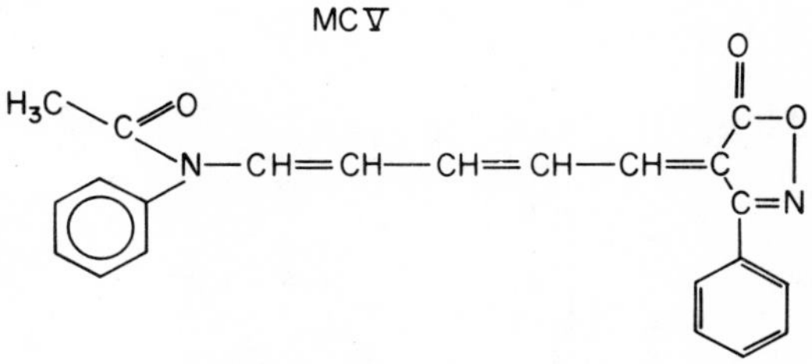

MC V

$H_3C-C(=O)-N(C_6H_5)-CH=CH-CH=CH-CH=C$ (isoxazolone ring: C(=O)–O–N=C–$C_6H_5$)

Figure 12. Component R-6 of Brooker (cf Reference 1)

---

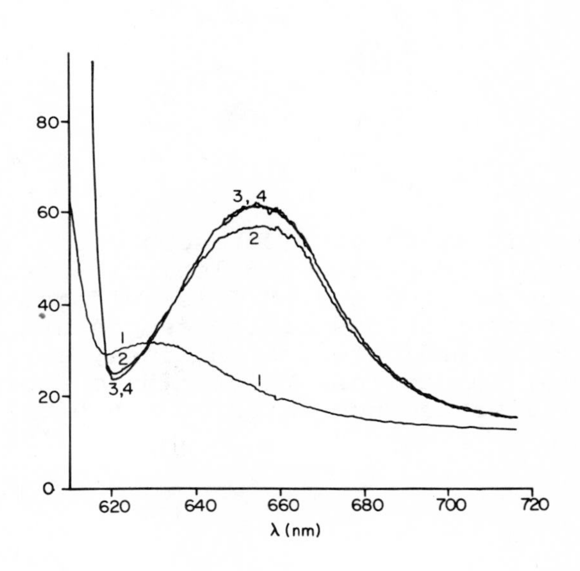


Figure 13. Emission spectra of MC-V in the presence and absence of dipalmitoyl lecithin vesicles. Trace #1, 1.7 μg of MC-V in 5 mM Tris-sulfate, pH = 7.4. Trace #2, in the presence of dipalmitoyl lecithin vesicles, 17 μl phospholipid/ml. Trace # 3, supplemented with 1.7 mM $K_2SO_4$. Trace #4, suplemented with 3.4 mM $K_2SO_4$. Excitation at 605 nm. (Data obtained by Ms. J.H. Owen)

(JHO-10)

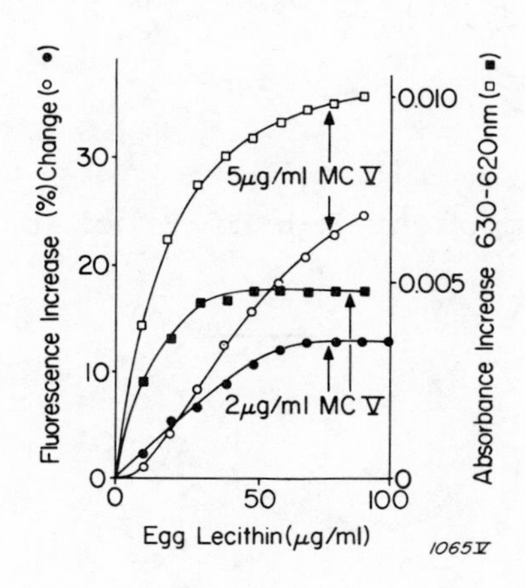


Figure 14. Titration of fluorescence and absorption change of MC-V with egg lecithin vesicles. The reaction medium is 5 mM Tris-sulfate, pH = 7.4. The concentrations of the components are indicated on the ordinate and abscissa, as are the experimental conditions. Fluorescence excitation at 577 nm; fluorescence emission > 640 nm. (1065-5)

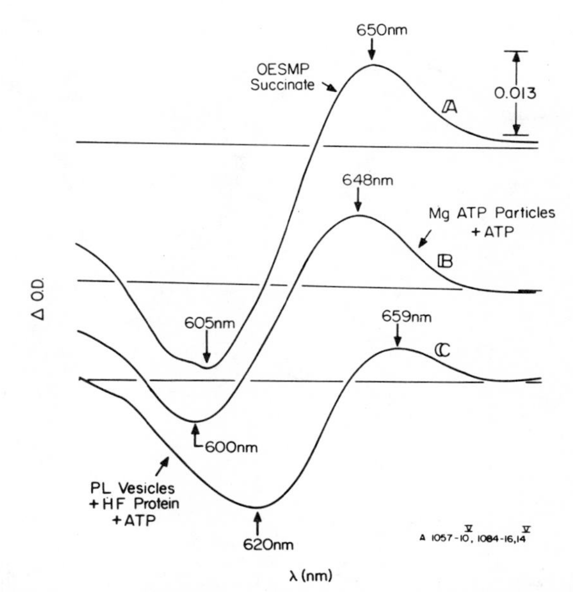


Figure 15. Responses of MC-V to membrane energization by electron transport in submitochondrial particles (A); by addition of ATP to $Mg^{++}$-ATP particles (B); by ATP addition to the reconstituted ATPase.

A. Top trace, oligomycin-supplemented submitochondrial particles, 0.2 mg of protein/ml, 7 μg MC-V/ml in 5 mM Tris-Cl, pH = 7.4. The baseline is memorized by the computer. Difference spectrum recorded 88 sec after the addition of 10 mM succinate. (1057-10)

B Middle trace. $Mg^{++}$-ATP submitochondrial particles, 0.5 mg of protein/ml, 7 μg of MC-V/ml, in 5 mM Tris-Cl, pH = 7.4. The baseline is memorized by the computer after which 0.5 mM ATP is added. (1084)

C. Bottom trace. 150 μl reconstituted ATPase, 7 μg MC-V/ml, 3 mM ATP. This baseline is memorized by the computer, and the deflected trace corresponds to 5 min after the addition of 7 mM $Mg^{++}$. The bulk of the red shift is eliminated by the addition of FCCP. (Thanks are due to Dr. E. Racker for the gift of the preparation). (1084)

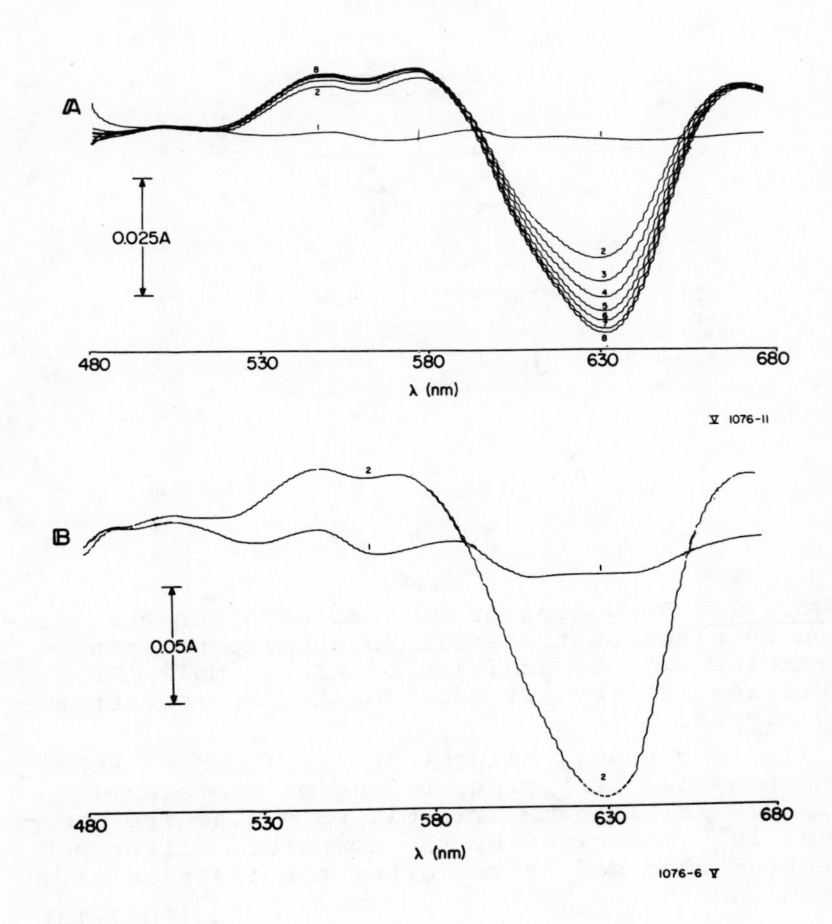


Figure 16. Membrane energization in chromatophores by pyrophosphate (A) or by ATP (B).

A) 11 μM bacterial chlorophyll; 7 μM of MC-V per ml; 6.7 mM $Mg^{++}$. The computer then memorizes the baseline and Traces #2 - #8 follow the addition of 330 μM pyrophosphate at 44 sec per scan.

B) The same conditions, except that 330 μM ATP is added, instead of pyrophosphate.

(1076-6,11)

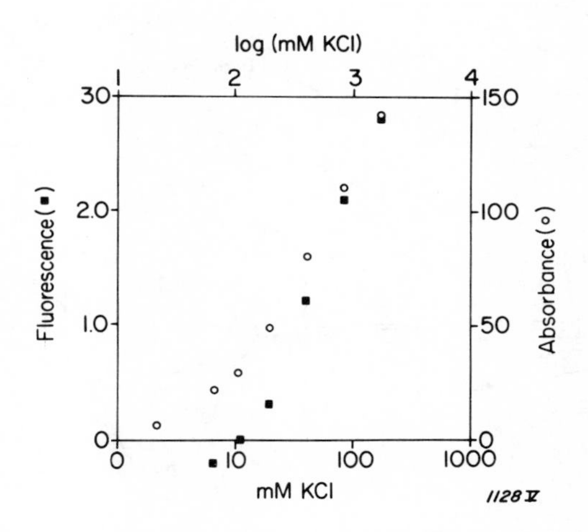


Figure 17. Effect of a valinomycin induced $K^+$ flux upon the fluorescence and absorption of the probe MC-V in *R. rubrum* chromatophores. The abscissae are in logarithmic coordinates indicating the concentration of added KCl. The ordinates represent the amplitudes of the fluorescence and absorption changes. 15 μM bacteriochlorophyll, 5 mM Tris-sulfate, pH = 7.5, 5 μM valinomycin, 7μg of MC-V per ml. 23°C.

This figure is reprinted from a previous paper (18).

(1158)

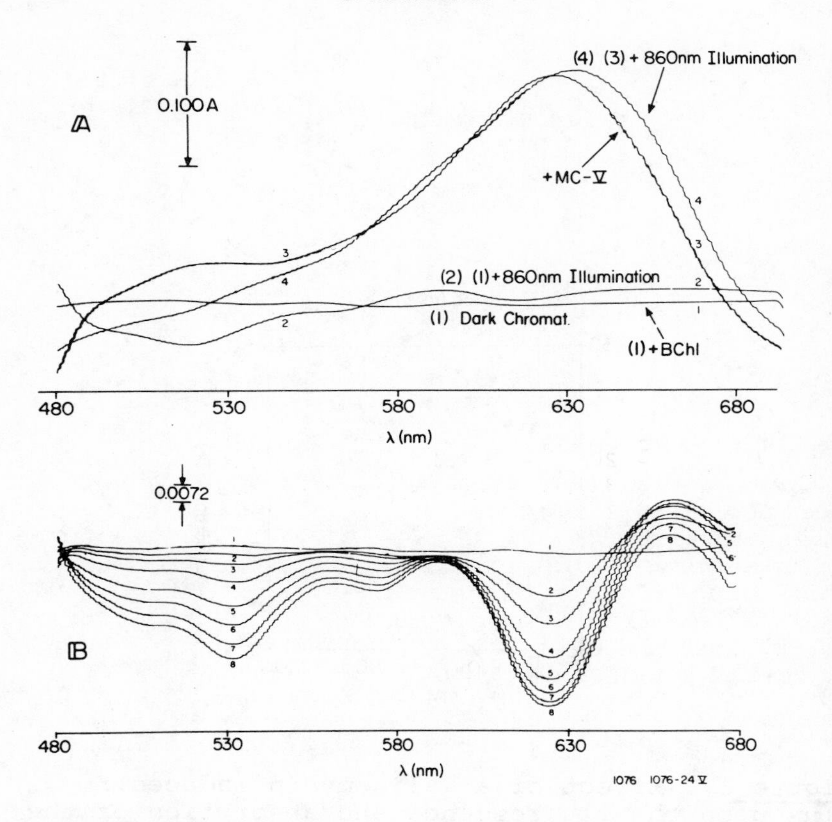


Figure 18. Illustration of the absolute and difference spectra of the response of the probe MC-V to illumination of R. rubrum chromatophores, 11 M bacteriochlorophyll, 7 g of MC-V per ml.

A) The baseline is memorized prior to the addition of the probe, and trace #2 indicates illumination of the chromatophores without the probe present. Trace #3 is following the addition of the probe, but without illumination at 860 nm. Trace #4 represents illumination following the addition of the probe. The red shift is apparent.

B) Difference spectra of essentially the same response, except that the trace equivalent to Trace #3 of 18A is read into the memory of the computer, giving flat baseline #1 of 18B. Traces #2 - #11 represent illumination at successively greater intensities.

(1076-18,24)

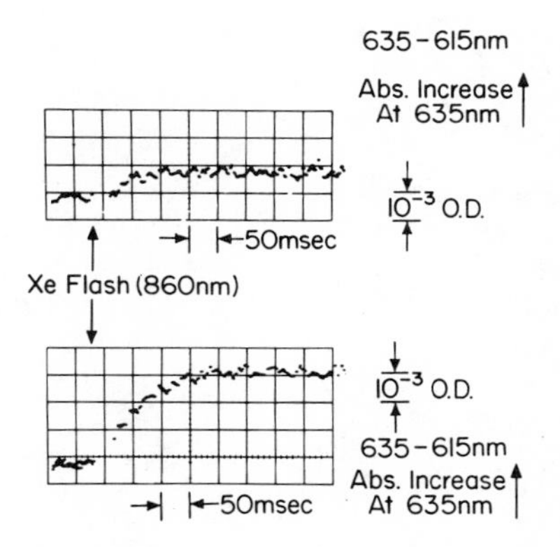


Figure 19. Illustration of the response of MC-V to a light pulse, applied to a suspension of *R. rubrum* chromatophores. A comparison of the flash response of MC-V in membranes of wild *R. rubrum* (lower trace), 90 μM bacteriochlorophyll, and the blue-green mutant of *R. rubrum* (upper trace), 18 μM bacteriochlorophyll. 14 μg of MC-V per ml, 5 mM Tris-sulfate, pH = 7.2; other conditions indicated on the diagram. (1059-7,8)

---

FLUORESCENCE LIFE-TIME DATA (M.C. V)

(nanoseconds)

| Substrate: | Substrate 0 | Succinate | Succinate |
|---|---|---|---|
| Deconvolution: | 0 | 6 min | 12 min |
| one exponential | 0.4 | 0.31 | 0.27 |
| two exponentials | 1.2 | 0.83 | 0.53 |
| | 0.33 | 0.21 | 0.06 |

M.T. - 362

# IV

# LIGHT MEDIATED PHENOMENA IN MEMBRANES

# THE ELECTRON TRANSPORT SYSTEM AS A $H^+$ PUMP IN PHOTOSYNTHETIC BACTERIA

A. R. Crofts

Department of Biochemistry
Medical School
University of Bristol
University Walk
Bristol, BS8 1TD

Studies on ion and substrate transport have led many in the field, whether working on mitochondria, chloroplasts or bacteria, to conclude that the primary driving force for transport is an electrochemical $H^+$ gradient generated at the expense of free energy made available through electron transport or ATP hydrolysis. The reasoning and experimental evidence leading to this conclusion are amply and excellently reviewed elsewhere (1-6).

Although the existence, and utility of the $H^+$ gradient in the context of ion transport is now fairly generally accepted, controversy still exists as to the relation between the $H^+$ gradient and the high energy state. The crux of this controversy is the experimental difficulty of distinguishing between a primary coupling role for the $H^+$ gradient as envisaged by Mitchell in his chemiosmotic hypothesis (1,7,8), and a secondary role in which the transmembrane gradient is considered to be in equilibrium with a high energy state which is chemical (9), or conformational (10,11) in nature, or due to changes in proton activity in a localised environment within the membrane (12), and represents the primary channel for coupling (13, 14). If the equilibration envisaged in these latter models is considered to be rapid, then no experimental criteria can be devised to test the effects of $H^+$ gradient and high energy state separately. Attention has therefore most fruitfully been directed towards establishing the free energy relation between the $H^+$ gradient and high energy state, with a view to testing the validity of the hypotheses of equilibration or equivalence. This has necessarily involved measurements of energetic potentials and the appropriate stoichiometric relationships, and has led to an exposure of some of the difficulties which these measurements entail. Indeed, until a more general agreement among workers in this field can be achieved, the outsider will continue to see this as an area of complete confusion, and the specialist proceed by judicious selection among the data available, with the danger that this will be directed by wishful thinking.

A more direct approach, and one which Mitchell (1,7,8,15) has been foremost in proposing, is an investigation of the mechanisms of the $H^+$ pumps driven by electron flow or ATP hydrolysis. The hypotheses outlined above can be expressed in mechanistic terms which allow an experimental distinction between them. In the case of the role of the electron transport system as envisaged in the chemiosmotic hypothesis, the following general criteria must be met:

i) The electron transport system must drive a $H^+$ pump.

ii) The free energy changes associated with electron flow must be accounted for in terms of work done in maintaining a $H^+$ gradient.

iii) The generation and decay of the $H^+$ gradient must be consistent with the kinetics of electron transport and of reactions which utilise the energy of the gradient.

In terms of these criteria, a variety of mechanism by which electron transport could drive a $H^+$ pump have been suggested, and some of these could be compatable with either of the general hypotheses of energy coupling above. In order to distinguish between the two hypotheses, it is necessary to postulate an additional criterion (which is not necessary to the chemiosmotic hypothesis in its most general expression).

iv) The electron transport system must be directly involved in the pumping of $H^+$.

Mitchell has suggested a number of possible mechanisms by which an electron transport system could act as a $H^+$ pump and has pointed out certain general features which might be thought of as supplementary criteria.

v) Electrogenic and neutral (H-carrying) spans of electron flow must occur across the coupling membrane.

vi) The components of the electron transport system must be arranged anisotropically in the membrane so as to correspond to these spans.

Taking these supplementary criteria, we may consider complementary kinetic and thermodynamic criteria.

vii) The separate development of electrical and chemical components of the $H^+$ gradient should be kinetically compatable with particular reactions of the electron transport system.

viii) In the coupled steady state the free energy differences between carriers defining electrogenic or H-carrying spans of the electron transport system should approach equivalence respectively with free energy of electrical or chemical components of the $H^+$ gradient.

In this paper I shall discuss our attempts to test some of these criteria using photosynthetic bacteria as experimental material.

## Photosynthetic Bacteria as Tools for Research in Bioenergetics

It is an article of faith in the field of bioenergetics that the mechanisms by which mitochondria, chloroplasts, bacteria, etc. couple electron transport and phosphorylation, are basically similar. It is therefore worth considering which of the different systems available offers the most favourable features for experimental attack on a particular problem. The photosynthetic bacteria offer several advantages as experimental material, and these are summarized below:

1) Reproducible growth of cultures either photosynthetically or aerobically.
2) Electron transport pathways inducible by growth under different conditions.
3) Availability of mutants.
4) Ease of preparation of chromatophores or respiratory particles.
5) Stability of chromatophore preparation.
6) Simplicity of the cyclic electron transport system.
7) Illumination provides both oxidising and reducing equivalents, no substrate addition.
8) Flash illumination provides a single equivalent of oxidant and reductant extremely rapidly.
9) Energy linked phenomena (carotenoid change and delayed fluorescence) as'indicators of the highenergy state.

These last few features are particularly convenient for experimental approaches based on spectrophotometric measurements of the rapid kinetics of electron flow and related phenomena.

## The Electron Transport Systems of *Rhodopseudomonas Spheroides* and *Rhodopseudomonas Capsulata*

The components contributing to the visible and electron paramagnetic spectra of chromatophores from Rps. spheroides and Rps. capsulata have recently been elucidated in some detail (16-24). The results of this work are summarised in Fig. 1 (for Rps. spheroides, 16-20) and 2 (for Rps. capsulata, 20-24); the data are derived from redox titrations in the dark of the components indicated, using the techniques developed largely by Dutton and co-workers (16,25). It is likely that the pathway of cyclic photosynthetic electron flow is similar in both species, and involves the primary photochemical reactants, P870 and X, called photoredoxin by Dutton and Leigh (26), and an electron transport chain containing ubiquinone, cytochrome b (of $E_m \sim 60$ mV), cytochrome $c_2$ (of $E_m \sim 300$ mV), and other components which have not been spectrally characterised but may include the high potential non-haem irons. The role of the other components present is not yet clear, but presumably some are associated with dark electron transport at the level of substrates and NAD(P), and with aerobic pathways (16-30). A simplified scheme for cyclic electron flow in Rps. capsulata is shown in Fig. 3, with sites of inhibitor sensitivity and half-times for the reactions also indicated. These latter refer to the maximal rates measured under coupled conditions following a single short saturating flash. The experimental evidence on which some of these values are based will be discussed more fully below.

## Rapid $H^+$-uptake by Chromatophores, and its Relation to Electron Flow

The kinetic characteristics of rapid $H^+$-uptake by chromatophores from a variety of photosynthetic bacteria and effects of inhibitors and ionophores, have been extensively reported elsewhere (31-35). In Rps. spheroides and Rps. capsulata the reaction elicited by a single turn-over flash has the following characteristics.

i) The half-rise time is 250-350 μs.

ii) The reaction is stimulated by valinomycin, which induces a second phase of equivalent extent but slower (2-10 ms) rise time.

iii) Antimycin inhibits the valinomycin stimulation but has no effect on the $H^+$ uptake in the absence of ionophore.

iv) Ortho-phenanthroline inhibits $H^+$ uptake in parallel with the inhibition of electron flow from primary to secondary acceptor.

v) Uncoupling agents, or ionophores of the nigericin type, accelerate the decay of the change but do not effect it's onset.

vi) In the absence of inhibitors or valinomycin, just over one $H^+$ is taken up per photochemical reaction centre.

vii) The rapid $H^+$-uptake attenuates on lowering the redox potential of the suspension, with an $E_m \sim 0$ at pH 7, but this value varies by $\sim$ 60 mV/unit as the pH is changed.

These characteristics show that the rapid phase of $H^+$-uptake is associated with reduction of a secondary acceptor which is a H-carrier, and have led us to suggest that the $H^+$ taken up represents the $H^+$ bound on reduction of the secondary acceptor by an electron from the primary acceptor. The characteristics of the acceptor identified are very similar to those of ubiquinone (36).

The features of the valinomycin stimulated, antimycin sensitive slower phase of $H^+$ uptake are less easily interpreted, but may be taken to indicate that a second reaction of $H^+$ uptake occurs close to the site of antimycin inhibition.

## Rapid $H^+$ Uptake on Repetitive Flash Illumination

The relation between the flash elicited $H^+$ uptake induced by continuous illumination is obviously of central importance in the elucidation of the mechanism of the $H^+$ pump. The condition of continuous illumination can be approached by repetitive flash illumination (37, 38). By varying the flash frequency, it is possible to study the initial events contributing to the overall change, and the way in which these depend on the rate of turnover of the electron transport system. On repeated illumination with single turn-over flashes at relatively low frequency, a series of discreet reactions of rapid $H^+$ uptake occur which accumulate to give an extent of $H^+$ uptake approaching that induced by continuous illumination. Superimposition of successive traces showed that each followed a time course similar to that of the rapid change induced by a single flash. Valinomycin stimulated the extent of each event so that the cumulative slope of the change was twice as steep as that in the absence of ionophore. These and similar experiments at varying frequencies of repetitive illumination show quite unambiguously that the reactions of rapid $H^+$-uptake are an elementary and integral component of the overall reaction of $H^+$-uptake induced by continuous illumination (38).

## Rapid $H^+$ Uptake by Reaction Centre Preparations

Our previous results show that rapid $H^+$-uptake is associated with reduction of a H-carrying secondary electron acceptor (33-35 and above) More recently (39) we have been able to demonstrate in photochemical reaction centres prepared by the method of Clayton and Wang (40) an uptake of $H^+$ on illumination stoichiometric with endogenous ubiquinone. On flash excitation, a rapid $H^+$ uptake stoichiometric with the concentration of reaction centres was observed, but only when reduced cytochrome c was present to re-reduce the photo oxidised primary donor (Fig. 4). This rapid $H^+$-uptake was inhibited by orthophenanthroline, but was completely insensitive to uncoupling agents, ionophores or antimycin. On excitation with repeated flashes the cumulative extent of $H^+$ -uptake in the absence of added acceptor was equal to the oxidising equivalents initially available in the ubiquinone pool. On addition of 1,4-naphthoquinone as an acceptor, the $H^+$-uptake on continuous illumination or the cumulative $H^+$-uptake on repeated flash excitation was much more extensive. These results are summarised in Table I.

## Rapid $H^+$-Uptake and the $H^+$-Pump

The rapid $H^+$-uptake by reaction centre preparations confirms and extends our previous observations on the relation between $H^+$-uptake and electron transport. However some major difference between flash induced $H^+$-uptake by chromatophores and reaction centres are worthy of note.

a) The reaction centre changes were insensitive to FCCP and nigericin, which accelerate the decay of the chromatophore rapid $H^+$-uptake.

b) No stimulation of $H^+$-uptake by valinomycin, or inhibition by antimycin, was observed with reaction centres.

These differences can be readily explained in terms of a further reaction of $H^+$-uptake associated with electron transport through the antimycin sensitive site, and a vectorial arrangement of the chromatophore electron transport system in the membranes so that hydrogen ions taken up from the external medium are ultimately released within the vesicles. The differences and similarities between $H^+$-uptake by chromatophores and reaction centres may be taken as further evidence of the involvement of electron transport in the mechanism of the $H^+$-pump, and together with the other results discussed above, lead us to conclude that the binding of a $H^+$ from the external medium on reduction of a H-carrier (probably ubiquinone) is an essential part of the mechanism of the pump. This strongly suggests that the electron transport chain is itself involved in the vectorial transport of $H^+$ (1,7,8), and that the protons involved in electron transport at this level are not localised (12).

## Photochemical Reaction Centre Preparations and the Reaction With Cytochrome $c_2$

We have recently studied the reaction between mammalian cytochrome c or cytochromes $c_2$ from a variety of photosynthetic bacteria and photochemical reaction centres prepared from cells of *Rps. spheroides* R26 (40) and wild type (41) or *Rps. capsulata* A1A pho+ (42). In every case the reaction was found to be second order, in contrast to the zero order kinetics reported by Ke et al (43) for the photooxidation of mammalian cytochrome c by reaction centres prepared from *Rps. spheroides* R26 by using Triton. The reaction was sensitive to increased ionic strength indicating that collision between the ionic species was involved. The reaction with mammalian cytochrome c varied with pH at all values of ionic strength with broad optimum between pH 7 and 9, and a negligible rate of reaction at values of pH below 4.5 and above 11.5. The reaction with native cytochrome $c_2$ was unaffected by pH, and this differential sensitivity to pH reflects the different isoelectric points of the two cytochromes.

In order to obtain information about the location of the cytochrome and reaction centres we have, in collaboration with Dr. G. Hauska of Bochum University, prepared antibodies to cyto chrome $c_2$ and reaction centres with a view to studying their reactivity with the antigens *in situ*.

## Structural Aspects - The Location of Cytochrome $c_2$

Photosynthetic bacteria are structurally similar to other bacteria, possessing a cell wall, separated by a periplasmic space from the cell membrane - which encloses the cytoplasm. In photosynthetically grown cells, the cell membrane is greatly increased in area by invaginations which may take the form either of small spherical vesicular structures, the chromatophores, or flattened stacked thylakoids of larger size. The cell wall is relatively impermeable to reagents, and this limits the experimental utility of intact cells. The cell wall can be readily digested by treatment with lysozyme (44); when this is done under suitable conditions, a preparation of spheroplasts (or protoplasts) is obtained in which the naked cell membrane remains intact, so that all cytoplasmic constituents are retained. By comparing the properties of such preparations with those of a suspension of isolated chromatophores, much information can be obtained about the structural relationship between the two preparations; in particular differential vectorial characteristics of the $H^+$ pump and reactivity with specific antibodies has made it possible to locate the site of the reaction between cytochrome $c_2$ and the photochemical reaction centre, and to show that the electron transport pathway near the photochemical reaction is so arranged in the membrane as to act as a $H^+$ pump (45). These results are summarised below, and shown diagramatically in Figs. 5 and 6.

1) Cytochrome $c_2$ is almost completely lost on preparation of spheroplasts from cells of Rps. spheroides or Rps. capsulata. The cytochrome is therefore a periplasmic protein. Cytochrome $c_2$ is retained on preparing a suspension of chromatophore from intact cells presumably because these are rapidly pinched off and reseal on mechanical disruption.

2) Spheroplasts readily oxidise externally added cytochrome $c_2$ or mammalian cytochrome c.

3) Antibodies to purified cytochrome $c_2$ or to purified reaction centres have no inhibitory effect on photooxidation of endogenous cytochrome $c_2$ in washed chromatophores. On treatment with cholate (or other detergents), both antibodies inhibit cytochrome $c_2$ oxidation.

4) Both antibodies inhibit photooxidation of added cytochrome $c_2$ by intact spheroplasts, but photooxidation of added mammalian cytochrome c is inhibited only by reaction centre antibody.

5) The $H^+$ pump is outwardly directed in spheroplasts (as in intact cells), in contrast to chromatophores where $H^+$ uptake occurs.

6) The antibody to the reaction centre does not inhibit the photochemical reaction itself in spheroplasts, chromatophores or isolated reaction centres.

These results show unambiguously that cytochrome $c_2$ is reacting from the aqueous phase which is outside the spheroplasts (equivalent to the periplasmic space of the intact cell), and inside the chromatophore, and that a site on the photochemical reaction centre is available for reaction from the same aqueous phase with a specific antibody, and this reaction prevents the reaction with cytochrome $c_2$.

Taken together with the information on rapid $H^+$ uptake (see above) these results lead to the following conclusions.

a) The secondary donor (cytochrome $c_2$) and secondary acceptor ($H^+$, ubiquinone) react with the primary photochemical reactants on opposite sides of the membrane.

b) The photochemical reaction centre must be so arranged in the membrane that oxidising equivalents ("holes") and reducing equivalents (electrons) are available to the secondary reactants on opposite sites of the membrane.

c) The photochemical reaction must there fore occur effectively across the membrane, and involve the performance of electric work in transporting charge across the dielectric.

d) The changes in charge must be effective in the aqueous phases on either side of the dielectric at least as rapidly as the secondary reactions occur.

These conclusions are summarised in the scheme in Fig. 6. We had previously suggested (following Mitchel (1, 8) and Junge & Witt (46) on the basis of work on the carotenoid change (47-50) and delayed fluorescence (51-54)) that the photochemical reactions involved charge transfer across the membrane. Our present results show that vectorial movement of charge occurs across the full width of the membrane, and is not localised within the membrane (12, 37). The photochemical reaction may therefore be regarded as the electrogenic arm of a Mitchellian loop in which the secondary acceptor (probably ubiquinone) is a H- carrier potentially completing the loop. The results discussed above may also be taken as supporting the view that the carotenoid change is an indicator of membrane potential (47-50), a possibility we discuss further below.

## The Carotenoid Change as an Indicator of Membrane Potential

The spectra of the carotenoid and bacteriochlorophyll pigments of cells or chromatophores of a variety of photosynthetic bacteria undergo a shift to the red under conditions in which the coupled system is in an 'energized' state (47-51, 55-57). Similar spectral changes can be induced by imposing diffusion potentials across the chromatophore membrane, using ionic gradients operating through ionophores (47). The wide variety of ions which can be used, under conditions where the increased permeability of the membrane to the ion can be inferred from direct physico-chemical measurements on model systems, lends support to the conclusion that the pigments are responding more or less directly to the electrical gradient imposed across the membrane. The response appears to be linear with potential. Junge and Witt (46) on rather less direct evidence, have ascribed the similar spectral changes of chloroplasts to a direct response of the pigments to the electrical field generated across the chloroplast membrane. They have suggested that an electrochromic shift (58) of the pigment spectrum occurs, and have mimicked the spectral change in a model system. However, certain difficulties arise when a similar explanation is attempted for the spectral changes of the bacteria.

1) Carotenoid and bacteriochlorophyll pigments in a wide range of bacteria show a response or lack of response which follows no consistent pattern (56).

2) In <u>Rps. spheroides</u>, where the change has been best characterised, the difference spectrum can closely be fitted assuming a shift of ∿ 10 nm of ∿ 10% of the carotenoid pigment (59,60).

3) The induction of diffusion potentials of opposite sign (-ve inside) to the light induced membrane potential gives rise to a carotenoid difference spectrum which is a mirror image of the light induced change (47).

4) Sherman and Clayton have shown that the carotenoid change is absent in a mutant of Rps. spheroides (61) which lacks the reaction centre complex, but has an otherwise normal complement of pigments.

The difficulty of interpretation has been compounded by the presence in wild-type Rps. spheroides (and other species) of up to 6 different carotenoid types, several of which are present in sufficient amounts to contribute significantly to the change observed. Two of these, spheroidene and OH-spheroidene have identical spectra.

In an attempt to simplify the problem, we have been looking for mutants of Rps. spheroides with a reduced compliment of carotenoid types. A green mutant recently isolated by Dr. Venetia Saunders of this department and designated Rps. spheroides G1C, has been found to contain only a single major carotenoid species (⩾ 99% neurosporene, Table II, N. Holmes, unpublished observations), and this looks a promising material for an exhaustive analysis of the carotenoid change. Our preliminary results show that the changes induced by light (Fig. 7) or diffusion potentials (Fig. 8) follow a similar pattern with regard to kinetics, sensitivity to inhibitors and linearity with respect to electrical gradient to the changes observed in chromatophores from the wild type cells (47) or cells of the Ga mutant (35). Absolute spectra of the chromatophores in the dark and in the light, and the difference between them are shown in Fig. 7. It is clear from these traces that, as suggested by previous workers (59, 60), the change can be approximated by a shift of ∿ 10 nm of ∿ 10% of the pigment. However, in contrast to previous studies, no ambiguity exists as to the species undergoing a change in these experiments. Difference spectra of the change induced by diffusion potentials of several magnitudes are shown in Fig. 9A. It can be seen that no true isobestic point for the change is observed in the cross over region, but rather that the point of null change is displaced further towards the red with increasing potential. Difficulties arise in interpreting the difference spectra quantitatively because of the contribution of multiple transitions to the overall change, and of a possible shift in base line with increasing potential. Nevertheless they appear to represent a qualitative demonstration that the spectral change is a true red shift, and that the degree of shift increases with increasing electrical gradient (see Fig. 9B).

We believe that an extension of our present observations, with a computer analysis of the spectral data, will provide very precise information about the spectral change of a single chemically defined carotenoid, neurosporene.

Already we may note a number of interesting features.

1) Only ∿ 10% of the pigment responds, suggesting that a proportion of the carotenoid is in a different environment from that of the bulk of the pigment.

2) Neurosporene (see formula below) is a neutral molecule in which asymmetry is limited to the difference of one double bond (cf. positions 7, 8 and 8', 7'). There are no polar groups on the molecule.

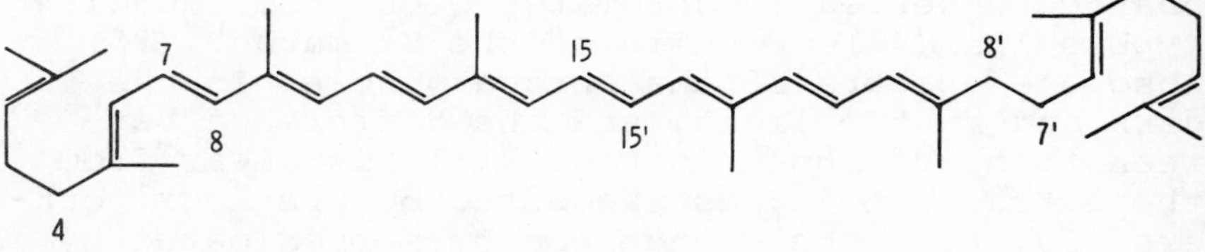

neurosporene

3) The shift of 10 nm observed is greater than would be expected from a simple electrochromic effect (62).

4) Despite the lack of marked molecular asymmetry, the responses of the carotenoids to electrical gradients show a characteristic polari-

ty, indicating that the pigments *in situ* are oriented anisotropically in the membrane.

Possibly some of the above features may be related to the asymmetry of the membrane itself, resulting from the curvature consequent upon the small size of the chromatophore vesicles. Some of the electrical consequences of such curvature have been recently discussed by Israelachvili (63), while the asymmetric distribution of charged and neutral lipids in small vesicles has been previously noted by Michaelson et al (64).

The charged lipid molecules tend to predominate in the outer layer of a curved membrane, the asymmetry of distribution increasing with increased curvature and decreased ionic strength. A consequence of the higher surface charge density of the outer layer is that the lipid molecules would be expected to pack less tightly in this layer; this in addition to effects arising from molecular packing which would also be expected to give rise to a less tight packing in the outer layer (63). In this context, the molecular asymmetry of neurosporene might be seen as imposing a greater rigidity on the unsaturated end of the molecule than on the saturated end, and this might in turn give rise to a degree of orientation due to packing requirements in the asymmetrical membrane. This possibility offers several lines of investigation both at a physico-chemical and a biological level which have yet to be explored.

Clearly, if the carotenoid change could be adequately explained on a theoretical basis in terms of a response to electrical field, the value of the change as an indicator would be greatly enhanced, and many lines of investigation at the level of both energy coupling, and ion and metabolite transport in intact cells, would become attractive.

## The Carotenoid Change and the Photochemical Reactions

The extremely rapid rise time of the 515 nm change in chloroplasts ($<10^{-8}$ s) led Witt and his coworkers (6, 46) to suggest that the photochemical reaction itself occurred across the membrane, and that the 515 nm change reflected the field generated in the photoact. We have followed this interpretation in discussing the close association between the rapid phase of onset of the carotenoid change and the photochemical reactions in chromatophores (48-50). In particular, the rapid phase of the carotenoid change was shown to be present under all conditions of ambient redox potential in which the photochemical reactions occurred, and to attenuate at extremes of the potential range with those reactions (35, 37). At potentials in the middle of the range, the rapid phase of the carotenoid change was followed by a slower, antimycin sensitive phase, which we attributed (35) to a further span of electrogenic electron flow between cytochrome b and c. However, Dutton et al (16) and Jackson and Dutton (37) have suggested as an alternative possibility that the rapid phases of the change may reflect a response to local fields generated in the photoact, but that these are effective across the membrane only after serial contributions to the vectorial displacement of charge from slower electron transfer reactions, which are reflected in the slower phases of the carotenoid change.

The results from the more biochemical approach discussed above preclude this latter explanation but leave open the hypothesis of a second electrogenic site. However the possibility still exists that the carotenoid changes are in response to local fields, and that the rapid phase of the change does not faithfully reflect the kinetics of development of a transmembrane potential.

We have recently approached this problem from another point of view, that of the dependence on the high energy state of delayed fluorescence.

## Delayed Fluorescence and Its Relation to Membrane Potential

Possible mechanisms for the dependence of the intensity of delayed fluorescence on the high energy state have been discussed elsewhere (51-54, 65-66), and recently reviewed by Lavorel (67). Many lines of evidence suggest that delayed fluorescence occurs as a result of a reversal of the photochemical reactions, and represents that small fraction of events in which sufficient energy is available, with an additional contribution from the environment, to excite chlorophyll (or bacteriochlorophyll) bulk pigment back to the singlet level. The activation energy derived from glow curves suggests that more energy is available than would be expected from the redox free energy difference between reactants (68). We have previously suggested that, if the photochemical reactions are arranged in the membrane so that charge separation occurs across the membrane and they thus perform electrical as well as chemical work, then the electrical free energy stored should be available to drive the reactions backwards to give delayed fluorescence (51-54). Consideration of this possibility using transition state theory shows that the intensity of delayed fluorescence should depend logarithmically on the membrane potential

$$L \propto \exp(\Delta \Psi)$$

(where L is the intensity of delayed fluorescence, and $\Delta \Psi$ is the membrane potential), and this predicted relationship has been found to occur in several different systems (51,54,69). From the above relationship (and ignoring complications introduced by other energetic terms) a simple proportionality would be expected between the change in intensity of delayed fluorescence, and change in membrane potential:

$$\Delta(\Delta \Psi) = 2.303 \frac{k\,T}{n} \log_{10} \frac{L\,1}{L\,2}$$

where n, the number of charges involved in the reaction, would be expected to be one. We have attempted to correlate changes in membrane potential as indicated by the carotenoid change, with changes in the intensities of delayed fluorescence under similar conditions. Fig. 10 shows the time course of the onset and decay in the light, of delayed fluorescence intensity plotted on a logarithmic scale, and of the carotenoid change plotted on an arbitrary linear scale. The close similarity in kinetics of the two changes is obvious. The relation between extent of carotenoid change and potential can be separately calibrated from the change induced by diffusion potentials on addition of KCl in the presence of valinomycin (47). Fig. 11 shows the relation between delayed fluorescence intensity and membrane potential (measured as above), when the changes were measured under similar conditions, and induced by a variety of changes in conditions. Two points are of immediate interest.

a) The logarithmic relationship is obeyed over the full range tested.

b) The slope of the curve suggests that a change in potential of ∿ 140 mV was required to give a ten fold change in intensity of delayed fluorescence. This latter observation is noteworthy; the expected slope of the curve would be ∿ 60 mV for a value of $n = 1$.

We do not wish to lay too much emphasis on this value, since our apparatus did not enable us to measure the carotenoid and delayed fluorescence changes under identical conditions. However, the results may indicate that the electrical field available to drive the photochemical reaction in reverse is less than that indicated by the carotenoid change. If the carotenoid change indicated a transmembrane potential, this would suggest that the photochemical reaction does not occur across the full width of the membrane. Clearly there are several possibilities as yet unexplored theoretically which could reconcile this situation with that apparently indicated by the carotenoid change (47-50). Apart from the suggestions of Witt et al. (6, 46) and of Jackson and Dutton (77), it will be necessary to consider tunnelling effects, and the extent to which the local field changes generated in the photoact are capacitatively or resistively coupled to the aqueous ionic phases on either side of the dielectric (41,42). Although some ambiguity remains in the interpretation of both the carotenoid change and delayed fluorescence and their dependence on the high energy state, the concept of the photochemical reactions orientated in the membrane so as to separate charge within or across the dielectric, and perpendicular to the plane of the membrane, provides a working hypothesis which explains with some consistency many observations on the two phenomena.

## A Model for the Electron Transport System as a $H^+$ Pump

I have discussed above the evidence which suggests that the electron transport system near the photochemical reactions is arranged as a $H^+$ pump. I have also mentioned some indications of other reactions both of electrogenic electron flow, and of $H^+$ uptake, closely associated with the region of the chain near the site of inhibition by antimycin. The evidence, summarised below, suggests that a second span of electron flow between cytochromes b and c acts as a $H^+$ pump.

1) Antimycin sensitive slow phase of the carotenoid change.

2) Antimycin sensitive slow phase of flash induced $H^+$ uptake in the presence of valimycin.

3) Uncoupling agents stimulate electron flow between cytochrome b and c as indicated by

a) changes in the steady state redox levels of the cytochromes,
b) reversibility of the redox changes on repetitive flash excitation (73),
c) increased rate of cytochrome b oxidation after a single flash(24).

4) Antimycin reverses the effect of uncoupling agents on the steady state redox levels. The site of action of antimycin can be unambiguously located between cytochromes b and c on

the basis of effects on both flash induced redox changes and the steady state redox level of the cytochromes.

Space does not permit a full discussion of this second site of energy conservation in the cyclic chain of photosynthetic electron flow. However it is worth considering some features to be expected from the site if it follows a pattern similar to the site associated with the photochemical reaction centre.

1) The pathway of electron flow between cytochrome b and cytochrome c must be organised as a $H^+$ pump.

2) The site may be expected to contain an electrogenic span, electron flow through which would account for the slow phase of the carotenoid change

3) The site may be expected to include a H-carrier whose reduction would account for the antimycin sensitive rapid $H^+$ uptake.

Several features of coupled electron transport in this region suggest that the process may be more complex than the simple linear scheme we have previously proposed (35), and the complexity of the similar site in the mitochondrial electron transport chain (74, 75) underlines this possibility. In particular the possible involvement of a H- carrier, the reduction and oxidation of which occur by a two stage mechanism with a stable semiquinone or free radical form, may account for some of the anomalies we have noted before (34).

I have incorporated the results discussed above into a somewhat speculative scheme (Fig. 12) in which the two sites of coupling are shown arranged as $H^+$ pumps. The evidence for the arrangement of the second site is obviously inadequate, and the dotted line is an expression of ignorance as to the precise mechanism.

## Concluding Remarks

I believe that in terms of the criteria (iv-vii) discussed in the introduction to this paper, we have established that the electron transport system in the neighbourhood of the photochemical reactions in Rps. spheroides and Rps. capsulata is arranged as a $H^+$ pump as envisaged by Mitchell in his chemiosmotic hypothesis. If this is the case, we may expect that coupled electron transport in other systems is similarly arranged, as was already strongly indicated by the work of Racker and his coworkers on reconstituted systems (76, 77). Our present results make it seem unlikely that either cryptic molecular $H^+$ pumps driven indirectly by electron flow, or mechanisms dependent solely on localised protons, play a major role in energy coupling. We cannot preclude a system of mixed mechanisms, and some features of coupling in the cytochrome b-c region may reflect a lack of free equilibration of protons between membrane sites and aqueous phase (73). Such a lack of equilibration can be demonstrated for the primary acceptor, where the pH dependence of the mid point potential suggests the involvement of a proton, but no $H^+$ change is seen corresponding to kinetic participation of a proton in the reaction (31-35). In this case at least, the failure of equilibration is unlikely to be associated with energy coupling since the rate of reduction of the secondary acceptor is completely insensitive to uncoupling agents. With these reservations, it seems likely that a chemiosmotic approach to the problems of energy coupling will prove more fruitful than a search for hypothetical intermediates, or high energy states distinct from the electrochemical $H^+$ gradient.

## Acknowledgements

The research reviewed above has been done in cooperation with Dr. J.B. Jackson, Dr. C.A. Wraight, Dr. R.J. Cogdell, Dr. E.H. Evans, R.C. Prince and N.G. Holmes of the University of Bristol. In addition we are grateful to Dr. P.L. Dutton of the Johnson Research Foundation for enthusiastic collaboration on the characterisation of EPR signals in Rps. capsulata, and for support and travel funds for one of us (R.C.P.), and Dr. Günter Hauska, and Dr. B.A. Melandri who collaborated on the work on antibody preparation and reactivity, and to Dr. Venetia Saunders for the G1C mutant of Rps. spheroides. We would like to thank the Science Research Council and the Royal Society for grants towards support and equipment, and NATO for a travel grant to support a joint project with Dr. G. Hauska and Dr. B.A. Melandri. We are grateful to Mr. G. Tierney and Mrs. Janet Fielding for skilled technical assistance.

## References

1. Mitchell, P. 1966. In: Chemiosmotic Coupling in Oxidative and Photosynthetic Phosphorylation. Glynn Research Ltd., Bodmin, Cornwall.
2. Greville, G.D. 1969. In: Current Topics in Bioenergetics Vol. 4., p. 185. ed. Sanadi, D. R. Academic Press, New York & London.
3. Walker, D.A. and Crofts, A.R. Ann. Rev. Bio-Chem. 39 (1970) 389.
4. Chance, B. and Montal, M. 1971. In: Current Topics in Membranes and Transport. Vol. 2, p. 99, eds. Bronner, F. and Kleinzeller, A. Academic Press. New York.
5. Harold, F. M. Bacteriological Rev. 36 (1972) 172.
6. Witt, H.T. Quarterly Reviews of Biophysics 4 (1971) 365.
7. Mitchell, P. Nature 191 (1961) 144.
8. Mitchell, P. 1968. In: Chemiosmotic Coupling and Energy Transduction. Glynn Research Ltd. Bodmin, Cornwall.
9. Slater, E.C. Quarterly Reviews of Biophysics 4 (1971) 35.
10. Boyer, P.D. 1968. In: Biological Oxidations p. 193, ed. Singer, T.P. Interscience Publishers, Inc. New York.
11. Green, D.E. and Ji, S. J. Bioenergetics 3 (1972) 159.
12. Williams, R.J.P. 1969. In: Current Topics in Bioenergetics Vol. 3, p. 79, ed. Sanadi, D.R. Academic Press, N. Y.

13. Chappell, J.B. and Crofts, A.R. Biochem. J. 95 (1965) 393.
14. Skulachev, V.P. (1973) Abst. 9th Internatl. Cong. Biochem. Stockholm, No. 4Sb1.
15. Hinkle, P. and Mitchell, P. J. Bioenerg. 1 (1970) 45.
16. Dutton, P.L. and Jackson, J.B. Europ. J. Biochem. 30 (1972) 495.
17. Kuntz, I.D., Loach, P.A. and Calvin, M. Biophys. J. 4 (1964) 227.
18. Cogdell, R.J., Jackson, J.B. and Crofts, A.R. J. Bioenergetics 4 (1973) 211.
19. Dutton, P.L., Morse, S.D., Jackson, J.B. and Leigh, J.S. (1973) Abst. 9th. Internatl. Cong. Biochem. Stockholm. No. 4d3.
20. Dutton, P.L., Leigh, J.S. and Reed, D.W. Biochim. Biophys. Acta 292 (1973) 654.
21. Klemme, J.H. 1969. In: Progress in Photosynthesis Research. Vol. 3, p. 1492, ed. Metzner H. Tübingen, Germany.
22. Crofts, A.R., Evans, E.H., Cogdell, R.J. and Jackson, J.B. (1972) Abst. VI. Internatl. Cong. on Photobiology. Bochum 1972, No. 37.
23. Crofts, A.R., Evans, E.H. and Cogdell, R.J. 1974. In: Proc. N.Y.A.S. Conf. on Mechanism of Energy Transduction in Biological Membranes. ed. Green, D.E. In press.
24. Evans, E.H. 1973. Dissertation for Ph.D. degree. University of Bristol, Bristol, U.K.
25. Dutton, P.L. Biochim. Biophys. Acta 226 (1971) 63.
26. Dutton, P.L. and Leigh, J.S. Biochim. Biophys. Acta 314 (1973) 178.
27. Connelly, J.L,, Jones, O.T.G., Saunders, V. A. and yates, D.W. Biochim. Biophys. Acta. 292 (1973) 644.
28. Marrs, B. and Gest, H. J. Bacteriol. 114 (1973) 1045.
29. Melandri, A., Zannoni, D. and Melandri, B.A. 1973. Abst. Symp. on Prokaryotic Photosynthetic Organisms, Freiburg, p. 56.
30. Drews, G., Leutiger, I. and Ludwig, R. Ark. Mikrobiol. 76 (1971) 349.
31. Crofts, A.R. and Jackson, J.B. 1970. In:

Electron Transport and Energy Conservation. p. 383, eds. Tager, J.M., Papa, S,, Quagliariello, E. and Slater, E.C. Adriatica Editrice, Bari.
32. Chance, B., Crofts, A.R., Nishimura M. and Price, B. European J. Biochem. 13 (1970) 364.
33. Crofts, A.R., Cogdell, R.J. and Jackson, J. B. 1973. In: Mechanisms in Bioenergetics. p. 337, Academic Press, N.Y.
34. Cogdell, R.J., Jackson, J.B. and Crofts, A.R. J. Bioenergetics 4 (1973) 211.
35. Crofts, A.R., Jackson, J.B., Evans, E.H. and Cogdell, R.J. 1972. In: Proc. IInd. International Cong. on Photosynthesis, Stresa 1971, p. 873. Junk, The Hague.
36. Reed, D.W., Zankel, K.L, and Clayton, R.K. Proc. Nat. Acad. Sci. U.S.A. 63 (1969) 42.
37. Jackson, J.B. and Dutton, P.L. Biochim. Biophys. Acta 325 (1973) 102.
38. Crofts, A.R., Cogdell, R.J., Evans, E.H., and Prince, R.C. 1974. In: Festschrift for Britton Chance, Proc. Symp. Bioenerg. 9th. Internatl. Cong. Biochem. Elsevier Publisher Co. Amsterdam. In press.
39. Cogdell, R.J., Prince, R.C. and Crofts, A.R. FEBS Letters 35 (1973) 204.
40. Clayton, R.K. and Wang, R.T. 1971. In: Methods in Enzymology. Vol. 234, p. 696, ed. San Pietro, A. Academic Press, New York.
41. Slooten, L. Biochim. Biophys. Acta 256 (1972) 452.
42. Prince, R.C. and Crofts, A.R. FEBS Letters 35 (1973) 213.
43. Ke, B., Chaney, T.H. and Reed, D.W. Biochim. Biophys. Acta. 216 (1970) 373
44. Karunairatnam, M.C., Spizizen, J. and Gest,H. Biochim. Biophys. Acta 29 (1958) 649.
45. Prince, R.C., Hauska, G. and Crofts, A.R. (1974) Abst. 546th. Meeting, Biochim. Soc. York.
46. Junge, W. and Witt, H.T. Z. Naturforsch. 23 (1968) 244.
47. Jackson, J.B. and Crofts, A.R. FEBS Letters 4 (1969) 185.

48. Jackson, J.B. and Crofts, A.R. European J. Biochem. 18 (1971) 120
49. Crofts, A.R. and Jackson, J.B. 1970. In: Electron Transport and Energy Conservation p. 383, eds. Tager, J.M., Papa, S., Quagliariello, E. and Slater, E.C. Adriatica Editrice, Bari.
50. Fleischmann, D.E. and Clayton, R.K. Photochem. Photobiol. 8 (1968) 287
51. Fleischmann, D.E. Photochem. Photobiol. 14 (1971) 277.
52. Wraight, C.A. and Crofts, A.R. Europ. J. Biochem. 19 (1971) 386.
53. Crofts, A.R., Wraight, C.A. and Fleischmann, D.E. FEBS Letters 15 (1971) 89.
54. Evans, E.H. and Crofts, A.R. Biochim. Biophys. Acta. (1974) In press.
55. Vredenberg, W.J. and Amesz, J. Brookhaven Symp. Biol. 19 (1967) 49.
56. Baltscheffsky, M. Arch. Biochem. Biophys. 130 (1969) 646.
57. Amesz, J., 't Mannetje, A. and de Grooth, B. G. 1973. Abst. Symp. on Prokaryotic Photosynthetic Organisms, p. 34.
58. Labhart, J. 1967. In: Advances in Chemical Physics. Vol. XIII, p. 179, ed. Prigogine, I. Interscience Publishers, N.Y.
59. Amesz, J. and Vrendenberg, W.J. 1966. In: Currents in Photosynthesis. p. 75, eds. Thomas, J.B. and Goedheer, J.C. Donker, Rotterdam.
60. Okada, M., Murata, N. and Takamiya, A. Plant and Cell Physiol. 11 (1970) 519.
61. Sherman, L.A. and Clayton, R.K. FEBS Letters 22 (1972) 127.
62. Schmidt, S., Reich, R. and Witt, H.T. Naturwissenschaften 8 (1971) 414
63. Israelachvili, J.N. Biochim. Biophys. Acta 323 (1973) 659.
64. Michaelson, D.M., Horwitz, A.F. and Klein, M. P. Biochemistry 12 (1972) 2637.
65. Mayne, B. Brookhaven Symp. Biol. 19(1967)460.
66. Kraan, G.P.B. 1971. Ph. D. Thesis, Leiden.
67. Lavorel, J. 1972. In: Bioenergetics of Photosynthesis, ed. Govindjee. In press.

68. Arnold, W. and Azzi, J.R. Proc. Nat. Acad. Sci. U.S.A. 61 (1968) 29.
69. Barber, J. Biochim. Biophys. Acta 275 (1972) 105.
70. Fleischmann, D.E. and Cooke, J.A. Photochem. Photobiol. 14 (1971) 71.
71. Kuhn, H. (1972) Abst. VI Internatl. Cong. on Photobiology. Bochum. No. 2.
72. Ullrich, H. M. and Kuhn, H. Biochim. Biophys. Acta. 266 (1972) 584.
73. Dutton, P.L., Morse, S.D. and Wong, A.M. 1974. In: Festchrift for Britton Chance. Proc. Symp. Bioenerg. 9th. Internatl. Cong. Biochem. Stockholm. In press.
74. Slater, E.C. Reviews on Bioenergetics, Biochim. Biophys. Acta 301 (1973) 129.
75. Wikström, M.K.F. Ibid. 155 (1973).
76. Racker, E. 1974. In: Festchrift for Britton Chance. Proc. Symp. Bioenerg. 9th. Internatl. Cong. Biochem. Stockholm. In press.
77. Ragan, C.I. and Racker, E. Fed. Proc. 32 (1973) 516 abs.
78. Norris, K.H. and Butler, W.L. IRE Trans. Bio-Med. Electronics 8 (1961) 153.
79. Jensen, S.L., Cohen-Bazire, G., Nakayama, T. O.M. and Stanier, R.Y. Biochim. Biophys. Acta. 29 (1958) 477.

---

TABLE I

| Change | Extent (nmoles $H^+$) | mole/mole reaction centres |
|---|---|---|
| Flash induced $H^+$ uptake with cytochrome *c* | 8.0 | 0.75 - 0.85 |
| with cytochrome *c* + napthoquinone | 11.0 | 1.0 - 1.1 |
| with cytochrome *c* + napthoquinone + orthophenanthroline | 2.1 | 0.25 - 0.30 |
| Total $H^+$ uptake induced by Repetitive xenon flashes | 24.0 | 2.25 - 2.55 |
| Total content of ubiquinone | | 1.5 |

TABLE II

Carotenoid content of Rhodopseudomonas spheroides, mutants Ga and G1C

| | G1C | Ga |
|---|---|---|
| Neurosporene | 99% | 63% |
| Lycopenes | trace | 1% |
| Chloroxanthin | - | 33% |
| ζ-carotenes | 1% | 3% |

Method as Jensen et al (79)

" - " indicates undetectable.

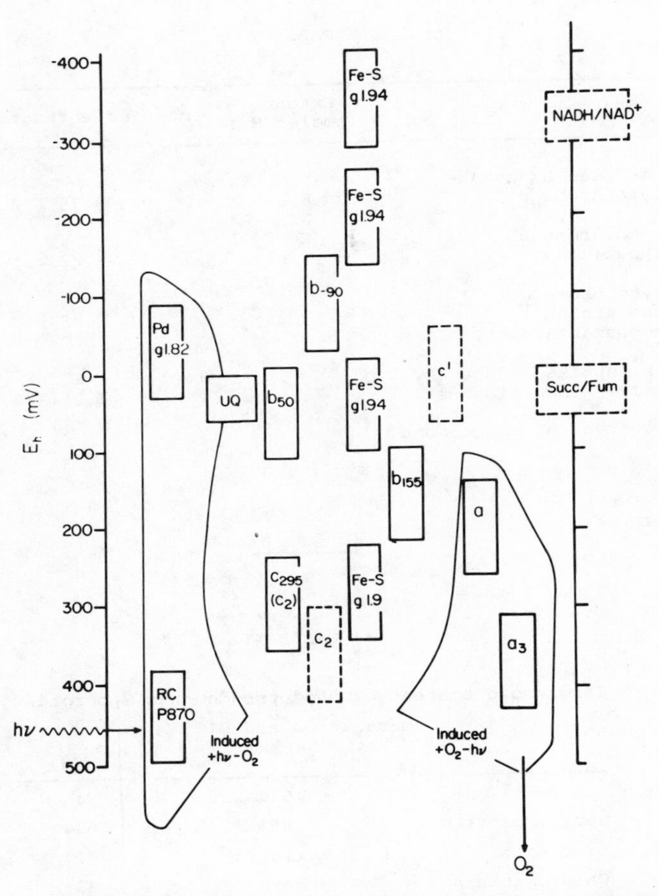


Figure 1. Redox components of Rhodopseudomonas spheroides. The components identified by spectrophotometry or EPR spectroscopy are arranged according to oxidation reduction potential at pH 7. Data from Dutton and Jackson (16) and Dutton et al (19,20) and Dutton, Lindsay and Leigh (unpublished data).

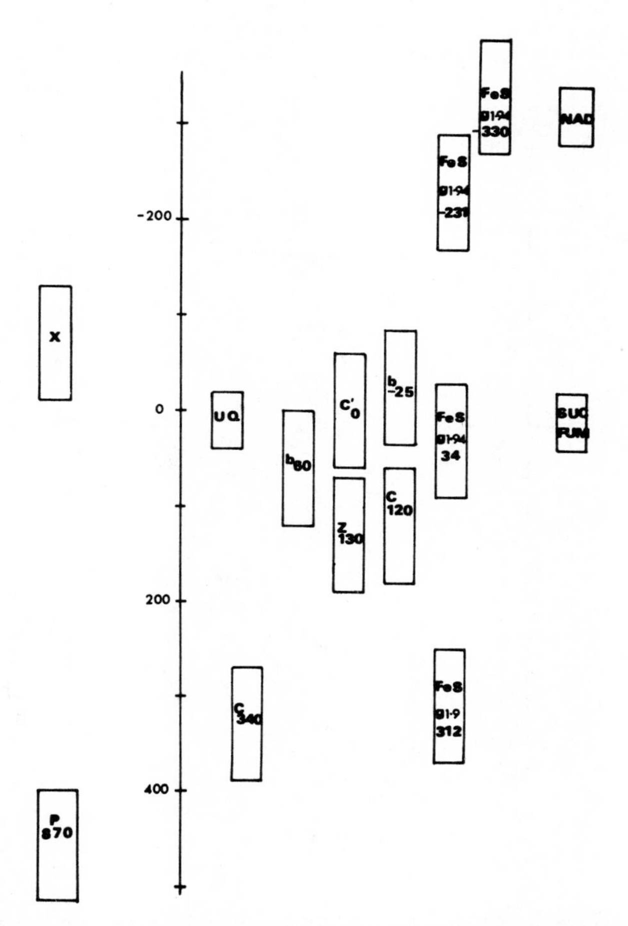


Figure 2. Redox components of Rps. capsulata. As for Fig. 1. Data from Crofts et al (22) Evans (24), and Prince, Dutton and Crofts (unpublished observations)

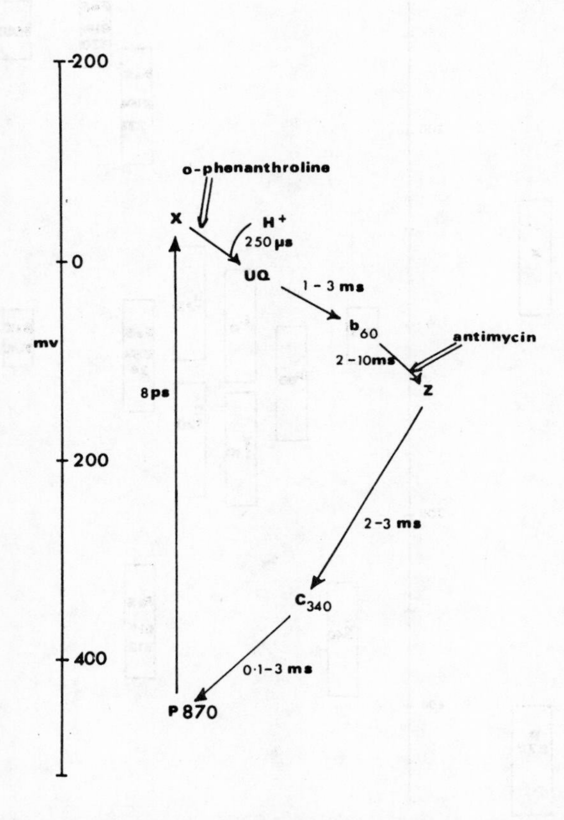


Figure 3. Scheme for pathway of cyclic photosynthetic electron flow in *Rps. capsulata*.

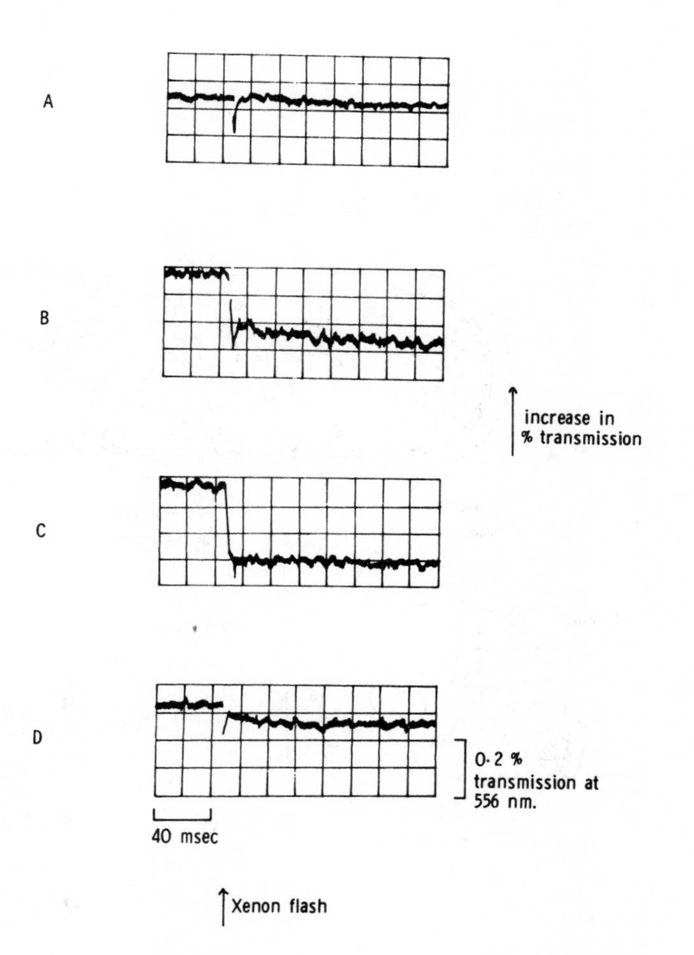


Figure 4. Flash induced $H^+$ uptake by reaction centre preparation from Rps spheroides R26. The $H^+$ uptake was monitored from the flash induced absorbancy change due to phenol red at 566 nm.

The anaerobic cuvette contained 8 $cm^3$ of 10 mM KCl at pH 7.5, with 1.3 µM reaction centres, and 25 µM reduced mammalian cytochrome c.

A - with no further additions. B - with 52 µM phenol red. C - as B, but 22 µM 1,4-naphthoquinone. D - as C, but with 3 mM o-phenanthroline. Addition of 20 nmoles HCl gave a deflection of 5.2% transmission change.

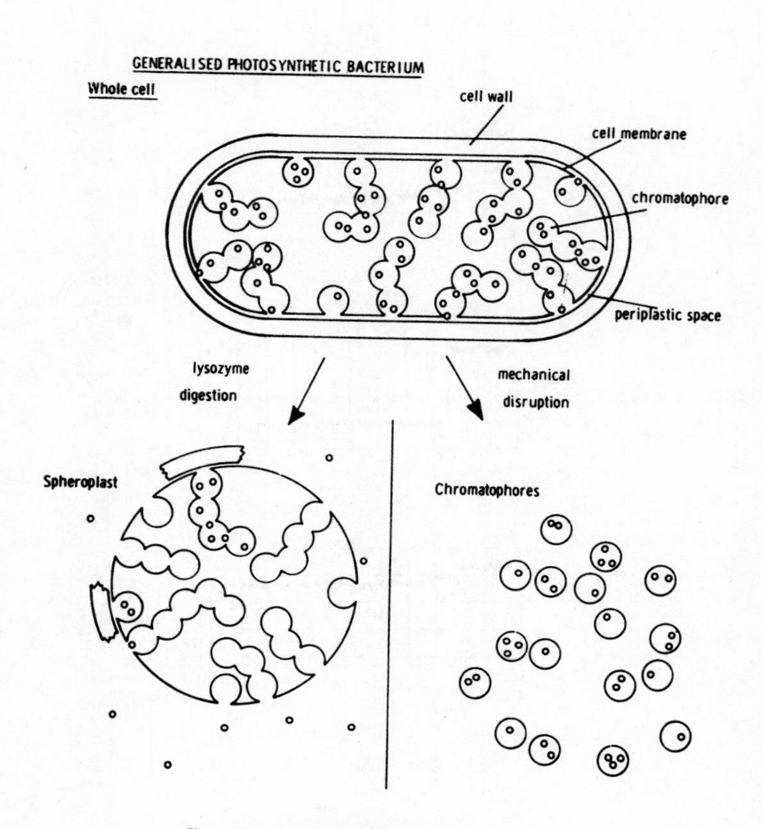


Figure 5. Diagramatic photosynthetic bacterial cell showing relation to spheroplast and chromatophore preparations. The filled circles represent periplasmic proteins, including cytochrome c.

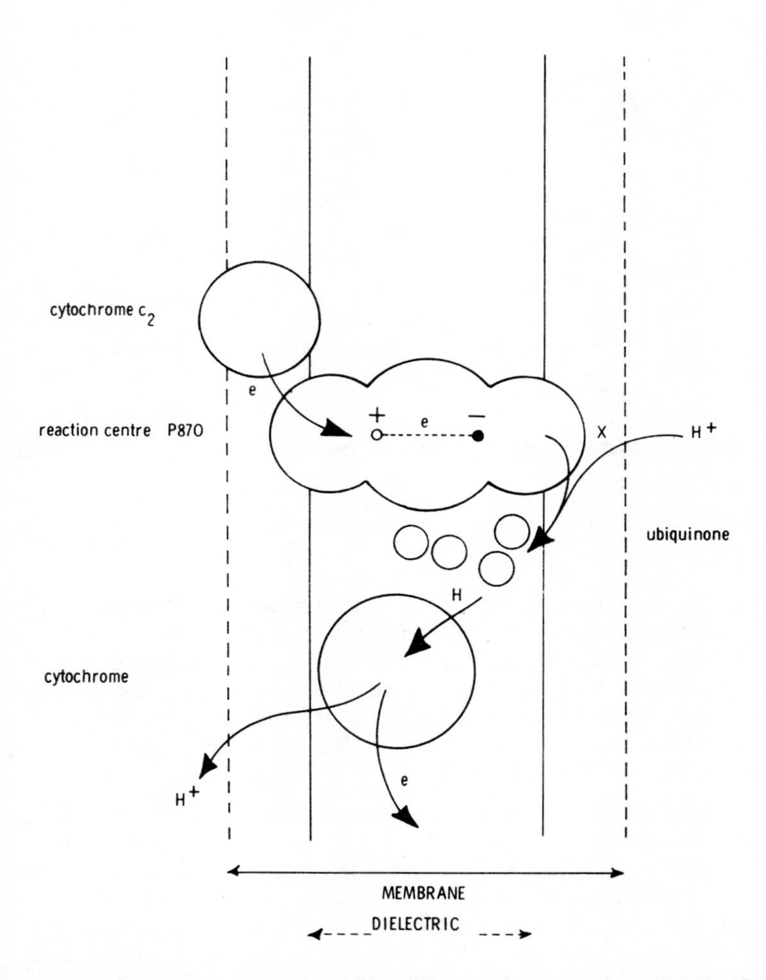


<u>Figure 6.</u> Schematic diagram showing a possible arrangement of the electron transport system close to the reaction centre as a $H^+$ pump.

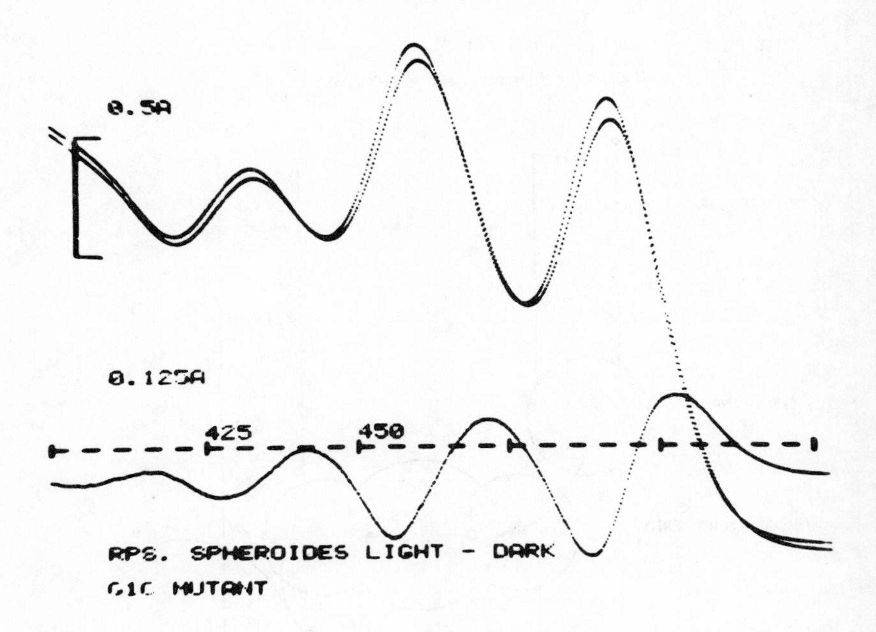


Figure 7. Spectrum of the carotenoid change in Rps. spheroides G1C. Chromatophores (39 μg/ml) were suspended in 2.5 ml of 100 mM Choline Cl, 20mM MES, pH 6.8. Illumination was provided by a 55W quartz halogen lamp focussed diffusely on the cuvette by a suitable lens system, and screened by a Wratten 87C filter.

Difference spectra were measured by accumulating uncorrected spectra in a mini-computer (PDP 11/10, Digital Equipment Company, Ltd., Reading, England) interfaced with a Hilger D330 monochromator (Rank Precision Industries Ltd.,) in which the wavelength drive unit was replaced by a stepper motor (G. Berger, Lahr, Germany, type RDM 50/A). The monochromator was incorporated into a simple single beam spectrophotometer, in which the current from the measuring photomultiplier (Type 9695B, EMI Ltd., Hayes, Middlesex) was logarithmically amplified (Analog Devices Ltd., operational amplifier type 755P) before being read by the analogue to digital conversion system (LPS 11-S, Digital Equipment Company Ltd.). Base line corrected difference spectra were derived by subtraction of the uncorrected spectra as suggested by Norris and Butler (78), and plotted on a CRT display (Tektronix 603 storage monitor ) as a 512 or 1024 point trace, with appropriate software-generated labels and scales. The spectrophotometer cuvette housing could accommodate either a conventional cuvette, or an anaerobic redox cuvette with overhead stirring, and actinic lamp for continuous illumination (as above), and a xenon flash lamp (Wotan X1E40) for flash illumination.

Absolute spectra were generated by subtracting from the uncorrected absorption spectrum the trace accumulated with a cuvette containing medium without chromatophores.

The photomultiplier was screened by a Corning No. 9782 glass filter.

---

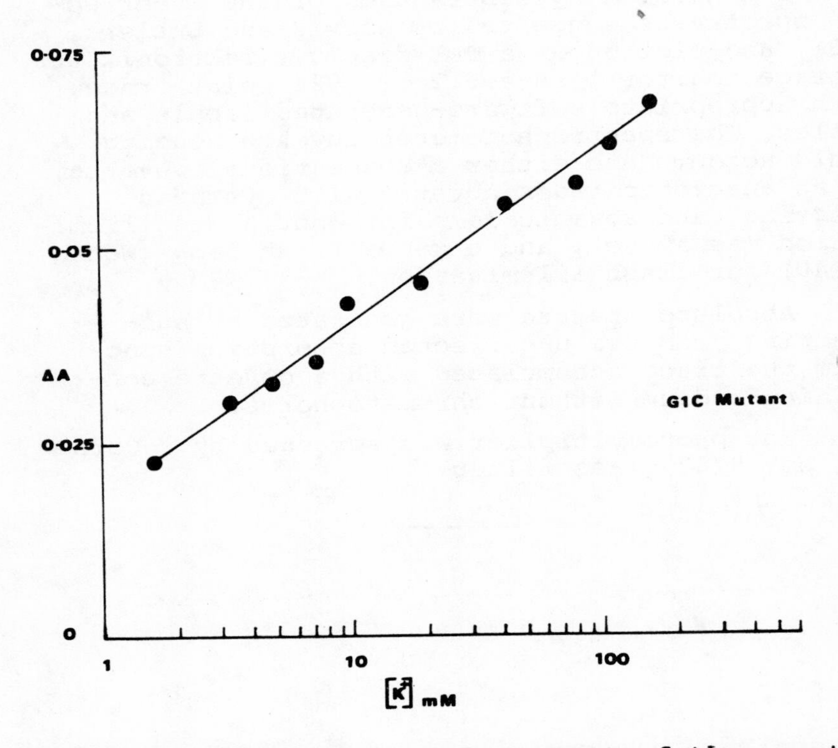


Figure 8. Dependence of the extent of the carotenoid change on diffusion potential. Chromatophores (15.5 μg/ml) from Rps. spheroides G1C were suspended in 100 mM choline chloride, 20 mM MES, pH 6.8. Diffusion potentials were generated and the carotenoid changes was measured in a dual wavelength spectrophotometer as previously described, (47), using the wavelength pair 524-480 nm. The chromatophores were prepared in a choline chloride medium from cells previously washed in a potassium free medium.

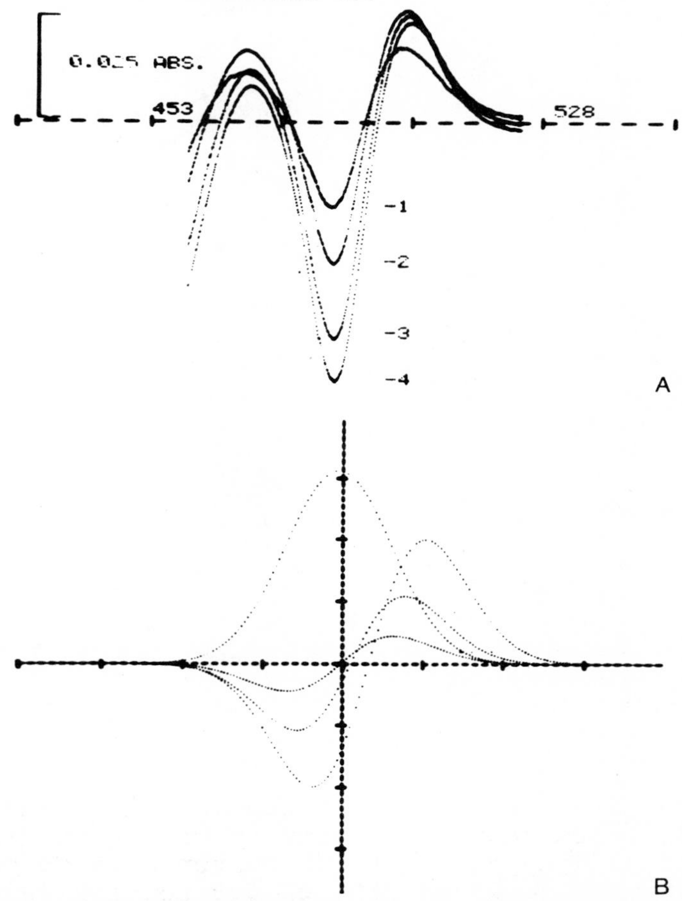


Figure 9. Spectra of carotenoid changes induced by diffusion potentials. A) Conditions as for Fig. 8, with 1) 3.2; 2) 19.2; 3) 80; 4) 112 mM KCl. A full spectrum (as shown) was accumulated in 10s, 15s after addition of KCl to the final concentration indicated. B) Theoretical traces showing differences between the Gaussian curve shown, and the same curve shifted to the right 0.1, 0.25, and 0.5 of the half width. The curves were generated using the computing system above and a Focal programme which included routines to display points and scales on the oscilloscope.

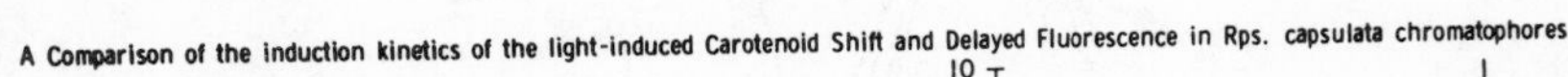

A Comparison of the induction kinetics of the light-induced Carotenoid Shift and Delayed Fluorescence in Rps. capsulata chromatophores.


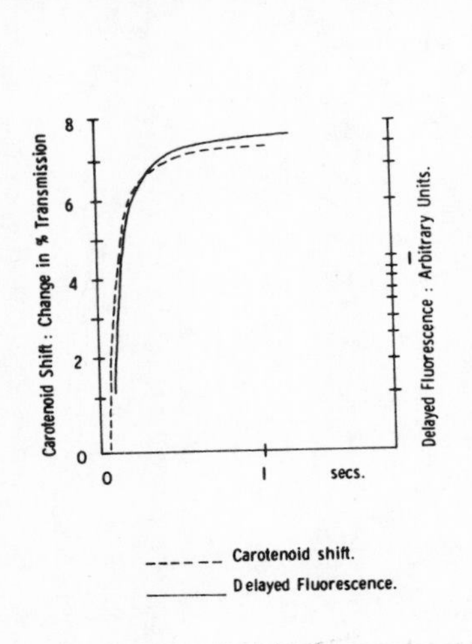


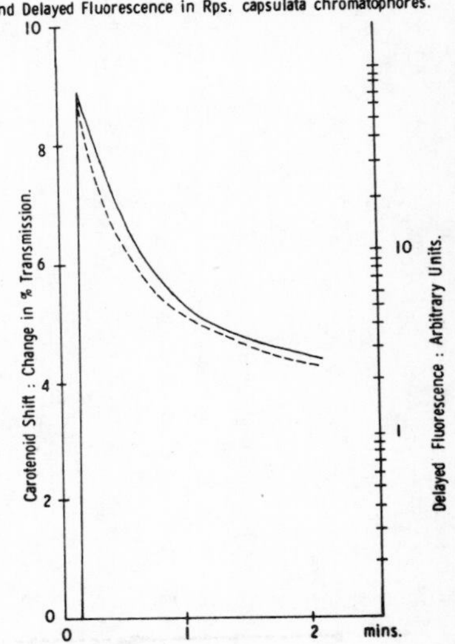


Figure 10. Kinetics of carotenoid change and delayed fluorescence intensity on continuous illumination. Rps. capsulata chromatophores suspended in 100 mM KCl, 20 mM MES, pH 6.8 to give 48 μg $ml^{-1}$ BChl for delayed fluorescence measurements (1 ms delayed fluorescence followed in a phosphoroscope), and 24 μg $ml^{-1}$ BChl for the carotenoid shift.

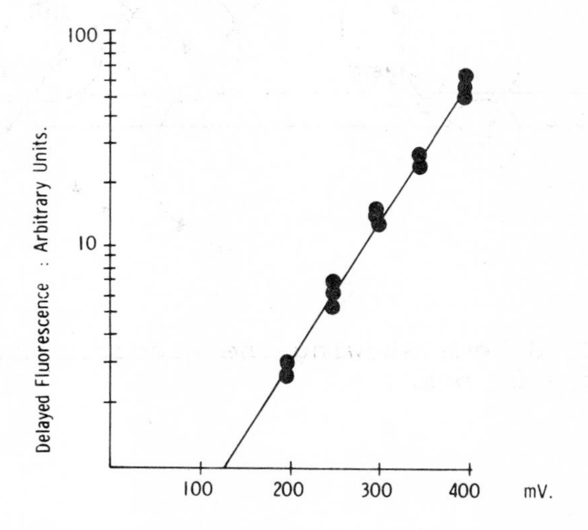


Figure 11. Relation between decayed fluorescence intensity and membrane potential. Conditions as in Fig. 3. Transients in delayed fluorescence and the carotenoid change were induced by additions of 0.2 μM nigericin at various points in the time course of the light induced change, and the levels before and after addition compared with reference to the zero level.

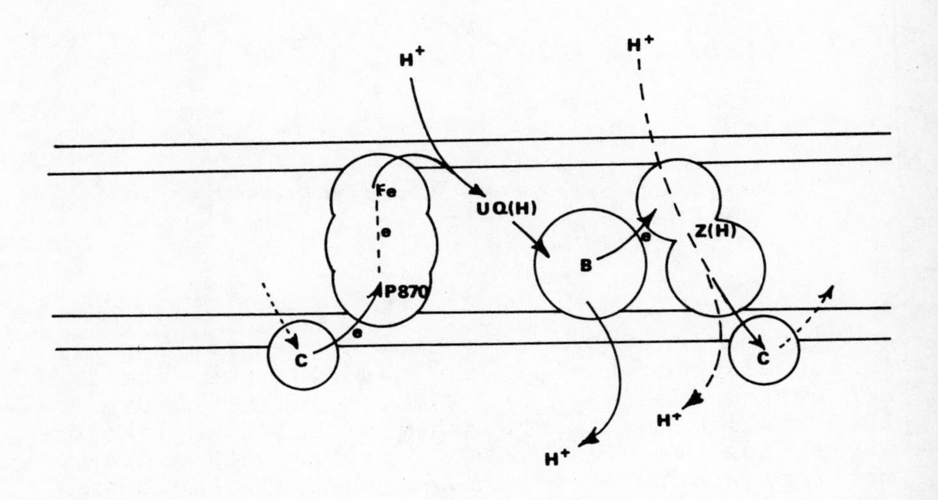


Figure 12. Scheme showing the electron transport system as a $H^+$ pump.

# PRIMARY ACTS OF ENERGY CONSERVATION IN THE FUNCTIONAL MEMBRANE OF PHOTOSYNTHESIS

H.T. Witt

Max-Volmer-Institut für Physikalische
Chemie und Molekularbiologie
Technische Universität Berlin
West Germany

The basic events of photosynthesis take place in subcellar organella, the chloroplasts. The inner system of the chloroplasts contains about 1000 small compartments, the so-called thylakoids. Thylakoids are disk shaped vesicles with a diameter of about 5000 Å. In the thylakoid membrane about $10^5$ pigment molecules are embedded. In green plants the pigments are chlorophylls-a and -b and carotenoids. Light absorbed in the bulk of these pigments is channelled by excitation energy migration to the photoactive chlorophylls (0,5% of the bulk chlorophyll). The excited centers transfer electrons from $H_2O$ via intermediates to $NADP^+$. This electron transfer is accompanied by phosphorylation, i.e. by synthesis of ATP from ADP and P. With the help of NADPH and ATP absorbed $CO_2$ can be reduced into sugar ($CH_2O$) and "everything else". The biochemical dark processes of the $CO_2$ -cycle are known as the Calvin-cycle. The problem of photosynthesis today is the question of the molecular mechanism of the transformation of light energy into the chemical energy of NADPH and ATP. Information on the mechanism of the molecular machinery can be obtained by direct measurement of the molecular events. For the characterization of the molecules which are reported in this lecture the change in optical properties has been used. For excitation of the whole process light pulses are necessary which are shorter than the reaction time of the considered event. This requirement has been realized in photosynthesis by using normal flashes, ultra short flashes or laser giant pulses. For registration of the small signals within short times the repetitive pulses spectroscopic method is most suitable. The extremely high sensitivity of this technique enables the detection of very small signals. The high time resolution of this technique extended the analysis of biological events down to the 10 nano-second range. These techniques are summarized elsewhere (1,2).

With the pulse spectroscopic methods the following events have been analysed under single turnover conditions in the range between 10 nsec and 1 sec:

1). Formation of metastable states of carotenoids.

2). Primary reactions of the active chlorophylls.

3). Electron transfers during the redox reactions of electron carriers as plastoquinone, cytochromes,etc. The events (1) - (3) have been measured by their characteristic intrinsic absorption changes.

4). Generation of electrical fields or potential differences respectively across the thylakoid membrane (measured by absorption changes due to electrochromism, i.e. by the shift of absorption bands in an electric field.

5). Ion fluxes across the membrane (measured by the derivative of the electrochromic absorption changes with regard to time).

6). Proton transfer across the membranes (measured by pH indicators in the outer space of the thylakoid. As indicator preferentially umbelliferone has been used which changes from a nonfluorescing into a fluorescing state during deprotonation).

7). Formation of proton gradients ΔpH across the membrane (measured by an intrinsic indicator for protons in the inner space of the thylakoids; the $H^{+}_{in}$ depending rate of reduction of Chl-$a^{+}_{I}$ by

$PQH_2$ has been used for this purpose).

8). Generation of ATP (measured by the irreversible fluorescence change of the pH indicator umbelliferone; during the ATP generation from ADP and P protons are consumed).

Details and relationships which have been evaluated between the eight events are discussed elsewhere (3,4).

This report is focussed on the experimental evidence that the cooperation of two photoactive chlorophylls and plastoquinone leads to a vectorial electron transfer from $H_2O$ (inside) to $NADP^+$ (outside). In this way, $NADP^+$ is reduced and the functional membrane electrically charged. It is shown that the discharging of the electrically energized membrane is coupled with the generation of ATP.

In photosynthesis two light reaction centers have been tentatively assumed (5) and postulated to operate in series (6,7). Such a coupling has been demonstrated (8,9,10). The two photoactive chlorophylls within the coupled systems have been observed spectroscopically as Chlorophyll-$a_I$ - 700 (11) and Chlorophyll-$a_{II}$ - 680 (12,13). These two centers transfer in their excited states one electron from $H_2O$ to $NADP^+$. A pool of plastoquinone is the link between the two light centers (14). Excitation of Chlorophyll-$a_I$ and Chlorophyll-$a_{II}$ leads to the generation of an electrical potential difference $\Delta\psi$ (15). In a single turnover each of the two light reactions generates one-half of the electrical potential change (16):

$$h\,\nu_I \sim \Delta\psi/12 \quad \text{and} \quad h\,\nu_{II} \sim \Delta\psi/12.$$

The generation of $\Delta\psi$ takes place together with the photoact at the two chlorophylls in $\leq$ 20 nsec (17).

$\Delta\psi$ has been measured by field indicating absorption changes. These have been assumed to be caused by electrochromism (s. above). This interpretation has been proved in the last seven years by different lines of evidences:

1). As expected the absorption bands of all

bulk pigments - chl-a, chl-b and carotenoids - are shifted in the field (18,19).

2). The shape of the spectral changes are characteristic for changes caused by electrochromism (18).

3). The spectrum of the changes is in agreement with the spectrum induced in the dark by electrical fields across artificial multilayers built up with chl-a, chl-b and carotenoids (19,20).

4). The relaxation of the changes is sensitive to artificial ionophores in contrast to the relaxation of changes caused by e-transfer, metastable states, etc.

5). The relative extent and the relaxation of the changes correspond to changes measured electrically by electrodes (22,23,24).

6). Field indicating absorption changes were established also across the inner membrane of photobacteria. The spectrum of these changes is in agreement with the spectrum induced in the dark by artificial electrical fields across the pigment bearing membrane (21).

From the further analysis of the field indicating absorption changes the following five properties have been derived:

1). The field indicating absorption changes can be regarded as a self-operating intrinsic molecular volt- and ammeter.

2). The indication of $\Delta\psi$ and $\Delta\psi \sim i$ is prompt, i.e. without restriction to time at least down to about 10 nsec (17).

3). The indication of $\Delta\psi$ is linear, i.e. $\Delta\psi$ is proportional to the field indicating absorption changes $\Delta A$ (16,25,24).

4). The indication of $\Delta\psi$ and i has been calibrated as follows (16):

$$\Delta\psi \simeq 50\ \text{mV} \cdot \Delta A / {}^{1}\Delta A$$

$$i \simeq 1/^{\text{uF}} \cdot 50\ \text{mV} \cdot d\,\Delta A/dt \cdot 1/{}^{1}\Delta A$$

$\Delta A/^{1}\Delta A$ = absorption change in relation to that produced in a saturated single turnover flash.

5). The indication of $\Delta\psi$ is due to potential differences perpendicular to the membrane and not to local potential differences in the plane of the membrane (15,22,23,24).

The last result has been explained by an electron shift at each light reaction from inside of the membrane to the outside (15).

In consecutive reaction steps at each of the two light reactions a proton translocation of 1 $H^+$ into the inner phase of the thylakoid has been observed (16):

$$h\nu_{I} \sim 1\ H^+ \qquad\qquad h\nu_{II} \sim 1\ H^+$$

This can be explained by four protolytic reactions which are coupled with the electron transfer from $H_2O$ to $NADP^+$ (26). At the outer surface of the membrane 1 $H^+$ is accepted by 1/2 $PQ^=$ and 1 $H^+$ by 1/2 $NADP^+$. At the inner surface 1 $H^+$ is released by the oxidation of $H_2O$ and 1 $H^+$ by the reoxidation of $PQH_2$. This is equivalent to a translocation of 1 $H^+$ to the inside of the thylakoid.

It has been shown that the discharging of the membrane occurs by the field driven efflux of protons (27).

The discussed experimental results - vectorial electron flow and proton circulation - are formulated in the zigzag scheme in Fig. 1. The scheme is additionally supported by the following relations between potential change $\Delta\psi$, proton translocation $\Delta H^+$ and reduction of plastoquinone PQ:

1) The electrical potential change can be increased in a long flash proportionally to the number of electrons (up to seven) injected into the pool of PQ (25,28):

$$\Delta\psi \sim \{PQ^{2-}\}$$

This substantiates the proposed transmembrane shift of electrons from $H_2O$ (inside) to PQ (outside).

2) The proton uptake can be increased in a long flash proportionally to the number of electrons which are injected into the pool of PQ (25):

$$\Delta H^{+} \sim \{PQ\ H_2\}$$

This substantiates the proposed $H^{+}$ translocation by $H^{+}$ uptake at PQ (outside) and $H^{+}$ release from $H_2O$ (inside).

3) The identity of the recovery time of $\Delta\psi$ (induced particularly by Chl-$a_{II}$) with the electron transfer time from $H_2O$ to the PQ pool (0,6 msec) supports the orientated electron shift from $H_2O$ (inside) to PQ (outside) (24).

The discharging of the electrically energized membrane is coupled with the generation of ATP: The amount of ATP generated is under optimal conditions and constant value of the proton gradient, $\Delta pH$, proportional to $\Delta\psi$, i.e. proportional to the number of charges Q (protons) by which the membrane has been loaded (27, 3):

$$\Delta\ ATP \sim \Delta\ \psi\ .\ C \quad \text{or} \quad \Delta\ ATP \sim \Delta\ H^{+}.$$

For the stoichiometry ATP : Q it has been found that the generation of 1 ATP is probably coupled to four protons (29):

$$1\ ATP \sim 4\ H^{+}.$$

The actual rate of ATP formation follows the actual rate of the electrical potential decay and proton efflux (27, 3):

$$\dot{ATP} \sim \dot{\Delta\psi}_{H^{+}} \quad \text{or} \quad \dot{ATP} \sim i_{H^{+}}\ .$$

The proton efflux occurs probably through the ATPase.

Titration experiments with ionophores indicate that the functional unit of phosphorylation as well as the unit of the electrical events is the membrane of one thylakoid (27):

$$ATP\ -\ unit\ =\ \Delta\psi\ -\ unit\ =\ 1\ thylakoid.$$

The results indicate that the electron transfer and phosphorylation are primary coupled by $\Delta\psi$ on the membrane of a thylakoid. This result is consistent with the hypothesis of Mitchell (26). Generation of ATP by a chemical intermediate directly produced by the e-transfer, is not in accordance with our experimental results. The action of $\Delta\psi$ can of course subsequently induce in the ATPase special configurations between $H^+$, ADP, P, $H_2O$ and other components, i.e. special intermediates and conformations are probably formed before ATP is synthesized. However, these reactions would be a result of the action of $\Delta\psi$. For a more detailed discussion see references 3 and 4.

## References


1. Witt, H.T. 1967. In: Fast Reactions and Primary Processes in Chemical Kinetics, Nobel Symp. V., 81 and 261, ed. Claesson, S. Stockholm: Almqvist and Wiksell; New York, London, Sydney: Interscience.
2. Rüppel, H. and Witt, H.T. 1969. In: Methods in Enzymology, Fast Reactions, Vol. 16, pp. 316-380, ed. Colowick, S.P. and Kaplan, N.O. New York:Academic Press.
3. Witt, H.T. Quart. Rev. Biophys. 4 (1971) 365.
4. Witt, H.T. 1974. In: Bioenergetics of Photosynthesis, ed. Govindjee. New York: Academic Press. In press.
5. Emerson, R. Ann. Rev. Plant. Physiol. 9 (1958) 1.
6. Hill, R. and Bendall, F. Nature (Lond.) 186 (1960) 136
7. Kautsky, H., Appel, W. and Amann, H. Biochem. Z. 332 (1960) 277.
8. Kok, B. Biochim. Biophys. Acta 48 (1961) 527.
9. Duysens, L.N.M., Amesz, J. and Kamp, B.M. Nature (Lond) 190 (1961) 510.
10. Witt, H.T., Müller, A. and Rumberg, B. Nature (Lond) 191 (1961) 194.
11. Kok, B. Biochim. Biophys. Acta 48 (1961) 527.
12. Döring, G., Renger, G., Vater, J. and Witt,

H.T. Z. Naturforsch. 24b (1969) 1139.

13. Gläser, M., Wolff, Ch., Buchwald, H.-E. and Witt, H.T. FEBS Letters (1974) In press.

14. Stiehl, H.H. and Witt, H.T. Z. Naturforsch. 23b (1968) 220; Z. Naturforsch. 24b (1969) 1588.

15. Junge, W. and Witt, H.T. Z. Naturforsch. 23b (1968) 244.

16. Schliephake, W., Junge, W. and Witt, H.T. Z. Naturforsch. 23b (1968) 1571.

17. Wolff, Ch., Buchwald, H.-E., Rüppel, H., Witt K. and Witt, H.T. Z. Naturforsch 24b (1969) 1038.

18. Emrich, H.M., Junge, W. and Witt, H.T. Z. Naturforsch. 24b (1959) 1144.

19. Schmidt, S., Reich, R. and Witt, H.T. Naturwiss. 58 (1971) 414.

20. Schmidt, S., Reich, R. and Witt, H.T. 1972. Proc. 2nd Intern. Congr. Photosynthesis Res. Stresa 1971, pp. 1087-1095, ed. Forti, G., Avron, A., Melandri, A. The Hague:Junk Publ.

21. Jackson, J.B. and Crofts, A.R. Eur. J. Biochem. 18 (1971) 120.

22. Kok, B. 1972. In: VI. Intern. Congr. Photobiol. Bochum Ed. Schenck, O. In preparation.

23. Witt, H.T. and Zickler, A. FEBS Letters 38 (1973) 307.

24. Witt, H.T. and Zickler, A. FEBS Letters 39 (1974) 205.

25. Reinwald, E., Siggel, U. and Rumberg, B. Naturwiss. 55 (1968) 221.

26. Mitchell, P. Biol. Rev. 41 (1966) 445.

27. Boeck, M. and Witt, H.T. 1972. Proc. 2nd Intern. Congr. Photosynthesis Res. Stresa 1971. pp. 903-911, ed. Forti, G., Avron, S., Melandri, A. The Hague: Junk Publ.

28. Witt, H.T., Rumberg, B., Schmidt-Mende, P., Siggel, U., Skerra, B., Vater, J. and Weikard, J. Angew. Chem. 4 (1965) 799. Intern. Ed.

29. Rumberg, B. and Schröder, H. 1973. In: VI. Intern. Congr. Photobiol. Bochum 1972. Ed. Schenck, O. Abstr. 036.

---

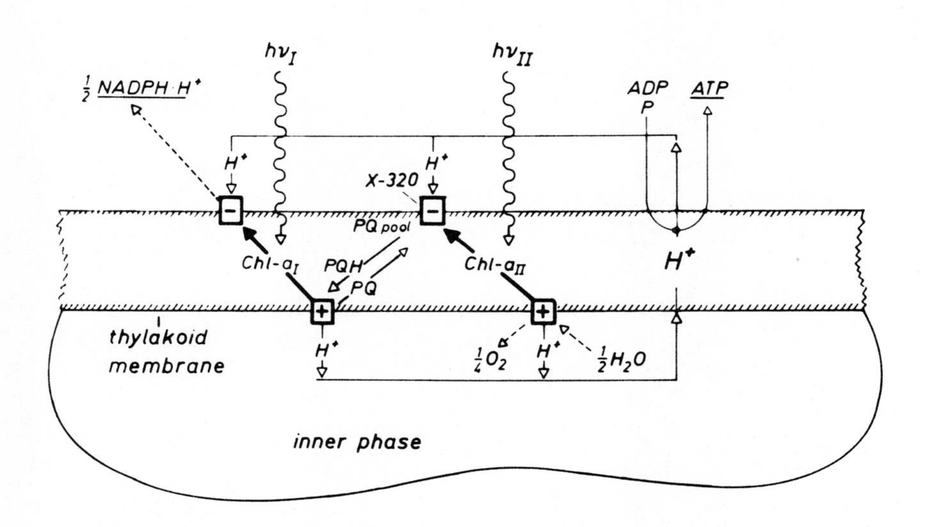


Figure 1. Zigzag pathway of electrons and cycling of protons in the primary acts of photosynthesis derived from pulse spectroscopic results. (X-320 is probably plastosemiquinone). For details see text.

# RHODOPSIN, VISUAL EXCITATION, AND MEMBRANE VISCOSITY

Richard A. Cone

Department of Biophysics
The Johns Hopkins University
Baltimore, Maryland 21218

In this talk I will try to describe some of the characteristics of the visual excitation mechanism which I expect will help form a basis for future research on visual receptors, and I will then discuss characteristics of rhodopsin which may reveal general properties of cell membranes.

Photoreceptors in both vertebrate and invertebrate eyes can detect the absorption of a single photon. In many of these receptors it appears that the rhodopsin molecule, after catching the photon, in some way regulates numerous $Na^+$-channels located elsewhere on the membrane. The $Na^+$-channels are often located several microns away from the rhodopsin molecule, and in the vertebratebrate rod, the $Na^+$ channels are in a completely separate membrane. The problem, then, is to find out how rhodopsin regulates these distant $Na^+$- channels. We know that in some invertebrate receptors rhodopsin opens many $Na^+$ channels, while in vertebrate rods and cones it closes them (for recent discussion see refs. 1 and 2). The mechanism is more obscure in invertebrate receptors than in rods and cones and I will confine this discussion to rods and cones even though many invertebrate receptors are large and easily impaled with microelectrodes, offering many advantages for future work.

In rods and cones, the rhodopsin is located

in the disc membranes which are formed from lamellar invaginations of the plasma membrane. In all rods (and some cones), the invaginations pinch off to become "free-floating" flattened vesicles, completely detached from the plasma membrane (3,4). In darkness, both rods and cones produce a steady $Na^+$ current which is pumped out of the cell in the inner segment region by pumps which are almost certainly driven by ATP produced by the mitochondria packed into the inner-segment of the receptor call (5,6). The $Na^+$ current then flows extracellularly to the outer segment, enters through the plasma membrane, and returns to the pumps through the ciliary connection between the outer segment and the rest of the cell. The effect of light is to transiently reduce this steady $Na^+$ current: a flash which delivers about 1 photon per rod reduces the $Na^+$ current about 1% for a fraction of a second. The reduction in $Na^+$ current is caused by a transient reduction in the $Na^+$-permeability of the plasma membrane surrounding the outer segment and results in a hyperpolarizing receptor potential.

The light-initiated reduction in $Na^+$-permeability was first inferred from intra- and extracellular microelectrode observations (see ref. 2), and can be directly inferred from osmotic shock experiments on freshly isolated rod outer segments (3,7). When a retina is shaken gently, the outer segments break off at the ciliary connection, and the plasma membrane reseals, making the isolated outer segment a good osmometer. Surprisingly, the plasma membrane of the outer segment is impermeable to $K^+$: when the outer segments are hyperosmotically shocked with KCl they shrink and remain shrunken, indicating KCl cannot leak in through the plasma membrane. However, following a hyperosmotic NaCl shock, the outer segments shrink and then recover in volume, returning to their original volume at a rate which depends directly on the $Na^+$ permeability of the plasma membrane. Light transiently reduces this rate of volume recovery by reducing the rate of $Na^+$ influx. The transient decrease in $Na^+$ influx is comparable to that observed in the perfused retina: in both cases, the entry of some $10^7$ $Na^+$ ions is blocked per photon absorbed.

Since the mechanism of visual excitation remains intact in freshly isolated outer segments, the osmotic technique, being both simple and direct, offers much promise for future work. The excitation mechanism is clearly confined in this case to a simple two-membrane organelle in which a rhodopsin molecule in the disc membrane must "reach out" and transiently block some 1% of the $Na^+$-channels in the plasma membrane.

## The $Ca^{2+}$ Hypothesis

One way rhodopsin might regulate the $Na^+$-channels would be for it to release messenger particles which could diffuse to and block the channels, and Yoshikami and Hagins have proposed that $Ca^{2+}$ ions might serve this function proposal is based on the compelling idea that rhodopsin could in this case operate by the same mechanism in both rods and cones. Since in cones the disc membranes are usually continuous with the plasma membrane, the space within a disc is in fact extracellular space. Hence there are very few candidates for messenger particles; rhodopsin in cone discs can release into the cytoplasm those particles available in the extracellular fluid, for example, $Na^+$, $K^+$, $Mg^{2+}$, $Ca^{2+}$, and $Cl^-$ ions.

The evidence now available which supports the $Ca^{2+}$ hypothesis is: 1) Yoshikami and Hagins have shown that the $Na^+$ current in rat rods becomes extremely sensitive to extracellular $Ca^{2+}$ as soon as a $Ca^{2+}$-ionophore (X537A) is introduced into the solution which perfuses the retina They report that in the absence of the $Ca^{2+}$-ionophore upwards of 10 millimolar $Ca^{2+}$ is required to block the $Na^+$-channels, but as soon as the $Ca^{2+}$-ionophore is added to the perfusion solution, the $Ca^{2+}$ concentration must be dropped to below $10^{-7}$ molar before the $Na^+$-channels begin to open, and they are not completely open until the $Ca^{2+}$-

concentration drops below $10^{-9}$ molar. Accounting approximately for the membrane potential, this in dicates cytoplasmic $Ca^{2+}$ appears to block $Na^{+}$-channels when its activity is on the order of $10^{-7}$ $-10^{-6}$molar.

2) In my laboratory, Ete Szuts has recently found that light causes the release of a large quantity of $Ca^{2+}$ from rod discs but has no effect on their content of $Na^{+}$, $K^{+}$, and $Mg^{2+}$ 10, 11. This work is still in progress but the results to date indicate that each photoactivated rhodopsin molecule can initiate the release of on the order of 1000 $Ca^{2+}$ ions under the conditions of the experiment. The experiment is performed as follows: a suspension of rod outer segments, isolated by gently shaking a freshly dissected frog retina, is divided into two equal aliquots which are treated identically and in total darkness except that one is exposed to a dim flash of light just before both aliquots are hyposomotically shocked to break the plasma membrane but not the disc membranes. The shock occurs in an EDTA-solution which chelates all accessible $Ca^{2+}$ and $Mg^{2+}$ ions. The shocked fragments are quickly sedimented and their ion contents determined by atomic absorbance. The $Na^{+}$, $K^{+}$, and $Mg^{2+}$ content of both pellets is found to be the same, but the flash-illuminated fragments loose over half their $Ca^{2+}$. In some experiments, Szuts has found this loos of $Ca^{2+}$ occurred even when the flash bleached only 0.01% of the rhodopsin. In this case, on the order of 1000 $Ca^{2+}$ ions were released per bleached molecule of rhodopsin.

A crucial step in this experiment is that the flash is delivered to the rod outer segments at a time when the excitation mechanism is intact, as shown by the osmotic shock technique. For a few months last fall, Szuts found to his dismay that the results of this experiment, though routine during the summer months, no longer showed any light-activated release of $Ca^{2+}$. However, during this time the isolated outer segments were also found to be unresponsive to light when tested with the osmotic shock technique. The cause of this lapse in sensitivity to light is obscure, perhaps due to some seasonal variation in frogs, but the parallel loss of $Ca^{2+}$ release and light sensitivity tends to confirm the idea that $Ca^{2+}$ may act as an internal messenger.

There are several crucial tests which the $Ca^{2+}$ hypothesis must pass before it can be considered established. In particular, observations must be made of the time-course of $Ca^{2+}$ release, the cytoplasmic activity of $Ca^{2+}$ must be measured (at least in the resting state), and the effective mobility of $Ca^{2+}$ in the cytoplasm must be measured. With such measurements it should be possible to demonstrate whether or not enough $Ca^{2+}$ is released in the required time to block the $Na^+$ -channels, and whether or not $Ca^{2+}$ can actually diffuse to the $Na^+$-channels in the time available. Finally, an appropriate $Ca^{2+}$ uptake mechanism must also be demonstrated.

## $Ca^{2+}$ Release Mechanism

By what mechanism might rhodopsin release $Ca^{2+}$ through the disc membrane? The simplest idea, of course, is that rhodopsin may be a light-activated $Ca^{2+}$-ionophore. The rate of $Ca^{2+}$ release is probably consistent with either a pore or a diffusional carrier mechanism: in the experiments by Szuts, if all the $Ca^{2+}$ is released within a fraction of a second (a time comparable to the duration of the late receptor potential) the rate of release per photoactivated rhodopsin molecule need not exceed $10^3$-$10^4$ $Ca^{2+}$ ions per second. As a light-activated pore rhodopsin could probably permit some $10^4$-$10^6$ $Ca^{2+}$ ions per second to pass through the membrane. As a diffusional carrier, given its size and the viscosity of the membrane (see below), rhodopsin could probably transport as many as $10^3$ $Ca^{2+}$ ions per second. Thus the transport rates for both mechanisms appear sufficient to account for the $Ca^{2+}$ release observations to date. However, even though the transport rate of a diffusional carrier might be adequate, it is far from clear whether a molecule like rhodopsin can tumble or bob in the membrane with a motion which would permit it to carry ions through the membrane. Clearly, a small molecule such as valinomycin can tumble in the membrane, but what about large membrane proteins? The enzyme-like specificity of many transport systems could be easily accounted for if large membrane proteins could undergo such tumbling and/or bobbing motions but to date there is no clear evidence available. It seems apparent that one of the most important advances to be made in the next few years will be to establish whether or not large membrane proteins can mediate transport by tumbling or bobbing in a way which permits them to act as diffusional carriers. The key point of course is to demonstrate that a given selective binding site on the molecule alternately appears on opposite sides of the membrane.

It is possible that rhodopsin initiates the release of $Ca^{2+}$ but does not itself transport $Ca^{2+}$. It is interesting to pursue, briefly, the question of whether all types of rhodopsin molecules operate by essentially the same mechanism. Recent experiments by Stoeckenius, Oesterhelt, and Racker (see the paper by Racker in this volume) have shown that bacteriorhodopsin is almost certainly a light-driven proton pump. In a photoreceptor, instead of using the energy of the photon to build up an ion gradient, rhodopsin probably triggers the release of energy previously stored in ion gradients across the membrane. If rhodopsin releases protons as messenger particles, calculations indicate that the release of one proton into the disc lumen does not constitute an adequately reliable signal for the rest of the excitation sequence[1]. In our present state of ignorance, it is at least possible, though unlikely, that vertebrate rhodopsin might passively translocate many protons (10-100) which would in turn initiate the release of $Ca^{2+}$ from the disc. However, recent experiments by Lisman and Brown have shown that an invertebrate rhodopsin in the ventral eye of _Limulus_ almost certainly does not release protons as internal messengers: when the ventral eye photoreceptor is loaded with pH buffers (using a microelectrode) there is little or no effect on the response of the cell to light (12). Moreover, if the photoreceptor is loaded with a $Ca^{2+}$ buffer, as the $Ca^{2+}$ activity in the cytoplasm rises above normal, the $Na^{+}$-channels become blocked, much as in vertebrate rods. But in these *Limulus* photoreceptors, the effect of light is *to open* $Na^{+}$-channels, not close them. Thus *Limulus* rhodopsin must act on the $Na^{+}$-channels with *something* other than $Ca^{2+}$ or $H^{+}$ ions. It appears, therefore, that each type of rhodopsin probably operates by a different mechanism: a proton pump in bacteriorhodopsin, perhaps as a $Ca^{2+}$ ionophore in vertebrate photoreceptors, and by an unknown mechanism in invertebrate photoreceptors.

It is apparent that the role of each type of rhodopsin will be difficult to establish until the rhodopsin can be purified and incorporated into a model membrane system, much as has been done for bacteriorhodopsin. In this way the effects of light in the intact receptor may be compared with the effects observed in the model system. Thus if a purified rhodopsin is found to release $Ca^{2+}$ with a rate and time-course similar to that observed in the intact receptor there will be little reason to invoke an intermediary process such as proton release. Instead, the simpler $Ca^{2+}$-ionophore hypothesis will be adequate and it will be relatively easy to distinguish a pore from a carrier mechanism. For example, the transport rate of a diffusional carrier should depend reciprocally on membrane viscosity, whereas the transport rate for a pore can in some cases be essentially independent of membrane viscosity.

This brings me to one of the most important properties of rhodopsin for studying membrane structure and function, namely, that it has an intrinsic, dichroic, and bleachable chromophore which makes possible direct observations of both the orientation and the motions of rhodopsin in the membrane. In addition, rhodopsin occurs in high concentration in the disc membrane and is nearly the only protein present in these membranes(13). Finally, the entire rhodopsin content in the receptor can be synchronously activated. This permits one to observe the molecule "in action" with numerous techniques: its spectral transitions can be observed with flash photometry, its charge configuration with dielectric and electrophysiological techniques, and its conformation and disposition in the membrane can be studied with X-ray, electron microscopy, circular dichroism, birefringence, neutron diffraction, ESR, and NMR. Surprisingly, although research on rhodopsin is being actively pursued by all of these as well as other techniques, we still do not know the disposition of rhodopsin within the membrane. However, the most recent evidence makes it highly likely that at least some portion of rhodopsin spans the membrane and that a major fraction of the molecule, including its sugar moiety, resides on the cytoplasmic face of the membrane (for a recent discussion see ref. 10). In addition, there is suggestive evidence that rhodopsin changes its disposition in the membrane as it undergoes the metaI → metaII spectral transition, the last spectral transition to occur before the $Na^{+}$-current is blocked.

Rhodopsin in isolated outer segments offers one of the best preparations for quantitatively studying diffusion of proteins in cell membranes, and hence elucidate the interactions between lipids and proteins. The diffusion of rhodopsin can be directly observed in a microspectrophotometer simply by bleaching the pigment on one side of an isolated outer segment and monitoring the redistribution of the remaining pigment as it diffuses laterally in the discs to achieve uniform distribution(10,14). The diffusion constant for this lateral diffusion of rhodopsin in the membrane is $4x10^{-9}$ $cm^2$/sec in both frog and mudpuppy (Necturus) rods at 20°C. The diffusion rate slows by about a factor of 3-4 with a 10°C drop in temperature, indicating the viscosity of the membrane increases by this same factor (an apparent energy of activation of about 20-25 kcal/mole). If the effective diameter of rhodopsin in the plane of the membrane is 45 Å, as suggested by its molecular weight and by x-ray evidence (15), then the Stokes-Einstein relationship for the translational diffusion of a sphere in a homogenous medium implies the approximate viscosity of the disc membrane is about 2 poise. Thus the membrane viscosity for lateral motion of large molecules is about 200 times the viscosity of water, a viscosity similar to that of a light oil such as olive oil.

The rotational diffusion of rhodopsin can also be directly observed, though with considerably greater difficulty since such a measurement approaches the experimental limits of flas photometry. For reasons which in retrospect seem unconvincing, observations of rotational diffusion were undertaken before the much simpler experiments on lateral diffusion were pursued. Rhodopsin was found to undergo rotational diffusion at a rate which agrees remarkably well with its rate of lateral diffusion (16). The relaxation time of rhodopsin in both frog and rat rods is about 20 μsec at 20°C and the relaxation time for frog rhodopsin increases by about a factor of 3-4 for a 10°C drop in temperature (16,17). Again using the Stokes-Einstein relation, in this case for the rotational diffusion of a sphere about one axis, if rhodopsin has an effective diameter of about 45 Å, the approximate viscosity of the membrane is again found to be about 2 poise at 20°C.

{The question is often raised whether "viscosity" is a satisfactory term for describing the interactions which determine the diffusion rates of molecules embedded in a structure such as a membrane. It seems to me that as long as one is not interested in minor correction factors the term "viscosity" is exactly right, especially if one confines the term to refer to motions which require lateral displacements and slippage in the membrane, in other words, to lateral diffusion and to rotational diffusion about an axis perpendicular to the membrane surface. The idea that a small molecule might experience a "microviscosity" altogether different than the macroscopic viscosity of a solution is not true. In a solution, little molecules experience much the same viscosity as big molecules, and big molecules experience the macroscopic viscosity of the solution, as was first demonstrated by Einstein who showed that Stokes' law correctly predicts the viscous drag experienced by a sugar molecule in an aqueous solution (18). Later experiments have shown that Stokes'law is nearly correct even for molecules smaller than the solvent molecules: using the Stokes-Einstein relation, the viscosity predicted by the rate of self-diffusion of water is only off by a factor of 2, and the viscosity predicted by the diffusion of $H_2$ in water is off by a factor of 3. Hence Stokes' law, though derived entirely with macroscopic concepts, appears to hold reasonably well even for the smallest of molecules. Thus there is good reason to expect that regardless of the size or shape of the molecule used to "probe" a membrane, the rate of diffusion, when properly interpreted to account for the disposition of the molecule, should yield essentially the same viscosity. Hence diffusion constants can be directly compared by predicting the viscosity, and if a wide variety of probe molecules should all predict the same viscosity then it would seem fitting to call this the "viscosity of the membrane". It is perhaps worth noting that the term "fluidity" as used in spin-label studies is a mixed parameter which depends not only on diffusion rates but also on the range of possible orientations - a variation in fluidity need not imply the viscosity varies.}

The viscosity of the disc membrane, as predicted by both the rotational and lateral diffusion of rhodopsin, is comparable to the viscosity predicted by several other observations based on the diffusion rates of proteins, phospholipids, and small lipid soluble molecules embedded in the lipid phase of various cell and model membranes (see refs. 14, 15 and 19). Indeed, cell membranes in general appear to have viscosities within an order of magnitude of 1 poise, as long as the lipid phase is in a liquid state. Thus it is clear that large membrane proteins can experience much the same viscosity as the lipids, and it is apparent that the proteins will be free to undergo rapid diffusion unless they are hindered by interactions with other proteins (aggregation) or by interactions with molecular structures outside the bilayer (e.g., with actin or spectrin).

It is somewhat surprising that the viscosity of the disc membrane is not higher. Rhodopsin molecules are highly concentrated in this membrane, and occupy about 50% of the area of one face of the membrane. This is equivalent to a 50% v/v solution. If rhodopsin had any tendency to aggregate the effective viscosity would increase markedly, especially for rotational diffusion which depends on the cube of the diameter of the diffusing "particle" whereas lateral diffusion only varies with the first power of the diameter. The agreement in the viscosity predicted by the rotational and lateral diffusion rates of rhodopsin suggests that rhodopsin probably diffuses as a monomer. This inference is strengthened by the x-ray analysis by Blasie and Worthington (15) in which they conclude rhodopsin is distributed in a liquid-like array, but a firm basis for concluding that rhodopsin diffuses as a monomer awaits future work on model membrane systems in which the concentration of rhodopsin and other membrane components can be varied to determine their effect on diffusion rates. This will be especially useful for theoretical analysis of the results since at present only dilute solutions can be fully interpreted.

The rate at which rhodopsin diffuses in the membrane implies rhodopsin-rhodopsin collisions occur with very high frequency, each molecule undergoes about $10^6$ such collisions/sec (14). This points to what I think is a fundamental characteristic of cell membranes: most are constructed in such a way that rapid reactions can occur between proteins, and between proteins and their coreactants. For example, even though the viscosity of the membrane is on the order of 100 times that of water, molecules in the membrane are ordered, and this ordering can greatly speed reaction rates. Moreover, substrates which partition into the lipid phase, or onto the oil-water interface, may reach much higher effective concentrations than in the aqueous phase. Together, such effects can more than compensate for the slower diffusion rates in the membrane. I am sure that as we get to know more about membrane proteins and their interactions we will find many reactions take place in the lipid phase of the membrane, and I expect an area of considerable future interest in membranes will be to characterize such lipid phase reactions.

## References

1. Cone, R.A. 1973. Biochemistry and Physiology of Visual Pigments, pp. 275, ed. H. Langer, Springer-Verlag: New York.
2. Hagins, W.A. Ann. Rev. Biophys. Bioeng. 1 (1972) 131.
3. Korenbrot, J.I., Brown, D.T. and Cone, R.A. J. Cell Biol. 56 (1973) 389.
4. Cohen, A.I. J. Cell Biol. 48 (1971) 547.
5. Penn, R.D. and Hagins, W.A. Biophys. J. 10 (1972) 1073.
6. Yoshikami, S. and Hagins, W.A. 1973. Biochemistry and Physiology of Visual Pigments. pp. 245, ed. H. Langer, Springer-Verlag:New York.
7. Korenbrot, J.I. and Cone R.A. J. Gen. Physiol. 60 (1972) 20.
8. Yoshikami, S., and Hagins, W.A. Abstr. 14th Ann. Meet. Biophys. Soc. WPM-13 (1970).
9. Yoshikami, S. and Hagins, W.A. Exp. Eye Res. 18 (1974) In press.
10. Poo, M-M. and Cone, R.A. Exp. Eye Res. 17 (1973) 503.
11. Szuts, E.Z. and Cone, R.A. Abstr. 18th Ann. Meet. Biophys. Soc. (1974).

12. Lisman, J.E. and Brown, J.E. J. Gen. Physiol. 59 (1972) 701 (Also, personal communication).

13. Bownds, D., Gordon-Walker, A., Gaide-Huguenin, A.D. and Robinson, W.E. J. Gen. Physiol. 58 (1971) 225.

14. Poo, M-M. and Cone, R.A. Nature 247(1974) 438

15. Blasie, J.K. and Worthington, C.R. J. Mol. Biol. 39 (1969) 417.

16. Cone, R.A. Nature New Biol. 236 (1972) 39.

17. Poo, M-M., Ph. D. thesis, The Johns Hopkins University, Baltimore (1973).

18. See discussion and Fig. 3.2 in chapter 3 of Stein, W.D., The Movement of Molecules Across Cell Membranes (1962) Academic Press: New York.

19. Edidin, M. Ann. Rev. Biophys. Bioeng. 3 (1974) In press.

# V

# ROLE OF THE MEMBRANES IN GENOME EXPRESSION

# HORMONE RECEPTORS AND THEIR FUNCTION IN CELL MEMBRANES

P. Cuatrecasas[1] and V. Bennett[2]

Department of Pharmacology & Experimental Therapeutics and Department of Medicine
The Johns Hopkins University
School of Medicine
Baltimore, Maryland 21205

## General Comments Regarding Detection of Membrane Receptors

*General Criteria.* During the past few years there has been a large burst of interest in membrane receptors for hormones, and considerable progress has been made in the identification and study of receptors for such peptide hormones as insulin, glucagon, drenocorticotropin, thyrotropin, angiotensin, calcitonin, growth hormone, prolactin, follicle stimulating hormone, luteinizing hormone, chorionic gonadotropin, oxytocin and vasopressin, as well as nonpeptide hormones such as prostaglandins and acetylcholine (reviewed in ref. 1). The general approach in these studies has been to measure the interaction (binding) of a radioactively labelled hormone with intact target cells or with isolated membrane preparations derived from such cells. The binding is surmised to "specifically" reflect receptor interactions if it demonstrates: a) strict structur-

---

[1]P. Cuatrecasas is the recipient of a United States Public Health Service Research Career Development Award AM31464
[2]Postgraduate student supported by the Insurance Medical Scientist Scholarship Fund.

al and steric specificity; b) saturability, which reflects a finite and limited number of binding sites; c) tissue specificity in accord with biological target cell sensitivity; d) high affinity, in harmony with the physiological concentrations of the hormone; and e) reversibility kinetically consistent with the reversal of the physiological effects observed upon removal of the hormone from the medium.

A number of problems and pitfalls can be encountered in the kinds of studies described above, and considerable caution must be exercised in interpretation of data (reviewed in ref. 1 and 2). Because physiological membrane receptors are present in extremely small quantity in membranes, the hormone must be labelled to very high specific activity (e.g., with $^{125}I$ or $^{131}I$ at 1 to 2 Ci/µmole) without destroying the biological activity of the hormone. Since virtually all chemical compounds used as binding ligands (hormones) exhibit some nonspecific adsorptive or binding properties to a variety of inert as well as nonreceptor biological materials, since such "binding" may be of extremely high affinity, and since the number ("infinite", by definition) of such nonspecific binding sites greatly exceeds that of specific receptors, great difficulties may be encountered in detecting such specific receptors. These problems may be compounded by the fact that numerous kinds of heterogenous nonspecific binding sites (differing in affinity) may coexist, the sum of which can for example result in Scatchard plots which at high concentrations of bound ligand give the appearance of a "second" class of receptors. Furthermore, if nonspecific adsorptive materials are present in very small quantity, binding to these materials can exhibit saturability and even stereospecificity; for examples, such saturable binding has been observed with peptide hormones to certain kinds of filters and to glass materials, and the differential binding of D and L steroisomers of tryptophan to albumin is well known. In addition, cells and biological membranes may contain "specific" functional structures (e.

g., degradative or metabolizing enzymes) which are strictly "nonspecific" with respect to receptor interactions but which may masquerade as receptors because they satisfy many of the binding criteria described above.

The binding data (specificity, affinity, number of sites, reversibility) must therefore be scrupulously evaluated by careful and detailed comparisons with the biological activity of the hormone. It is desirable that the initial preparation used for binding studies be a simple, intact system (e.g., isolated, homogenous and viable cells) so that binding and biological responses can be measured in the same system, before disruptive procedures are performed on the cells.

Examples of Hormone-Receptor Interactions

*Insulin and fat cells.* In some systems, such as with insulin, glucagon and some other peptide hormones, it is reasonably certain that specific receptor interactions can be detected with $^{125}I$-labelled hormones.

The binding of insulin to a variety of cells and membranes has been studied in detail in various laboratories (reviewed in refs 1,3,4). The binding observed is highly specific for this hormone and chemical derivatives compete for binding in accord with their relative effectiveness as agonists. A saturable binding site of high affinity (about $10^{-10}$ M) can be discerned in fat cells which correlates well with the known biological effects of the hormone in the same cells. Fat cells have very few receptors for insulin, about 10,000 per cell or about 10 per $\mu m^2$ of surface area. The binding is spontaneously reversible, and the rates of dissociation and association can be measured independently. The insulin molecules which dissociate are presumably chemically intact since their biological potency is unaltered. Thus, interaction of this hormone with receptors does not involve stable covalent bonds and it is not accompanied by degradation or inactivation of the hormone.

This data suggests that receptor occupancy and activation by insulin may be simply dictated by the concentration of the hormone in the extracellular spaces, equilibration occurring at rates dependent primarily on the rate of dissociation of the hormone-receptor complex. It is not known, however, whether special mechanisms exist in peripheral tissues for specifically internalizing or otherwise degrading insulin.

The insulin binding sites of fat cells are probably localized exclusively on the external surface of these cells since virtually no binding is detected in the total particulate fraction obtained from cells previously digested with trypsin-agarose (reviewed in ref 1,4). Localization to the external aspect of the cytoplasmic membrane has been established by using inside-out membrane vesicles. Trypsin digestion of these vesicles does not destroy the binding which is detected upon subsequent rupture of the vesicles, and if $^{125}$I-insulin is bound to the intact cell before endovesiculation and preparation of the inside-out vesicles, the hormone cannot dissociate unless the vesicles are disrupted physically or by detergent or phospholipase treatment. These studies also indicate that very little or no "flip-flop" of these receptors occurs between the outer and inner aspects of the membrane.

The receptor structures for insulin are probably glucoprotein in nature since sequential digestion of cells or membranes with neuraminidase followed by β-galactosidase decreases the affinity of the insulin-receptor complex and since certain plant lectins appear to bind directly to these structures. These binding sites are very susceptible to proteases; mild digestion affects primarily the affinity of binding whereas more drastic digestion appears to cause more drastic destruction. One of the most interesting but unexplained features is the unmasking of considerably quantities of "new" binding sites by digesting cells or membranes with certain phospholipases or by exposure to high concentrations (2N) of NaCl. The possible physiologic role of these cryptic sites is not understood.

The insulin binding macromolecules have been solubilized from cell membranes with nonionic detergents, and the binding properties of the solubilized material are remarkably similar to those seen in the intact membrane. The molecular parameters of the receptors indicate an asymmetric shape and a M.W. in the range of 300,000. It has not yet been possible to obtain forms of lower molecular weight (subunits) which are still capable of binding insulin. The receptors have been purified nearly to homogeneity by the combination of conventional procedures and affinity chromatography on insulin-agarose and wheat germ agglutinin-agarose derivatives. The major current problem faced in this area is the purification of these molecules in sufficient quantity to perform chemical characterization studies. The paucity of receptors in biological materials can be appreciated by the fact that purification must be approximately 500,000-fold using liver membranes. In contrast, purification of acetylcholine receptors from electric tissues of fish requires a prufication of about 100- to 500-fold.

*Insulin and lymphocytes and fibroblasts.* Human peripheral white blood cells possess specific binding sites for insulin (5). However, it has been shown (6,7) that whereas some leukocytes (e.g. macrophages) can bind considerable amounts of insulin, nylon column-purified lymphocytes free from macrophages, polymorphs and platelets have less than one binding site per cell. Virtually no binding is seen in these cells compared to in vitro transformed lymphocytes, or to permanent cell line lymphocytes (RPMI 0237), maintained in tissue cultures. The binding observed to unfractionated human lymphocytes (5) could be explained by a 0.3% contamination with cells such as macrophages or transformed lymphocytes. Recently it has been observed that during mitogenic blast transformation of human lymphocytes there is a dramatic de novo appearance of cell surface receptors for insulin (6, 7).

Although it has been known for some time that insulin can act as a serum-substitute to support the growth of cells in cultures, it is recognized that insulin does this only at concentrations far above those which occur in vivo. Compared to epidermal growth factor, for example, insulin is indeed a poor mitogen for human fibroblasts (2,8). It may be significant that the affinity of insulin (from binding studies) for receptors in fibroblasts and lectin-transformed lymphocytes is about an order of magnitude lower than the physiological concentration of the hormone and of the affinity of receptors in fat cells or liver membranes. The possibility must be entertained that these binding sites may normally be intended for another as yet unidentified serum peptide with growth-promoting and insulin-like properties, and that sufficient cross-reactivity of these "receptors" can occur with insulin for them to pose as low-affinity, specific "receptors" for this hormone (2,7). There is reason to speculate that certain insulin-like peptides in serum (e.g. somatomedin, nonsuppresible insulin-like activity, multiplication stimulating activity) may be capable of interacting with insulin receptor structures. Some precedents for overlapping receptor specificity may be found with vasopressin and oxytocin (antidiuresis and myometrial contraction) (9), human growth hormone, prolactin and human placental lactogen (10), and secretin and vasoactive intestinal polypeptide (11).

*Catecholamines and membrane structures disguising as receptors.* The potential perils of using binding studies to detect biological receptors can be illustrated with the case of the catecholamine hormones. The binding of $^3$H-labelled (5 to 10 Ci/mmole) catecholamines (norepinephrine, isoproterenol) to various mammalian cells and membrane preparations shows saturability, an apparent affinity not too dissimilar from that which is presumed to operate for the intact receptor system, and specificity for the catechol moiety of the ligand (12, 13). However, close scrutiny demonstrates that the binding is totally lacking in stereospecificity, that the ethanolamine portion of the molecule is not required for binding (e.g., pyrocatechol is as good as (-)-norepinephrine as a competing ligand), that the binding is essentially irreversible, and that the number of binding sites, compared to the corresponding number for peptide hormones, is enormously high (13). In addition, it has been established that catechol substances (e.g.,pyrocatechol, 3,4-dihydroxymandelic acid, the (+)-isomer of norepinephrine) which compete for binding indistinguish ably from the labelled hormone are themselves biologically inert and they do not inhibit the biological effects of the active hormone in intact cells or the stimulation of adenylate cyclase activity in membrane preparations (13). Furthermore, the non-catechol, m-methanesulfonamide derivative of (-)-isoproterenol, which is a potent β-adrenergic agonist, does not compete for binding. It is evident that the measured membrane binding of $^{3}H$-norepinephrine cannot represent direct interactions with β-adrenergic receptors. Some evidence has been presented (13) which suggests that the binding measured with $^{3}H$-norepinephrine represents binding to an altered form of a membrane-bound enzyme, catechol-O-methyl transferase. The number of true adrenergic receptors in these tissues must be very small since in the presence of a large excess of pyrocatechol (or other catechols) the residual binding of $^{3}H$-(-)-norepinephrine is extremely small and it is still not stereospecific. Thus, this is an example of a "specific" but non-receptor membrane-localized component which is present in very large excess compared to the true receptor components, and which because of some cross-specificity with the radioligand used, feigns initially as a receptor structure. Detection of specific β-adrenergic receptors will probably require the use of binding ligands of much higher specific activity which either lack a 3,4-dihydroxy phenolic group or which must be used together with substances which suppress the non-receptor catechol-binding components.

## How Do Hormone-Receptor Complexes Modify Membrane Functions?

One of the most exciting and important areas of future research in the field of membrane receptors for hormones is the elucidation of the precise mechanisms by which hormone-receptor complexes, once formed, modify the activity of specific membrane-localized enzymes or transport structures. In the past it has been assumed that the receptors themselves possess specific and separate functions (e.g., ionophores for acetylcholine receptors), or that the receptors are structurally contiguous with other molecules endowed with specific functions (e.g., adenylate cyclase for peptide hormones). It is in fact not necessary to make such assumptions, especially since there may be special advantages to having receptors which in their uncomplexed form are totally separate from other membrane macromolecules (1).

In this theory (1) the receptor assumes new properties upon binding of the specific ligand; one of these new properties is a special affinity for binding to and thus perturbing other membrane structures such as adenylate cyclase. This is, then, essentially a two-step mechanism (Fig. 1) which basically reflects the currently developing view that biological membranes are essentially fluid structures which permit relatively free diffusion of molecules along the plane of the membrane (Fig.2). Sequential specific interactions could therefore occur within the membrane in a fashion analogous to the well known behavior of molecules in aqueous solutions, except that different diffusion properties and special constraints would exist in the former. In the case of adenylate cyclase, hormone-receptor complexes of a stimulatory or inhibitory nature could complex separately or competitively with the enzyme (Fig. 3).

This two-step membrane hypothesis would help to explain many apparent anomalies in the action of hormones, including how insulin may modify separate and perhaps independent membrane processes by acting on a single receptor, and how several hormones acting through independent receptors may modify the same adenylate cyclase (Fig. 4) in a given cell (1). In addition, such considerations have important consequences on understanding and predicting the properties and kinetics of the interaction of hormones with membrane receptors.

In the case of insulin, recent studies have shown that this hormone, in addition to lowering intracellular levels of cyclic AMP, can cause very prompt and marked elevations of cyclic GMP in fat cells, liver slices and lymphocytes (14). Since insulin can directly inhibit in isolated membrane preparations the activity of adenylate cyclase (15) as well as guanylate cyclase (16), it has been postulated (14) that a unique insulin-receptor (or cholinergic homone-receptor) interaction at the cell membrane may act by transforming the substrate specificity of the membrane-localized adenylate cyclase into a form which now expresses preference for GTP rather than ATP. The same membrane-localized enzyme system would be responsible for the synthesis of both nucleotides, and the "balance" of ATP- or GTP-utilizing forms of the enzyme would depend on the relative occupancy of receptors which favor the ATP form (adrenergic hormones, glucagon, etc.) compared to the occupancy of those receptors that favor the GTP form (insulin, cholinergic hormones). This could explain the simultaneous "inhibition" of adenylate cyclase activity and elevation of cyclic GMP levels (by enhanced synthesis) through the action of a single, unique hormone-receptor interaction. This hypothesis, which can be envisioned to fit in the broader two-step mechanism described above, is attractive because it could explain the effects of insulin (and cholinergic hormones) on a "unitary" basis and it would place both cyclic nucleotides as "initial" chemical mediators.

## Cholera Toxin as a Model System For Hormone Receptors in Membranes

Cholera toxin, a protein which is extremely potent in stimulating adenylate cyclase activity in nearly every tissue which has been examined, binds specifically to certain glycolipids ($GM_1$ gangliosides) in membranes (17-21). There is good evidence that these are the true chemical receptors, and that the interaction of the toxin with these leads to the stimulation of the enzyme (18). Exogenous gangliosides can be incorporated spontaneously into cell membranes, and these can bind the toxin in an active form.

One of the most unique features of the action of the toxin is the characteristic lag phase (about 60 minutes) which transpires between binding of the toxin and activation of the enzyme. This lag is highly temperature dependent. Various studies on the nature of this latency have led to the suggestion that the initial toxin-ganglioside (or receptor) complex is inactive, and that after binding the complex must undergo a special temperature and time dependent transition which involves a spontaneous relocation of the complex within the two dimensional structure of the membrane (22) such that complexation with and modification of adenylate cyclase occurs (Fig. 5).

This transition could involve a translocation of part of the toxin molecule to the inner portion of the membrane, where adenylate cyclase is presumably localized. Choleragenoid, a structural analog of cholera toxin which binds to cells indistinguishably from the toxin but is biologically inactive, and is thus a competitive antagonist (20) of toxin action, would form the initial receptor complex but would be unable to undergo the required subsequent transition (19). In this scheme the toxin is binding stoichiometrically with adenylate cyclase, and it is possible from binding studies to estimate the number of these enzyme molecules in cells. Studies have shown that the nature and extent of dissociation of the toxin-receptor complex changes with time and temperature of incubation (19), and that the temperature transition of enzyme activation differs between erythrocytes of different species (23).

These observations are very consistent with the two-step mechanism for hormone receptor action described above, and it may serve as an excellent model system with which to study and test these hypotheses. For example, recent studies using fluorescein-labelled toxin indicate that there is indeed a redistribution and migration of the toxin on the lymphocyte surface after initial binding (24).

## References

1. Cuatrecasas, P. Ann. Rev. Biochem. 43. In press
2. Hollenberg, M.D. and Cuatrecasas, P. Biochem. Action of Hormones. In press.
3. Roth, J. Metabolism 22 (1973) 1059.
4. Cuatrecasas, P. Fed. Proc. 32 (1973) 1838.
5. Gavin J.R. III, Gorden, P., Roth, J. Archer, J.A. and Buell, D.N. J.Biol.Chem. 248(1973)2202.
6. Krug, U., Krug,F. and Cuatrecasas, P. Proc. Nat.Acad.Sci. USA 69 (1972) 2604.
7. Hollenberg, M.D. and Cuatrecasas, P. 1974. In: Control of Proliferation in Animal Cells. pp. 214, ed. Clarkson, B., Baserga, R. Cold Spring Harbor: New York.
8. Hollenberg, M.D. and Cuatrecasas, P. Proc.Nat. Acad. Sci. USA 70 (1973) 2964.
9. Soloff, M., Swartz, T., Morrison, M., and Saffran, M. Endocrinology 92(1973)104; Bockaert, J., Imbert, M., Jard, S. and Morel, F. Molec. Pharmac. 8 (1972) 230.
10. Shiu, R.P.C., Kelly, P.A., and Friesen, H.G. Science 180 (1973) 968.
11. Desbuquois, B. Personal communication.
12. Lefkowitz, R.J., Sharp, G., and Haber, E. J. Biol. Chem. 248 (1973) 342.
13. Cuatrecasas, P., Tell, G.P.E., Sica, V. Parikh,

I., and Chang, K-J., Nature. In press.

14. Illiano, G., Tell, G.P.E., Siegel, M.I. and Cuatrecasas, P. Proc. Nat. Acad. Sci. USA 70 (1973) 2443.
15. Hepp, K.D. FEBS Letters 20 (1972) 191; Illiano, G. and Cuatrecasas, P. Science 72 (1972) 906; DeAsua, L.J., Surian, E.S., Flawia, M.M. and Torres, H.M., Proc. Nat. Acad. Sci. USA 70 (1973) 1388; Flawia, M.M. and Torres, H.N. J. Biol. Chem. 248 (1973) 4517.
16. Siegel, M.I. and Cuatrecasas, P. Manuscript in preparation.
17. Cuatrecasas, P. Biochemistry 12 (1973) 3547.
18. Cuatrecasas, P. Biochemistry 12 (1973) 3558.
19. Cuatrecasas, P. Biochemistry 12 (1973) 3567.
20. Cuatrecasas, P. Biochemistry 12 (1973) 3577.
21. Cuatrecasas, P., Parikh, I. and Hollenberg,M. D. Biochemistry 12 (1973) 4253.
22. Singer, S.J. and Nicolson, G.L. Science 175 (1972) 720.
23. Bennet, V. and Cuatrecasas, P. Manuscript in Preparation.
24. Craig, S. and Cuatrecasas, P. Unpublished. Siegel, M. and Cuatrecasas, P. Manuscript in preparation.

a. $H + R \rightleftharpoons HR$

b. $HR + A.C. \rightleftharpoons HR - A.C.$

Figure 1. Sequential two-step hypothesis for the action of membrane hormone receptors (1) involves the discrete steps of a) binding of hormone (H) to the receptor (R) followed by b) binding of the hormone-receptor complex (H-R) to another separate membrane structure, such as adenylate cyclase.

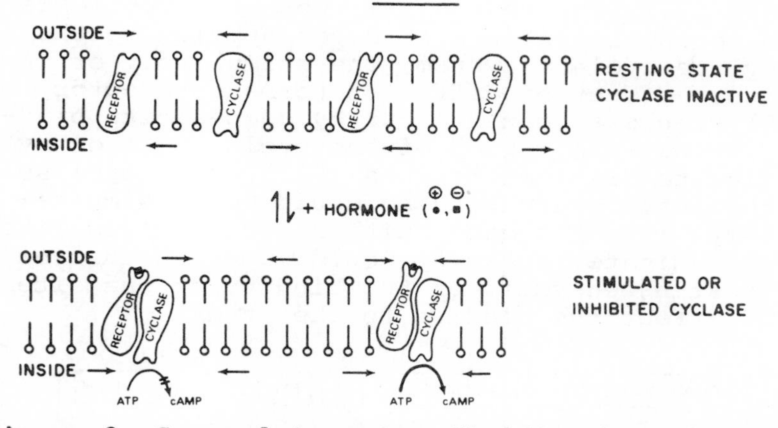


Figure 2. General two-step fluidity hypothesis for the mechanism of modulation of adenylate cyclase activity of cell membranes by hormone receptors (1). The central feature is that the receptors and the enzyme are discrete and separate structures which acquire specificity and affinity for complex formation only after the receptor has been occupied by the hormone. These structures can combine after binding of the hormone because of the fluidity of cell membranes. The hormone binding sites of the receptor are on the external face, exposed to the acqueous medium, and the catalytic site of the enzyme is facing inwardly toward the cytoplasm of the cell.

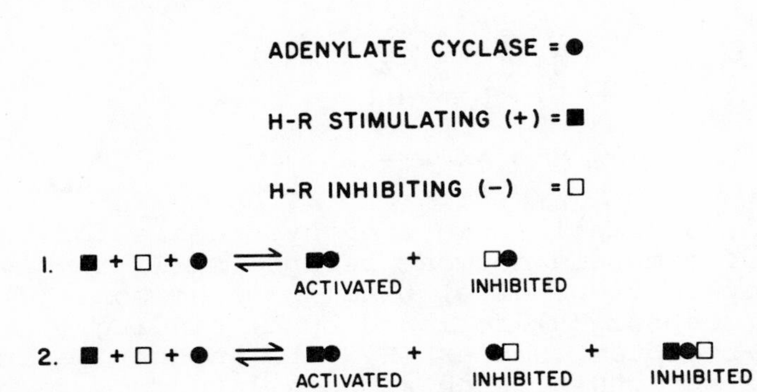


Figure 3. In the sequential two-step hypothesis of membrane receptor action, as applied to the adenylate cyclase system, the stimulatory or inhibitory effect of different hormone-receptor (H-R) complexes could occur by competitive binding to the same region of the cyclase (1) or by binding to distinct regions (2) of this cyclase. In the latter case binding could occur simultaneously, in which case neither effect would likely predominate (activity would approach that of the resting enzyme), or the binding at one site would affect the binding to the other.

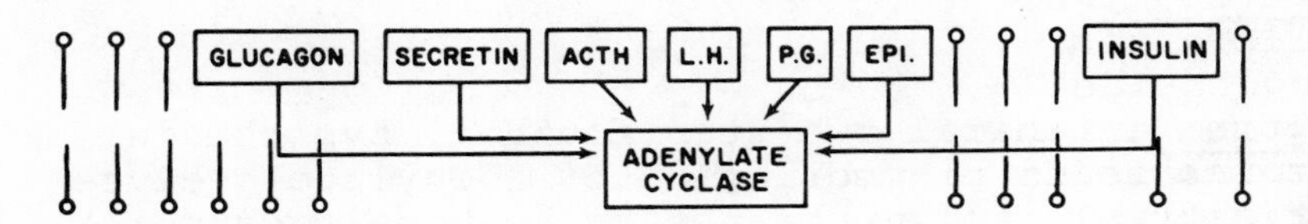


Figure 4. In certain cells a variety of hormones are believed to affect the same adenylate cyclase molecules through the action of separate receptors for each hormone. In the conventional theories structural contiguity is envisioned, which poses considerable geometric and spatial problems and requires structures (or complexes) of great size. In the two-step hypothesis (Fig. 1-3) no necessary connection is assumed to exist between the enzyme and the unoccupied receptor.

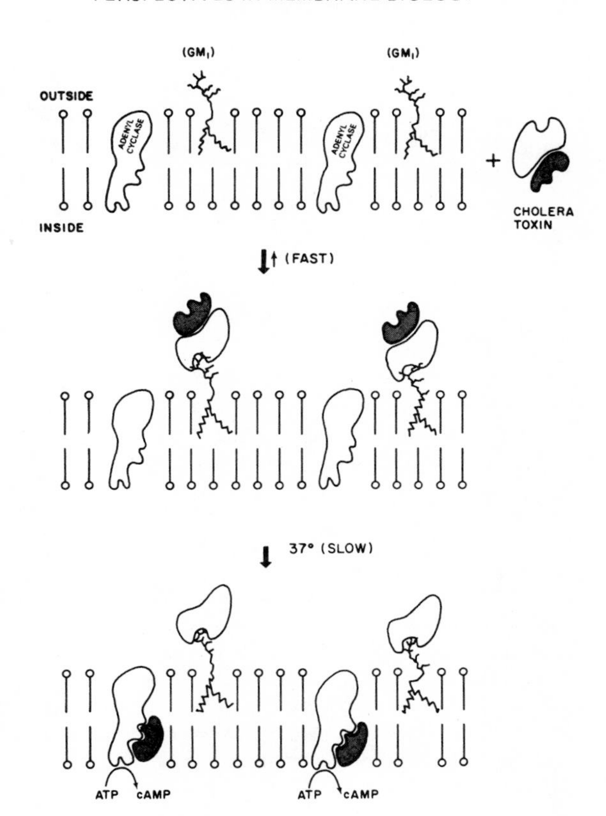


Figure 5. Postulated mechanism of action of cholera toxin. The toxin binds initially to ganglioside $GM_1$ receptors in the cell membrane to form an inactive toxin-receptor complex. This complex is not associated initially with adenylate cyclase, but after a special time and temperature transition which involves movement within the structure of the membrane, the complex associates with and perturbs the enzyme. One possible mechanism for this transition, suggested in this figure, involves a dissociation of the toxin subunits with translocation of one of these (which alone is inactive, and which is highly water insoluble) to the inner region of the membrane where the catalytic unit of the cyclase is localized.

# INITIAL REACTIONS OCCURRING IN THE CELL PLASMA MEMBRANE FOLLOWING HORMONE BINDING

G.V. Marinetti, L. Lesko and S. Koretz

Department of Biochemistry
University of Rochester
School of Medicine and Dentistry,
Rochester, New York

The binding of polypeptide hormones such as insulin, glucagon, ACTH, and growth hormone has been shown to occur with receptors localized in the cell plasma membrane (1-27). The binding of the catecholamine hormones epinephrine, norepinephrine and isoproterenol has also been demonstrated to occur on membrane-bound receptors (28-33). Hormone binding is a necessary but not a sufficient event for the ultimate physiological action of the hormone. There can occur coupled or uncoupled binding, thus the binding may be productive or non-productive. This is analogous to substrate binding to enzymes.

Some progress has been made on the isolation and purification of the receptor for insulin (13-15), glucagon (34) and epinephrine (28,30). To date no membrane receptor for any hormone has been isolated in sufficient quantity for detailed biochemical analysis.

The molecular mechanism of action of hormones on cell membranes represents a challenging problem in biology. This chapter will cover the initial events which occur on binding of the hormone to the membrane. Space limitations do not allow a comprehensive coverage and discussion of this important biological process.

## Methods for Measuring Hormone Binding

Various laboratories use different methods for measuring hormone binding to isolated membranes (1,6,7,8,9b,11,13,19,20,28,29,30,31,34,35). The advantages, disadvantages and limitations of these methods will not be discussed in this paper. The conditions of binding must be specified. The labeled hormone must be biological active. Hormone uptake into membrane vesicles can offer difficulties and be confused with true binding (specific binding). The question of specific versus non-specific binding has not been adequately resolved.

## Hormone Concentration

One chooses to use a hormone concentration which is "physiological". This is not easy to ascertain since some hormones are produced at nerve endings and act directly on target cells without necessarily entering the vascular circulation. For those hormones entering the blood one can consider serum levels of the hormone as being "physiological". The levels can vary with diet and emotional stress and also will vary depending on whether portal blood levels or peripheral blood levels are considered. We have used hormone concentrations which Exton et al (36) have shown to give half maximal response on cAMP production in perfused rat liver. For epinephrine and glucagon these concentrations are $10^{-7}$M and $10^{-9}$M respectively.

## Hormone Modulation of Adenylate Cyclase and cAMP Levels

The most frequent initial event correlated with hormone binding is the activation of membrane bound adenylate cyclase (as with catecholamines, glucagon and ACTH ) or the inhibition of adenylate cyclase (as with insulin). The time course and dose response of hormone binding to isolated plasma membranes can be correlated with the time course and dose response of adenylate cyclase activation. These studies have shown that with glucagon and epinephrine, hormone binding precedes adenylate cyclase activation. Hormone binding and adenylate cyclase activation should in theory show saturation kinetics.

The important question which still remains unresolved is how is hormone binding coupled to adenylate cyclase activation or inhibition.

With isolated whole cells the binding of hormone can be correlated with a metabolic process in the cell which the hormone modifies. Thus insulin bindins has been correlated with lipogenesis or with the oxidation of glucose to $CO_2$ in isolated fat cells (11,37). Epinephrine and norepinephrine binding can be correlated with lipolysis in isolated fat cells. With isolated cells it is also possible to correlate hormone binding with adenylate cyclase activity by measuring cyclic AMP levels.

## Catecholamines. Correlation of Binding with Adenylate Cyclase Activation

The selective binding of epinephrine to isolated rat liver plasma membranes with the concomitant stimulation of membrane bound adenylate cyclase was demonstrated by Marinetti et al (1,19). The time course studies showed that epinephrine binding preceded adenylate cyclase activation. The epinephrine stimulation of adenylate cyclase was small relative to the stimulation produced by glucagon, a finding confirmed in isolated perfused liver by Exton et al (36)* and in liver cell membrane by Rodbell et al (9).

Schramm et al (29) have studied catecholamine binding to turkey erythrocytes and its activation of adenylate cyclase. The concentration of epi-

---

* It is of interest that although Exton et al. found a 60 fold maximum stimulation of cAMP by glucagon as compared to a four fold stimulation by epinephrine, both hormones gave nearly identical stimulation of glucose production. Thus it appears that cAMP is made in large excess over that required for glycogenolysis.

nephrine for half maximal binding was 30 μM which was the same as the epinephrine concentration required for half maximal activation of adenylate cyclase. Epinephrine binding was nearly complete within 1 minute at 37°. The beta adrenergic agent propranolol inhibited both epinephrine binding and epinephrine stimulation of adenylate cyclase. These workers reported that some analogs of epinephrine which were effective in displacing epinephrine binding caused only a small activation of adenylate cyclase but were able to inhibit the epinephrine stimulation of adenylate cyclase. The authors concluded that hormone binding was an essential but not sufficient condition to elicit a hormone response.

The correlation of catecholamine binding to its activation of adenylate cyclase in turkey erythrocytes was examined by Bilezikian and Aurbach (33). They found that the affinity of the hormone receptor was identical with the Km for adenylate cyclase activation by isoproterenol. Propranolol was shown to be a potent inhibitor of the activation of adenylate cyclase by isoproterenol but was a weak inhibitor of hormone binding. These workers concluded that the catechol ring of the catecholamine was required for binding and that the side chain bearing the secondary hydroxyl group and an amine or substituted amine was required for adenylate cyclase activation. The hormone concentration for half maximal activation of adenylate cyclase was in close agreement with the hormone concentration giving half maximal stimulation of adenylate cyclase. This concentration was approximately 4-6 μM which was appreciably lower than the value of 30 μM found by Schramm et al (29). Bilezikian and Aurbach found that hormone binding reached a plateau between 5-10 minutes in contrast to the 1 minute reported by Schramm et al (29).

The action of catecholamines on fat cells has been investigated by several laboratories (9,38, 39,40). The general finding is that isoproterenol is a more potent stimulator of adipose tissue lipase than epinephrine. AH-3365, an analog of epinephrine in which the meta hydroxyl group is hydroxymethyl and in which the $CH_3$-N is replaced by t-butyl-N was shown to be less potent than epinephrine and isoproterenol in stimulating fat cell lipolysis (39). The catecholamine receptor in the fat cell has the properties of a β-receptor.

Gorman et al (41) found that with fat cell ghosts norepinephrine and isoproterenol were as effective as epinephrine in stimulating adenylate cyclase and gave nearly complete inhibition of epinephrine binding to fat cell ghosts. On the other hand DOPA gave no stimulation of adenylate cyclase but inhibited epinephrine binding by 85%.

Tomasi et al (42) studied the binding of epinephrine and glucagon to isolated liver plasma membranes. They found a relatively high localization of binding to the plasma membrane as compared to other cell particles. Sulfhydryl agents PCMB and DTNB at $10^{-4}$M gave a marked inhibition of both epinephrine binding (by 90-98%) and glucagon binding (by 40-49%). A study of epinephrine analogs showed that analogs which possessed the catechol ring were the most potent inhibitors of binding and consequently this part of the epinephrine molecule appears to be essential for binding. A similar study of epinephrine binding to fat cells (Table I) and isolated fat cell plasma membranes (Figure 1) confirmed the findings on rat liver membranes. The most potent inhibitors of epinephrine binding were those containing a catechol ring. The side chain does not appear to play an important role in binding but is believed to play a role in coupling the binding to the hormone response.

To answer the question whether hormone binding is coupled or uncoupled to its physiological action, epinephrine binding to isolated fat cells was examined in the presence and absence of α-adrenergic and β-adrenergic blocking agents. The data in Table II show that α-adrenergic agents do not influence epinephrine binding nor do they influence significantly the epinephrine stimulation of lipolysis. However the β-blocking agent propranolol had no influence on epinephrine binding but did inhibit the epinephrine stimulated lipolysis. These observations are consistent with an uncoupling effect of propranolol. On the other hand, the β-blocker dichloroisoproterenol which has a structural resemblance to epinephrine, inhibited both the binding of epinephrine and the epinephrine stimulation of lipolysis to the same extent.

A study of the binding of the specific stereoisomers of norepinephrine is shown in Table III. The l-isomer was bound 5 fold greater than the d-isomer in isolated fat cells. With isolated liver plasma membranes the l-isomer was bound about 6 fold greater than the d-isomer. It was previously shown (1,28) that the 3-methoxy-4-hydroxyphenylethylamine was bound far less than the 3,4-dihydroxyphenylethylamine, stressing the importance of the catechol ring for binding. Since the d-isomer of norepinephrine is bound to the fat cell and to the rat liver plasma membrane to an appreciable extent, although less than the l-isomer, it seemed that this binding was in excess of that predicted from its weak physiological response. Hence we assume that the d-isomer has uncoupled or unproductive binding to a large extent.

To examine the influence of the stereoconfiguration of the side chain and whether the d- and l-isomers of norepinephrine were bound to the identical receptor, we studied the effect of unlabeled l-norepinephrine on the binding of radioactive l- and d-norepinephrine in isolated fat cells and liver plasma membranes. The data in Table IV shows that at a molar ratio of 10/1 of unlabeled isomer to labeled isomer the unlabeled l-isomer gave a 24-26% inhibition of binding of the labeled d-isomer in liver plasma membranes and fat cells. In the same system the unlabeled l-norepinephrine gave an 87% inhibition of binding of the labeled-l-norepinephrine in liver and a 42% inhibition in fat cells. Hence the unlabeled l-isomer is more effective in inhibiting the binding of labeled l-norepinephrine than in inhibiting the binding of labeled d-norepinephrine. This difference is believed due to the specific stereoconformation of the side chain. One cannot rule out the possibility that these isomers may bind to some extent to different receptors.

Our studies with catecholamine analogs are in agreement with similar studies on turkey erythrocytes (29,33) and cardiac muscle (30,43).

The general conclusion to be derived from these studies is that catecholamine binding to the membrane receptor requires the catechol ring but that the side chain binding is responsible for coupling directly or indirectly to adenylate cyclase, in order for the hormone to exert its physiological action. Hormone binding is thus a necessary but not sufficient event for the hormone action. In the fat cell, β-blocking agents are believed to inhibit catecholamine binding to the receptor by competing at the catechol binding site or they also may act to uncouple the binding by interfering with the catecholamine side chain interaction. Marley (44) earlier had suggested that both the side chain and the catechol ring were essential for the activity of catecholamines.

## Insulin Binding. Correlation with Its Physiological Actions

A large number of studies has shown that insulin binds to "specific" receptors on the plasma membrane of liver, fat cells, thymocytes, and lymphocytes (7,8,11,13,14,15,16,17,18,19,21,23, 25,26,37). This section will not attempt to enumerate the many studies on insulin binding. Primary emphasis will be concerned with those studies which correlate binding with the action of insulin.

Freychet et al (11) found that eight insulin derivatives with biological potencies varying over a 100 fold range gave inhibition of binding of radioinsulin to liver plasma membranes in direct proportion to their ability to stimulate glucose oxidation in isolated fat cells. Inactive insulin chains, glucagon, ACTH and GH had no effect on insulin binding.

Goldfine et al (45) studied insulin binding to isolated rat thymocytes and correlated the binding with α-aminobutyric acid (AIB) transport. They found a half maximum response on AIB transport at an insulin concentration of 40 nM and a maximum response at 1 μM. The concentration of cold insulin required for half maximal displacement of labeled insulin was 20 nM with complete displacement occurring at 1 μM. Insulin was found to increase AIB influx by raising the maximum influx capacity ($J_{max}$) and lowering the concentration of AIB at which influx was half maximal (Km). Cycloheximide inhibited AIB basal influx and reduced the response to insulin. The authors suggest that (a) insulin stimulation of AIB influx correlates in sensitivity and specificity with binding of $I^{125}$-insulin to the cell and (b) insulin stimulation of AIB influx may be mediated by a change in the synthesis and degradation of the AIB transport system.

Goldfine and Sherline (46) also studied the relationship of insulin action on rat thymocytes to cAMP. It was observed that dibutyryl cAMP, PGE, and cyclic AMP stimulated AIB influx. The maximum effect of dibutyryl cyclic AMP was seen at 1 mM and cyclic AMP had only 1% the activity of the dibutyryl analog. Whereas insulin increased the J max and lowered the Km for AIB influx, dibutyryl cAMP increased the J max but had no effect on the Km. The addition of insulin and dibutyryl cAMP gave additive effects. The authors concluded that insulin increases AIB influx by a mechanism not involving changes in cyclic AMP levels in the cell.

In adipose tissue insulin stimulates glucose oxidation to half its maximal value at 200 pM (37). Attempts to correlate this with insulin binding have been complicated since the binding studies on fat cells have varied. Dissociation constants for insulin binding have ranged from 0.1 pM (8) to 8 nM (37). Kono and Barham (37) have estimated that only 2.4% of the total insulin receptors are occupied at the time insulin exerts its maximal effect on stimulation of glucose oxidation in isolated fat cells.

## Glucagon-Binding and Adenylate Cyclase Activation

Birnbaumer and Pohl (20) have found that des-histidine glucagon added 5 minutes after incubation of glucagon with liver plasma membranes, inhibits completely the glucagon stimulation of adenylate cyclase but only displaces 10-20% of the bound glucagon. These authors conclude that only 10-20% of the glucagon binding sites in liver plasma membranes participate in the activation of adenylate cyclase. The function of the other binding sites was not specified; therefore it is not safe to assume they necessarily represent excess receptors associated with adenylate cyclase.

The evidence cited above indicates that a relatively small number of the total glucagon binding sites in the liver plasma membrane are associated with adenylate cyclase activation. The question of how glucagon binding to these sites correlates with the hormone response is important. Moreover, are the other binding sites involved in other actions of glucagon or do they represent uncoupled receptors or spare receptors?

Dunnick and Marinetti (28) reported a 50% inhibition of glucagon binding to liver plasma membranes by SH agents such as DTNB or pCMB. Storm and Dolginow (47) recently showed that basal activity, glucagon stimulated activity, and fluoride stimulated activity of adenylate cyclase in rat liver plasma membranes was inhibited by the SH agent indoacetamide. Glucagon (1-2 μM) stimulated the SH reactivity of the membrane. Since other workers have demonstrated the inhibition of basal and glucagon stimulated adenylate cyclase activity by SH agents the authors conclude that glucagon alters the conformation of the adenylate cyclase complex to make available more SH groups which are crucial for adenylate cyclase activity.

Rodbell et al (10) studied a variety of glucagon analogs with respect to their ability to inhibit glucagon binding to liver plasma membranes and their ability to stimulate basal adenylate cyclase or the glucagon stimulated adenylate cyclase. They found that des-histidine glucagon did not activate adenylate cyclase but did inhibit glucagon binding. However glucagon 1-21, glucagon 1-23, glucagon 20-29 and glucagon 22-29 failed to activate adenylate cyclase or to inhibit glucagon binding. Since des-histidine glucagon has no biological activity it appears clear that this N-terminal amino acid is essential for biological activity and also is necessary for the glucagon stimulation of adenylate cyclase. It therefore appears to be required for the coupling of glucagon binding to adenylate cyclase activation.

## Role of Phospholipids in the Hormone Stimulation of Adenylate Cyclase

Inasmuch as the adenylate cyclase of mammalian cells is localized in the plasma membrane and the plasma membrane contains a high content of lipid, in particular phospholipid, it is not surprising that phospholipids play a role in the action of this enzyme. Whether phospholipids play a specific role or whether they play a non-specific structural role in helping to maintain a particular conformation of the enzyme complex in the membrane is not easy to unravel. Birnbaumer (21) in a recent review article takes the view that phospholipids may play a specific role in coupling the hormone-receptor interaction to adenylate cyclase activation. Evidence cited to support this view was based on the finding from different laboratories that treatment of fat cell membranes or liver plasma membranes with high concentrations of phospholipase A gives a total loss of adenylate cyclase activation but that lower concentrations of phospholipase A produce a selective loss of glucagon stimulation but not fluoride stimulation of the enzyme. Furthermore, with digitonin-treated membranes of rat liver, the loss of glucagon stimulation of adenylate cyclase was partially restored by the addition of membrane phospholipids with PS being more effective than PC or PE.

Birnbaumer states that evidence for a more specific role of phospholipids on the hormone-receptor interaction was provided by the work of Levey (48,49) and Lefkowitz and Levey (43). Treatment of cat heart membranes with Lubrol-PX gave a loss in the response of adenylate cyclase to norepinephrine and glucagon. Addition of phosphatidyl inositol (PI) selectively restored the response to norepinephrine whereas the addition of phosphatidylserine (PS) restored the response to glucagon. This response of PI has been confirmed by Rethy et al (50). These workers found that PI was more effective than PS in partially restoring the epinephrine stimulation of adenylate cyclase from delipidized liver plasma membranes. Rethy et al (51) later showed that PS added to delipidized liver plasma membranes gave near complete restoration of the epinephrine stimulation of adenylate cyclase and only partially restored the glucagon response. However, they found that PI, phosphatidyl choline (PC) and phosphatidyl ethanolamine (PE) had no effect on the system. This lack of effect of PI with liver differs from the finding of Levey (48, 49) with cat heart.

Dunnick and Marinetti (28) showed that liver plasma membranes extracted in the cold with organic solvents bound epinephrine to the same extent as control membranes. However phospholipases A,C, and D were found to inhibit the binding of norepinephrine to rat liver plasma membranes (Table V). Pronase and trypsin also inhibited hormone binding. Thus it can be inferred from the above work that phospholipids are not essential for binding of this catecholamine to the membrane receptor but rather phospholipids may play a role in either maintaining the membrane conformation to stabilize adenylate cyclase or may be required for coupling the hormone binding to activation of adenylate cyclase.

Working with turkey erythrocytes, Bilezikian and Aurbach (52) found that phospholipase, Triton X-100, SDS and Lubrol PX destroyed the catecholamine activation of adenylate cyclase but did not effect the hormone binding. Lubrol was shown to liberate the receptor from the membrane. Hence in this case the uncoupling by detergent may be due solely to a physical separation of the receptor from the adenylate cyclase complex in the membrane.

At the present time one cannot define the precise role of phospholipids in the membrane-bound adenylate cyclase system. Phospholipids have long been known as essential structural components of cell membranes. Recent work has shown that phospholipids are asymmetrically localized in the erythrocyte membrane (53-55) with PS and PE being located predominantly on the inner membrane surface and PC and sphyngomyelin (SPH) being localized predominantly on the outer membrane surface. Since the hormone receptors are localized on the outer membrane surface they may interact primarily with PC and SPH. However since adenylate cyclase (at least the catalytic subunit) is located on the inner membrane surface it may interact primarily with PS and PE.

The localization of PI in the membrane remains to be determined. The asymmetry of the phospholipids is in all probability a secondary consequence of protein asymmetry since phospholipid flip-flop, although slow (56), is not so slow as to prevent complete randomization of the membrane phospholipids unless they were prevented from doing so by interacting with proteins which could not undergo flip-flop. Marinetti et al (57) have shown that up to 20% of PE plus PS becomes cross linked to membrane proteins by difluorodinitrobenzene (57). These phospholipids are thus closely associated with membrane proteins and are the lipids most likely to influence the structure and function of these proteins.

## Modulation of the Hormone Sensitive Adenylate Cyclase by Nucleotides

A role for nucleotides in the action of hormones has been shown in liver (22,58), pancreatic beta cells (59) and platelets (60). It has been proposed by Rodbell (22,58) that the glucagon stimulation of adenylate cyclase is dependent on GTP and ATP which are presumed to bind reversibly to "allosteric" sites on the enzyme complex. ATP and GTP have been found to increase both the rate of dissocation of bound glucagon and the rate at which glucagon binding reaches equilibrium. It is suggested that only a fraction of the regulatory sites are occupied by glucagon and these are sufficient to explain the activation of adenylate cyclase. A model is proposed in which the regulatory component of the cyclase exists in two states in equilibrium with each other and which bind glucagon differently . Only one state is competent for activating adenylate cyclase. Nucleotides are believed to act at topographically distinct sites from glucagon and shift the equilibrium in favor of the competent state.

Harwood et al (61) found that GTP markedly inhibits basal adenylate cyclase activity in fat cells ghosts or fat cell plasma membranes, giving a maximal effect at $10^{-6}$M. The fluoride activation of adenylate cyclase was less inhibited by GTP. The terminal phosphate group of GTP was postulated to be required for the effect. The authors suggest that the terminal phosphate is transferred to the catalytic component of adenylate cyclase resulting in an inhibited enzyme in the basal state. They found that GTP converts the enzyme to a state which is thermally stable at 45° and which shows marked increased sensitivity to hormones. Indeed GDP enhances 2 fold the activation of adenylate cyclase by ACTH. The authors conclude that the inhibitory effect of GTP on basal adenylate cyclase activity combined with the stimulation by GDP of hormone activation of the enzyme play a regulatory role in the production of cAMP. If GTP leads to a phosphorylation of the enzyme it no longer can be considered an allosteric modifier and one must consider another enzyme (a GTP dependent kinase) in the regulation of the enzyme and a phosphatase to restore the enzyme to its original state.

Leray et al (62) found that purified liver

plasma membranes from adrenalectomized rats have a high sensitivity to glucagon, epinephrine and fluoride. GTP appears to be essential for both hormones since it was found to enhance the glucagon and epinephrine stimulation of adenylate cyclase, and had no effect on the fluoride activation of the enzyme. GTP increased the basal activity of adenylate cyclase in contrast to the findings of Harwood et al (61) using fat cell membranes.

Birnbaumer (21) in a review article on hormone sensitive adenylate cyclase points out the complex effects of GTP and ATP on this system. This author postulates that with renal medullary adenylate cyclase, ATP may interact at 3 different sites on this enzyme complex.

The effects of nucleotides on the membrane bound adenylate cyclase system are poorly understood to date. Since the catalytic subunit of adenylate cyclase is localized on the inner membrane surface, this unit is the most likely site of interaction with GTP and ATP which are produced inside the cell. A direct action of these nucleotides on the regulatory subunit of adenylate cyclase seems unlikely, since ATP and GTP are localized inside the cell and are quite impermeable to the plasma membrane and hence are not available to the receptor subunit which is on the exterior surface of the cell membrane.

The asymmetric arrangement of receptors on cell membranes is supported by the findings of Cuatrecasas (23) that inside-out vesicles of fat cell ghosts do not bind insulin unless they are disrupted by sonication. Recent work by Kant et al (63) has also shown that cyclic AMP binds only to the inside surface of the turkey red cell membrane.

Do Hormones Modulate Phosphodiesterase (PDE)?

This enzyme hydrolyzes cyclic AMP to 5'-AMP and is believed to be involved in regulating cyclic AMP levels in the cell. It is widely distributed throughout cells occurring both in the cytosol and bound to membranes. The methyxanthines are inhibitors of the enzyme.

Vaughan (64) found about 50% of the low Km (1-5 μM) phosphodiesterase is localized in the particulate fraction of fat cells. The high Km (30-50 μM) enzyme was about 75% localized in the soluble portion of the cell. ATP, GTP, AMP and GMP (at 5 mM) inhibited only the high Km enzyme. Insulin was found to increase primarily the activity of the low Km phosphodiesterase but only if the activity was measured in the presence of cAMP below 5 μM. These workers suggest that insulin lowers the cell concentration of cyclic AMP by stimulating membrane bound PDE . However relatively high insulin levels (100-1000 μU/ml) were required to produce an effect on PDE and the effect was not very large. Loten and Sneyd have also found an activation of fat cell PDE by insulin (64a).

Recently Manganiello and Vaughan (65) have confirmed their original finding that insulin (1 mU/ml) increases phosphodiesterase activity of fat cells. The PDE activity was increased 71% but primarily in a $P_2$ fraction which was enriched with the low Km (0.2 μM) phosphodiesterase.

Thompson et al (26) found that liver phosphodiesterases have apparent Kms of 63 and 73 μM for cAMP, show negative cooperative kinetics and appear to be membrane bound. Thirty minutes after injection of insulin (3U/100 mg) into streptozotocimized diabetic rats or injection of bovine growth hormone (1 mg/100g) into hypophysectomized rats, the separated membrane-bound high affinity PDE was activated but no effect was seen on the cytosol PDE. The PDE from hormone treated rats had the same Km but increased Vmax as compared to PDE from control animals. It is of interest that no effect of these hormones was seen in vitro systems.

Hepp (17a) found no effect of insulin on mouse liver PDE (using particulate preparations in vitro). They rather found that insulin inhibited the glucagon stimulation of adenylate cyclase. Diabetes induced by streptozotocin led to a marked increase in sensitivity of a adenylate cyclase to glucagon stimulation. Insulin treatment of diabetic mice in vivo led to a normalization of the response. Hepp suggests that in liver insulin directly influences adenylate cyclase by interfering with the transmission of the glucagon signal between receptor and the catalytic subunit.

House et al (65a) have isolated a plasma membrane fraction from rat liver. This fraction contains PDE which is rapidly stimulated by insulin with a half maximum response at 2-3 nM insulin.

## Do Hormones Inhibit Adenylate Cyclase?

Hepp (17a, 17b) found that insulin inhibited the glucagon stimulation of adenylate cyclase in mouse liver membranes. Cuatrecasas (18,23) also has reported that in both liver and fat cell membranes of the rat, insulin ($10^{-10}$ - $10^{-11}$M) markedly inhibits adenylate cyclase activity when stimulated either by ACTH, glucagon or epinephrine. The insulin effect was not seen at high insulin concentration (above 50 μU/ml or $10^{-9}$M).

Ray et al (5) found both an inhibition of basal adenylate cyclase activity and glucagon stimulated adenylate cyclase activity in isolated rat liver plasma membranes. However, since these effects were seen only at $10^{-5}$ - $10^{-6}$M insulin they may not be physiological significant.

## Are Membrane-Bound Enzymes Other Than Adenylate Cyclase and Phosphodiesterase Influenced by Hormones?

Hadden et al (66) have reported a stimulation by insulin of the membrane bound ATPase of human lymphocytes. The insulin stimulation of ATPase was small. The authors report that norepinephrine also stimulates the ATPase by an action mediated via an alpha receptor. They suggest that the stimulation of ATPase by these hormones is direct and not mediated by an action on adenylate cyclase. The relationship between ATPase activity and glucose transport was not resolved.

## Modulation of Membrane Phosphorylation by Cyclic AMP Dependent Kinases

Inasmuch as cyclic AMP activates cytosol protein kinases (67-71) and these kinases in turn use ATP to phosphorylate a variety of enzymes (70,71) it was reasonable to assume that cAMP might also influence the phosphorylation of the plasma membrane. The data in Table VI show that protein kinase isolated from rat liver and added to a rat liver plasma membrane system does phos phorylate the plasma membrane via ATP and that $10^{-6}$M cAMP gives a two fold stimulation of the membrane phosphorylation (72). Moreover, it has been shown that the phosphorylated membrane is capable of binding more $Ca^{++}$ than the control membrane (72). Phosphorylation increased the number of sites of lower affinity for $Ca^{++}$.

## Hormone Stimulation of Membrane Phosphorylation

Since glucagon and epinephrine stimulate the membrane bound adenylate cyclase to produce cAMP, it would be expected that these hormones could replace cAMP for the protein kinase phosphorylation of the membrane (72). The results in Table VII, confirm this expectation and show that glucagon and much less so, epinephrine at $10^{-7}$M (concentrations which stimulate adenylate cyclase) were able to stimulate the protein kinase phosphorylation of the membrane. Calculations show that 1 molecule of glucagon ($10^{-7}$M) causes the uptake of 100 molecules of phosphate covalently bound to the membrane. Zahlten et al (73) have shown the in vivo stimulation of cell membranes by glucagon administration to rats.

## Hormone Modulation of $Ca^{++}$ Binding to Plasma Membranes

It was reasonable to determine if hormones could perturb the membrane structure directly and thereby modify the membrane structure and lead to altered permeability characteristic with enhanced ion transport. A relationship between hormones, cAMP and calcium ions has been discussed by Rasmussen (70,71). He proposes a model in which certain hormones enhance $Ca^{++}$ influx into cells by a cAMP dependent process. To test this model we studied $Ca^{++}$ binding to the liver plasma membrane with and without added hormones and cyclic AMP (74). cAMP at $10^{-3}$M influences the $Ca^{++}$ binding sites by making available more binding sites of lower affinity. A study of four hormones (Figure 2) showed that hydrocortisone, epinephrine and glucagon stimulated $Ca^{++}$ binding but insulin inhibited $Ca^{++}$ binding (74). Calculations showed that at $10^{-8}$M hormone concentration, one molecule of hydrocortisone effects the uptake of 3000 molecules of $Ca^{++}$. This large amplification suggests that the hormone modifies a fairly large domain of the membrane. Since these hormone effects are produced without added ATP or added protein kinase, the effects are not believed due to the hormones stimulating the membrane-bound adenylate cyclase with an increase in cAMP which then activates a membrane-bound protein kinase. These hormone effects are believed to be due to a direct perturbation of the membrane with exposure of cryptic sites for $Ca^{++}$ binding. The effect of a hormone on the membrane conformation is supported by the work of Sonnenberg (27). He showed that growth hormone at very low concentrations produced a large alteration of the red cell membrane conformation.

## Modulation of Ion Transport by Hormones

Gardner et al (75) have recently shown that catecholamines stimulate $Na^{+}$ flux by 2-4 fold in isolated turkey erythrocytes. The potency of the catecholamine analogs for stimulation of $Na^{+}$ transport correlated directly with their potency to activate adenylate cyclase. It remains to be determined whether cAMP exerts a direct effect on the system or whether cAMP stimulates membrane bound protein kinases which phosphorylate the membrane and that the phosphorylated membrane leads to the observed change in ion flux. The role of membrane bound ATPase in this process must be considered. Recent work has indicated that certain hormones can modulate membrane bound ATPase (66). Himms-Hagen (40) has proposed a model in which adenylate cyclase and ATPase are closely linked in the membrane.

Friedman and Park (76) have observed the cAMP stimulation of $Ca^{++}$ efflux in rat liver. Shlatz and Marinetti (72) found that glucagon ($10^{-8}$M) stimulated $Ca^{++}$ binding to isolated plasma membranes. Friedman et al (77) later found that glucagon (as well as cAMP, cGMP and isoproterenol) hyperpolarized perfused rat liver cells. The hyperpolarization followed a time course similar to an increase in $K^{+}$ efflux which was preceded by an increase in $Ca^{++}$ efflux. Glucagon (and other hormones and cAMP) therefore can influence cation binding and cation flux across cell membranes. Shlatz and Marinetti (74) found a stimulation of $Ca^{++}$ binding by glucagon in washed plasma membranes not fortified with ATP (which would be required to form cAMP). Unless the membranes used contained endogenous bound ATP, it was concluded that the glucagon effect on $Ca^{++}$ binding was not dependent on cyclic AMP production. Indeed, added ATP (1 mM) inhibited the glucagon stimulation of $Ca^{++}$ binding. In another experiment Shlatz and Marinetti (72) found that cAMP added to a system containing rat liver plasma membranes and fortified with rat liver protein kinase stimulated phosphorylation of the membrane and that the phosphorylated membrane bound more $Ca^{++}$ ions than control membranes. The addition of glucagon or epinephrine to this membrane system replaced the requirement for added cyclic AMP in the protein kinase phosphorylation of the membrane. Hence it was concluded that hormones such as glucagon and epinephrine can directly modify the plasma membrane so as to enhance $Ca^{++}$ binding or they can act indirectly by stimulating cAMP production which activates protein kinase which then can phosphorylate the membrane. It appears from these studies that some of the excess glucagon binding sites not correlated with adenylate cyclase activation may be involved in the glucagon modulation of cation binding to the cell membrane and possibly to cation flux.

## Models for Hormone Action on Cell Membranes

A model system depicting the action of epinephrine on the membrane bound adenylate cyclase is shown in Figure 3. The adenylate cyclase is hypothesized as consisting of a regulatory subunit (R) and a catalytic subunit C (Fig. 3a). The regulatory subunit R has 2 binding sites (A and B) for the epinephrine. The catechol ring of the hormone binds at site B and the side chain binds at site A. Binding at site B alone is a necessary but not sufficient condition for hormone action. Site A is postulated to be the coupling site. Productive binding requires both sites to be occupied.

The catalytic subunit also is envisioned as having two binding sites, one for ATP binding (site H) where hydrolysis of ATP occurs and one for binding of modifiers such as GTP (site M). Binding of the modifier M may induce an allosteric change in the complex enzyme altering the conformations of subunits C and R to C' and R' (Fig. 3b). This can modify the binding of the hormone to the regulatory subunit (decrease or increase) and hence allosterically modulate the cyclase in this fashion. The modifier may influence to some extent the catalytic site H of the subunit C.

The action of the epinephrine is shown in Fig. 3c: The epinephrine is shown binding the sites B and A of the regulatory subunit in the R" conformation. This binding now induces a conformational change in the catalytic subunit to C" activating the catalytic site H so that it can effectively bind and convert ATP to cAMP.

The model also shows the asymmetric arrangement of the phospholipids in the membrane with PE, PS and PI being localized on the inner surface and PC and SPH on the outer surface. Cholesterol is omitted for sake of convenience. The PS and PI are shown closely juxtaposed and interacting with the catalytic subunit. It is assumed that the PS and PI are required to hold the catalytic subunit in the membrane and that PC and SPH are required for holding the regulatory subunit in the membrane.

A more generalized model to explain the multiple effects of hormones on cell membranes is shown in Figure 4. The model is invoked to explain the multiple effects of hormones on $K^+$-$Na^+$ flux, $Ca^{++}$ binding and flux, substrate transport, and on adenylate cyclase. Subunits R and C represent the regulatory and catalytic subunits of adenylate cyclase. The hormone stimulation of adenylate cyclase increases the intracellular level of cAMP which in turn binds to the regulatory unit of protein kinase (s) (PKi) which is in the inactive state and converts it by dissociation to the active protein kinase (PKa). The protein kinase (or kinases) then phosphorylate (via ATP) a variety of enzymes (E) and converts them either to active or inactive forms (E'). The enzymes E' are converted back to the original state (E') by specific phosphatases.

The protein kinase(s) in the cytosol are also able to phosphorylate membrane proteins (protein F). Thus the membrane can exist in a different state of phosphorylation. The phosphorylated state is envisioned as having altered permeability properties for binding of cations such as $K^+$, $Na^+$ and $Ca^{++}$ and therefore ion flux can be modulated in this way. Phosphatases would be required to dephosphorylate the membrane proteins to convert the membrane to its original state.

The hormones may also directly influence substrate and cation flux across the membrane by binding to specific receptors (such as A-D) which binding induces a conformational change in the membrane enhancing cation flux. In the case of insulin, these receptors (C and D) may be associated with transport of glucose, amino acids and nucleosides.

Whether membrane ATPase is coupled to the hormone action by interacting with adenylate cyclase or whether the ATPase plays a role in the hormone stimulated flux of $K^+$, $Na^+$, or $Ca^{++}$ remains to be elucidated. Components E and F can represent an ATPase complex.

The effect of nucleotides (78, 58) phospholipids (79,80,48,49,50,51,52), detergents (13,14,15, 43,52), lipases (42,52) and proteases (4,42,52) on the adenylate cyclase system can be explained in terms of the models presented. Proteases can inhibit the system in isolated cells by hydrolyzing the hormone receptor whereas in isolated membranes they can inhibit by hydrolyzing either the hormone receptor or the catalytic unit of adenylate cyclase.

Phospholipases might inhibit the adenylate cyclase system in several different ways. They can hydrolyze the membrane phospholipids and disrupt the membrane to such a degree that the hormone receptor is physically separated from the catalytic subunit or is uncoupled from this subunit. The phospholipids may act merely as structural cementing agents to hold the adenylate cyclase complex in the membrane. It is possible that certain phospholipids, in particular PS and PI which have been shown to be effective in restoring hormone sensitivity to delipidized membranes, may play a functional role. Birnbaumer (21) has suggested that phospholipids may act as coupling agents between the hormone receptor and the catalytic subunit of adenylate cyclase. However, in view of recent work (53-55) which shows an asymmetric arrangement of phospholipids in the plasma membrane of the red cell (and assuming this asymmetry prevails in the liver and fat cell plasma membrane) one can postulate that these acidic phospholipids such as PS might interact specifically with the catalytic subunit since both PS and the catalytic subunit are situated on the same inner surface of the membrane.

Detergents are considered to produce effects similar to phospholipases in that they disrupt lipid-lipid, lipid-protein interactions. They can also disrupt protein-protein interactions and thus can be more disruptive than lipases. Moreover, because of their ability to cause extensive alteration of the membrane and their ability to penetrate the membrane, they can exert their effects on isolated cells as well as on isolated membranes.

A unitary hypothesis specifies that all the actions of a hormone are mediated by a primary single event. It appears unlikely that polypeptide hormones and catecholamines which are known to bind to the plasma membrane and which apparently can exert some of their biological effects without penetrating the membrane, exert their multiple actions via a single event on the membrane. It is appealing to assume that all the actions of these hormones are mediated directly or indirectly by increasing or decreasing cellular cyclic AMP levels by interacting with the membrane adenylate cyclase. Although one cannot rule out this unitary mode of action, the evidence does not appear to support his hypothesis. The models (Figures 2 and 3) are provisional and speculative and are aimed to explain the experimental evidence cited in this chapter dealing with hormone effects on isolated plasma membranes or isolated cells. It is hoped that they aid rather than deter productive thinking in this important area of molecular biology.

## Hormone Binding - Hormone Function - Summary of Major Observations

The properties of the hormone-membrane system and evidence that high affinity insulin binding has biological significance are enumerated below:

1. The hormone receptors are localized on the plasma membrane of target cells.
2. Hormone binding is rapid and the onset of the biological response is equally rapid.
3. Continuous occupation of the receptor appears to be essential for sustained action of the hormones.
4. The concentration dependence curves for binding and for the biological response are similar.
5. The hormone binding and its response occur at "physiological" concentrations of the hormone.

6. The Km for hormone binding and for the biological response are similar.
7. The native unlabeled hormone displaces the labeled bound hormone.
8. There is a correspondence between the ability of hormone analogs to initiate the biological response and to compete with binding of the native hormone.
9. Hormone binding is a necessary but not sufficient event for eliciting the hormone response. The binding can be coupled or uncoupled.
10. Part of the hormone molecule appears to be required for binding to the receptor and another part of the hormone molecule appears to be required for coupling.
11. Other agents such as nucleotides (ATP, GTP) modify the hormone response.
12. An intact membrane is essential for coupled binding. Phospholipases and detergents disrupt this coupling.
13. The full biological response is observed when apparently only a small number of the total receptors are occupied.
14. Hormones can produce multiple effects on the membrane, such as modifying adenylate cyclase, phosphodiesterase, ATPase, $K^+$, $Na^+$ flux, $Ca^{++}$ binding and efflux, substrate transport and membrane phosphorylation.
15. Hormone degradation, inactivation or removal must occur to terminate the hormone response. Hormone inactivation by the membrane can be rapid.

## Perspectives for Future Studies on Hormone Receptors

*Isolation of Hormone Receptors and Adenylate Cyclase. Reconstitution Experiments.* In order to elucidate the molecular mechanism of action of hormones on cell membranes, in particular the mechanism of stimulation of adenylate cyclase, it seems inevitable that the isolation and purification of these components must be achieved. Although considerable progress has been made in the isolation of the glucagon (34) and insulin receptor (13) the amounts isolated have been too small to measure chemically. The isolation and purification of membrane-bound adenylate cyclase has not been achieved. The adenylate cyclase appears to be very labile to freezing whereas the hormone receptors are quite stable. Both of these components occur in very small concentration in cell membranes (about 5000 -10,000 molecules per cell). Therefore large scale mass production methods will have to be developed for their isolation and purification. Once this has been achieved the properties of each component can be studied and attempts can be made to reconstitute them by addition of specific phospholipids to obtain a hormone sensitive system.

*Hormone receptors-abnormal states.* In abnormal states such as insulin resistance, obesity and certain forms of diabetes are the hormone receptors defective, uncoupled or altered in amount?

How does one hormone (thyroxine) influence the action of another hormone (epinephrine or glucagon)? The permissive action of one hormone on another remains to be elucidated.

Are the physiological responses of a single hormone due to one primary interaction on the cell or does one hormone have several different receptors each coupled to a specific function.

*Hormone Inactivation by the Membrane.* In order for hormones to act in a cyclic fashion the hormone must be removed from the cell membrane or it must be inactivated by some process. It has been shown that glucagon is rapidly inactivated by isolated plasma membranes (22) and that insulin is also degraded by isolated plasma membranes (16) although not at a very rapid rate. Epinephrine is not appreciably degraded by isolated plasma membranes. An interesting and important observation with all three hormones is that the bound hormone is stable and protected from degradation. Thus the hormone must dissociate from the receptor in order for degradation to occur. Glucagon and insulin, which appear not to penetrate the fat cell plasma membrane readily, if at all, have much higher binding constants ($10^9$-$10^{10} M^{-1}$) than catecholamines ($10^6$-$10^7M^{-1}$), which apparently can penetrate the membrane. To date the problem of how hormone synthesis and release is balanced by hormone degradation remains to be elucidated.

This work was supported in part by a grant HL02063 from the Heart and Lung Institute, The National Institutes of Health, and U.S. Public Health Grant 2T07 AM 01004-09.

References

1. Marinetti, G.V., Ray, T.K. and Tomasi, V. Biochem. Biophys.Res.Comm. 36(1969)185.
2. Pohl.,S.L., Birnbaumer,L. and Rodbell, M. Science 164 (1969)566.
3. Bar, H. and Hechter, O. Proc. Nat. Acad. Sci. U.S.A. 63 (1969) 350.
4a. Birnbaumer, L.,Pohl, S.L. and Rodbell, M. J. Biol. Chem. 244 (1969) 3468.
4b. Rodbell, M., Birnbaumer, L. and Pohl, S.L. J. Biol. Chem. 245 (1970) 718.
4c. Birnbaumer, L. and Rodbell, M. J. Biol. Chem. 244 (1969)3477.
5. Ray, T.K., Tomasi, V. and Marinetti, G.V. Biochim. Biophys. Acta 211 (1970) 20.
6. Emmelot, P. and Bos, C.J. Biochim. Biophys. Acta. 249 (1971) 285.
7. Freychet, P., Roth, J., and Neville, D.M. Jr. Biochem. Biophys. Res. Comm. 43 (1971) 400.
8. Cuatrecasas, P. Proc. Nat. Acad. Sci. U.S.A. 68 (1971) 1264.
9. Pohl, S.L., Birnbaumer, L. and Rodbell, M. J. Biol. Chem. 246 (1971) 1849.
9b. Rodbell, M., Krans, H.M.J., Pohl, S.L. and Birnbaumer, L. J. Biol. Chem. 246(1971)1861.
10. Rodbell, M., Birnbaumer, L., Pohl, S.L. and Sundby, F. Proc.Nat.Acad.Sci. USA 68(1971)909.
11. Freychet, P., Roth, J. and Neville, D.M.Jr. Proc. Nat. Acad. Sci. U.S.A. 68 (1971)833.
12. Birnbaumer, L., Pohl, S.L. and Rodbell, M., J. Biol. Chem. 246 (1971) 1857.
13. Cuatrecasas, P. Proc. Nat. Acad. Sci. U.S.A. 69 (1972) 318.
14. Cuatrecasas, P. J. Biol. Chem. 247 (1972)1980.
15. Cuatrecasas, P. Proc. Nat. Acad. Sci. U.S.A. 69 (1972) 1277.

16. Freychet, P., Kahn, R., Roth, J., and Neville D.M. Jr. J. Biol. Chem. 247 (1972) 3953.
17a. Hepp, K.D. Europ. J. Biochem. 31 (1972) 31.
17b. Hepp, K.D. FEBS Letters 12 (1971) 263
18. Illiano, G. and Cuatrecasas, P. Science 175 (1972) 906.
19. Marinetti, G.V., Shlatz, L., and Reilly, K. 1972. In: Insulin Action p. 207-276, ed. I.B. Fritz, New York: Academic Press.
20. Birnbaumer, L. and Pohl, S.L. J. Biol. Chem. 248 (1973) 2056.
21. Birnbaumer, L. Biochim. Biophys. Acta 360 (1973) 129.
22. Rodbell, M. Fed. Proc. 32 (1973) 1854.
23. Cuatrecasas, P. Fed. Proc. 32 (1973) 1838.
24. Birnbaumer, L., Pohl, S.L. and Rodbell, M. J. Biol. Chem. 247 (1972) 2038.
25. DeMeyts, P., Roth, J., Neville, D.M. Jr., Gavin, J.R., and Lesniak, M.A. Biochem. Biophys. Res. Comm 55 (1973) 154.
26. Thompson, W.J., Little, S.A. and Williams, R. H. Biochem. 12 (1973) 1889
27. Sonnenberg, M. Proc. Nat. Acad. Sci. U.S.A. 68 (1971) 1051.
28. Dunnick, J.K. and Marinetti, G.V. Biochim. Biophys. Acta 249 (1971) 122.
29. Schramm, M., Feinstein, H., Naim, E., Long, M. and Lasser, M. Proc. Nat. Acad. Sci. U.S.A. 69 (1972) 523.
30. Lefkowitz, R.J., Haber, E. and O'Hara, D. Proc. Nat. Acad. Sci. U.S.A. 69(1972)2828.
31. Lefkowitz, R.J., Sharp, G.W.G. and Haber,E. J. Biol. Chem. 248 (1973) 342.
32. Lefkowitz, R.J., O'Hara, D.S. and Warsham,J. Nature New Biol. 244 (1973) 79.
33. Bilezikian, J.P. and Aurbach, G.D. J. Biol. Chem. 248 (1973) 5577.
34. Krug, F., Desbuquois, B. and Cuatrecasas, P. Nature, New Biol. 234 (1971) 268.
35. Cuatrecasas, P. Proc. Nat. Acad. Sci. U.S.A. 63 (1969)450.
36. Exton, J.H. Robison, G.A., Sutherland, E.W. and Parks, C.R. J.Biol.Chem. 246(1971)6166.
37. Kono, T., and Barham, F.W. J. Biol. Chem. 246 (1971) 6210.

38. Fain, J.N. Pharmacol. Rev. 25 (1973) 67.
39. Fain, J.N. Fed. Proc. 29 (1970) 1402.
40. Himms-Hagen, J. Fed. Proc. 29 (1970)1388
41. Gorman, R.R., Tepperman, H.M. and Tepperman, J. J. Lipid Res. 14 (1973) 279.
42. Tomasi, V., Koretz, S., Ray, T.K., Dunnick, J.D. and Marinetti, G.V., Biochim. Biophys. Acta 211 (1970) 31.
43. Lefkowitz, R.L. and Levey, G.S., Life Sci. 2 (1972) 821.
44. Marley, E. Adv. Pharmacol. 3 (1964) 167.
45. Goldfine, I.D., Gardner, J.D. and Neville, D.M. Jr. J. Biol. Chem. 247 (1972) 6919.
46. Goldfine, I.D. and Sherline, P. J. Biol. Chem. 247 (1972) 6927.
47. Storm, D.R. and Doginow, Y.D. J. Biol. Chem. 248 (1973) 5208.
48. Levey, G.S. Biochem. Biophys. Res. Comm. 43 (1971) 108.
49. Levey, G.S., J. Biol. Chem. 246 (1971) 7405.
50. Rethy, A., Tomasi, V. and Trevisani,A.Arch. Biochem. Biophys. 147 (1971) 36.
51. Rethy, A., Tomasi, V., Trevisani, A. and Barnabei, O. Biochim. Biophys. Acta 290 (1972) 58.
52. Bilezikian, J.P. and Aurbach, G.D. J. Biol. Chem. 248 (1973) 5584.
53. Gordesky, S., and Marinetti, G.V. Biochem. Biophys. Res. Comm. 50 (1973) 1027.
54. Verklij, A.J., Zwaal, R.F.A., Roelofsen, B., Comfuruis, P., Kastelyn, D. and VanDeenen, L.L. M. Biochim. Biophys. Acta 323 (1973) 178.
55. Zwaal, R.F.A., Roelfosen, B., and Cooley, C.M. Biochim. Biophys. Acta. 300 (1973) 159.
56. McConnell, H.J. and McFarland, B.C., Quart. Rev. Biophys. 3 (1970) 91.
57. Marinetti, G.V., Baumgarten, R., Sheeley, D. and Gordesky, S. Biochem. Biophys. Res. Comm. 53 (1973) 302.
58. Rodbell, M., Birnbaumer, L., Pohl, S.L., and Krans, H.M.J., J. Biol. Chem. 246 (1971) 1877.
59. Goldfine, I.D., Roth, J. and Birnbaumer, L. J. Biol. Chem. 247 (1972) 1211.
60. Krishna,G., Harwood, J.P., Barber, A.J. and Jamieson, G.A. J. Biol. Chem. 247(1972)2253.
61. Harwood, J.P., Low, H., and Rodbell, M. J. Biol. Chem. 248 (1973) 6239.

62. Leray, F., Chambaut, A.M. and Hanoume, J. Biochem. Biophys. Res. Comm.48(1972) 1385.
63. Kant, J.A. and Steck, T.L. Biochem. Biophys. Res. Comm. 54 (1973) 116.
64. Vaughan, M. 1972, In: Insulin Action. p. 297-318, ed. Fritz, I.B. New York, Academic Press.
64a.Loten, E.G. and Sneyd, G.T. Biochem. J. 120 (1970) 187.
65. Manganiello, V. and Vaughan, M. J. Biol. Chem. 248 (1973) 7164.
65a.House, P.D.R., Poulis, P. and Weidemann, M.J. Europ. J. Biochem. 24 (1972) 429.
66. Hadden, J.W., Hadden, E.M., Wilson, E.E., Good, R.A. and Coffey, R.G. Nature New Biol. 235 (1972) 174.
67. Kuo, J.F. and Greengard, P. Proc. Nat. Acad. Sci. U.S.A. 64 (1969) 1349.
68. Maeno, H., Johnson, E.M. and Greengard, P. J. Biol. Chem. 246 (1971) 134.
69. Kumon, A., Yamamura, H. and Nashizuka, Y. Biochem. Biophys. Res. Comm. 41 (1970) 1290.
70. Rasmussen, H. Science 170 (1970) 404.
71. Rasmussen, H., Goodman, D.P.P. and Tenenhouse, A. Critical Rev. Biochem. 1 (1972) 95.
72. Shlatz, L. and Marinetti, G.V. Biochem. Biophys. Res. Comm. 45 (1971) 51.
73. Zahlten, R.N., Hochberg, A.A., Stratman, F.W. and Lardy, H.A. Proc.Nat.Acad.Sci. USA 69 (1972) 800.
74. Shlatz, L., and Marinetti, G.V. Science 176 (1972) 175.
75. Gardner, J.D., Klaeneman, H.L, Bilezikian, J.P. and Aurbach, G.D. J. Biol. Chem. 248 (1973) 5590.
76. Friedman, N. and Park, C.R., Proc. Nat. Acad. Sci. U.S.A. 61 (1968) 504.
77. Friedman, N., Somlyo, A.V. and Somlyo, A.P. Science 171(1971)410.
78. Rodbell, M, Krans, H.M.J., Pohl, S.L. and Birnbaumer, L. J. Biol. Chem. 246 (1971) 1872
79. Pohl, S.L., Krans, H.M.J., Kozyreff, V., Birnbaumer, L.and Rodbell, M. J. Biol. Chem. 246 (1971) 4447.
80. Cuatrecasas, P. J. Biol. Chem. 246(1971)6532.
81. Laurell, S. and Tibbling, G. Clin. Chem. Acta 13 (1966) 317.

## TABLE I

The effect of Epinephrine Analogs on the Binding of $^{3}$H-Epinephrine to Rat Epididymal Fat Cells

| Structure | | 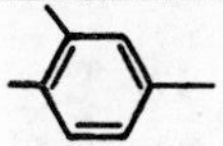 | β CH | α CH | NH | Hormone analog added | % of the control |
|---|---|---|---|---|---|---|---|
| 3OH | | 4OH | OH | H | $CH_3$ | unlabeled 1-epinephrine | 7.4 ± 1.6 |
| $3OCH_3$ | | 4OH | OH | H | $CH_3$ | metanephrine | 24.1 ± 6.8 |
| 3OH | | 4OH | OH | H | $CH(CH_3)_2$ | isoproterenol | 3.2 ± 2.4 |
| 3OH | | 4OH | OH | H | H | 1-norepinephrine | 15.1 ± 2.4 |
| 3OH | | 4OH | H | COOH | H | 1-3,4-dihydroxyphenylalanine | 7.8 ± 3.3 |
| 3OH | | 4OH | OH | OH/ | | 1-3,4-dihydroxyphenylglycol | 8.5 ± 4.0 |
| 3OH | 6OH | 4OH | H | H | H | 6-hydroxydopamine | 6.3 ± 1.8 |
| 3OH | | 4OH | OH COOH/ | | | d,1-3,4-dihydroxymandelic acid | 1.3 ± 1.3 |
| $3OCH_3$ | | 4OH | OH COOH | | | d,1-3-methoxy-4-hydroxy-mandelic acid | 68.4 ± 5.8 |
| $2OCH_3$,$5OCH_3$ | | | OH | $CH_3$ | H | methoxamine (vasoxyl) | 107.1 ± 6.5 |
| | | | H | COOH | H | 1-phenylalanine | 97.9 ± 3.3 |
| | | 4OH | H | COOH | H | 1-tyrosine | 85.8 ± 5.2 |
| 1OH | 2OH | $4NO_2$/ | | | | 4-nitrocatechol | (0-5 %) |

/ *indicates end of the molecule*

Fat cells were added in aliquots of 50-90 μg fat cell protein to 2.0 ml of KRB-BSA-glucose pH 7.3 containing 1 x $10^{-4}$M concentration of analog. Following a 10 minute incubation at 37°C, $^{3}$H-epinephrine was added to a final concentration of 1 x $10^{-7}$M (1.0 μC). The incubation proceeded for an additional 10 minutes; the suspension was applied to the Millipore filtration apparatus. Each value represents the mean percent ± S.E. of triplicate analyses of three separate experiments. KRB=Krebs-Ringer bicarbonate, BSA = bovine serum albumin. The preparation of fat cells and method of hormone binding have been described previously (19). Taken from the Ph.D. thesis of L. Lesko, University of Rochester, 1973. The d,1-epinephrine L-bitartrate (7 - $^{3}$H) spec. act. 9.45 Ci/m mole was obtained from New England Nuclear, Boston, Mass.

TABLE II

The Effect of Alpha and Beta Adrenergic Blocking Agents On Epinephrine Binding to Rat Epididymal Fat Cells and Glycerol Release

| Blocker Concentration | Epinephrine Binding | Epinephrine-stimulated Glycerol Release |
|---|---|---|
| | % of Control | % of Control |
| *Alpha Blockers* | | |
| Phentolamine | 103.0 ± 8.4 | 89.0 ± 2.6 |
| Phenoxybenzamine | 95.6 ± 8.1 | 94.2 ± 2.2 |
| *Beta Blockers* | | |
| Propranolol | 98.9 ± 8.9 | 44.4 ± 5.6 |
| Dichloroisoproterenol | 74.6 ± 6.1 | 77.5 ± 2.4 |

Hormone Binding: Fat cells (50-90 μg fat cell protein) isolated from rat epididymal fat pads were incubated in 2.0 ml of KRB-BSA-glucose buffer, pH 7.3, containing $1 \times 10^{-7}$M $^3$H-epinephrine (1.5 μC) and alpha and beta blocking agent at a concentration of $1 \times 10^{-5}$M for 10 minutes at 37°C Control samples without blocker were incubated simultaneously. The amount of epinephrine bound was determined by direct Millipore filtration (19). Each value for $1 \times 10^{-5}$M blocker represents the mean ± S.E. of triplicate samples from eight separate experiments.

Glycerol Determination: Fat cells (50-90 μg fat cell protein) from the same preparations used for binding were incubated in 1.0 ml of KRB-BSA-glucose buffer, pH 7.3, containing $1 \times 10^{-5}$M 1-epinephrine and blocking agent at a concentration of $1 \times 10^{-5}$M for 20 minutes at 37°C. Controls in the absence of blocking agent and/or epinephrine were incubated simultaneously. The incubations were terminated by the addition of 1.0 ml of 3% PCA and 1.0 ml aliquots of the centrifuged supernatant were neutralized with 1.0 ml of saturated $KHCO_3$ solution. A 0.5ml aliquot of this supernatant and 0.2 ml of the enzyme mixture were incubated together for 60 minutes at room temperature. Glycerol was determined by the method of Laurell and Tibbling (81). Fluorescence was measured on a Perkin-Elmer Model MPF-3 Fluorescence Spectrophotometer (350 nmeters exciting wavelength and 458 nmeters emitting wavelength). Standards were run simultaneously. Each value represents the mean ± S.E. of duplicate determinations from eight separate experiments.
Taken from the Ph.D. thesis of L. Lesko, University of Rochester, 1973.

TABLE III

d-and l-$^{14}$C-Norepinephrine Binding to Rat Liver Plasma Membrane and Epididymal Fat Cells

| Tissue | Hormone Concentration | Hormone Bound pmoles/mg Protein measured | corrected |
|---|---|---|---|
| *Liver* | | | |
| d-$^{14}$C-norepinephrine | 4 x $10^{-6}$ M | 109.3 ± 4.3 | 24.6 ± 0.9 |
| l-$^{14}$C-norepinephrine | 0.9 x $10^{-6}$ M | 151.8 ± 11.8 | 151.8 ± 11.8 |
| *Fat Cells* | | | |
| d-$^{14}$C-norepinephrine | 4 x $10^{-6}$ M | 121.8 ± 12.3 | 27.4 ± 2.8 |
| l-$^{14}$C-norepinephrine | 0.9 x $10^{-6}$ M | 115.3 ± 4.9 | 115.3 ± 4.9 |

Rat liver plasma membranes (90 - 120 μg membrane protein) and fat cells (50 - 90 μg fat cell protein) were incubated in 2.0 ml of KRB buffer, pH 7.3, (KRB-BSA-glucose buffer, pH 7.3 for fat cells) containing either 4 x $10^{-6}$M d-$^{14}$C-norepinephrine (0.17μC) or 0.9 x $10^{-6}$M l-$^{14}$C-norepinephrine (0.1 μC) for 10 minutes at 37°C. Binding was measured by direct Millipore filtration. Each value represents the mean ± S.E. of triplicate samples from three separate experiments. The methods of preparing plasma membranes and fat cells and the Millipore filtration procedure are given elsewhere (19). Taken from the Ph.D. thesis of L. Lesko, University of Rochester, 1973. The d-norepinephrine (methylene - $^{14}$C) D-bitartrate, spec. act. 21.2 mC/m mole and l-norepinephrine-D-bitartrate, spec. act. 57 mC/m mole were obtained from Amersham/Searle, Arlington Heights, Ill.

TABLE IV

The Stereochemical Aspects of d and 1-$^{14}C$-Norepinephrine Binding To Rat Liver Plasma Membranes and Epididymal Fat Cells

| Tissue | Molar ratio of unlabeled isomer to labeled isomer | % Inhibition |
|---|---|---|
| *Liver* | | |
| d-$^{14}C$-Norepinephrine + 1-Norepinephrine | 10/1 | 26.5 ± 3.4 |
| 1-$^{14}C$-Norepinephrine + 1-Norepinephrine | 10/1 | 86.9 ± 2.0 |
| *Fat Cells* | | |
| d-$^{14}C$-Norepinephrine + 1-Norepinephrine | 10/1 | 24.0 ± 17.3 |
| 1-$^{14}C$-Norepinephrine + 1-Norepinephrine | 10/1 | 41.6 ± 10.5 |

Rat liver plasma membranes (90-120 μg membrane protein) and fat cells (50-90 μg fat cell protein) were incubated for 10 minutes as indicated in Table III. To test the influence of unlabeled 1-norepinephrine on the binding of radioactive isomers, membranes and fat cells were incubated in the assay mixture described in Table III with the addition of $4 \times 10^{-5}$M 1-norepinephrine. Binding was performed by direct Millipore filtration. The values represent the mean ± S.E. of triplicate analyses from three separate experiments. Taken from the Ph.D. thesis of L. Lesko, University of Rochester, 1973.

TABLE V

The Effect of Phospholipases and Proteases on Epinephrine Binding to Rat Liver Plasma Membranes

| Agent | % of Control |
|---|---|
| Phospholipase A | 20.8 ± 4.6 |
| Phospholipase C | 18.8 ± 3.0 |
| Phospholipase D | 40.7 ± 8.3 |
| Trypsin | 3.3 ± 0.0 |
| Pronase | 53.3 ± 16.3 |

Phospholipase A (Naja naja) (5 μg of PLA/100 μg of membrane protein), Phospholipase C (Cl. welchii) (5 μg PLC/100 μg of membrane protein) of Phospholipase D (5 μg PLD/100 μg membrane protein) was added to 4.0 ml of plasma membrane suspension (450-600 μg membrane protein) in KRB-buffer, pH 7.3. The membranes were incubated at 37°C for 45 minutes and then centrifuged and washed 3 times with 4.0 ml of KRB-buffer, pH 7.3. Aliquots of the phospholipase-treated membrane (90-120 μg membrane protein) were incubated in 2.0 ml of KRB buffer, pH 7.3, containing 1 x $10^{-7}$M $^3$H-epinephrine (1.0-1.5 μC) for 10 minutes at 37°C. The direct Millipore filtration method was employed to measure epinephrine binding. Each value represents the mean ± SE of triplicate determinations of three separate experiments.

Trypsin (50 μg of trypsin/ 100 μg membrane protein) or pronase (50 μg of pronase/ 100 μg membrane protein) were added to 4.0 ml of membrane suspension (500-600 μg membrane protein/ml) in KRB-buffer, pH 7.3. The membrane suspension was incubated for 1 hour at 37°C and then centrifuged and washed three times with 4.0 ml of KRB-buffer, pH 7.3. Aliquots of the treated membranes (90-130 μg of membrane protein) were then incubated in 2.0 ml of KRB-buffer, pH 7.3 containing 1 x $10^{-7}$M $^3$H-epinephrine (1.0-1.5 μC) for 10 minutes at 37°C. Binding was determined by direct Millipore filtration. Each value represents the mean ± SE of triplicate determinations of three separate experiments.

Control membranes were incubated simultaneously with the treated membranes. Taken from the Ph.D. thesis of L. Lesko, University of Rochester, 1973. Phospholipase A, from Miami Serpentarium, phospholipase C and D from Sigma Chem. Co. Trypsin from Worthington Biochem. and pronase from Cal. Biochem.

TABLE VI

Cyclic-AMP-Dependent
Protein Kinase Phosphorylation of the Plasma Membrane

| Additions | nmoles of Pi per mg membrane protein |
|---|---|
| No kinase | 3.0 ± 0.8 |
| Kinase (300 μg) | 7.4 ± 0.7 |
| Kinase (300 μg) + C-AMP ($10^{-8}$M) | 9.4 ± 1.0 |
| Kinase (300 μg) + C-AMP ($10^{-7}$M) | 12.3 ± 0.5 |
| Kinase (300 μg) + C-AMP ($10^{-6}$M) | 17.6 ± 1.4 |

Rat liver plasma membranes (60-70 μg protein) were incubated in 0.1 M Tris buffer pH 7.5 in a total volume of 1.0 ml containing 300 μg of rat liver protein kinase, 1 mM $MgCl_2$, 1 mM γ-$P^{32}$-ATP (0.1 μC) and varying amounts of cAMP. After 10 min. incubation at 37°C the membranes were collected on HA 0.45μ Millipore filters and washed twice with 5 ml of buffer. The filters were counted on a Packard Tri Carb Scintillation Spectrometer. Taken from Shlatz and Marinetti (72).

TABLE VII

The Effect of Hormones on the Protein Kinase
Mediated Phosphorylation of the Plasma Membrane

| Additions | $P^{32}$ Binding nmoles/mg protein | $C^{14}$-ATP Binding nmoles/mg protein |
|---|---|---|
| Control | 7.4 ± .70 | 3.5 ± 1.02 |
| Glucagon $10^{-7}$M | 15.4 ± .20 | 3.0 ± .56 |
| Epinephrine $10^{-7}$M | 8.5 ± .65 | 3.0 ± .85 |
| Glucagon + Epinephrine $10^{-7}$M each | 15.5 ± 1.22 | 3.6 ± .75 |
| Epinephrine $10^{-6}$M | 11 ± 0.6 | |

Hormones were added at $10^{-7}$M except where indicated otherwise and NaF was added at 10 mM. Rat liver membranes (70-80 μg protein) suspended in 0.1 M Tris-HCl buffer, pH 7.5 were incubated with 1.0 mM ATP containing either 0.1 μC of γ-$P^{32}$-ATP or 0.05 μC of adenosine-8-$C^{14}$-triphosphate, 1.0 mM $MgCl_2$, 300 μg of protein kinase and additions noted for 10 min at 37°C. Samples were filtered by Millipore filtration. Results represent the mean ± standard deviation of duplicate determinations of four separate experiments. Taken in part from Shlatz and Marinetti (72).

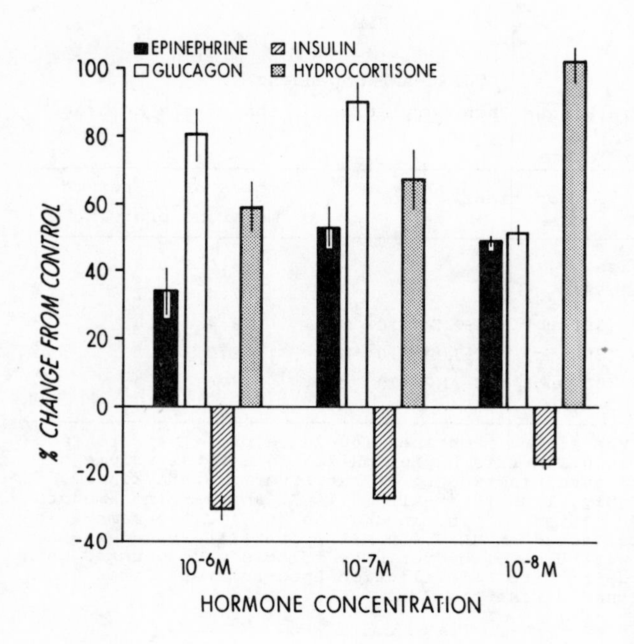


Figure 1. The inhibition of binding of $^3$H-1-Norepinephrine to isolated fat cell plasma membranes by catecholamine analogs. Fat cell plasma membranes (60-80 μg protein) were incubated at 37°C for 10 min in 2.0 ml of KRB buffer pH 7.4 containing 2% BSA, $1x10^{-7}$M 7-$^3$H-1-norepinephrine (0.5 μC, Amersham/Searle) and varying amounts of analog as shown in the Figure. The suspensions were chilled in ice and spun at 3000 rpm for 15 min. The membrane pellets were resuspended in 2.0 ml of ice cold KRB buffer and immediately filtered through Millipore filters (EGWP-02500, Millipore Corp.) and washed three times with 5 ml of cold KRB buffer. The filters were put into vials containing 10 ml of Beckman Biosolv-TLA cocktail and the radioactivity determined. Each point represents the mean ± S.D. of triplicate analyses of two experiments. Three rats were pooled in each experiment for the isolation of the fat cell plasma membranes. From top to bottom the structures are: 1-phenylephrine, d,1-metanephrine, dihydroxymandelic

acid, l-dihydroxyphenylalanine, and l-norepinephrine. The ordinate represents the percent of control binding (without added analog).

---

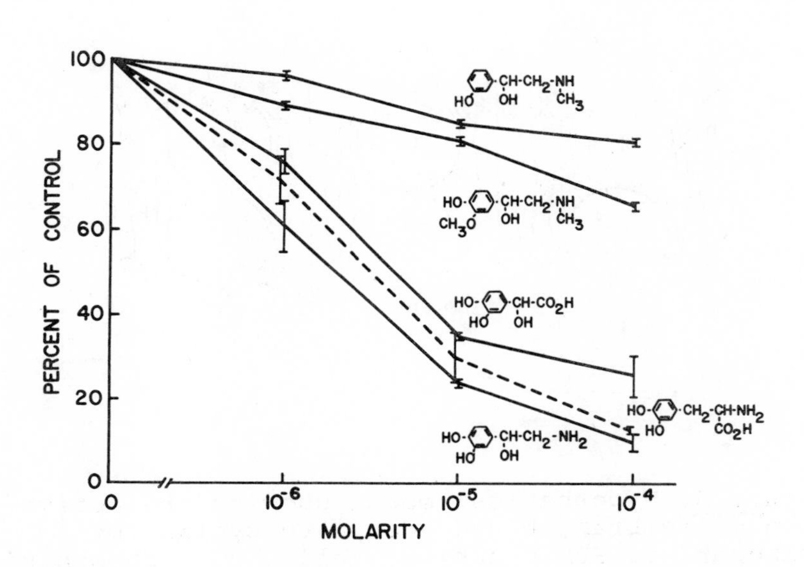


Figure 2. The effect of hormones on the calcium binding to isolated rat liver plasma membranes. Membranes (60-70 μg protein) were incubated with 1 mM $CaCl_2$ (containing 0.5 μC of $^{45}Ca^{++}$) for 10 min at 37° in a total volume of 1.0 ml containing 0.1 M Tris buffer pH 7.5 and varying amounts of hormone. $Ca^{++}$ binding was determined by Millipore filtration. Taken from Shlatz and Marinetti (74). (Science, Vol. 176. pp.175-177, 1972. Copyright by the American Association for the Advancement of Science).

---

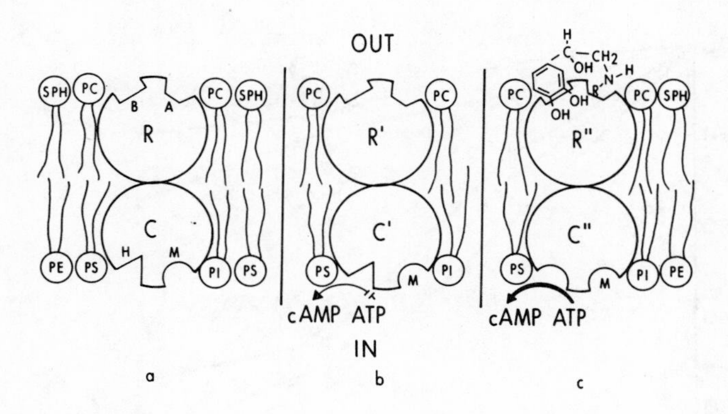


Figure 3. Hypothetical model showing the activation of membrane-bound adenylate cyclase by epinephrine. SPH = sphingomyelin, PC = phosphatidyl choline, PE = phosphatidyl ethanolamine, PS = phosphatidyl serine, PI = phosphatidyl inositol, R = hormone receptor protein. This subunit binds epinephrine at sites A and B. C = the catalytic subunit of adenylate cyclase. It has two binding sites. Site H is the catalytic site for binding of ATP. Site M is the modifier site for allosteric agents such as guanyl nucleotides. ATP = adenosine -5'-triphosphate, cAMP = 3',5'-cyclic adenosine monophosphate.

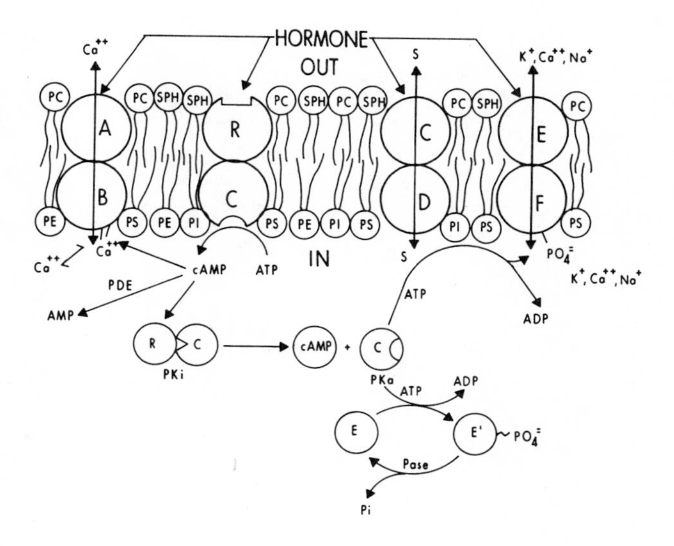


Figure 4. Hypothetical model showing the multiple effects of hormones on the plasma membrane. R = hormone receptor coupled to the catalytic subunit C of adenylate cyclase, A = a different hormone receptor (independent of ATPase) involved in calcium binding and $Ca^{++}$ flux through the membrane. This receptor is coupled to a calcium transport protein B. C and D represent substrate (S) transport proteins for sugars, amino acids and nucleosides. They may be in close approximation to either adenylate cyclase (R,C) or to proteins E,F which may be ATPases involved in the transport of cations. The protein F is capable of being phosphorylated by the cAMP dependent protein kinase(s) PKa = active catalytic unit of protein kinase(s), PKi is the inactive protein kinase(s) consisting of a catalytic subunit C and a regulatory subunit R which binds cAMP. E and E' represent a family of enzymes activated or inactivated by PKa. Pase = specific phosphatases. PDE = phosphodiesterase. The other symbols are explained in Figure 3.

# SURFACE ULTRASTRUCTURE OF NORMAL AND PATHOGENIC CULTURED CELLS

Adolfo Martínez-Palomo and
Pedro Pinto da Silva

Departamento de Biología Celular, Centro de Investigación y de Estudios Avanzados del IPN, Apartado Postal 14-740, México. and The Salk Institute for Biological Studies, San Diego, California, U.S.A.

## Introduction

Changes in the surface components of tumoral cells have been extensively investigated in recent years in view of the possible significance of surface properties in the control of cell-to-cell interactions and cell growth. A particularly promising area was established with the finding of a selective agglutination reaction of tumoral cells induced by a variety of lectins (1,2,3).

In general, cultured tumoral cells have been found to be susceptible to agglutination with concanavalin A (con A) or with wheat germ agglutinin, while normal cells will only agglutinate in the presence of these lectins if previously treated with low concentrations of trypsin (3,4). At the time these differences were observed, it was assumed that the sensitivity of intact tumoral cells to lectin-induced agglutination was due to the presence of exposed surface receptors in tumoral cells, while these receptors were assumed to be in a cryptic location at the surface of normal cells (4).

## Distribution of Concanavalin A Receptors in Normal and in Transformed Cells

As a continuation of our studies on the surface ultrastructure of normal and transformed cells (5,6) we initiated in collaboration with Drs. Bernhard and Wicker, a study on the surface distribution of con A receptors in normal and in tumoral cells. We have studied secondary cultures of golden hamster embryos as normal cells, since the use of normal established culture cell lines as controls may be open to question. Hamster cells transformed by polyoma virus *in vivo* (T. Py XV line) were used, after their oncogenic potential was confirmed. Binding sites for con A were detected by means of the electron microscopical technique described by Bernhard and Avrameas (7). When living normal and polyoma transformed cells are treated with con A (100 μg/ml) during 15 minutes, the distribution of the label produced by the reaction con A - peroxidase - benzidine is different in both types of cultures. In normal hamster embryo cells (Fig. 1) the reacting layer appears as a uniform and continuous line of dense precipitate all over the cell surface, when viewed in vertical sections. Observations of similar sections of polyoma transformed cells (Fig. 2) demonstrates, in about one third of the cells, the presence of a discontinuous surface layer formed by dots and short traits, leaving in between areas of the plasma membrane devoid of label. The difference between a uniform labeling at the cell surface of normal cells and a patchy distribution of surface receptors in transformed cells is only apparent in cultures that were reacted with con A before fixation with glutaraldehyde. These differences disappear when con A is added after cultured cells have been fixed with glutaraldehyde; both normal and polyoma transformed cells (Fig. 3) show then a uniform distribution of con A receptors. The homogeneous distribution of con A receptors in normal cells and the clustered arrays of these receptors in transformed cells treated with con A before fixation becomes more apparent with the study of tangenctial sections. In grazing sections of normal cells, the reacting surface layer appears as a continuous sheet formed by the confluency of globular deposits approximately 30 nm in diameter (Fig. 4). The layer of polyoma transformed cells studied in surface views was found to be represented by similar, but more isolated globular deposits, which formed dense patches between essentially unlabeled regions of surface membrane. This patchy distribution of electron dense precipitate observed in tangenctial sections gives the appearance of isolated dots and discontinuous lines in vertical sections of polyoma transformed cells.

The clustered topological distribution of con A molecules at the surface of transformed cells was observed and reported independently by Nicolson (8) and ourselves (9,10). Basically similar differences were reported by Bretton et al (11) and Rowlatt et al (12). Although it was initially assumed both from our studies (9,10) and those of Nicolson (8) that the clustered distribution of con A receptors was inherent to the plasma membrane of transformed cells, it became evident that the initial distribution is similarly uniform in both normal and in transformed cells. The clustered distribution of con A receptors in transformed cells treated with con A before fixation is now interpreted as the result of a reaggregation of surface receptors induced by con A (13). Fixation of transformed cells will inhibit the con A induced aggregation of surface receptors, as shown in Fig. 3.

## Significance of the Redistribution of Concanavalin A Receptors in Transformed Cells

From the above mentioned ultrastructural studies on the distribution of con A receptors at the surface of normal and transformed cells, and from the lack of significant quantitative differences in the binding of radioactively labeled con A to normal and tumoral cells surfaces (14,15,16) it has been concluded that the main difference between normal, non agglutinable cells, and tumoral, agglutinable cells, lies in the susceptibility of con A sites at the tumoral cell membrane to aggregate upon interaction with con A.

Taken together, the results of the abovementioned studies seem to indicate that the selective agglutination of tumoral cells induced by con A is not the result of quantitative differences in the number of specific receptors at the surface of normal and transformed cells. The possibility that these receptors are exposed at the surface of tumoral cells and are located in a cryptic position in normal cells has also been ruled out. At the present time the most plausible explanation for the higher sensitivity of tumoral cells to be agglutinated by lectins appears to be the lectin-induced redistribution of surface receptors. However, it seems premature to accept this explanation as final or to freely extrapolate results to other cellular systems, as exemplified by the recent report of De Petris and Raff (17) on the induction of redistribution of con A surface sites both in normal and in transformed cells. The reasons for the latter results which are clearly in discrepancy with a number of other studies (8,9,10,11,12,13,16) are not evident at this time. The requirement of a certain degree of mobility of con A receptors in the plane of the plasma membrane of tumoral cells is supported by the inhibition of the agglutination by prefixation of cells in glutaraldehyde. Whether or not this higher mobility of con A sites in tumoral cells has any bearing on the establishment or maintenance of the malignant transformed state is still a matter of pure speculation.

## A More Suitable Cell System to Study Lectin Induced Agglutination

In order to inquire into the significance of the agglutination process with respect to the pathogenic properties of certain cultured cells, we needed a cellular system that would allow us to study the agglutination process induced by lectins under better conditions than those provided by cultures of transformed fibroblastic cell lines. The use of fibroblastic cultures requires treatment of the cells with chelating substances or with proteases in order to detach the cells from the culture substrate, in order to form a cell suspension prior to testing the agglutinability of a given cell culture. There is little doubt that these chemical treatments considerably modify cell surface components. The use of cells that multiply normally in suspension, or at least, that can be removed from culture tubes by gentle agitation, would overcome the problem. For these reasons we attemped to study the agglutinability of Entamoeba histolytica trophozoites.

## Agglutination of Pathogenic E. Histolytica Trophozoites

We have found, in collaboration with Arturo Gonzẩlez and Margarita de la Torre (18) that E. histolytica trophozoites of the pathogenic strain HK9 cultured under axenic conditions are extremely sensitive to agglutination with con A. The specificity of the agglutination reaction was determined by the inhibitory action of α-methyl-D-mannopyranoside and α-methyl-D-glucoside. When the agglutinability of various pathogenic strains of E. histolytica was compared with that of trophozoites from cultures isolated from asymptomatic carriers, a clear cut difference was observed; pathogenic strains are much more sensitive to clumping with con A than "non pathogenic" cultures (18).

Among the various cellular factors which might influence the agglutination of pathogenic cells, cytoplasmic microtubules seem to be necessary for agglutination of tumoral cells (19). Treatment of E. histolytica cells with microtubule disrupting drugs such as colchicine and vinblastine will not affect agglutination with con A. In addition, con A-induced agglutination of pathogenic E. histolytica is unaffected by cell pretreatment with cytochalasin B. Only incubation at 4°C will abolish the agglutination of pathogenic amebae with con A.

## Distribution of Concanavalin A Receptors in E. Histolytica

Ultrastructural examination of con A-peroxidase labeled E. histolytica pathogenic trophozoites does not reveal a clustering of con A receptors sites, as seen in tumoral cells. A uniform distribution of the label is observed both in pathogenic cells prefixed with glutaraldehyde before treatment with con A, or in cultures where fixation was carried out after con A was added. Therefore, in E. histolytica HK9 strains, which are very sensitive to agglutination with con A, no redistribution of surface receptors is needed for the agglutination process to take place. In order to test whether con A receptors move at all in the plane of the surface membrane, we treated HK9 trophozoites first with con A (10 μg/ml, 15 minutes, room temperature), followed by fixation with glutaraldehyde. In trophozoites treated with the con A-peroxidase-glutaraldehyde sequence, con A receptors show a striking accumulation at one pole of the cell where the uropod is located (Fig. 7), in a manner similar to the induction of "cap" formation described in lymphocytes (20). These results tend to demonstrate that con A receptors may show a prominent displacement over the cell surface of pathogenic amoeba; however, redistribution of these receptors does not appear to be a prerequisite for the occurrence of agglutination.

## Relationship Between Cell Surface and Membrane Structure

One of the puzzling aspects in the study of membrane structure is the extent to which the mobility of surface antigens and other peripheral receptors reflect the translational movement of protein components embedded within the plasma membrane matrix. In fact, mobility of surface receptors and surface antigens has been taken by Singer and Nicolson (21) as an evidence of fluidity of the plasma membrane. However, a direct correlation between mobility of surface receptors and aggregation of intercalated membranes particles revealed by the freeze-fracture and freeze-etch techniques has been observed only in red blood cell ghosts where it has been demonstrated that membrane intercalated particles are the exclusive sites at the surface which bear A and B antigens (22), wheat germ agglutinin (23), influenza virus, and anionic sites (24).

To study the relationship between intramembranous particles and surface components in _E. histolytica_, we have induced reaggregation of intercalated particles by incubation of living cells in glycerol, followed by fixation and freeze-fracture (25). Glycerol-induced aggregation of membrane particles in _E. histolytica_ is more prominent than that previously reported for lymphoid cells (25). Aggregation of membrane particles induced by glycerol provides a clear distinction between smooth and particulate regions of _E. histolytica_ fracture faces (Fig. 6). We have investigated to what extent the induced redistribution of structural components within the plane of the plasma membrane is accompanied by topographical modifications of surface receptors, such as con A receptors or negative sites.

Glycerol treatment of living _E. histolytica_ cells is not followed by a redistribution of con A receptor sites, revealed by the con A - peroxidase technique (Fig. 8). When _E. histolytica_ cells are fixed with 1.5% glutaraldehyde before immersion in glycerol, the negative surface sites revealed by the colloidal iron technique (26) have a uniform distribution at the cell surface (Fig. 9). However, when cells are treated with glycerol before fixation with glutaraldehyde and subsequently, the negative sites are labeled with colloidal iron, a clear-cut patchy distribution of the label is observed (Fig. 10). These results tend to demonstrate that the glycerol-induced redistribution of intramembranous particles revealed by freeze-fracture is accompanied by a clustering of negative sites labeled with colloidal iron, whereas con A surface sites do not redistribute when membrane particles are induced to aggregate. Thus, the relationship membrane structure-surface components is a complex one in living cells; whereas in red blood cells several surface receptors will reaggregate at the same time as intramembranous particles redistribute, in *E. histolytica* the relationship between membrane particles and surface receptors is not straightforward. From these observations, it becomes apparent that the redistribution of surface receptors and surface antigens may not be necessarily an evidence of the translational movement of intramembranous components.

## References

1. Aub, J.C., Tieslau, C. and Lancaster, A. *Proc. Nat.Acad.Sci. USA*. 50 (1963) 613.
2. Burger, M.M. and Goldberg, A.R. *Proc. Nat. Acad. Sci. USA* 57 (1967) 359.
3. Inbar, M. and Sachs, L. *Proc. Nat. Acad. Sci. USA* 63 (1969) 1418.
4. Burger, M.M. *Proc. Nat. Acad. Sci. USA*. 62 (1969) 994.
5. Martínez-Palomo, A. and Brailovsky, C. *Virology* 34 (1968) 379.
6. Martínez-Palomo, A., Brailovsky, C. and Bernhard, W. *Cancer Res*. 29 (1969) 925.
7. Bernhard, W. and Avrameas, S. *Exp. Cell Res*. 64 (1971) 232.
8. Nicolson, G.L. *Nature New Biol*. 233 (1971) 244.
9. Martínez-Palomo, A. and Wicker, R. *Proc. Amer. Soc. Cell Biol*. (1971) 179.
10. Martínez-Palomo, A., Wicker, R. and Bernhard, W. *Int. J. Cancer*. 9 (1972) 676.
11. Bretton, R., Wicker, R. and Bernhard, W. *Int. J. Cancer*, 10 (1972) 397.
12. Rowlatt, C., Wicker, R. and Bernhard, W. *Int. J. Cancer*. 11 (1973) 314.
13. Rosenblith, J.Z., Ukena, T.E., Yin, H.H., Berlin, R.D. and Karnovsky, M.J. *Proc. Nat. Acad.*

Sci. USA 70 (1973) 1625.

14. Cline, M.J. and Livingstone, D.C. Nature New Biol. 232 (1971) 155.

15. Ozanne, B. and Sambrook, J. Nature New Biol. 232 (1971) 156.

16. Noonan, K.D. and Burger, M.M. J. Cell Biol. 59 (1973) 134.

17. De Petris, S., Raff, M.C. and Mallucci, L. Nature New Biol. 244 (1973) 275.

18. Martínez-Palomo, A., González-Robles, A. and de la Torre, M. Nature New Biol. 245 (1973)186.

19. Yin, H.H., Ukena, T.K. and Berlin, R.D. Science 178 (1972) 867.

20. Taylor, R.B., Duffus, W.Ph., Raff, M.C. and de Petris, S. Nature New Biol. 233 (1971) 225.

21. Singer, S.J. and Nicolson, G.L. Science 175 (1972) 720.

22. Pinto da Silva, P., Branton, D. and Douglas, S.D. Nature 232 (1971) 194.

23. Tillack, T.W., Scott, R.E. and Marchesi, V.T. J. Exp. Med. 135 (1972) 1209.

24. Pinto da Silva, P., Moss, P.S. and Fudenberg, H.H. Exp. Cell Res. 81 (1973) 127.

25. McIntyre, J.A., Gilula, N.B. and Karnovsky, M.J. J. Cell Biol. 60 (1974) 192.

26. Gasic, G., Berwick, L. and Sorrentino, M. Lab. Invest. 18 (1968) 63.

---

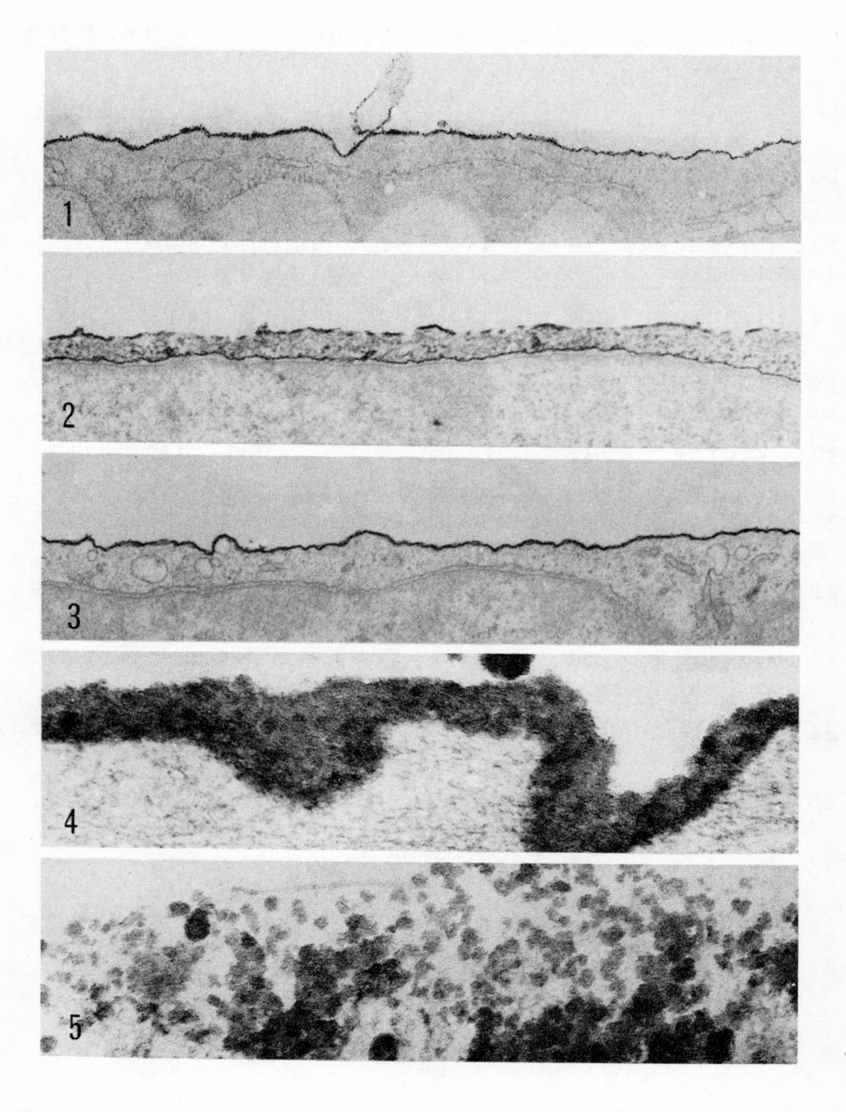


Figure 1. Hamster embryo cell, secondary culture. Vertical section. The culture was reacted with con A, fixed with glutaraldehyde, and treated with peroxidase and benzidine. X 30,000.

Figure 2. Polyoma transformed cell. Vertical section. Clustered distribution of con A receptors present in cultures treated with con A before fixation. X 30,000.

---

Figure 3. Polyoma transformed cell. Vertical section. When transformed cell are fixed with glutaraldehyde before the addition of con A, lectin surface receptors show a uniform distribution similar to that seen in normal cells. X 30,000.

---

Figure 4. Hamster embryo cell, secondary culture. Tangenctial section. The uniform distribution of electron dense precipitate induced by the con A-peroxidase reaction becomes more evident in tangenctial sections. X 100,000.

---

Figure 5. Polyoma transformed cell. A clustered distribution of the label is seen in tangenctial sections of transformed cells, in contrast to the uniform distribution observed in normal cells X 100,000.

---

Figure 6. Freeze-fracture replica of *E. histolytica* cell treated with 25% glycerol before fixation. Face A. Notice the pronounced redistribution of intramembranous particles. X 60,000.

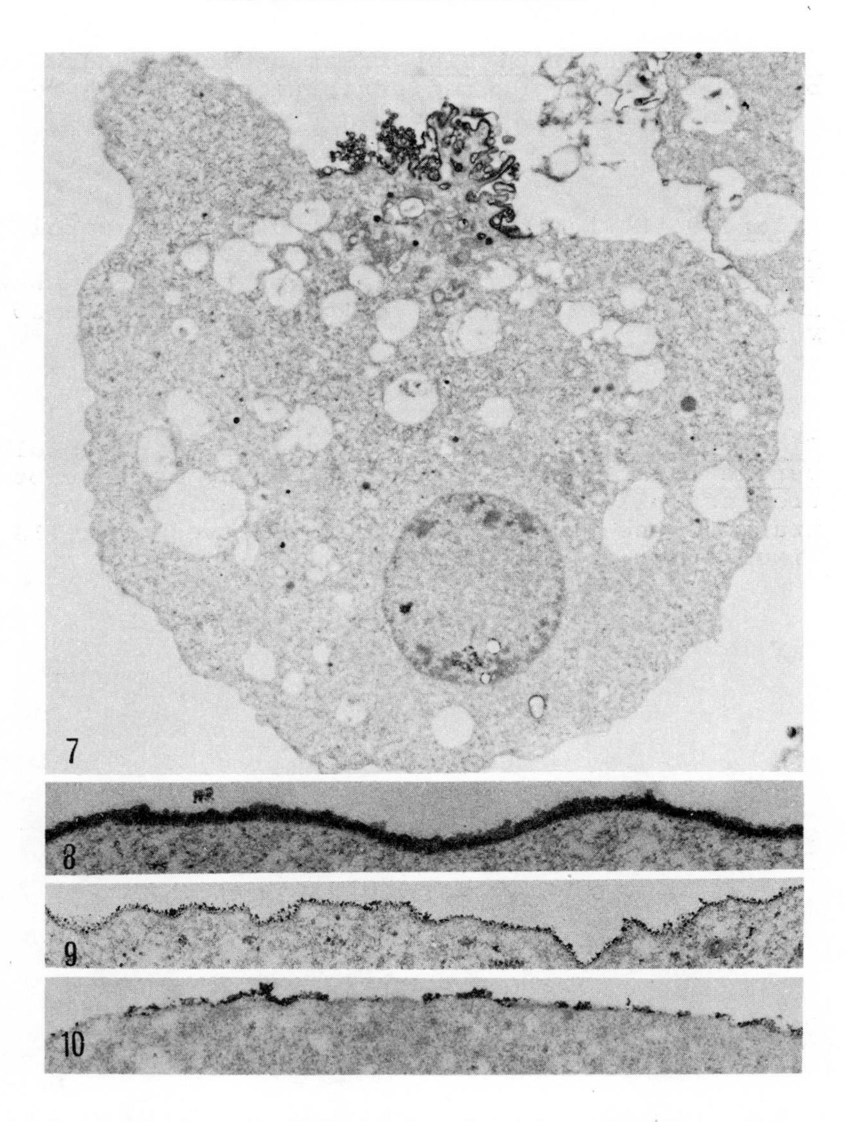


Figure 7. *E. histolytica* cell treated with con A and peroxidase before fixation. Con A receptors sites show a prominent redistribution "capping" at one pole of the cell where the uropod is located. Uncontrasted section. X 9,000.

Figure 8. *E. histolytica* treated with glycerol before fixation. No redistribution of con A receptors is seen. X 30,000.

---

Figure 9. *E. histolytica* treated with glycerol *after* fixation with glutaraldehyde. Negative sites labeled with colloidal iron hydroxide show a uniform distribution. Compare with Figure 10 X 30,000.

---

Figure 10. *E. histolytica* treated with glycerol *before* fixation. A clear-cut patchy redistribution of colloidal iron label is seen, in contrast to the uniform labeling observed in cells fixed prior to glycerol. X 30,000.

---

# THE ISOLATION OF SURFACE COMPONENTS INVOLVED IN SPECIFIC CELL-CELL ADHESION AND CELLULAR RECOGNITION

Max M. Burger

Department of Biochemistry
Biocenter of the University of Basel
Klingelbergstrasse 70
CH 4056 Basel, Switzerland.

Little attention has been given to the biochemistry of cell surface components that are not involved in membrane structure or transport systems, but are involved in recognition or adhesion functions between cells.

Some form of cell-cell interaction is a necessary condition for the formation of the intricate cellular architecture of an organ, as well as for the organisation and interconnection of organs. Such interactions have to occur, since many cells have a carefully defined position during embryonal development, and since they behave according to their immediate environment, if transplanted elsewhere in the embryo (regulative development). Furthermore, close cellular contacts seem to be necessary for the induction of differentiation, in which information as well as a feedback control of information are probably passed back and forth from cell to cell.

If cell-cell recognition is involved in the proper alignment of cells during enbryogenesis, then the following theoretical possibilities exist for the establishment of specific cell-cell contacts (1): For single cells, the attraction stimulus could be chemical and could be transmitted over large distances, a process generally called chemotaxis (2), (see Fig. 1a). Alternatively, the attraction could be directed by a specific surface mechanism involving a guideline of extracellular material, e.g. mucopolysaccharides, collagen (3), elastin, calcium or silica salts (Fig. 1b). Either single whole cells or networks of cells could serve as guidelines (4) for the directed movement of cells to their final destination (Fig. 1c). Finally cells can migrate randomly and subsequently form attachments only with appropriate (recognized) cells, attachments that could entrap a cell in a particular location by virtue of specific surface interactions (Fig. 1d).

The same theoretical possibilities exist for cell clusters, cell sheets and whole organs. In this case, the interacting cell in Fig. 1 can be visualized either as a leading cell followed by many other non-interacting cells or it could be substituting for a group of cells, all of which interact with the neighboring cells. Furthermore, in this situation information could also pass through such groupings of cells, since most cells seem to be electrically coupled after a certain stage in development and can thereafter exchange small molecules, i.e. possible "information" through the same coupling.

Before leaving the above theoretical argument, it should be pointed out that we have not even considered the question of the original spatial distribution of information in the egg and its relevance to development, namely whether the original information for organogenesis and morphogenesis could already be distributed specifically in the egg cytoplasm and cortex, or alternatively this information is successively read out by the DNA during the growth and differentiation of the organism.

Cell-cell interactions have been studied at many levels in recent years. In addition to the immune response (5), which is probably the most popular area of investigation within the field, nerve muscle interactions (6), fertilization (2), nerve regeneration (7) and neuronal development (4,8) have been attracting more and more attention, and in a short time these will be subjected to a multi-disciplinary approach. New techniques developed in all of these areas will certainly benefit both the biochemistry and cell biology of cell-cell interactions during development. The dissociation of organ rudiments with proteolytic enzymes and chelators by the Mosconas (12) has shown that organ-specific reaggregation results when cell suspensions isolated from several different early chick embryo organs are swirled gently. The relevance of such "sorting out" experiments is not yet clear. Since many of these sorting specificities are present only during a certain time span in the developing embryo and not in the adult, they are thought to be involved in the centripetal forces that are perhaps responsible for the formation of the organ in certain instances. They may be required, however, only up to the time when the cells will be fixed together in the proper organ architecture by intercellular bridges, or by a solid capsule surrounding the organ.

In most cases, aggregation in vitro unfortunately occurs in two discrete stages: first, unspecific aggregation ascribed to the damage inflicted on the cell surface while the proteolytic dissociation of the embryonal tissue takes place, and then histiotypic sorting out follows as a second stage. This problem with surface damage was later corrected by a modification of the assay (13). Furthermore aggregation has since been studied in several invertebrate organisms where dissociation is generally less damaging (14-17).

Vertebrate rudiments can also be aligned close to one another to form small balls that subsequently engulf each other and lead to onion-like structures that maintain a constant external-internal relationship. Steinberg's results with such experiments led him to his "differential adhesion" hypothesis which offers a purely physical explanation for *in vitro* cellular arrangements. Although his hypothesis is independent of any consideration of the chemistry of the surface constituents involved, he by no means excluded a biochemical receptor mechanism (11, 18).

Lilien (19) and Garber and Moscona (20) have demonstrated that chick retinal and mouse cerebrum cells can release soluble components which specifically enhance the rate and/or degree of reaggregation of dissociated cells from the tissue of origin. Recently Merell and Glaser (21) have introduced a novel procedure for measuring specific reaggregation. Labeled, partially purified plasma membrane preparations were incubated with homologous cultured cells and the amount of label that could not be washed off under mild rinsing conditions was determined. Since even in this system questions can still be raised concerning the degree of damage inflicted on one of the two parts in the reaggregation system (the membrane), and since in all reaggregation assays used to date one still has to ask whether cells that were left (even under optimal conditions) to repair their surface have really regained the same state of differentiation in vivo as before, we can still expect further improvements in the assays.

## Sponge Reaggregation as a Biochemical Model

Investigations of the reaggregation of sponge cells have several advantages over similar studies on vertebrate cells. First, the sponge tissues can be dissociated simply with calcium-magnesium free seawater and do not require proteolytic enzymes, that unfortunately have to be used for all of the work with vertebrate tissue. This then allows experiments on essentially intact cells. Second, the aggregation factor can be isolated more reliably and in sufficient amounts for large scale biochemical analyses.

Guided by suggestions made by Andrews (22) that a viscous surface substance could enhance the establishment of contacts between "like" cells, Wilson (23) chemically dissociated two different sponge species collected at Woods Hole, mixed them and found that aggregates that subsequently formed only contained cells from a single species. This clever experiment, using color as a marker for the two species, instead of isotopes as one would use today, suggests that two different sponge species must have a way of recognizing their own from different cells - presumably via cell surface components (Fig. 2).

These findings were later confirmed by Galtsoff(24), as well as extended to include the observation that reaggregation is dependent on the presence of calcium. Humphreys (25) and Moscona (26) made good use of this information by developing a gentle dissociation procedure consisting of a calcium-magnesium free seawater treatment which releases a large molecular weight factor from the sponge tissue. This material called aggregation factor (AF) from now on, promoted species-specific aggregation of the dissociated cells (Fig.3).

However, the specificity of the aggregation factor does not seem to be absolute, as was first pointed out by MacLennan and Dodd (27) by showing that a close taxonomic relationship of 3 sponges decreases their specificity barriers, an observation which we have quantitated and confirmed (Turner and Burger, 28). Taxonomic considerations were also brought forward in recent autoradiographic analyses of reaggregated, mechanically dissociated cells (29). Since sponge taxonomy is far from being well established, and furthermore since the technique for determining specificity may have to be reevaluated (30), the details of such questions remain to be resolved.

The chemical nature of the species-specific aggregation factor was first analyzed by indirect procedures as for example with the help of degradative enzymes (10), as well as by other procedures (31), and it was found to be a proteoglycan. A very careful and detailed analysis of a specific Microciona (M. parthena) aggregation factor comes to the conclusion that it is a large heterodisperse 70 S complex with a molecular weight averaging $2 \times 10^6$ Daltons (32). After removal of $Ca^{++}$ and irreversible dissociation, 9-10 S subunits with a molecular weight of about $2 \times 10^5$Daltons could be isolated (32) that seem to have about the composition of the native aggregation factor. 47% consisted of protein and 49% of carbohydrate, a fifth of which was uronic acid (33). Microciona prolifera, the factor used in our study, similarly had a ratio of protein to carbohydrates of 1:1 and equally contained some uronic acid although somewhat less (10% of the total carbohydrate, Turner and Burger, unpublished).

## Indication for the Involvement of Carbohydrate in a Species-Specific Sponge Aggregation Process

If carbohydrates are involved in the specific reaggregation process in a manner analogous to an antigen in an antibody-antigen reaction, then one could expect that certain carbohydrates and their derivatives should be able to inhibit reaggregation in the manner classical haptens do.

A few years ago we found (34) that the reaggregation of Microciona prolifera cells induced by the homologous aggregation factor could be inhibited with glucuronic acid, as well as by cellobiuronic acid, a disaccharide derivative of glucuronic acid. This inhibition turned out to be quite specific, insofar as none of the other available sponges could be inhibited by this particular sugar, and since no other carbohydrate displayed any inhibitory activity towards this particular sponge. Since $Ca^{++}$ was important for the stability of the aggregation factor and since the aggregation did not occur in the absence of $Ca^{++}$ the criticism could be raised that glucuronic acid might simply sequester $Ca^{++}$ due to its ability to chelate $Ca^{++}$. Such an artefact can be ruled out in view of the fact that galacturonic acid does not inhibit reaggregation at all, although it binds $Ca^{++}$ at least as well, if not better than glucuronic acid. Furthermore the addition of excess $Ca^{++}$ did not overcome the inhibition by glucuronic acid (28).

Based on Gasic and Galanti's (35) observations that disulfide groups in the aggregation factor are crucial and that the factor loses its activity when incubated with proteases, it was generally assumed that factor activity resides in the proteinaceous moiety, although the presence of the carbohydrate moiety of the aggregation factor has never been subject to doubt. In view of the carbohydrate inhibition data cited above, however, we interpreted these observations differently. Agents that destroy the protein portion of the aggregation factor might, in many cases, dissociate the protein core of the carbohydrate carrying subunits, thereby leading to the formation of monovalent or non-functional oligovalent pieces of aggregation factor. Small amounts of such pieces would not only have lost their aggregation capability for cells but they could impede aggregation by binding as "monovalent" pieces to their respective binding sites on the cells, thereby preventing the remaining active aggregation factor molecules from binding to and aggregating the cells.

This then led us to the concept (1) that the cell surface might carry a receptor molecule that probably is a protein capable of recognizing carbohydrates on the aggregation factor which have antigenic, rather than antibody-like function, as was generally assumed previously.

Evidence that _Microciona prolifera_ aggregation factor carries the antigenic carbohydrate essential for the recognition process can be summarized by the following (28, 34):

1. A _Helix pomatia_ mixture of glycosidases containing, among others, glucuronidase is capable of destroying aggregation factor activity after preincubation at 37°, but not at 0°. This inactivation can be prevented with high concentrations of glucuronic, but not galacturonic acid. Such product inhibition of glycosidases is well known, and is spread quite widely within this class of enzymes. Pure β-glucuronidase preparations do not inactivate aggregation factor, an observation which is interpreted to mean, among other possibilities, that glucuronic acid can be

protected by some other sugars at the end of the chain and that these groups would first have to be removed before glucuronic acid can be removed.

2. Preincubation of glucuronic acid with the dissociated cells is much more efficient in preventing aggregation than is preincubation of the aggregation factor with glucuronic acid.

3. Preliminary studies by Kuhns (36,31) with aggregation factors isolated from several different sponge species indicates that the aggregation factor from Microciona prolifera can be inhibited by several plant lectins specific for more than one type of sugar. This suggests that those particular carbohydrates with which the lectin interacts, or other sugars in close proximity, might be involved in the aggregation process. Since several other aggregation factors either cannot be inactivated or can be inactivated poorly, the question arises as to whether the appropriate lectins for the functional carbohydrates of those factors have not yet been found, or whether these aggregation factors, in fact, operate via a different type of aggregation mechanism.

4. Preliminary and unpublished studies by Turner have shown that oligosaccharide pieces can be isolated from the Microciona prolifera aggregation factor that turn out to be powerful inhibitors of Microciona prolifera reaggregation.

The model developed therefore predicts the presence of a carbohydrate-recognizing component of the Microciona aggregation factor. After the execution of a number of procedures that are known to release surface components, the treatment of $Ca^{++}$-$Mg^{++}$ free seawater-dissociated sponge cells with hypotonic NaCl removes some components and makes the cells refractory to the subsequent addition of aggregation factor (Fig. 4). When the supernatant of that hypotonic treatment was preincubated with such cells with a defective surface, and the cells were then rinsed, their sensitivity to the addition of aggregation factor was restored (38). The technique for the release of the surface material, which for the time being we will call "baseplate", still requires several standardizations for each batch of cells and needs improvement. To date, we have very little data on the biochemical characterization of the baseplate, since its functional aspects will have to be worked out first. However, most tests previously carried out confirm the notion that the hypotonic supernatant containing the baseplate has the predicted characteristics (see Table I).

As is the case for most tests on whole and intact cells, they unfortunately include a number of uncertainties. Cell-free model systems can avoid many of them. With that aim, we insolubilized the proteins from the hypotonic shock medium by covalently coupling them to solid beads (Fig. 5 and ref. 38). Such beads aggregate as soon as aggregation factor from the homologous cell is added (Table II). They also attach to mechanically disrupted cells, that still carry aggregation factor, but not, however, to $Ca^{++}$-$Mg^{++}$ free seawater treated cells that therefore were free of functioning aggregation factor. The aggregation of baseplate beads induced by aggregation factor can be inhibited with glucuronic acid (Fig. 5 and Table II). Inhibition was primarily seen when beads were preincubated with glucuronic acid and not when the aggregation factor alone was preincubated (see Table II). Such experiments promise a quantifiable study of surface interactions between cells using in vitro systems. It is unlikely that all types of cell surface interactions can be faithfully mimicked with such coated bead systems, because many properties of the cell surface membrane are not included in the bead. For example, the lipophilic core structure of the membrane, furthermore membrane fluidity, and in particular, the potential mobility of a glucoprotein within the natural membrane, are quite obviously absent when membrane components are covalently attached to a rigid bead, and consequently such properties cannot be studied in this system.

Figure 6a summarizes our present working model for the reaggregation of Microciona prolifera cells. For other cell pairs, as well as for organ-specific cell-cell recognition in vertebrate organisms, we still have to consider inverted polarities where the recognizing moiety lies between the cells and the recognized site of the surface (6b), and particularly a model where each surface carriers male and female parts, i.e. the antibody-like groove that defines the steric or chemical specificity of the antigen on the same surface complex (6c).

Although sponge reaggregation, the biological function of which incidently is not clear so far, may turn out to be a very fruitful field of study to elaborate several different molecular mechanisms for cell-cell recognition as a general phenomenon, we have to be aware that vertebrate cell-cell interaction may be restricted to only one of those found in invertebrates or may have evolved their own type.

Some of the basic questions in the field of morphogenetic migration and reshaping during embryogenesis are still the following: do they take place and are they guided and controlled entirely from within the cell or are they also guided by cell-cell communications? If they are controlled by cell-cell communications, are those provided for by intercytoplasmatic bridges, as gap junction for instance, or are chemical surface components involved in recognition as well as in guidance? The next questions, which we have to ask, are more familiar to the developmental biologist and molecular biologist: how does the information for the distribution of such surface recognition sites over the early embryo take place, how is the formation of such cell-cell recognition systems turned on, and how is it turned off in late embryo and adult cells? Finally, we would like to know how the proper cell distribution is fixed and stabilized. It would be unlikely, if within this catalogue of questions cell surfaces would not play an important role somewhere, and provide the membrane biochemist some formidable conceptual and technical problems.

## Acknowledgements

Almost all of the work was carried out by the author, Drs. R.S. Turner, G. Weinbaum and W.J. Kuhns, as well as Mr. S.M. Lemon and R. Radius, over the last 8 years at the Marine Biological Laboratory at Woods Hole, Mass. I thank Miss M.E. Hatten, Dr. R. Mannino and Dr. R.S. Turner for reading the manuscript and the Swiss National Foundation for support.

## References

1. Burger, M.M. 1974. In: The Neurosciences: 3rd study program, pp. 773, ed. Schmitt, F.O. Cambridge: MIT Press.
2. Metz, C. and Monroy, A. 1969. In: Fertilization. New York: Academic Press.
3. Hauschka, S.D. and Konigsberg, I.R. Proc. Nat. Acad. Sci. USA 55 (1966) 119.
4. Sidman, R.L. 1971. In: Third Lepetit Colloquium on Cell Interactions. pp. 1, ed. Silvestri, L. Amsterdam: North Holland.
5. Mitchison, N.A. Immunpathology 6 (1971) 52.
6. Fischbach, G.D. Develop. Biol. 28 (1972) 407.
7. Hunt, R.K. and Jacobson, M. Proc. Nat. Acad. Sci. U.S.A. 69 (1972) 2860.
8. Weiss, P.A. 1970. In: The Neurosciences: 2nd study program. pp. 1068, ed. Schmitt, F.O. New York. Rockefeller Univ. Press.
9. Roseman, S. Chem. Phys. Lipids. 5 (1970) 270.
10. Moscona, A.A. Develop. Biol. 18 (1968) 250.
11. Steinberg, M.S. J. Exp. Zool. 173 (1970) 395.
12. Moscona, A.A. and Moscona, M.H. J. Anat. 86 (1952) 287.
13. Roth, S.A. and Weston, J.A. Proc. Nat. Acad. Sci. U.S.A. 58 (1967) 974.
14. Sconzo, G., Pirrone, A.M., Mutolo, V. and Giudice, G. Biochim. Biophys. Acta. 199 (1970) 441.

15. Hynes, R.O., Raff, R.A. and Grass, P.R. Develop. Biol. 27 (1972) 150.

16. Gierer, A., Berking, S., Bode, J., David, C. N., Flick, K., Hansmann, G., Schaller, H., Trenkner, E. Nature New Biol. 239 (1972)98.

17. Antley, R.M. and Fox, A.J. Neurosci. Res. Progr. Bull. 10(3)(1972) 304.

18. Wiseman, L.S., Steinberg, M.L. and Phillips, H.M. Develop. Biol. 28 (1972) 498.

19. Lilien, J.E. 1969. In: Current Topics in Devel. Biol. Vol. 4, pp. 169, eds. Moscona, A. A. and Monroy, A., New York: Academic Press.

20. Garber, B.B. and Moscona, A.A. Develop. Biol. 27 (1972) 235.

21. Merrell, R. and Glaser, L. Proc. Nat. Acad. Sci. U.S.A. 70 (1973) 2794.

22. Andrews, G.F. J. Morphology 12 (1897) 367 and suppl. 12 (1897) 1.

23. Wilson, H.V. J. Exp. Zool. 5 (1907)245

24. Galtsoff, P.S. J. Exp. Zool. 42(1925)223.

25. Humphreys, T. Develop. Biol. 8 (1963)27.

26. Moscona, A.A. Proc. Nat. Acad. Sci. U.S.A. 49 (1963) 142.

27. MacLennan, A.P. and Dodd, R.Y. J. Embryol. Exp. Morphol. 17 (1967) 473.

28. Turner, R.S. and Burger, M.M. Nature New Biol. 244 (1973) 509.

29. Mc Clay, D.R. Biol. Bull. 141 (1971) 319.

30. Curtis, A.S.G. and Vyver, G. van de. J. Embryol. Exp. Morph. 26 (1971) 295.

31. Margoliash, E., Schenck, J.R., Hargie, M.P., Burokas, S., Richter, W.R., Barlow, G.H. and Moscona, A.A. Biochem. Biophys. Res. Commun. 20 (1965) 383.

32. Cauldwell, C.B., Henkart, P. and Humphreys, T. Biochemistry. 12 (1973) 3051.

33. Henkart, P., Humphreys, S. and Humphreys, T. Biochemistry. 12 (1973) 3045.

34. Lemon, S.M., Radius, R. and Burger, M.M. Biol.. Bull. 141 (1971) 380.

35. Gasic, G.J. and Galanti, N.L. Science 151 (1966) 203.

36. Kuhns, W.J. and Burger, M.M. Biol. Bull. 141 (1971) 393.

37. Kuhns, W.J., Weinbaum, G., Turner, R.S. and Burger, M.M. Humoral Control of Growth and Differentiation, eds. LoBue, J. and Gordon, A.S. New York: Academic Press. In press.

38. Weinbaum, G. and Burger, M.M. Nature 244 (1973) 510.


---

TABLE I

Whole Cell Aggregation Studies

| Expt. No. | Cell Type Used | Aggreg. Factor Added | Baseplate Added | Glucuronic Acid | Degree of Aggregation |
|---|---|---|---|---|---|
| 1 | CMF | - | - | - | 0 |
| 2 | CMF | + | - | - | 4+ |
| 3 | CMF | Second* | - | First* | 0 |
| 4 | CMF + HY | + | - | - | 0 |
| 5 | CMF + HY | - | + | - | 0 |
| 6 | CMF + HY | First** | Second** | - | 0 |
| 7 | CMF + HY | Second** | First** | - | 3+ |

Where present 0.05 ml Microciona prolifera aggregation factor, 0.3 ml baseplate (about 700 μg protein per ml), $5 \times 10^{-2}$M neutralized glucuronic acid were used in a final volume of 2 ml seawater. The reaction was started by the addition of another ml containing $5 \times 10^{7}$ cells/ml Microciona prolifera cells. CMF: calcium magnesium free seawater dissociated M. prolifera cells. CMF + HY: the same cells after an additional treatment with a hypotonic NaCl solution. 15 minutes after mixing the cells at room temperature the samples were scored according to an earlier description (ref. Turner and Burger, Nature).

* First reagent was in contact with the cells for 5 minutes and then only was the reaction started without rinsing in between.

**After preincubation of the cells with the first agent the cells were spun and resuspended in the second reagent.

Data partially from Weinbaum and Burger, 1973.

TABLE II

Bead Aggregation Studies with Immobilized Aggregation Factor and Baseplate

| Expt. No. | Bead Type | Aggreg. Factor Added | Baseplate Added | Glucuronic Acid | Degree of Aggregation |
|---|---|---|---|---|---|
| 1 | AF | - | - | - | 3+ |
| 2 | AF | - | - | + | 3+ |
| 3 | BP | - | - | - | 0 |
| 4 | BP | - | - | - | 4+ |
| 5 | BP | P r e i n c u b a t e d | | - | 0 |
| 6 | BP | Second* | - | First* | 0 |
| 7 | BP | First* | - | Second* | 4+ |
| 8 | AF+BP | - | - | - | 4+ |

0.05 ml of a suspension of beads substituted with aggregation factor (AF) or baseplate (BP) was added to the 2 ml calcium magnesium free seawater. Aggregation factor, baseplate and glucuronic acid solution as in Table I.

* First reagent was in contact with beads for 2 minutes. Only then was the second added and the degree of aggregation of the beads occurred almost immediately after additon of the calcium containing seawater and the reactions were scored as described earlier. Data partially from Weinbaum and Burger, 1973.

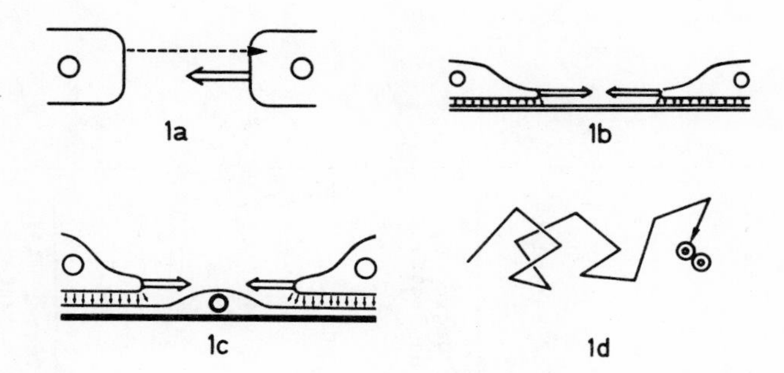


Figure 1. Mechanisms by which specific cell-cell contacts can be established. (1a) Chemotaxis. (1b) Attraction mediated and guided by an extracellular matrix or (1c) by another cell. (1d) Random movement and entrapment due to specific cell-cell interaction (from Burger, 1974).

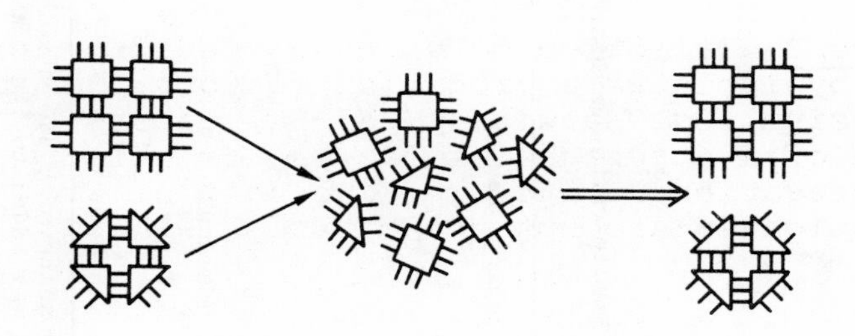

Figure 2. Specific sorting out of sponge cells. When two unrelated species of sponge tissue are mechanically dissociated into single cell suspensions and mixed, they will in due time sort out into two separate sponge cell clumps of the original species. The rectangular species can be e.g. Microciona prolifera and the triangular species Haliclona occulata while the spikes represent the hypothetical surface material necessary for specific reaggregation (from Burger, 1974).

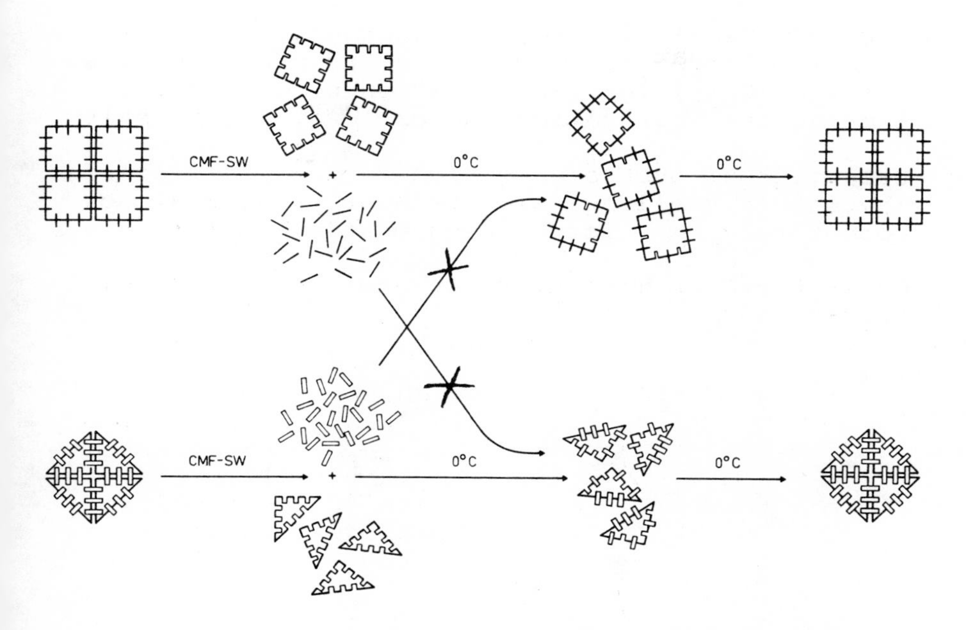


Figure 3. Promotion of species specific aggregation by isolated aggregation factors. Calcium-magnesium free seawater (CMF-SW) causes the release of a species-specific factor which promotes reaggregation of its own kind only if the species investigated are not too closely related.

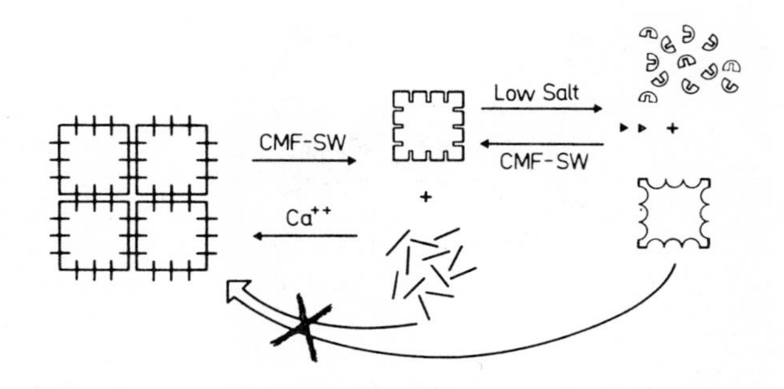


Figure 4. Requirement of a membrane-bound baseplate for the specific sorting-out process. Sponge cells dissociated in $Ca^{++}$-$Mg^{++}$ free sea-

water (CMF-SW) will release aggregation factor (\⌐/\/ \//) that is required in the presence of calcium ($Ca^{++}$) for reaggregation of the cells lacking the aggregation factor . Low salt treatment of such aggregation factor- less cells (hypotonic swelling) will release a baseplate or receptor for the aggregation factor. Cells that are missing baseplate (From Burger, 1974).

---

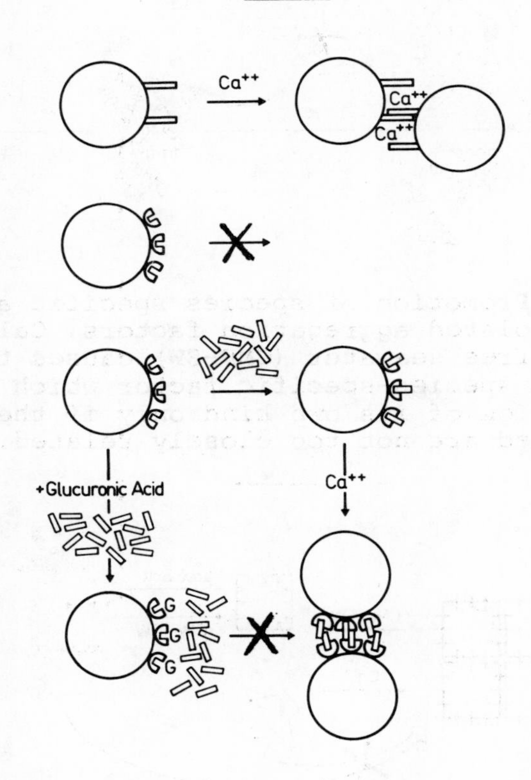


Figure 5. Aggregation of factor or baseplate substituted beads. On the first line beads which carry the aggregation factor (bars) are shown to aggregate in the presence of calcium ($Ca^{++}$). This unspecific reaggregation could not be inhibited

with glucuronic acid. Beads substituted with base-plate (kidney-shaped receptors with groove for aggregation factor) do not aggregate. If, however, aggregation factor (bars) is presented (third line) and then calcium is added, aggregation takes place (third to fourth line on the right). This aggregation could be inhibited by glucuronic acid only (third to fourth line on the left and bottom).

---

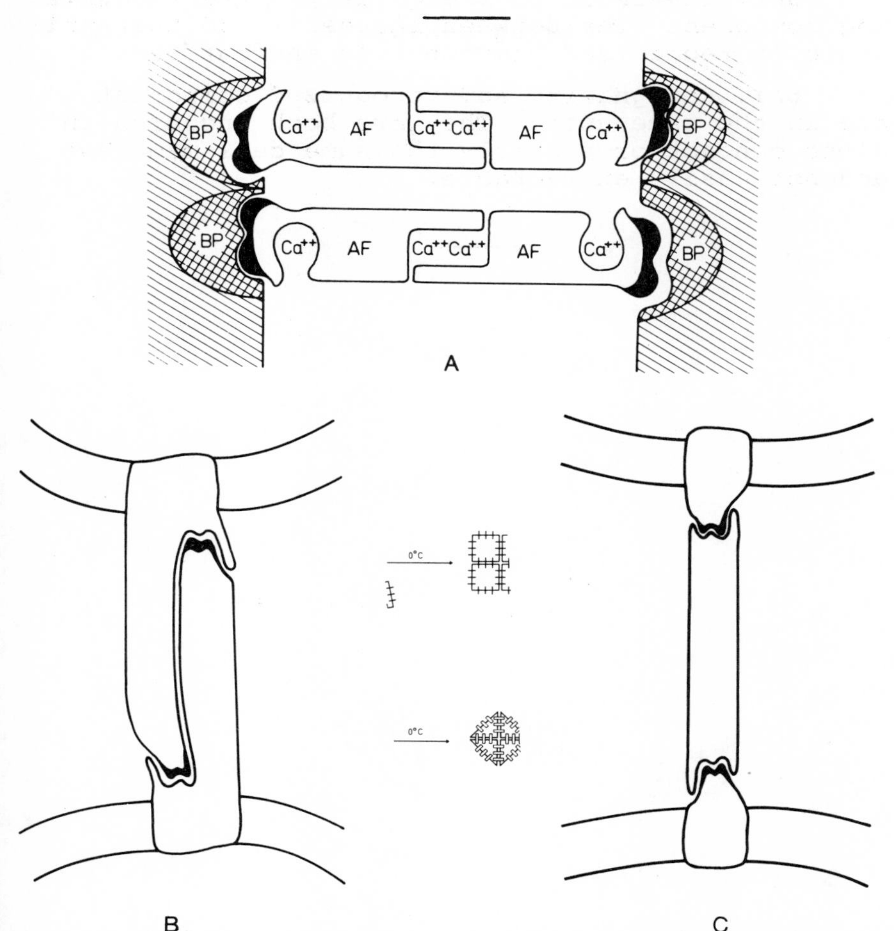


Figure 6. Three possible molecular mechanisms for surface specific cell-cell recognition.

6a). Tentative model for the recognition observed in the sponge *Microciona prolifera*. Two macromolecular aggregation factors (AF) are illustrated each consisting of at least two subunits. The black termini at each pole carry the carbohydrates that are recognized by the baseplate (BP) anchored in adjoining cell surfaces (from Weinbaum and Burger, 1973).

6b). Inversion of the polarity: the recognizing component lies between the cells and the antigenic or recognized component in the surface.

6c). Recognizing and recognized component are in the same unit. They are both anchored in close contact or possibly the same cell membrane and not easily extractable.

---

# VI

# MEMBRANE AND THE IMMUNOLOGICAL RESPONSE

# LYMPHOCYTE MEMBRANES IN LYMPHOCYTE FUNCTIONS

Fritz Melchers

Basel Institute for Immunology
Grenzacherstrasse 487
CH-4058 Basel,
Switzerland

Lymphocytes originate from the bone marrow. During differentiation they arrive in the secon dary lymphoid organs (e.g. the spleen and lymph nodes) either directly (bone-marrow-derived, B-lymphocytes) or by way of the thymus (thymus-derived, T-lymphocytes). In the secondary organs these lymphocytes rest prior to the invasion of antigen. They rest in the $G_o$-phase of the cell cycle in which they do not synthesize DNA and do not divide. Lymphocytes exist as single, motile cells in all these organs, some of them even circulate through the body. This makes it possible to correlate the quantity of a measured effect to single cells.

Lymphocytes can be stimulated to synthesize DNA, to divide and to secrete immunoglobulins (Ig). Antigens do so, since they are recognized as "foreign" structures by receptor molecules on lymphocytes which are believed to be Ig molecules. It is one of the distinct advantages of lymphocytes that the structure of the Ig molecule is known and well understood. Ig receptors on lymphocytes possess a wide variety of specificities to a wide variety of different "foreign" structures, antigens. One lymphocyte, however, possesses only one type of Ig molecule with a given recognizing capacity for an antigen. Thus, the binding of different antigen stimulates different recognizing cells. One antigen stimulates only a very small portion of all lymphocytes in the body (~0.02%). Biochemical studies on changes occurring in lymphocytes after antigen stimulation are therefore very difficult, if not impossible to conduct with a normal lymphocyte population.

Stimulation of lymphocytes by antigen is furthermore complicated by the involvement of more than one type of cell in this triggering process. Antigen stimulates T- and B-lymphocytes to DNA synthesis and proliferation. It has been shown, however, that only B-lymphocytes synthesize and secrete Ig, i.e. differentiate into the Ig-secreting function needed for the so-called humoral antibody response. T-cells help the B-cells to do so. If T-lymphocytes are missing in the body (thymectomy, genetic defect) the humoral antibody response is generally abolished. It is not clear at present exactly how T-cells cooperate with B-cells. Still another type of cell appears needed to bring about the response of a B-lymphocyte to an antigen: adherent cells (A-cells) or macrophages, at least "in vitro", help.*

From this, lymphocytes appear interesting because of their Ig receptors having such a wide range of recognizing capacities for different antigenic structures, and because of their cell-cell cooperations in the regulation of their antigen-stimulated growth and differentiation. For a biochemist searching for structural changes measurable by his techniques in a homogeneous population of cells or molecules, lymphocytes are homogeneous neither as cells nor in the antigen-recognizing Ig-molecules mediating the stimulatory signal to the cell.

---

* A vast literature exists to substantiate the aforementioned picture of the diversity of antigen recognition and the regulation of the response of lymphocytes. The reader is referred to text books and review articles on general immunology.

In this situation it could be considered a break-through that mitogens were found which could be classified as reacting either specifically with T-cells or specifically with B-cells. Lipopolysaccharides (LPS), the purified protein derivative of tuberculin (PPD), certain components in fetal calf serum, dextransulfate, poly AU and other polyanions are some of these mitogens which specifically stimulate B-cells (1). These substances stimulate a large part of lymphocytes to synthesize DNA and to divide. B-cell-specific mitogens also induce the differentiation to Ig-secreting cells. With mitogens we can study, with biochemical methods, changes in lymphocytes which occur after stimulation.

I will describe reactions of B-cells only. B-lymphocytes from secondary lymphoid organs can be obtained devoid of T-cells and A-cells from genetically athymic ("nude") mice by a sequence of purification steps involving the removal of A-cells through adherence to plastic surfaces and to iron fillings, followed by enrichment of surface membrane Ig-positive B-cells in free-flow electrophoresis. These B-lymphocytes can finally be enriched for small, resting cells from a minor (~1%) contamination of large, actively Ig-secreting cells by sedimentation over Ficoll gradients under earth's gravity. These small purified B-lymphocytes can be cultured in serum-free medium and stimulated by B-cell mitogens to synthesize DNA, divide and differentiate into Ig-secreting cells.

Structure, synthesis and turnover of Ig are taken as parameters to characterize the functional state of the B-lymphocyte. Resting small lymphocytes before stimulation will be described first. In these resting cells Ig is deposited in the surface membrane where it is thought to serve as a receptor for antigen. Then the molecular principle of action of two B-cell mitogens, lipopolysaccharide (LPS) and locally concentrated Concanavalin A (Con A), is described. Changes in structure, synthesis and turnover of Ig is followed with time of mitogenic stimulation of B-lymphocytes. Early changes are distinguished from late changes. After long-term mitogenic stimula tion actively Ig-secreting plasma B-cells develop, in which the mechanism of active secretion involving the membranous components of the endoplasmic reticulum can be studied. The actively secreted Ig is no longer inserted into the surface membrane. Ig therefore changes its amphipathic nature (2) during stimulation and differentiation to active secretion. Finally, it is shown that stimulation of B-lymphocytes may balance between reactions leading to proliferation and reactions leading to differentiation. Manipulation of B-lymphocytes by "factors", some of them interacting on the surface membrane with the cell, will unbalance them to only proliferate or only mature to Ig-secreting cells.

Much is done to clarify the functions of lymphocytes. Little, however, has been accomplished yet to correlate these functions to structures of lymphocyte membranes. I will emphasize those functions of lymphocytes which can be expected to be "membrane-bound".

## Resting B-lymphocytes Before Stimulation

*Synthesis of Ig.* One small resting B-lymphocyte has been estimated to contain 3-15x$10^4$ Ig molecules bound in the surface membrane (3). We follow synthesis of Ig in small B-cells and its release from the cells by the incorporation of radioactive leucine during 4 hours into material which is precipitable by Ig-specific antisera in cellular lysates and extracellular fluids (4,5,6). Synthesis of radioactive Ig continuously increases during the 4-hour-period of labelling. Ig does not reach a constant specific activity inside the cells. This indicates that the intracellular pool of Ig molecules has not been equilibriated with radioactive molecules during these 4 hours.

Small B-cells from spleens of "nude" mice synthesize IgM, but not IgG or IgA. Thus, only Hμ-chain-specific antisera, but not Hγ- or Hα-chain-specific antisera precipitate radioactive material from cells labelled for 4 hours with radioactive leucine. Reduced and alkylated radioactive polypeptide chains contained in the Hμ-chain-specific serological precipitates show mobilities in SDS-urea-polyacrylamide gels coinciding with those of reference Hμ- and L-chains (Melchers and Andersson, 1973).

Small $G_0$-lymphocytes, up to 4 hours of labelling, keep over 90% of their radioactive IgM bound to the cells. Active secretion of IgM is absent. Biosynthetic incorporation of radioactive leucine into B-cells labels many proteins. IgM represents only 1-3% of the total incorporated radioactivity.

We can estimate how many IgM molecules have been made in 4 hours within one quiescent B-cell. For this estimation it is assumed that all cells have the same synthetic activity and the specific radioactivity of each of the estimated 120 leucine-residues in 7-8 S IgM (see below) synthesized during the 4 hours equals the specific radioactivity of ($^3$H) leucine in the incubation medium (38 Ci/mM) (7,8). From the amount of radioactivity incorporated into IgM within 4 hours, we can calculate that one small B-cell has synthesized between 1000 and 2000 7-8 S IgM molecules in that time. It will therefore take a small B-cell 40 to 200 hours to synthesize half of the estimated $3\text{-}15 \times 10^4$ receptor IgM molecules (3) found on the surface of that cell. The estimated 1000-2000 IgM molecules represent the lower limit for the number of molecules which one small B-cell may synthesize within 4 hours, since the specific activity of the cellular leucine pool is unknown and since not "all nude" spleen lymphocytes may synthesize Ig.

*Turnover of IgM.* For turnover measurements small B-lymphocytes are labelled for 4 hours with radioactive leucine, then transferred into non-radioactive chase medium. Turnover of IgM is then monitored by the disappearance of leucine-labelled IgM from the cells and by the appearance of this radioactive IgM in the supernatant nonradioactive chase medium. The time of median disappearance of radioactive IgM from the small cells and of medium appearance in the supernatant medium varies between 20 and 80 hours in different experiments. Other laboratories have found much shorter turnover times for surface-located Ig (9,10,11). This discrepancy will be dealt with below.

The time of median disappearance of IgM from small B-cells, 20 to 80 hours, agrees reasonably well with the time it takes to synthesize half of the 3-15x$10^4$ surface receptor Ig molecules on a small cell (calculated above). Turnover of IgM molecules in resting $G_o$-lymphocytes constitutes shedding from the cells into the supernatant medium. IgM molecules are not taken back inside the cells and are intact in their polypeptide structure, i.e. not degraded by proteases, when they appear in the supernatant medium (see below for the size of the molecules). It is tempting to speculate that different membrane proteins may turn over differently. Some, such as H-2 proteins (2) are degraded, others such as Ig, are shed from the cell surface. 10% of the radioactive proteins shed from B-cells are IgM. Other membrane proteins appear therefore shed from B-lymphocytes. Synthesis and shedding of membrane-bound IgM exemplifies the dynamic state of the lymphocyte membrane in which components of the membrane are continually replaced as they decay by degradation or shedding.

*Stability of IgM synthesis in the presence of Actinomycin D.* Actinomycin D selectively inhibits DNA-dependent RNA synthesis. Actinomycin D is expected not to influence the capacity of small cells to synthesize IgM, if all of the factors involved in synthesis of IgM, including the mRNAs for Hμ-chain and L-chain are stable. Actinomycin D is expected to decrease the capacity to synthesize IgM if any one of the factors involved in IgM synthesis is unstable in the presence of the inhibitor.

It was found (6) that synthesis of IgM in small B-cells is unstable in the presence of Actinomycin D. In fact, IgM synthesis decays even more rapidly than the sum of syntheses of all proteins in the cell, evident in the drop of the values for ratios of synthetic rates of IgM over those of total protein. These experiments therefore do not permit conclusions about the life span of mRNA for Hμ-chain in small cells. We conclude that continued DNA-dependent RNA synthesis is requisite in small, resting B-lymphocytes for the maintainance of IgM synthesis. Lerner et al. (12) have reached the opposite conclusion for the stability of the mRNAs for Hμ- and L-chains in unstimulated lymphocytes. I will deal with this discrepancy below.

*Size of IgM*. Resting $G_o$-lymphocytes contain over 90% of their radioactive IgM, labelled within 4 hours, as 7-8 S subunits ($H\mu_2$-$L_2$) or as Hμ-L precursor material. These polypeptide chains are all associated and linked by disulfide bridges.

Radioactive IgM shed into the supernatant medium from cells labelled for 4 hours, then chased in nonradioactive medium for 6 hours, is to 80% 7-8 S subunits and to 10% Hμ-L precursor material. Again, the polypeptide chains are not only associated, but also linked by disulfide bridges (Melchers and Andersson, 1973; and in preparation). These size analyses are in agreement with results obtained by Marchalonis, Cone and Atwell (11) and by Vitetta and Uhr (10).

Our analyses indicate no difference in the size of Hμ- and L-chain (other than that attributable to carbohydrate differences - see below) between intracellular, surface-membrane-bound, shed and secreted IgM. If, therefore, hydrophobic parts of the polypeptide chains of IgM should exist which hold the receptor IgM molecule in the membrane and which may be cleaved off when IgM leaves the cell such hydrophobic parts cannot account for more than 10% of the molecular weights of the chains. Since quite small hydrophobic pieces have been found anchoring membrane proteins in the lipid bilayer (2), it remains to be investigated with more sensitive methods, whether surface membrane-bound IgM contains additional hydrophobic polypeptides which are absent in secreted IgM.

*Carbohydrate composition of IgM.* Radiochemical analysis of IgM molecules labelled by radioactive precursors of carbohydrate moieties such as those found in the glycoprotein IgM (13) permit the qualitative analysis of the presence or absence of certain sugar residues in IgM (4,7,8, 14).

We find that radioactive mannose labels both intracellular and shed IgM of resting B-lymphocytes. Radioactive fucose and galactose do not however label either the intracellular nor shed IgM molecules during a 4-hour-labelling period (6). From such results we conclude in analogy to earlier reports with other types of B cells (quoted above) that IgM inside of and shed from small B-lymphocytes contains the core sugars mannose and glucosamine, but not the penultimate galactose and the terminal fucose residues common to branched carbohydrate moieties of the 19 S form of IgM-actively secreted by plasma cell-like B-cells (see below). The Hμ chains, but not the L-chains, of subunit IgM carry the carbohydrate moieties which are labelled by radioactive mannose.

The roles of carbohydrate groups in membrane glycoproteins is generally poorly understood. Carbohydrate groups may determine the polarity and position of glycoproteins in lipid bilayers of membranes (2), carbohydrate groups containing different sugars ("core" vs. "branch" sugars, charged vs. noncharged sugars, etc.) and attached at different sites within the polypeptide chains may fix different conformations of the polypeptides and thereby determine the interactions with other components of the membrane. They may also interact directly with glycolipids in the membrane (2). Immunoglobulins in lymphocyte surface membranes may be a powerful model system to study the role of a membrane glycoprotein in defined cell functions.

*Quantitation of surface-bound IgM molecules.* Two methods are used to quantitate surface-bound immunoglobulin. Both employ the binding of antibodies with specificities for mouse Ig to viable B-lymphocytes.

In method 1, $I^{125}$-labelled Fab fragments with specificities for mouse Ig are bound in saturating concentrations to B-cells. With the known specific radioactivity of the $I^{125}$-Fab-preparation we can quantitate that between 5 and $15 \times 10^5$ molecules of Fab are bound to one B-cell.

In method 2 (6) surface-bound Ig is complexed on intact cells with whole rabbit antibodies having specificities for mouse Ig. Prior to the binding of the (anti-mouse Ig) antibodies the distribution of Ig molecules on the surface membrane of resting B-lymphocytes is diffuse (15,16). Immunofluorescent studies have shown that interaction with polyvalent (anti-Ig) antibodies induces aggregation of these Ig molecules first into small aggregates ("spots"), then into larger aggregates ("caps") found over the uropod of the lymphocyte. Aggregation, in analogy to the prozone effect in immunoprecipitation, is inhibited at increasing concentrations of (anti-Ig) antibodies. This suggests that lattice formation, such as in solution, takes place in the surface membrane between the Ig molecules and the polyvalent (anti-Ig) antibodies. Diffusion of Ig molecules in the surface membrane is compatible with the view that the surface membrane of animal cells is quasi-fluid (17,18).

We use this precipitation reaction in the surface membrane between mouse Ig molecules and (anti-mouse Ig) antibodies to distinguish surface IgM from labelled intracellular IgM in B-lymphocytes. For this we label B-cells through biosynthetic incorporation of radioactive leucine. (Anti-Ig) antibodies will bind to such labelled cells and precipitate only those labelled IgM molecules which face the outside of intact cells within the surface membrane. They will not be able to reach IgM molecules located in intracellular compartments.

On small B-cells which are labelled for 4

hours with radioactive leucine 45% of all cellular radioactive IgM can be precipitated by the (anti-mouse Ig) antibodies. During the first 6 hours after the pulse of radioactive leucine in nonradioactive chase medium most of the labelled IgM disappears from the intracellular compartments. Cells after the 6 hour chase period have mostly surface-bound-labelled IgM. From the surface membrane-labelled IgM molecules disappeared by shedding into the extracellular fluid (turnover) with a median disappearance time between 20 and 80 hours (see above). The median appearance time in the surface membrane of the 1000 to 2000 radioactive IgM molecules which are synthesized within the 4-hour-labelling period (see before) is therefore short compared to the median disappearance time of these IgM molecules from the cells. The surface membrane, with 3 to $15x10^4$ Ig molecules,therefore appears to be the main cellular pool of IgM molecules in small, resting B-lymphocytes. This pool appears freely mixable for all Ig molecules in the surface membrane, since IgM molecules labelled in a 4 hour pulse with radioactive leucine leave the cell in the subsequent chase period with no measurable lag period early and late in the chase.

## Mitogens for B-lymphocytes

All B-cell mitogens stimulate a large portion of all lymphocytes. If small, surface-membrane Ig-positive cells from the spleen are studied, more than 80% of the cells showed increase thymidine uptake as displayed by autoradiography of the labelled cells. Some mitogens are not known in their molecular structure (i.e. PPD) or are a mixture of many components, again unknown in molecular structure (i.e. FCS). I will take two mitogens, locally concentrated Concanavalin A (ConA) (19) and bacterial lipopolysaccharide (LPS) (20), of which the molecular structure is known (21), to discuss the possible modes of action of B-cell mitogens.

The mitogenic part of LPS resides in the Lipid A portion of the molecule (22). The mitogenic effect of Lipid A, consisting of phosphorylated glucosamine disaccharide units with ester- and amide-linked fatty acids, is lost after alkali treatment which removes the ester-linked fatty acids. The initial step in mitogenesis and activation may be an insertion of the lipid A portion of LPS into the lipid bilayer of the surface membrane of B-cells. Lipid A and all other forms of LPS can be viewed as glycolipids. Thus, while the insertion into the lipid bilayer may primarily be conducted by affinities displayed by the lipid portion, subsequent interactions in the membrane may also involve the carbohydrate portions of it.

Concanavalin A has 4 binding sites for mannopyranosyl-residues. It therefore can crosslink carbohydrate residues of either glycoproteins or glycolipids which carry the proper carbohydrate residues at terminal positions. In free form, Con A acts as a T-cell mitogen (1), while in its insolubilized form (Con A), bound either to plastic petri dishes or to Sepharose beads, it acts as a B-cell mitogen. This illustrates that the presentation of the reactive groups in form of a matrix to the cell surface is of crucial importance in B-cell triggering. Matrix presentation of repeating determinants of antigens has also been postulated as a necessary requirement of antigen-activated B-cell triggering (23). It appears that fixation of molecules reacting with the stimulating substances in an array in the surface membrane facilitates activation. By binding to carbohydrate group, Con A may indirectly or directly change interactions of molecules in the membrane. If a direct interaction is altered by Con A, carbohydrate groups are involved in these interactions (see above for possible roles of conformation of carbohydrate groups in biological reactions).

## Activation of B-lymphocytes. Early Events

*Changes in the number of surface bound IgM molecules.* Within 20 minutes after the addition of LPS 7-8 S IgM molecules on the surface membrane aggregate to complexes much larger than 19

S-pentameric IgM. These complexes may contain other cellular material, and it is not known at present whether they contain also the mitogen, bacterial lipopolysaccharide. Binding of $^{125}$I-labelled (anti-mouse Ig) Fab to such mitogen-treated cells decreases to 10% of the original binding capacity found before addition of the mitogen. Biosynthetically-labelled, surface-bound IgM can no longer be precipitated in the surface membrane of intact cells by (anti-Ig) antibodies (24).

The aggregated surface-IgM-determinants are progressively degraded, such that less and less intact Hμ-and L-chains can be detected by acrylamide gel electrophoresis of the reduced and alkylated serological precipitates of the aggregated material. An activation of surface-membrane-located proteases appears to occur early after stimulation. Thus, surface-membrane bound receptor-IgM appears to be used up after stimulation.

Levels of cyclic AMP inside B-cells do not change significantly within 4 hours nor later than up to 72 hours after stimulation (Bauminger, S., Andersson, J. and Melchers, F. in preparation)

Between 6 and 12 hours after the initiation of mitogenic stimulation binding capacities for $^{125}$I-(anti-mouse Ig) Fab reappear on the stimulated B-cells. Biosynthetically-labelled-IgM can be precipitated again on the surface of the intact cells by (anti-Ig) antibodies after 10 to 20 hours, i.e. at the time when an increase in intracellular "de novo" synthesis of IgM is observed after mitogenic stimulation. Between 10 and 20 hours after the initiation of stimulation the same amount of IgM appears on the surface as was originally present on the unstimulated small cells.

Beyond 20 hours of stimulation more and more IgM molecules appear on the surface membrane. This is concluded from the finding that a) higher concentrations of (anti-Ig) antibodies are needed to precipitate an optimal amount of IgM molecules on the surface of the cells, and b) that more $^{125}$I-labelled (anti-mouse Ig) Fab molecules are bound per cell. Exact estimates as to how much more IgM molecules appear on the surface are difficult to make because the quantitative relationship between the number of antibody molecules which bind IgM molecules arranged in certain patterns on the surface membrane is unknown.

The increased number of IgM molecules on the surface membrane of stimulated B-cells show a rapid rate of turnover. In B-cells stimulated for 44 hours with mitogen, 30% of all IgM molecules labelled within 4 hours by radioactive leucine are on the surface, while 70% are in intracellular compartments. During the first 4 hours in nonradioactive chase medium, labelled IgM molecules disappear from intracellular compartments with a median disappearance time of 2 hours. They appear in the surface membrane, since the amount of radioactive IgM there increases during the first 4 hours. The surface membrane is therefore an intermediate station of IgM molecules leaving the cell with a rapid rate. From the disappearance of labelled IgM molecules from the surface membrane in the time between 4 and 13 hours in chase medium, this rapid rate is estimated to have a median disappearance time around 5 hours. We estimate from the amount of radioactive IgM which has left the cells after 13 hours in chase medium that at least 60 to 70% of all surface-bound IgM in stimulated B-cells turns over rapidly. The observed increase in the number of surface bound IgM molecules after stimulation is therefore largely due to active secretion of IgM molecules via the surface membrane after mitogenic stimulation (24).

Much needs to be done. Levels of cyclic GMP should be determined in B-cells stimulated under different conditions (see later) with different antigens at different concentrations for different times. The chemical nature of the aggregates containing Ig has to be elucidated. Early changes in the surface membrane should be studied with isolated plasma membrane vesicles, as it is done with other reactions in bacterial membranes (2).

*Synthesis and turnover of IgM.* Small resting B-lymphocytes synthesize 10% of its total cellular pool of immunoglobulin M (IgM) molecules within 2 to 4 hours. They release those IgM molecules from the cells into the supernatant medium with a half disappearance time between 20 and 40 hours, mainly as 7-8 subunit IgM's (see before). LPS stimulates small resting B-lymphocytes within the first hour to a 2 to 3 fold increased rate of IgM synthesis. This stimulation is mitogen-dose dependent and can also be observed with other B-cell mitogens (PPD, fetal calf serum). IgM molecules synthesized with this increased rate are actively secreted from the cells with a median disappearance time of 2 to 4 hours, mainly as 19S-IgM-pentamers.

The initial change in the rate and type of IgM synthesis after mitogenic stimulation can also be observed in the presence of 5 μg/ml Actinomycin D. IgM synthesis which is more sensitive to inhibition by Actinomycin D than the sum of all cellular protein syntheses in small unstimulated B-lymphocytes, is rendered more resistant to this inhibitor immediately after mitogenic stimulation. Stimulation of small, resting B-cells, in the presence or absence of Actinomycin D, leads to a redistribution of ribosomes from monoribosomes to polyribosomes within the first hour of stimulation.

Stimulation of B-cells, therefore, by mitogen leads to stabilization of RNA-synthesis-dependent components of IgM synthesis from degradation through the formation of polyribosomes and reprograms IgM synthesis from a synthesis of membrane-associated IgM to a synthesis of actively secreted IgM (Melchers, F. and Andersson, J., manuscript in preparation).

## Activation of B-lymphocytes. Late Events

*Activation of DNA synthesis.* DNA-synthesis, as measured by the uptake of radioactive thymidine by B-cells, commences after a lag period of 12 to 14 hours after initiation of mitogenic stimulation

The rate of DNA-synthesis thereafter continues to increase up to 72 hours after initiation of stimulation (5).

*IgM synthesis and secretion.* The ratios of the rates of synthesis or of secretion of IgM over those of all proteins made in the cell increase with increasing time of stimulation. Other proteins than IgM are made and secreted by B-cells after activation, but IgM synthesis and secretion increases selectively over synthesis and secretion of other proteins in the cells. Since the number of plaque-forming, immunoglobulin-secreting cells increases with time of stimulation, we interpret the increase in the ratios of synthesis and secretion rates to say that more cells within the population change from non-secretors to secretors, rather than that a given cell in the population synthesizes and secretes more IgM and less other protein. These long-term stimulated B-cells now show IgM-synthetic rates which are stable up to 5 hours in the presence of Actinomycin D. In fact, IgM synthesis is more stable than syntheses of other cellular proteins. The change from Actinomycin D-sensitive to Actinomycin D-resistant IgM synthesis can therefore be taken as another sign that small B-cells have undergone stimulation to increased synthesis and to active secretion. IgM synthesized at the increased rate is secreted with a half time around 4 hours as a 19-S pentamer. Thus, pentamerization and a rapid turnover of IgM represents another sign of B-cell activation.

After 72 hours of mitogenic stimulation, the B-cell cultures produce and secrete as much IgM as does an IgM-producing and secreting mouse plasma cell tumor (14). This makes it likely that a large proportion of all cells in the culture actively secrete IgM after 72 hours of stimulation. The average secretory capacity of a B-cell stimulated for 72 hours can be calculated as the number of molecules of IgM secreted per unit time with the same assumptions made for the calculations of the synthetic capacity of small B-cells (see before). An average B-cell stimulated for 72 hours contains between 2 and $5 \times 10^6$ molecules 7-8 S IgM (see below for size) and secretes 3 to $5 \times 10^4$ molecules 19 S IgM per hour (see below for size). This synthetic capacity is comparable to that of IgG-producing plasma cells (4). Jerne (25) has calculated that between 5000 and 10000 mRNA molecules each for H- and for L-chains must exist in such a cell.

*Activation of the synthesis of "branch" carbohydrate portions of IgM molecules.* Radioactive mannose, galactose and fucose are used as precursors to label biosynthetically specific "core" or "branch" sugar residues of immunoglobulins (4,8, 14).

Radioactive mannose labels - as in resting lymphocytes - intracellular and extracellular IgM. In addition, however, radioactive galactose and fucose now also label IgM molecules in galactose respectively fucose residues of carbohydrate moieties which appear on secreted, but not on intracellular IgM after mitogenic stimulation (8). These secreted IgM molecules therefore contain the "branch" sugars galactose and fucose. Again, ratios of synthesis and secretion of fucose- and of galactose-labelled IgM and of those of all labelled carbohydrate-containing macromolecules secreted from cells increase with time of stimulation.

The increases in the ratios of synthetic and secreting rates for polypeptide and carbohydrate portions of IgM can be taken as parameters to monitor the stage of activation of a B-cell population. The appearance of secreted fucose- or galactose-labelled IgM is a further sign of B-cell activation to Ig-secretion, since IgM molecules shed from resting B-lymphocytes are not labelled by these two sugars (see before).

The rapid disappearance of IgM from our activated B-cells is reminiscent of rates measured by other laboratories (9,10,11) in unstimulated B-cells. Since their IgM appears outside the cell as 7-8 S subunits, while ours is polymerized to 19 S pentamers, when it shows the rapid rate of disappearance, comparisons of the various methods of turnover and size measurement will have to be made to clarify these discrepancies.

Furthermore, the observations of Lerner et al (12) that IgM synthesis is stable in Actinomycin D-treated, unstimulated lymphocytes may be expected if stimulated, large cells contaminate the unstimulated cell populations.

## Mechanism of Active Secretion of Ig

The work, shortly to be summarized here, has been undertaken in many laboratories (26-29).

Ig is synthesized on membrane-bound polyribosomes which sit on that side of the rough endoplasmic reticulum membrane which faces the intracellular compartments around the nucleus of the cell. Newly synthesized Ig then is released from the ribosomes and passes through the membranes into the channels ("cisternae") of the rough endoplasmic reticulum from which it migrates into the channels of the smooth endoplasmic reticulum before it is secreted.

Assembly of the chains into disulphide bound molecules is Ig-class and sub-class specific, begins on polyribosomes and continues while Ig traverses the cell. While IgM occurs as surface membrane-bound molecules even in stimulated, immature plasma cell (see below), it is not yet clear whether IgG exists as surface bound molecules in mature plasma cells. In mature plasma cells over 95% of the synthesized Ig is actively secreted. Appreciable differences exist in the size of the intracellular pools of Ig in cells synthesizing different classes of Ig: while IgG in smooth membranes represents a measurable amount of the total intracellular IgG, IgM can hardly, if at all, be detected in that compartment. Synthesis of the polypeptide portion of Ig takes only 2 to 4 minutes. Newly synthesized Ig appears secreted from the cells after a lag of 20 to 30 minutes. Hence almost all of the time Ig spends inside plasma cells is needed for its transport from the site of synthesis through and out of the cell. While only between 5% and 40% of of all the protein synthesized inside the plasma cells is Ig, 80% to 95% of the secreted material can be identified as Ig.

The question arises why Ig is the only protein in B-cells which is actively secreted, while in small, unstimulated B-cells, many other surface-bound proteins are shed. After stimulation, B-cells acquire the capacity to synthesize Ig at a higher rate than any other cellular protein and to actively secrete Ig.

During its migration through the cell, Ig acquires stepwise the residues of its carbohydrate moieties. At least four sequential precursor-product-relationships exist between different intracellular forms and the secreted form of Ig. Ig associated with polyribosomes contains very little, if any, glucosamine and/or mannose (30). Ig in the rough endoplasmic reticulum contains most of its glucosamine and mannose residues, but only traces of galactose and no fucose residues, while Ig in the smooth endoplasmic reticulum contains glucosamine, mannose, galactose and N-glycolyl-neuraminic acid residues, but only traces of fucose residues. Only the secreted form of Ig has the full complement of all carbohydrate residues.

These findings have suggested the hypothesis that the biosynthetic steps by which carbohydrate residues are attached to Ig, might be part of the process by which plasma cells transport and secrete Ig (30,31). Glycosyl-transferases, located in membranes of the rough and smooth endoplasmic reticulum, could attach sugar residues to Ig in an ordered fashion and thereby constitute an assembly line which directs Ig molecules out of the cells.

Intracellular IgM assembles only to the 7 S subunit (IgMs) form and polymerizes to 19 S IgM shortly before or during secretion. Mannose and glucosamine residues are added to intracellular IgMs at an early stage in the process of secretion, while galactose and fucose residues are added just before, or at the time, IgM leaves the cell (4,5, 8) (see below).

It is remarkable that IgM acquires 35% to 40% of its total carbohydrate content shortly before, or at the time of, secretion. This could indicate that carbohydrate residues determine different conformations of a glycoprotein. Such conformational differences were in fact concluded from hybridization experiments in which mixtures of intracellular 7 S and extracellular 19 S IgM from the same myeloma were reduced to 7 S-subunits, thereafter attempted to be reoxidized to 19 S IgM.

The results of these hybridization experiments suggested a difference of protein conformation in the $F_c$-portions of the molecules between intra- and extracellular IgM. Although at present these results have been obtained with only one myeloma IgM, they may be relevant to the stable inclusion of the $F_c$ portion of IgM receptor molecules in intracellular and surface membranes of lymphocytes.

These possible biological roles of carbohydrate moieties of Ig in selection for transport and secreting and in conformational changes of IgM during polymerization could be tested with an inhibitor of carbohydrate attachment to glycoproteins, 2-deoxy-D-glucose (32-34).

IgG-producing tumor cells and IgM-producing tumor cells and mitogen-activated cells were tested (5,35). 2-deoxy-D-glucose inhibited attachment of carbohydrate to Ig in tumor plasma cells and in mitogen-stimulated B-cells. In both types of cells 2-deoxy-D-glucose did not inhibit synthesis of μ-chains, respectively γ-chains, and L-chains, but inhibited active secretion of these newly made molecules. In polyacrylamide gel electrophoresis, mobility of the μ-chains synthesized in the presence of the inhibitor was that expected for a μ-chain with less or no carbohydrate attached to it.

Differences were observed in the active secretion of IgM molecules from 2-deoxy-D-glucose-inhibited cells, which had been synthesized _before_ the addition of the inhibitor. In tumor plasma cells around 30% of the intracellular IgM can be secreted as 19 S pentamers in the presence of the inhibitor (5). This indicates that sugar residues such as galactoses and fucoses normally added to these molecules shortly before secretion from the cells are not requisite for the pentamerization of IgM molecules. This makes unlikely that addition of these carbohydrate residues are primarily necessary to induce or stabilize any conformational changes between 7-8 S intracellular and 19 S extracellular IgM.

In contrast to tumor plasma cells, mitogen-stimulated B-cells did not secrete any IgM as 19 S pentamers in the presence of the inhibitor (5). Instead IgM molecules, which normally disappear from the cells with a rapid rate to be polymerized into 19 S pentamers, now appeared aggregated to molecules much larger than 19S. It is likely that mitogen, in the presence of the inhibitor, induces this aggregation of IgM subunits on the surface of the stimulated, 2-deoxy-D-glucose-inhibited B-cells, in ways analogous to the observed aggregation of surface IgM in B-cells in the initial phase of mitogen-stimulation. We take these findings as additional evidence that IgM may stay temporarily on the surface membrane during active secretion.

In IgG-producing and secreting tumor plasma cells, 2-deoxy-D-glucose inhibited the migration of newly synthesized IgG polypeptide chains from membrane-bound polyribosomes into the cisternae of the rough endoplasmic reticulum. It also inhibited the transfer of IgG molecules from rough to smooth membranes. It did not inhibit the transfer of $IgG_1$ molecules from the smooth membrane to the outside of the cells. 2-deoxy-D-glucose inhibited the attachment of sugar residues to $IgG_1$. Thus, attachment of galactose, fucose and N-glycolyl-neuraminic acid to $IgG_1$ molecules located in smooth membranes is not a prerequisite for their secretion from plasma cells. Glycosylation of $IgG_1$ molecules and/or to other intracellular carbohydrate moieties is however necessary to draw newly synthesized $IgG_1$ molecules into rough membranes and to transport them from there into smooth membranes.

Secretion of Ig-light chain subunits without detectable carbohydrate attached to them (7) and secretion of full $IgG_1$ molecules with incomplete carbohydrate moieties (36), could already be taken as evidence against the hypothesis that glycosylation of Ig-molecules is requisite for Ig-secretion. We have recently found (Melchers, F. and Andersson, J., to be published) that active secretion of the carbohydrate-free light chain of the plasma cell tumor MOPC 41 is inhibited by 2-deoxy-D-glucose. Since these tumor cells do not produce any detectable carbohydrate-containing Ig-chain, such as a heavy chain, we conclude from these experiments that glycosylation of intracellular structures other than Ig are requisite to draw newly synthesized Ig molecules into membraneous channels of the rough endoplasmic reticulum and from there into those of the smooth endoplasmic reticulum.

Immunoglobulin is primarily a membrane protein in small, resting lymphocytes. Thus we regard IgM synthesis, surface deposition, turnover and shedding in small B-lymphocytes as biochemical processes which are part of the overall process of surface membrane biosynthesis. IgM synthesis in small cells is in balance with the synthesis of other components (proteins, lipids, etc.) destined to be incorporated into the surface membrane of the cell. Turnover of surface IgM is just part of the turnover of the whole surface membrane. Many other proteins in the surface membrane turn over by shedding. IgM may assemble inside the cells with other membrane components to form premembrane complexes, replacing such structures in the surface membrane which have been lost by shedding.

After activation, IgM is synthesized at a higher rate than are other cellular proteins. The biogenesis of membranes becomes inbalanced: IgM displaces more and more other membrane proteins in lipid bilayers. IgM uses more effectively the assembly line for surface membrane biogenesis; active secretion is initiated. For this specialization for active secretion a membraneous apparatus is developed in B-cells which strikingly resembles that of other secretory cells from endocrine and exocrine glands. Polyribosomes, in electron microscopic pictures, now appear membrane-bound. They may synthesize less and less other membrane proteins and more and more IgM. Glycosylation of proteins of lipids other than Ig appear essential to initiate active secretion. Premembrane complexes may not be formed properly when certain parts of them are not glycosylated.

It appears possible that activation of a variety of exocrine and endocrine gland cells to maturation and active secretion involves the formation of a common cellular and molecular principle, morphogenetically the endoplasmic reticulum, with which different specific mRNAs are more efficiently translated into specific proteins which are characteristic for the activated cell secreting them. If one were to compare the synthesis of cellular components after stimulation to maturation of two different secretory cells, as e.g. a B-lymphocyte and a mammary gland cell (37), one should detect "common" as well as "special" macromolecules synthesized in these two cells.

## The Balance Between Proliferation and Maturation

B-lymphocytes are activated to proliferation and to maturation. Proliferation is measured by the incorporation of radioactive thymidine and by the appearance of cells in mitosis. It can be inhibited by inhibiting the biosynthesis of DNA through hydroxyurea (38) or cytosine arabinoside (39).

While DNA-synthesis is inhibited in 48 hours-mitogen-stimulated B-lymphocytes, protein and IgM synthesis and secretion and the plaque-forming capacity of those cells are unaffected. When hydroxyurea or cytosine arabinoside are added to small, resting B-cells at the time of initiation of mitogenic stimulation, these cells mature to 19 S IgM secreting, plaque-forming cells in the absence of DNA-synthesis and proliferation. Maturation in the absence of DNA-synthesis is also manifested by an increase in ratios of rates of synthesis and secretion of IgM over those of all proteins in the cell and by the attachment of the "branch" sugars, galactoses and fucoses, to 19 S IgM secreted by the cells. This maturation of B-cells in the absence of DNA-synthesis occurs within 12 to 30 hours after stimulation and is mitogen-dose-dependent. The inhibition of B-cells to synthesize DNA and to develop into clones of plaque-forming cells can be reversed by the removal of hydroxyurea for as long as 36 hours after mitogenic stimulation in the presence of the inhibitor.

In the presence of hydroxyurea, B-cells are stimulated to immature plasma blast-like cells containing surface-bound and little intracytoplasmic Ig within 16 to 24 hours of stimulation. Mature plasma cells containing no detectable surface-bound Ig, but abundant intracytoplasmic Ig, are only developed in the uninhibited, but not in the HU-inhibited mitogen-stimulated cells later in the response. Thus, two stages of B-cell maturation and differentiation can be distinguished: the first stage of an immature plasmablast develops in the absence of DNA-synthesis, while the later stage of the development of the mature plasma cell with membranous structures of the endoplasmic reticulum requires DNA-synthesis (40).

*Maturation.* The process of maturation of B-lymphocytes to active secretion can be inhibited by exposing small, resting B-cells before stimulation to anti-Ig-antibodies (Andersson, J. and Melchers, F., to be published). These antibodies do not, however, inhibit, but enhance at higher concentrations, DNA-synthesis in the subsequently stimulated cells. Bivalent pepsin-$(Fab')_2$-fragments, recognizing mouse Ig determinants, mediate the same effects at the same concentrations, while monovalent papain-Fab-fragments do not enhance DNA-synthesis and have a 100-fold lower capacity to inhibit maturation to plaque-forming cells. B-cells stimulated for 6 hours by the mitogen lipopolysaccharide can no longer be inhibited by anti-mouse-Ig-antibodies from maturing to plaque-forming cells.

Thus, resting B-cells can be activated by mitogen to maturation in the absence of DNA-synthesis and proliferation, and can be activated to DNA-synthesis and proliferation without maturing to antibody-secreting cells.

The molecular mechanism of action of the anti Ig-antibodies with receptor Ig on the B-cells is unclear. Although bivalent antibody structures are needed for the effects, it seems unlikely that it involves "capping" since no "prozone"-effect is observed and since our antibody preparations "cap" optimally with surface-bound Ig at much lower concentrations. A saturating occupancy of receptor Ig in the surface membrane with antibodies may leave one valency of the antibody molecules always free to react with any just dissociated receptor Ig determinants. This may effect an immobilization relative to each other of the antibody-bound Ig molecules in the fluid surface membrane. Such immobilization may suppress macromolecular syntheses inside the cells connected with the phenotypic expression of the Ig-genes.

From the results it appears that "mitogens", such as lipopolysaccharides, stimulate B-cells to two intracellular chains of reactions, <u>one</u> for the induction of DNA synthesis and cell proliferation, and <u>another</u> for the initiation of increased synthesis and active secretion of IgM. It might well be that different "mitogens" stimulate these two reactions with different strength. In an antigen-stimulated cell, growth-factors may play a role in the stimulation to these two reactions. Cells other than B-cells may supply these factors (see Introduction). In fact, antigen appears to prolong the period of proliferation in B-cells and therefore appears to have a modulating influence on the possible balance between proliferation and differentiation. It does so by interacting with receptor Ig on the surface membrane of the cells, not unlike anti-Ig-antibodies which also do not stimulate the B-cells by themselves, but modulate stimulation in favour of proliferation.

It is predictable that the regulation of the stimulatory signals leading to proliferation and to differentiation will involve a complex of several structures. From the activities of the structurally better known mitogens (Con A, LPS), I anticipate carbohydrate groups of glycoproteins or glycolipids (Con A) and glycolipids or other lipids (LPS) to be involved. Receptor Ig, a glycoprotein, must have modulating effects on the reactivity of the B-cells when bound to a matrix of repeating determinants (antigen anti-Ig antibodies, Con A?) such that proliferation is favoured over maturation in the clonal development of B-cells after stimulation.

## Acknowledgements

It is evident from the references to our own experimental work that much of it has been conducted in collaboration with Dr. Jan Andersson, now at the Salk Institute for Biological Studies, La Jolla, California, USA, some of it with the participation of Dr. Louis Lafleur at the Basel Institute. It is a pleasure to thank Misses Monica Gidlund, Dorothee Jablonski, Ingrid Möllegäard and Kerstin Wennberg for their expert technical assistance.

I would also like to thank the organizers of the Symposium, Drs. S. Estrada-O. and C. Gitler, for an exciting meeting and their overwhelming hospitality. It is tempting to compare the site of the symposium, the Hotel Victoria in Oaxaca, with a cell membrane. In fact, I consider it an ideal model membrane. The participating scientists could be taken as the protein components of the membrane who occurred arranged in different order and aggregation forms at different sites of this model membrane. In the tightly-packed form in the lecture room they served as receptors for light-mediated signals (i.e. slides, as psychedelic as some of them were) and sound-mediated signals (i.e. speakers, as varying as some of the English was). They were even sonicated by neighbouring hotel guests and by dance bands. The dance band in fact helped to aggregate some of them with homologous (wifes) or heterologous (other hotel guests) components. In many participants the mexican food changed their amphipathic character to a more hydrophobic nature. One of the most favoured sites certainly was the swimming pool, in which the highly liquid surrounding at comfortably elevated temperatures facilitated the interactions. I hope that the reaction of this model membrane to the input signals will be stimulation to differentiation of research in Mexico and, hopefully, also to proliferation into another such meeting in Mexico in the future.

## References


1. Articles in Lymphocyte activation by mitogens, ed. G. Möller, Transplant. Rev. 11 (1972).
2. Raff, M.C., de Petris, S. and Lawson, D. This volume.
3. Rabellino, E., Colon, S., Grey, H.M. and Unanue, E.R. J. Exp. Med. 133 (1971) 156
4. Melchers, F. Biochem. J. 119 (1970) 765
5. Melchers, F. and Andersson, J. Transpl. Rev. 14 (1973) 76.
6. Andersson, J., Lafleur, L. and Melchers, F. Eur. J. Immunol. 4 (1974) In press.
7. Melchers, F. Eur. J. Immunol. 1 (1971b) 330.
8. Andersson, J. and Melchers, F. Proc. Nat. Acad. Sci. U.S.A. 70 (1973) 416
9. Wilson, J.D., Nossal, G.J.V. and Lewis, H. Eur. J. Immunol. 2 (1972) 223.
10. Vitetta, E.S. and Uhr, J.W. J. Exp. Med. 136 (1972) 676.
11. Marchalonis, J.J., Cone, R.E. and Atwell, J.L. J. Exp. Med. 135 (1972) 956
12. Lerner, R.A., McConahey, P.J., Jansen, I. and Dixon, F. J. Exp. Med. 135 (1972) 136.
13. Shimizu, A., Putnam, F.W., Paul, C., Clamp, J.R. and Johnson, I. Nature New Biology 231 (1971) 73.
14. Parkhouse, R.M.E. and Melchers, F. Biochem. J. 125 (1971) 235.
15. Taylor, R.B., Duffus, P.H., Ra f, M.C. and

De Petris, S. Nature New Biology 233 (1971) 225.
16. Loor, F., Forni, L. and Pernis, B. Eur. J. Im munol. 2 (1972) 203.
17. Gitler, C. Ann.Rev.Biophys. Bioeng. 1(1972)51.
18. Singer, S.J. and Nicholson, G.L. Science 175 (1972) 720.
19. Andersson, J., Edelman, G.M., Möller, G. and Sjöberg, O. Eur. J. Immunol. 2(1972c)233.
20. Andersson, J., Sj¨oberg, O. and Möller, G. Eur. J. Immunol. 2 (1972a) 349.
21. Lüderitz, O., Galanos, C., Lehmann, V., Nurminen, M., Rietschel, E.T., Rosenfelder, G., Simon, M. and Westphal, O. J. Infect. Dis., Endotoxin Suppl. (1972) In press.
22. Andersson, J., Melchers, F., Galanos, C. and Lüderitz, O. J. Exp. Med. 137 (1973) 943.
23. Möller, G. Cell. Immunol. 1 (1970) 573.
24. Melchers, F. and Andersson, J. Eur. J. Immunol. 4 (1974) In press.
25. Jerne, N.K. Cold Spring Harbor Sympl Quant. Biol. 32 (1967) 591.
26. Uhr, J.W. Cell. Immunol. 1(1970)228.
27. Choi, Y.S., Knopf, P.M. and Lennox, E.S. Biochemistry 10 (1971) 668.
28. Bevan, M.J., Parkhouse, R.M.E., Williamson, A.R. and Askonas, B.A. Progr. Mol. Biophys. 2 (1972) 22.
29. Melchers, F. Histochem. J. 3 (1971) 389.
30. Melchers, F. and Knopf, P.M. Cold Spring Harbor Symp. Quant. Biol. 32 (1967) 255.
31. Swenson, R.M. and Kern, M. J. Biol. Chem. 242 (1967) 3242.
32. Farkas, V., Svoboda, A. and Baner, S. Biochem. J. 118 (1970) 755.
33. Liras, P. and Gascon, S. Eur. J. Biochem. 23 (1971) 160.
34. Gandhi, S.S., Stanley, P., Taylor, J.M. and White, D.O. Microbios. 5 (1972) 41.
35. Melchers, F. Biochemistry 12(1973)1471.
36. Melchers, F. Biochem. J. 125 (1971) 241.
37. Topper, Y.J. and Vondeshaar, B.K. 1974. In: Control of Proliferation in Animal Cells. Cold Spring Harbor Laboratory. In press.

38. Krakoff, I.H., Brown, N.C. and Reichard, P. Cancer Res. 28 (1968) 1559.
39. Skoog, L. and Nordenskjöld, B. Eur. J. Biochem. 19 (1971) 81.
40. Andersson, J. and Melchers. F. Eur. J. Immunol. (1974) In press.

---

# CELL MEMBRANE ANTIGENS DETERMINED BY THE MOUSE MAJOR HISTOCOMPATIBILITY COMPLEX — SOME DATA AND SOME SPECULATIONS

Stanley G. Nathenson

Departments of Microbiology and Immunology, and Cell Biology
Albert Einstein College of Medicine
Yeshiva University
Bronx, New York 10461
U.S.A.

## Introduction

Immunogenetic techniques have defined an increasing series of genetically determined, antigenically definable molecules on the surface of mammalian cells, especially lymphoid cells. Because of the availability of genetically defined strains in the mouse system, it has been used as the model in such studies. The antigens include those of the H-2 major histocompatibility complex (MHC), Θ, Ly-A, Ly-B, MSLA, PC-1, $G_{IX}$, etc. (1).

Each of these genetically defined antigens is an entity of its own. The structural, genetic, functional, and dynamic properties of each antigenic molecule must be worked out individually, and will not necessarily suggest the properties of another. However, these moeities share one important property, that of membrane integration. Thus, the structural basis of this property may be shared among these molecules.

Because our knowledge is most advanced on the H-2 MHC system, I should like to limit this discussion to the genetic and biochemical properties of these alloantigens. This system is of particular interest because of its central role in transplantation immunology. First described by Gorer in 1936 (2) the MHC system includes genes determining a number of functions, all apparently involving immunological processes, as well as cell surface membranes. The MHC chromosome segment (chromosome 17) is composed of 5 regions defined by recombination which map in the following order from the centromere: K, I, S, D and T (3). The K (H-2K) and D (H-2D) regions are separated by approximately an 0.5% recombination frequency. These genes control serologically detected cell membrane antigens referred to as the H-2 antigens. The S (Ss-Slp) region (4) contains a gene controlling the quantity of a serum protein thought to be involved in the complement system (5), and the I region contains genes involved in the immune response to certain polypeptide antigens (6), to lymphocytic proliferative responses (cf. reference 7 for review), and to susceptibility to certain mouse leukemias (8). The T region contains the Tla gene (1) which controls the TL antigen (thymus leukemia antigen). This gene is separated by an 0.5% recombination frequency from the H-2D gene of the D region.

## Overall Chemical Properties of The Products of The MHC

*H-2K and H-2D Antigens.* The glycoprotein nature of the H-2 alloantigens was first established by analysis of the H-2 alloantigen fragments released in water soluble form from spleen cell membranes by papain digestion (9). Two classes of fragments of different sizes and called Class I and Class II were recovered after the enzymatic solubilization step. The Class I fragments had a molecular weight of about 37,000 daltons, and the Class II fragments had a molecular weight of about 28,000. Both were glycoproteins of approximately 90% protein and 10% carbohydrate (Table I).

H-2 alloantigen glycoproteins could also be released from membranes in an intact form using the non-ionic detergent NP-40 (Nonident P-40), and isolated and recovered by an indirect immunoprecipitation method (10). The molecular size as judged by SDS (sodium dodecyl sulfate) polyacrylamide gel electrophoresis was approximately 43,000-47,000 daltons. When precipitates were not dissociated by a reducing agent in SDS, molecules of 90,000 daltons and of 45,000 daltons were found. The 90,000 M.W. species could be converted to the 45,000 daltons molecule by reduction and alkylation. Tentatively, this established a dimer and monomer relationship, presumably the dimer containing identical monomers.

Attempts to establish the in situ size of the H-2 alloantigen have suggested that the antigen isolated by NP-40 is larger than the antigen after reduction and alkylation in SDS. An approximate molecular size of the order of 300-400,000 could be obtained for NP-40 solubilized antigen as judged by its elution position during molecular seive chromatography on agarose 0.5 M columns in 0.5% NP-40 buffer. The larger molecular weight for the H-2 glycoproteins in the NP-40 non-ionic detergent suggests that they may exist in aggregates of possibly 8 to 10 monomeric units.

As shown in Table I the papain solubilized fragment (Class I) differed in M.W. from the native or intact NP-40 glycoprotein by approximately 3,000-6,000 daltons. It is reasonable to conclude, therefore, that the minimally cleaved Class I papain fragment represents a major part (approximately 80%) of the native glycoprotein. Thus, the loss of a relatively small portion of the intact antigen confers aqueous solubility on the papain fragment. It seems reasonable to assume that this antigenically silent small region is necessary for membrane integration of the molecule.

The carbohydrate content of the H-2 alloantigens includes galactose, glucosamine, mannose, fucose and sialic acid as the N-acetylneuraminic acid derivative. There is no glucose or galactosamine. The carbohydrate chains are approximately 3300 M.W. and there may be between one and two chains per peptide portion(11).

The antigenic activity of the H-2 glycoproteins is carried in the peptide structure. Stuies supporting such a conclusion come from experiments in which antigenic activity was found not to reside in the glycopeptide fraction, and studies in which enzymatic removal of much of the carbohydrate did not destroy the antigenic reactivity of the remaining antigen molecule.In addition, protein denaturation and specific modification of only certain amino acid residues detroyed the antigenic activity.

*TLA Antigen.* The thymus leukemia (TL)alloantigens are expressed on the cell surface of murine thymocytes and on certain leukemia cells (1). They are alloantigens of a special form and are more accurately called differentiation antigens since in the normal state they are expressed on one cell type, the thymocyte. However, they are also tumor specific since they can be found on leukemia cells of mouse strains whose thymocytes do not express them.

An important feature of the Tla locus is its close linkage (0.5%) to the H-2D gene, and the topographical relationship of the TL antigen to the H-2D antigen as shown by antibody blocking techniques (1).

The glycoprotein nature of the TL antigens was established by studies on radiolabeled products cleaved from the membrane by proteolytic digestion with papain and isolated by indirect immunoprecipitation. As pointed out in Table I, the M.W. of the TL papain fragment is similar to the H-2 glycoprotein fragment isolated by the same procedure.

The carbohydrate moiety of the TL antigen was isolated from TL antigen radiolabeled in its carbohydrate moiety after pronase digestion. By comparison to the glycopeptide from H-2 alloantigen (3300-3500 daltons) on Sephadex G-50 column chromatography, the molecular weight of TL glycopeptide was calculated to be 4500 (12). Thus the TL product is a glycoprotein with many similar biochemical properties to the H-2 glycoprotein;

for example, both can be released from the membrane by enzymatic digestion, the TL and H-2 glycoprotein fragments are very similar in molecular weight, and carbohydrate chains of both gene products have similar, though not identical molecular weights.

*Membrane Associated Molecules Determined by the I Region.* The Ia (I associated) antigens (13) have recently been defined as a system of cell surface antigens determined by genes in the I region mapping between the K and S region. The antigens are found only on lymphoid cells and are most easily detected with cytotoxic antisera using antibodies prepared from immunizations between mouse strains carefully selected to eliminate H-2K and H-2D gene differences. The antigens defined by appropriate antisera have been isolated using the techniques of indirect precipitation from NP-40 extracts of spleen cells (14).

Preliminary evidence suggests these antigens are glycoproteins, since they can be labeled with fucose, mannose or galactose. Their molecular size was estimated by electrophoresis in SDS on polyacrylamide gels. Again, as was seen with the H-2K and H-2D glycoproteins, these glycoproteins apparently have a dimer and monomer relationship. The dimers are approximately 60,000 M.W. and the monomers 30,000 M.W. The transition from the 60,000 M.W. dimer to the 30,000 M.W. monomer can be achieved by reduction and alkylation.

## Dynamic Properties of the H-2 Glycoproteins and TL Glycoproteins in the Membrane

Studies on the biogenesis of the membranes have established that different components of the cell membrane are not synthesized and inserted at the same rates. For example, different enzymes of the liver cell membranes turn over at different rates and lipids also show variable turnover properties (15). The H-2 glycoproteins have also been assessed in terms of their biosynthetic and turnover properties.

In one series of experiments tumor cells

bearing H-2 antigens were treated with papain until only about 40% of the H-2 antigen was left. The cells were then suspended in growth medium and the reappearance of antigenic activity was monitored. Antigenic activity on the papain treated cells began to increase after a lag of about 1-1/2 hrs and rose from a base level of 40% to 100% after 6 hrs. This 6 hr regeneration time is a much shorter time period than the 24 hr division time of the cells (16).

Turnover rates of radiolabeled H-2 antigen showed a half life of about 8-10 hrs. The doubling time of the cells was about 48 hrs and the half time of the H-2 antigen was 4-5 times faster than the rate of cell division. Thus, these studies, and others carried out in HL-A bearing human cell lines (17), suggests a very rapid biosynthesis of these glycoproteins relative to other membrane and intracellular components. Studies on the regeneration rates of the TL and Ia antigens have not been reported.

Studies on the antibody, lectin, or antigen induced movement (capping) of cell surface components are discussed in another paper in this series and will not be further considered here. However, it is important to comment that the TL antigens were the first for which antibody mediated disappearance was documented (18). Of interest is the concomitant observation that disappearance of TL antigen was accompanied by compensatory increase in the amount of H-2D antigen. This process, termed antibody mediated modulation, probably is a type of capping. H-2D and H-2K glycoproteins have also been shown to undergo capping, and they cap as idependent units (19).

Thus, H-2 glycoproteins have rapid dynamic properties in terms of turnover, and in addition are capable of undergoing rearrangement in the membrane plane.

## Peptide Comparison of H-2D and H-2K Gene Products

An understanding of the primary structure of the major histocompatibility glycoprotein alloantigens would provide the basis for defining their membrane affinity, their antigenic properties and their genetic parameters. While sequence data is not yet available, we have carried out analysis of the peptides of H-2K and H-2D glycoproteins produced by trypsin digestion (20). Such peptide comparisons provided information on two different questions. First, what were the similarities and differences in peptide composition among molecules determined by alleles of the same gene (e.g. H-$2K^b$ vs H-$2K^d$ or $K^b$ vs $K^d$)? Second, what were the similarities and differences in peptide composition between products of alleles of the H-2K and H-2D genes of the same haplotype (e.g. $K^d$ vs $D^d$)?

While so far only several different haplotypes and their alleles have been examined the following results have emerged from these studies. A great deal of diversity was present in the peptide profiles of different products. For example, comparison of products of alleles of the same gene (e.g. $K^b$ vs $K^d$, $D^b$ vs $D^d$) showed only 35% similarities. Peptide comparisons of the products of the alleles of the H-2K vs H-2D showed somewhat divergent results since $K^b$ vs $D^b$ showed 60% similarity, while $K^d$ vs $D^d$ showed about 40% similarities.

These results are quite striking since they show an extreme degree of diversity between the products of alleles of the same gene. In fact, differences between products of alleles of D and K genes are somewhat less than the differences of alleles of the same gene.

The finding of considerable uniqueness for the products of the alleles of the H-2 genes is possibly not unexpected, since the polymorphism of these genes has been their hallmark. In fact, the complex serological profiles associated with products of different alleles suggests that there must be a complex and variable structural basis.

## Conclusions and Speculations

The discovery of the MHC alloantigens has been due to their accessibility for antigenic stimulation (i.e. cell membrane location), and genetic variability. Thus, only because the techniques of immunogenetics selectively provided us the tools, were we able to identify these products. However, even this property sets these products and their genes apart from other membrane macromolecules which cannot be so studied.

The information about the H-2K and H-2D products is most advanced. And while still rudimentary, as compared to the massive structural literature available for immunoglobulins, or hemoglobins, such information can allow certain generalizations about other members of the MHC series and possible interrelationships. It must be emphasized that such speculations are often predicated on preliminary data. However, such speculation is the purpose of the chapter and allowances should therefore be made.

The H-2K, H-2D, Ia and TL antigens are immunologically detectable, cell surface located, and found primarily on cells of the lymphoid series. They are all glycoproteins. Further, it is indeed striking that the genes determining these products are so closely linked. Their glycoprotein nature and genetic linkage suggests a possible functional association.

However, in spite of obvious similarities, important differences in many properties are apparent. At present, only the H-2D and H-2K products seem to bear considerable immunological similarities, since both are very strong transplantation antigens. The TL antigen probably has no role directly in skin graft rejection properties. The present preliminary evidence suggests that the Ia antigens are not transplantation antigens and are not involved in graft rejection. Also the H-2K and H-2D antigens are most heavily concentrated on lymphocytes, but are detectable on nearly every cell type of the animal. However, the TL antigen, when expressed, is found only on thymocytes, although it may be found on some lymphoid cell tumors, even in mice carrying thymocytes which do not normally express that TL allele. Thus, as compared to the H-2D and H-2K products, TL is regulated quite separately, and in a much different manner.

The Ia antigens are even more puzzling. Mostly because of lack of data, we are not certain of their distribution. Certainly some of these antigens are expressed on B-lymphocytes, and possibly some on T-lymphocytes. They are apparently absent on most lymphoid tumor cells so far studied, whereas H-2 is strongly expressed on such cells. Therefore, such individual characteristics of these products, H-2, TL and Ia, suggests different functional properties.

Some of the physical properties of these antigens are of interest.

For example, present evidence suggests that the H-2 glycoproteins may exist in the cell membrane as aggregates. These aggregates may contain of the order of 8 to 10 monomeric 45,000 dalton units. In addition, the monomers may be linked by disulfide bonds to form dimeric units.

It is significant that the Ia glycoproteins may also be characterized by a monomer and dimer relationship. Furthermore, preliminary evidence suggests that Ia may exist in an aggregate at least as large as H-2 as judged by its chromatographic behavior on 0.5 M BioGel columns in NP-40 detergent. It is possible that the aggregation property is important for functional requirements.

The property of rapid biosynthesis and turnover, at least of the H-2 glycoproteins, is another important property. Such rapid dynamics, in addition to the capacity to aggregate or cap, may be related to function. For example, one can envision that certain mechanisms (e.g. receptor) might require association-dissociation for triggering signals at the cell surface. Rapid regeneration of new receptor would be important to re-arm the cell.

The major questions about the gene products of the mouse H-2 MHC are still unanswered. What are the biological functions of each? What are the mechanisms by which they function? Structural studies on the products will certainly play a central role in complementing biological studies on function. Thus, biochemical and structural characterization of each antigen molecule is necessary to describe unique characteristics and also those properties which some molecules may have in common. Such data will define the structural basis of membrane affinity, antigenic reactivity, and genetic variability. But in addition, such data will also provide crucial leads for the biologist and immunologist.

However, in terms of the cell, such structural descriptions only give information of the molecules in question. In addition, we need information of the membrane topography, such as gained by electron microscopic techniques, and on the dynamics and interrelationships of movement of molecules on the surface, as is studied by capping methods. We also need approaches at the somatic cell genetics level. Can we isolate non-cappable mutants?

Thus, the study of the cell membrane, and its components, requires multi-disciplinary approaches. Only in this way can we understand the nature of each component, and understand the relationships of these components with each other in dynamic as well as topographical terms.

## Acknowledgement

The work quoted by the author has been supported by United States Public Health Service grants, AI-07289 and AI-10702.

## References


1. Boyse, E.A. and Old, L.J. Ann. Rev. Gen. 3 (1969) 269.
2. Gorer, P.A. Brit. J. Exp. Path. 17 (1936) 42.
3. Klein, J., Demant, P., Festenstein, H., Mc Devitt, H.O., Shreffler, D.C., Snell, G.D. and Stimpfling, J.H. Immunogenetics (1974) In press.
4. Passmore, H.C. and Shreffler, D.C. Genetics 60 (1968) 210.
5. Hinzova, E., Demant, P. and Ivanyi, P. Folia Biol (Praha) 18 (1972) 237.
6. Benaceraff, B. and McDevitt, H.O. Science 175 (1972) 273.
7. Demant, P. Transpl. Rev. 15 (1973) 164.
8. Lilly, F. J. Natl. Can. Inst. 49 (1973) 927.
9. Shimada, A. and Nathenson, S.G. Biochemistry 8 (1969) 4048.
10. Schwartz, B.D., Kato, K., Cullen, S.E. and Nathenson, S.G. Biochemistry 12 (1973) 2157
11. Nathenson, S.G. and Muramatsu, T. 1971. In: Glycoproteins of Blood Cells. eds. G.A. Jamieson and T.J. Greenwalt. Lippincott Co., Philadelphia.
12. Muramatsu, T., Nathenson, S.G., Boyse, E.A. and Old, L. J. J. Exp. Med. 137 (1973) 1256.
13. Shreffler, D.C., David, C., Gotze, D., Klein, J., McDevitt, H.O. and Sachs, D. Immunogenetics (1974) In press.
14. Cullen, S.E., David, C.S., Shreffler, D.C. and Nathenson, S.G. Proc. Nat. Acad. Sci. (1974) In press.
15. Omura, T., Siekevitz, P., Palade, G.E. J. Biol. Chem. 242 (1967) 2389.
16. Schwartz, B.D. and Nathenson, S.G. Transpl. Proc. 3 (1971) 180.

17. Turner, M.J., Strominger, J.L. and Sanderson A.R. Proc. Nat. Acad. Sci. (U.S.) 69 (1972) 200

18. Old, L.J., Stockert, E., Boyse, E.A. and Kim J.H. J. Exp. Med. 127 (1968) 523.

19. Neauport-Sautes, C., Lilly, F., Silvestre, D. and Kourilsky, F.M. J. Exp. Med. 137 (1973) 511.

20. Brown, J.L., Kato, K., Silver, J. and Nathenson, S.G. Submitted for publication 1974.

21. Uhr, J. Vitetta, E. Fed. Proc. 32 (1973) 35.

TABLE I

Summary of Properties of MHC Products

| Product | Composition | Overall Molecular Weight Papain Solbl | NP-40 Solbl. | M.W. Carbohydrate |
|---|---|---|---|---|
| H-2K, H-2D | Glycoprotein (90% Protein 10% Carbohydrate) | 37,000 (Class I) | 45,000* | 3,500 |
| TL | Glycoprotein | 37,000 | 45,000** | 4,400 |
| Ia | Glycoprotein | | ∿30,000* | |

* *Can be found in dimers.*

** *Uhr and Vitetta (21).*

# LIGAND INDUCED REDISTRIBUTION OF MEMBRANE MACROMOLECULES: IMPLICATIONS FOR MEMBRANE STRUCTURE AND FUNCTION

Martin C. Raff, Stefanello de Petris* and Durward Lawson

Medical Research Council,
Neuroimmunology Project
Zoology Department,
University College London
London WC1E 6BT

It has been three years since the initial demonstration that when multivalent ligands, such as antibodies or lectins, bind to macromolecules on the surface of living cells, they can induce a large scale redistribution of the macromolecules in the plane of the cell membrane (1). In this paper, we will review the principle features of ligand-induced redistribution of membrane molecules as illustrated by the behavior of immunoglobulin (Ig) molecules on the surface of thymus-independent B lymphocytes, and then describe our more recent observations on other molecules on lymphocytes, fibroblasts, erythrocytes and neural cell membranes. Finally, we will consider some of the implications of these findings for membrane structure and function.

## Patching, Capping and Pinocytosis of Ig Molecules on Lymphocytes Induced by Anti-Ig Antibody

Ig is an important molecule in the membrane of a B lymphocyte, as it serves as receptor by

* Present address: Basel Institute of Immunology, Switzerland.

which the cell interacts with its specific antigen; the interaction of Ig receptor with antigen induces the cell to divide and differentiate into an antibody-secreting cell. Ig is demonstrated on the surface of a B cell by using anti-Ig antibody labelled with a radioactive isotope, or a fluorescent or electron dense marker, and the distribution one sees depends on the conditions of labelling. If the anti-Ig antibody is divalent it redistributes the membrane-bound Ig. Three distinct types of redistribution can be induced: patching, capping and pinocytosis.

When mouse B lymphocytes are labelled with anti-Ig antibody which has been digested with papain into its monovalent Fab fragments, and conjugated to fluorescein (1) or ferritin (2), Ig is seen to be diffusely and randomly distributed, presumably representing the normal distribution in the unperturbed membrane. B cells which have been trypsinized or activated for three days in tissue culture with lypopolysaccharide, show the same random distribution of Ig when labelled with fluorescent Fab anti-Ig antibody (our unpublished observations).

On the other hand, if B cells are labelled with divalent anti-Ig at 0-4°C and/or in the presence of a metabolic inhibitor, such as sodium azide, the Ig redistributes into clusters, leaving areas of unlabelled membrane between (1,2,3,4). Although this patchy distribution is not seen with monovalent Fab anti-Ig labelling, it can be induced by adding a second layer of divalent anti-Fab antibody (1,2,3) suggesting that it is caused by cross-linking of the Ig molecules by the divalent antibody. Patching implies that Ig molecules are able to move laterally in the plane of the membrane, and that the membrane is fluid. Since it cannot be inhibited by metabolic inhibitors (1,2,3) and occurs even at 0-4°C (although the rate is reduced (2) at low temperatures), it is probably a passive process, best visualized as a microprecipitation or agglutination reaction occurring in two dimensions. The slowed rate of patching at low temperatures is presumably related to increased viscosity of the membrane lipids, but the finding that it occurs at 0° indicates that the membrane does not freeze completely at 0° or above, although localized phase transitions may occur at these temperatures (5). The apparent absence of an abrupt freezing point is not surprising in view of the heterogeneity of the lipids and the presence of significant amounts of cholesterol (6). By measuring the packing density of ferritin molecules in the patches when B cells are labelled with anti-Ig conjugated to ferritin (anti-Ig-FT), one can estimate that the size of a mobile Ig unit is $<400\text{nm}^2$ (4), which would correspond to at most two or three 7S Ig molecules (7). Thus Ig molecules appear not to travel as large islands of protein.

If B lymphocytes are labelled with divalent anti-Ig and the cells are warmed up to 20°C or above in the absence of metabolic inhibitors, they rapidly move the anti-Ig-Ig complexes on their surface to one pole of the cell forming a 'cap' (1,2,3,4,8,9,10). Like patching, capping requires divalent binding of antibody, but unlike patching, it is an active process and does not occur at 0-4° and is readily (and reversibly) inhibited by metabolic inhibitors such as sodium azide or dinitrophenol, (1,2,3,4,8,9,10). Most B cells cap in less than two minutes at 37°(1) and the caps always form at that pole of the cell containing the cell organelles (4), which corresponds to the tail of a moving lymphocyte, but capping does not require that the cell move relative to a substrate, for it occurs readily when cells are maintained in suspension (1) or in silicone droplets (our unpublished observations). The mechanism of capping is not known. It is not affected by inhibitors of protein synthesis (8, our own unpublished observations), depletion of extracellular calcium and magnesium (1), depolarization of the membrane by high concentrations of potassium (2) or the microtubule-disrupting agents colchicine or vinblastine (1), although it is *partially* inhibited by cytochalasin B (1). Capping appears to be specific for those molecules that are cross-linked by the antibody, since when B cells are capped with anti-Ig, Ig molecules cap while the major histocompatibility antigens, and heteroantigens recognized by anti-lymphocyte serum remain diffusely distributed (1,3,8,9). It probably involves the interaction of contractile cytoplasmic components with the cross-linked membrane macromolecules resulting in their backward movement, with a countercurrent forward flow of uncross-linked membrane molecules in the plane of the membrane.

Metabolically active B cells labelled with divalent anti-Ig begin to pinocytose the anti-Ig-Ig complexes (and some adjacent unlabelled membrane) within minutes of being warmed to 37° (1, 4, 11). Although pinocytosis usually follows cap formation, the two processes can be dissociated. For example, cells labelled with monovalent anti-Ig can pinocytose at least some of the label without forming caps (2). The pinocytosis induced by divalent anti-Ig antibody can be massive at 37° and can lead to the complete disappearance of the anti-Ig and membrane Ig from the cell surface in less than 2 hours (1,3). Such cells no longer label with anti-Ig but label normally with antibodies directed against other surface antigens (1) indicating that, like patching and capping, the pinocytosis is largely specific for Ig. The phenomenon of antibody-induced specific disappearance of a cell-surface antigen is known as <u>antigenic modulation</u> (12) and these studies on B cell Ig indicate that pinocytosis is one mechanism whereby it can occur; in some instances active shedding of antibody-antigen complexes may also play a role (13). If B cells that have been modulated with anti-Ig are placed in culture, membrane Ig reappears within 6-20 hours (3, our unpublished observations).

## Random Distribution of Membrane Macromolecules on Dissociated Cells. A General Rule

The observations on the behaviour of Ig molecules on B cells when cross-linked by anti-Ig antibody provide strong support for the 'fluid-mosaic' model of the plasma membrane, proposed by Singer and Nicolson (14) and initially suggested by the experiments of Frye and Edidin (15) and Blaisie and Worthington (16). They suggest that in dissociated cells, membrane macromolecules are distributed at random as single molecules or small groups of protein subunits, which are free to move in the plane of the membrane. While the notion that biological membranes behave as two-dimensional fluids is now generally accepted, there has been considerable controversy about the distribution of membrane proteins, since a number of mapping experiments have demonstrated a non-random distribution of membrane components: (a) Aoki et al (17) and Stackpole et al (18) found a patchy distribution of alloantigens (Θ, H-2, TL) on mouse thymocytes using hybrid antibody immunoferritin labelling; (b) Mandel reported that intramembranous particles visualized by freeze-fracture electron microscopy were randomly distributed in mouse B lymphocytes but clustered in T lymphocytes (19); and (c) Nicolson found that Concanavalin-A (ConA) binding sites were randomly distributed on isolated membranes of normal 3T3 fibroblasts but were distributed in large aggregates on transformed (20) and trypsinized (21) 3T3 cells. There is now compelling evidence that all of these examples of non-randomness are artifacts induced by the conditions of the experiment.

*Alloantigens on thymus lymphocytes.* The patchy distributions of Θ, H-2 and Tl alloantigens that have been found on mouse thymocytes were initially interpreted as indicating that the various components of the lymphocyte surface were confined to specific areas of the membrane (17, 18) and recently it has been suggested that certain of these putative regions are chosen by viruses when they bud from the cell surface (22). Since all of these studies involved multivalent labelling procedures it could be argued in retrospect, that the patchy distributions seen may have resulted from ligand-induced redistribution of the surface antigens. Recently it has been shown that this is the case.

When Θ is demonstrated on mouse thymocytes by an indirect immunoferritin technique, using mouse anti-Θ antibody followed by rabbit anti-mouse Ig coupled to ferritin (anti-Ig-FT), the distribution of the ferritin labelling depends on the valency of the anti-Ig-FT (23). When it is divalent, a patchy distribution is seen, just as described by Aoki et al, however when monovalent Fab anti-Ig-FT is used the distribution is random and dispersed. This experiment shows that Θ, like Ig, is randomly distributed and free to move in the plane of the membrane. In addition, it indicates that anti-Θ antibody, even though it is divalent, cannot redistribute Θ on its own; it requires additional cross-linking by anti-Ig. We think the most likely explanation for this is that Θ is represented only once (or perhaps twice) on an individual macromolecule and therefore cannot form a lattice; for just as in a precipitation reaction, both the antigen and the antibody must by multivalent for a lattice to be formed. If this explanation is correct, then one can make a prediction about another alloantigenic system on mouse thymocytes, the Thymus-Leukemia (TL) antigens. Since there are three TL antigens, TL 1,2, and 3, which are thought to be present on the same molecule (24), antibody directed against all three antigens (i.e. anti-TL 1,2,3) should be able to redistribute TL on its own, without a second layer of divalent anti-Ig. In fact, that is the case, for when thymocytes are treated with anti-TL 1,2,3, and then monovalent Fab anti-Ig-FT, the ferritin is distributed in patches (23). Only if the thymocytes are prefixed in formaldehyde, and then labelled with anti-TL 1,2, 3 and anti-Ig-FT is a dispersed distribution seen (24). Similar studies have been reported by Davis, who found a patchy distribution of H-2 antigens on mouse lymphocytes when an indirect labelling procedure was used with divalent anti-Ig-FT, and a dispersed distribution when directly conjugated anti-H-2-ferritin was employed (25). In addition, studies with Concanavalin A conjugated to ferritin (ConA-FT) have shown a random distribution of ConA-binding molecules on the surface of lymphocytes which have been prefixed with glutaraldehyde (23). Since ConA binds to the majority of lymphocyte glycoproteins (26), it seems reasonable to conclude that glycoproteins in general are randomly dispersed on the lymphocyte surface.

*Lymphocyte intramembranous particles.* Although the relationship between intramembranous particles (IMPs), visualized by freeze-fracture, and lymphocyte surface antigens and receptors is far from clear (see below), the finding by Mandel that IMPs were distributed in a clustered manner in the membranes of mouse T cells (19) was difficult to reconcile with the apparently random distribution of the surface components. Recently, MacIntyre et al, demonstrated that the clustered distribution of IMPs in mouse lymphocytes is an artifact, induced by the glycerol (used as a cryoprotectant) and was not seen if the cells were fixed with glutaraldehyde prior to freezing (27). The mechanism by which glycerol clusters IMPs is unknown.

Unlike the situation in the erythrocyte membrane, where there is circumstantial evidence to suggest that at least some surface blood group antigens (28), and virus and lectin receptors (29) are associated with IMPs, experiments thus far have failed to demonstrate any relationship between IMPs and surface receptors and antigens in the lymphocyte membrane. When antibody (anti-H-2, anti-Ig) or lectins have been used to cap surface components on mouse lymphocytes, the distribution of IMPs has not been found to be appreciably changed (30, unpublished observations of S. de Petris). In the one report where this was not the case (31), unfixed cells were treated with glycerol prior to freezing, making interpretation difficult. In a preliminary experiment done in collaboration with N.B. Gilula, where mouse thymocytes were treated with 25% glycerol in order to cluster IMPs, then fixed in glutaraldehyde and labelled with Cona-FT, the distribution of ConA-binding sites remained randomly dispersed (Lawson, Gilula and Raff, unpublished).

On balance, the evidence to date suggests that if surface antigens and receptors are associated with IMPs, the association must be a labile one, or perhaps only certain surface components (ones not yet studied) are associated. Whether these findings indicate that the surface receptors and antigens studied thus far do not penetrate through the lipid bilayer is not clear.

*ConA receptors in fibroblasts*. It has been known for many years that normal fibroblasts require higher concentrations of lectins, such as ConA, to agglutinate them than do transformed (32) or trypsinized (33,34) fibroblasts, although the reason (s) for these differences in agglutinability remains unclear. The original hypothesis that these differences reflect excessive synthesis or unmasking (33) of lectin-binding sites on transformed and trypsinized cells has been undermined by the failure to find differences in the binding of radio-labelled lectins (35,36,37,38). The reports that ConA-FT binding sites are clustered on membranes of transformed (20) and trypsinized (21) 3T3 fibroblasts while being dispersed on normal 3T3 cells provided a possible explanation for differences in agglutinability. Although these findings were originally interpreted to mean that glycoproteins spontaneously aggregated in the membranes of transformed (20) and trypsinized (21) cells, it seemed possible that the clustered distributions were induced by the binding of the ConA, which is tetravalent at physiological pH.

To test this possibility, we, in collaboration with L. Mallucci, prefixed normal, trypsinized and polyoma-virus transformed 3T3 cells with glutaraldehyde and then labelled them with ConA-FT (39). In all cases the distribution was random and dispersed, excluding the possibility that ConA receptors spontaneously aggregate during the process of transformation or trypsinization. Similar observations have been made now by a number of investigators (40,41) including Nicolson (42). When we labelled unfixed 3T3 cells, the ConA-FT was distributed in a patchy manner, not appreciably different in trypsinized, transformed and normal cells, despite their striking differences in agglutinability (39). Since a number of laboratories have reported that ConA does not redistribute on normal 3T3 cells, (40,41,42) we have repeated these experiments with native, purified ConA. After incubating the cells at 20° and observing agglutination of transformed and trypsinized but not normal 3T3 cells, the cells were fixed with glutaraldehyde and the distribution of ConA visualized with anti-ConA antibody conjugated to ferritin. Here again, we observed a patchy distribution of ConA on all three cell populations (43). The reason(s) for the difference between our findings and those of others is still unclear, but differences in cell lines would seem the most likely explanation. No matter what the explanation turns out to be, it seems fair to conclude that transformation and trypsinization are not invariably accompanied by an appreciable increase in redistributionability of lectin-binding sites and that differences in agglutinability cannot always be related to differences in the redistributionability of lectin receptors.

## Redistribution of ConA Receptros on Myelin, Synaptosomes and Erythrocytes

Ligand-induced redistribution of membrane molecules has been demonstrated on a variety of different cell types in many laboratories, leaving little doubt that it is a general phenomenon. A few more examples are worth mentioning to emphasize the generality of these phenomena and to point out some possible exceptions.

Myelin is an unusual membrane in that it contains relatively little protein and it seemed of interest to see if the general rules of random distribution and mobility of membrane macromolecules applied. In experiments done in collaboration with Andrew Matus (44), we found that when myelin fragments in a P2 fraction from rat neocortex were prefixed in glutaraldehyde and then labelled with ConA-FT, the distribution was diffuse and random, while labelling of unfixed myelin followed by an additional layer of anti-FT, to increase the cross-linking, produced a strikingly patchy distribution. The same was true for synaptosomes in the P2 fraction, indicating that the nucleus does not influence this behaviour of membrane molecules.

Attempts to demonstrate antibody-induced redistribution of surface antigens on intact erythrocytes by immunofluorescence (3) or immunoferritin-electron-microscopy (D. Shotton, personal communication) have failed thus far, raising the possibility that the erythrocyte is the odd cell out in terms of macromolecule mobility. In addition, the various manoeuvres known to induce IMP aggregation in erythrocyte ghosts (45,46), in general do not work in intact erythrocytes, and there is evidence that even in ghosts, a large proportion of the spectrin, which is attached to the inside of the membrane (47), must be released before IMP aggregation can occur (48). These observations and others (49) have led to the suggestion that spectrin serves as an internal membrane skeleton holding the IMPs in a relatively fixed array. If most of the surface proteins are associated with IMPs then their failure to redistribute when cross-linked is nicely explained by this model. On the other hand, the reports of non-random distribution of H-2 antigens on mouse erythrocytes (17,50) are difficult to explain if redistribution did not occur. In addition, it has been shown that the exposure of intact human erythrocytes to sendai virus can induce clustering of IMPs (51).

In preliminary experiments, we have found that ConA-FT binds in a random and dispersed fashion to the surface of intact adult and newborn mouse erythrocytes. When a second layer of anti-FT antibody was added some patching of the ConA-FT was induced, more so on newborn erythrocytes than in adult cells. This difference between adult and newborn erythrocytes was difficult to interpret in view of the more marked agglutination and distortion of the newborn cells, but it is consistent with the finding that ferritin-conjugated anti-A antibodies bound in a random manner to adult human erythrocytes but in a patchy manner to newborn human erythrocytes (52). It is not clear whether the small degree of patching on adult erythrocytes in our studies reflects glycoprotein movement in the plane of the membrane or simply lectin-induced clumping of long glycoprotein side-chains on the surface of the cells. The larger scale patching on the distorted newborn erythrocytes, on the other hand, suggests redistribution of membrane glycoproteins, but at present we cannot exclude that this is dependent on cell damage induced by the ConA-binding and/or agglutination.

## Implications for Membrane Structure and Function

These studies on the distribution and induced-redistribution of cell-surface antigens and receptors suggest that, in general, proteins and glucoproteins exposed on the surface of dissociated nucleated somatic cells, are randomly distributed as single molecules, or small groups of like- or unlike-molecules which are free to move relative to one another in the plane of the fluid lipid bilayer. However, the concepts of randomness and fluidity require qualification. It is clear that when cells interact with each other to form organized tissues, an ordered arrangement of membrane proteins can be found at the area of cell to cell interaction. Thus acetylcholine receptors in striated muscle cells are confined to the region of the neuromuscular junction and not dispersed over the entire cell membrane (e.g. 53), unless the cell is denervated (e.g. 54), and certain membrane-bound enzymes appear to be restricted to specific regions of liver (55) and kidney (56) parenchymal cells. Thus interacting cells appear to be able to restrain the movement of certain membrane proteins, but the mechanism(s) involved remains to be determined. Even in dissociated cells, it is clear that the plasma membrane is not entirely fluid all of the time, otherwise cells would not be capable of directional movement. In fact, capping implies restricted fluidity in the region of the cap at the tail of the cell, or else the cross-linked components would eventually move away from the tail out over the rest of the cell surface (see ref. 4). There must be cytoplasmic structures which interact directly or indirectly with the cross-linked membrane proteins. One of the most pressing problems in membrane biology is to determine the nature of these cytoplasmic components, and the way in which they interact with membrane macromolecules. Although some of these components are probably actinomyosin-like microfilaments, the recent experiments of Edelman and Yahara and their colleagues (57) have suggested that microtubules, or other colchicine and vinblastine-sensitive proteins, may also be involved. They and Loor et al (3) have shown that the binding of tetravalent ConA to the lymphocyte surface at 37° can inhibit (partially or completely) the ability of divalent anti-Ig to Cap Ig on B cells, and this inhibition can be relieved by high concentrations of microtubule-dissociating drugs (57), despite the fact that these drugs appear not to influence the normal capping of Ig by anti-Ig (1). Since ConA appears to be able to bind directly to surface Ig units (Loor, personal communication, de Petris, unpublished observations), its ability to inhibit Ig capping may well be related to its cross-linking Ig to other membrane glycoproteins which themselves are unable to cap. The reasons for this inability to cap are unknown, but the release afforded by microtubule-dissociating drugs points to the possible importance of microtubules. In studies on polymorphonuclear leucocytes, Ukena and Berlin (58) have shown that phagocytosis does not normally alter membrane transport, whereas if cells are treated with low concentrations of microtubule-dissociating drugs, phagocytosis results in large decreases in transport. They have interpreted these results as indicating that microtubules may play an important role in maintaining the organisation of membrane proteins. Since microtubules are not found close to the plasma membrane whereas microfilaments are, it may be that the putative interaction of membrane proteins with microtubules is mediated via microfilaments.

Besides providing evidence concerning the distribution and mobility of membrane macromolecules, ligand-induced redistribution has provided a useful tool for determining the relationship between different determinants on the cell surface. After redistributing a particular membrane determinant with a specific multivalent ligand, one can determine whether other membrane components move with it (i.e. co-cap) or remain behind. Thus far, all the evidence suggests that, when exposed to ligands of well-defined specifity, only determinants on the same molecule co-cap (1, 8,59,60). In this way it has been shown that the antigenic specificities determined by the two ends of the major histocompatibility locus in mouse (H-2K and H-2D) (59) and man (LA and 4) (60) cap in an independent manner, providing compelling evidence that these specificities are carried on two distinct proteins. Similarly, it has been shown that a single polymeric antigen can cap virtually all of the Ig receptors on a B lymphocyte which specifically interacts with the antigen (61), providing the only direct evidence that all of the receptors on a single B cell have the same antigen-combining site (i.e. the same specificity). Co-capping experiments should prove useful in determining the relationship between T lymphocyte receptors, major histocompatibility antigens and products determined by the histocompatibility-linked immune response (IR) genes, as well as the relationship between oncogenic virus associated antigens and tumour specific antigens on the tumour cell surface.

Perhaps the most intriguing (although least substantiated) aspect of ligand-induced redistribution of membrane proteins is its possible relevance to problems of cell interactions and how ligands (hormones, lectins, antigens, etc.) interacting with cell-surface receptors, actually signal the cell. For example, when cells form gap-junctions in tissues or in tissue culture it is possible that randomly distributed and mobile putative gap junction proteins interact with corresponding or complementary proteins on adjancent cells causing their redistribution into organized clusters which provide the structural basis for the electrically low-resistance junctions (62,63). When multivalent antigens interact with their specific receptors on the surface of a lymphocyte, it has been suggested that the induced clustering of receptors provides at least part of the initial signal, leading to the cells decision to divide and differentiate into an effector cell (1,3,64). On the other hand, there is some evidence that excessive aggregation of receptors on basophils, induced by antigen or antibody, may inhibit the release of histamine from these cells (65). Although there is no direct evidence in any system for a causal relationship between surface receptor movement and cell signalling, the appreciation that receptors can and do redistribute when cross-linked by multivalent ligands has provided new ways of thinking about and studying the mechanisms that operate in cell-cell and ligand-cell interactions.

## References

1. Taylor, R.B., Duffus, W.P.H., Raff, M.C. and de Petris, S. Nature New Biol. 233(1971)225.

2. de Petris, S. and Raff, M.C. Nature New Biol. 241(1973)257.

3. Loor, F., Forni, L. and Pernis, G. Eur. J. Immunol. 2(1972)203.

4. de Petris, S. and Raff, M.C. Eur. J. Immunol. 2(1972)523.

5. Papahadjopoulos, D. Jacobson, K., Nil, S. and Isac, T. Biochim. Biophys. Acta, 311(1973)330.

6. Oldfield, E. and Chapman, D. FEBS Letters 23(1972)285

7. Valentine, R.C. and Green, N.M. J. Mol. Biol. 27(1967)615.

8. Kourilsky, F.M., Silvestre, D., Neauport-Santes,

C., Loosefelt, Y. and Dausset, J. Eur. J. Immunol. 2(1972)249.

9. Unanue, E.R., Perkins, W.D. and Karnovsky, M. J. J. Exp. Med. 136(1972)885.

10. Yahara, I. and Edelman, G.M. Proc. Nat. Acad. Sci. U.S.A. 69(1972)806.

11. Unanue, E.R., Perkins,W.D. and Karnovsky,M.J. J. Immunol. 108(1972)569.

12. Old, L.J., Stockert, E., Boyse, E.A. and Kim, J.H. J.Exp. Med. 127(1964)523.

13. Wilson, J. D. and Nossal, G.J.V. Eur. J. Immunol. 2(1972)225.

14. Singer, S.J. and Nicolson, G.L. Science 175 (1972)720.

15. Frye, L.D. and Echdin,M. J.Cell Sci. 7(1970)319.

16. Blaisie, J.K. and Worthington, C.R. J.Mol. Biol. 39(1969)407

17. Aoki, T., Hämmerling, U., de Harven, E., Boyse, E.A., Old, L.J. J. Exp. Med. 130(1969)979.

18. Stackpole, C.W., Aoki, T., Boyse, E.A., Old, L.J., Lumley-Frank, J. and de Harven, E. Science 172 (1971) 472.

19. Mandel, T. Nature New Biol. 239(1972)112.

20. Nicolson, G.L. Nature New Biol. 233(1971)244.

21. Nicolson, G.L. Nature New Biol. 239(1972)193.

22. Aoki, T. and Takahashi, T. J. Exp. Med. 135 (1972)443.

23. de Petris, S. and Raff, M.C. Eur. J. Immunol. In press.

24. Boyse, E.A. and Old, L.J. Ann. Rev. Genet. 3 (1969)269.

25. Davis, W.C. Science 175(1972) 1006.

26. Allan, D., Auger, J. and Crumpton, M.J. Nature New Biol. 236 (1972) 23.

27. MacIntyre, J.A., Giluta, N.B. and Karnovsky,

M.J. J. Cell. Biol. In press.

28. Pinto da Silva, P., Branton, D. and Douglas, S.A. Nature 232(1971) 194.
29. Tillack, T.W., Scott, R.E. and Marchesi, V.T. J. Exp. Med. 135(1972)1209.
30. Karnovsky, M.J. and Unanaue, E.R. Fed. Proc. 32(1973)55.
31. Loor, F. Eur. J. Immunol. 3(1973)112.
32. Aub, J.C., Sanford, B.H. and Cote, M.N. Proc. Nat. Acad. Sci. U.S.A. 50(1963)613.
33. Burger, M.M. Proc. Nat. Acad. Sci. U.S.A. 62(1969)994.
34. Inbar, M. and Sachs, L. Proc. Nat. Acad. Sci. U.S.A. 63(1969)1418.
35. Sela, B., Lis, H., Sharon, N. and Sachs, L. Biochim. Biophys, Acta 249(1971) 546.
36. Arudt-Jovin, D.J. and Berg, P. J. Virol. 8 (1971) 716.
37. Cline, M.J. and Livingston, D.C. Nature New Biol. 232 (1971) 155.
38. Ozanne, B. and Sambrook, J. Nature New Biol. 232 (1971) 156.
39. de Petris, S., Raff, M.C. and Mallucci, L. Nature New Biol. 244 (1973) 275.
40. Inbar, M. and Sachs, L. FEBS Letters 32 (1973) 124.
41. Rosenblight, J.Z., Ukena, T.E., Yin, H., Berlin, R.D. and Karnovsky, M.J. Proc. Nat. Acad. Sci. U.S.A. 70 (1973) 1625.
42. Nicolson, G.L. Nature New Biol. 243(1973)218.
43. de Petris, S., Mallucci, L. and Raff, M.C. Manuscript in preparation.
44. Matus, A., de Petris, S. and Raff, M.C. Nature New Biol. 244 (1973) 278.
45. Pinto da Silva, P. J. Cell Biol. 53 (1972)777.

46. Tillack, T., Scott, R.E. and Marchesi, V.T. J. Cell Biol. 47 (1970) 213.

47. Nicolson, G.L., Marchesi, V.T. and Singer, S. J. J. Cell Biol. 51 (1971) 265

48. Elgsaeter, A., Shotton, D. and Branton, D. Manuscript in preparation.

49. Nicolson, G.L. and Painter, R.G. J. Cell Biol. 59 (1973) 395.

50. Nicolson, G.L., Hyman, R. and Singer, S.J. J. Cell Biol. 50 (1971) 905.

51. Bächi, T., Aguer, M. and Howe, C. J. of Virol. 11 (1973) 1004.

52. Blanton, P.L., Martin, J. and Haberman, S. J. Cell Biol. 37 (1968) 716.

53. Peper, K. and McMahan, U.J. Proc. R. Soc. London Biol. Sci. 181 (1972) 43

54. Fambrough, D.M. and Hartzell, H.C. Science 176 (1972) 189.

55. Essner, E., Movikoff, A.B. and Masek, B. J. Biophys. Biochem. Cytol. 4 (1958) 711.

56. Goldfischer, J. Essner, E. and Novikoff, A.B. J. Histochem. Cytochem. 12 (1964) 72.

57. Edelman, G.M., Yahara, I. and Wang, J.L. Proc. Nat. Acad. Sci. U.S.A. 70 (1973) 1442.

58. Ukena, T.E. and Berlin, R.D. J. Exp. Med. 136 (1972) 1.

59. Neuport-Santes, C., Lilly, F. Silvestre, D. and Kourilsky, F.M. J.Exp.Med. 137(1973)511.

60. Bernoco, B.S., Cullen, S., Scudeller, G., Trinchieri, G. and Ceppellini, R. Proc. of the 5th International Histocompatibility Workshop Conference. Munksgaard, Copenhagen, 1973. In press.

61. Raff, M.C., Feldmann, M. and de Petris, S. J. Exp.Med. 137 (1973) 1024.

62. Johnson, R.G. and Sheridan, J. Science 174 (1971) 717.

63. Gilula, N.B., Reeves, D.R. and Steinbach, A. Nature 235 (1972) 262.

64. Metzger, H. Adv. Immunol. In press.

65. Becker, K.E., Ishizaka, T., Metzger, H., Ishizaka, K. and Grimley, P.M. J. Exp. Med. 138 (1973) 394.

---

# VII

# RECONSTITUTION OF SPECIFIC MEMBRANE FUNCTIONS

# LIPID-PROTEIN ASSEMBLY AND THE RECONSTITUTION OF BIOLOGICAL MEMBRANES

M. Montal

Departamento de Bioquímica
Centro de Investigación y de
Estudios Avanzados del I.P.N.
Apartado Postal 14-740
México, D.F.

<u>El camino es siempre mejor que la posada.</u>

*Cervantes*

It is now well accepted that biological membranes behave as reversible transducers of various energy-sources, mainly light, redox, electrochemical and phosphate-bond energies (1,2). Considerable effort has been directed towards the elucidation of the mechanisms underlying the phenomena of energy transduction but, with the exception of the nerve-excitable membrane, little has been accomplished. In this case, the analysis of the electrical behaviour of the membrane provided a sound basis for the proposal and test of several hypotheses (1,3). Unfortunately, such an approach is not applicable to most membranes, particularly those of subcellular organelles such as mitochondria, chloroplasts or photoreceptor discs, where the energy-transducing machinery is located in inner membrane compartments not amenable to electrical measurements. One approach intended to circumvent this problem is that of "model building", provided that a technical solution can be managed. The rationale is that an understanding of the model will ripen into insight concerning the biological mechanisms.

According to a current widely-held view, biological membranes are considered as fluid mosaic structures composed mainly of lipids and proteins, with a lipid bilayer as the basic structural element (1,4,5). Hence, it is within this context that resides the importance of the recognition between lipids and proteins, the nature of their interactions and the resultant assembly process.

In 1962 Mueller and Rudin (6) were able to form single planar bilayers separating two aqueous phases; by 1965 Bangham (7) and coworkers had established that lipid bilayers in the form of small vesicles were suitable for permeability studies. So it was that the basic structural framework required to begin the endeavour of building the model became available. In the past twelve years of bilayer research an impressive advance has been attained in the theoretical and experimental evaluation of the carrier and channel mechanisms of ion transport across membranes (1,8,9,10) and of electrical excitability (1). This progress has been possible due to the success in incorporating small peptides and membrane-active antibiotics into planar bilayers containing hydrocarbon solvents (1,8-10). In contrast, attempts to introduce large and defined membrane proteins in a functionally active state have met with considerable difficulties, although recently there have been reports of some promising advances (11-22). The failures may predominantly be attributed to the presence of hydrocarbon or to the inability of the proteins to interact with a preformed bilayer.

Aware of the obstacles that prevented the achievement of such a goal, we have first attempted to form planar bilayers in the absence of hydrocarbon solvents and, second, to employ both the protein and the lipid as building blocks of the bilayer. This is in contrast to traditional attempts in which the protein is added as a secondary adsorbate to a preformed bilayer.

Paul Mueller and I succeeded in forming bimolecular membranes from two lipid monolayers at an air-water interface by the apposition of their hydrocarbon chains through an aperture made in a hydrophobic partition which separates the two monolayers (23-25). This method of bilayer formation, overcomes the adverse effects of the presence of hydrocarbon solvents in black lipid films previously used in functional reconstitution studies. It allows us, for the first time, to attempt the simultaneous assembly of lipid and protein into a bilayer either after "both" components are spread as a monolayer at the air-water interface or after the proteins are allowed to penetrate the expanded lipid monolayers (26). The first of these two experimental possibilities requires the lipid-protein complex to be "soluble" in an organic solvent but still native and active. How can this be achieved?

Carlos Gitler and I (27,28) reasoned that the partition of a lipoprotein (i.e. a lipid-protein complex which shares solubility properties with proteins (29)) between an aqueous medium and a solvent of low dielectric constant would be strongly favored towards the non-aqueous phase if the overall complex were neutral. Since most membrane lipids are either neutral or negatively charged, it is likely that in a lipoprotein, ion-pairs between protein-cationic and lipid anionic groups exist (30). If the residual protein anionic groups are neutralized by ion-pair formation with suitable cations, thus rendering the overall complex neutral, the partition of the lipoprotein into solvents of low dielectric constant would be considerably enhanced.

By using a model lipoprotein composed of neutral and acidic lipids and cytochrome c, we established that these concepts were "operationally" valid, and we demonstrated the complete partition of the cytochrome c lipoprotein into hydrocarbon solvents in the presence of an adequate concentration of $Ca^{++}$ (the most effective counterion) (28). In other words, aqueous ion-pair formation between lipid and protein, with the concomitant charge neutralization of the complex, results in the formation of a proteolipid, (i.e., a lipid-protein complex which shares solubility properties with lipids (29)).

The next step in the logical sequence was to form a proteolipid with a "real" membrane protein. However, since membrane proteins are soluble only in detergents (cf. 31) it was necessary to obtain a detergent-free lipoprotein. The recent advances achieved particularly by Dr. Racker's group (31-35), as well as by several others (36-41), in preparing detergent-free phospholipid vesicles containing several membrane proteins which retain biological activity have permitted us to form proteolipids of membranes proteins. The techniques involve mixing detergent solutions of lipid and protein and removing the detergent slowly by dialysis (32,33,36-41), or sonicating a dispersion of lipid and protein in salt media (34, 35).

We have concentrated our efforts on two important membrane proteins, namely cytochrome c oxidase and rhodopsin.

## Cytochrome c oxidase

Cytochrome c oxidase (E.C. 1.9.3.1) is a mitochondrial inner-membrane enzyme which catalyzes the final reaction of respiration by reducing molecular oxygen to water with the electrons of the respiratory chain funelled through its substrate, cytochrome c. It is a dimer of two heme proteins: cytochrome a, which reacts with cytochrome c, and cytochrome $a_3$ which reacts with oxygen, cyanide and carbon monoxide (cf. 42). There is considerable evidence suggesting that the enzyme spans the entire width of the inner mitochondrial membrane, interacting with cytochrome c on the cristae side, and with oxygen on the matrix side (43,44). This electron-transfer reaction in "coupled" mitochondria is associated with the synthesis of one molecule of ATP. This orientation is a requirement in Peter Mitchell's hypothesis for energy coupling (45) in which the oriented protein allows electrons to flow across the mitochondrial membrane in a particular direction. As a consequence of this vectorial movement of electrons, charge separation occurs across the membrane: electrons on one side, protons on the other. Accordingly, such a transmembrane redox reaction can be thought of as a "proton-generator" (an electrogenic proton-pump). The energy of the resulting proton gradient, Mitchell postulates, is transformed into phosphate bond energy (ATP) by a proton-translocating membrane-bound ATPase (46).

Previous experimental evidence are very suggestive of the electrogenic nature of the redox reaction but they do not constitute direct proof (33,47,49). A move direct experiment would be to form planar bilayers with incorporated cytochrome oxidase accessible to electrical measurements.

Figure 1 illustrates the sequence of lipid-protein assembly in planar bimolecular membranes that we have followed. We prepared the lipoprotein vesicle according to Racker (34). The addition of $Ca^{++}$ results in the formation of a proteolipid as evidenced by the complete (over 90%) extraction or partition of the complex into n-hexane. No extraction at all is obtained in the absence of $Ca^{++}$. We have been able to assay the enzymatic activity of the proteolipid in three different interfacial organizations:

1. Lamellar, as proteolipid bilayer vesicles*
2. Proteolipid monolayers at the air water interface, and
3. Lamellar, as proteolipid planar bilayers.

*Proteolipid bilayer vesicle**. This is prepared from the proteolipid in hexane by solvent evaporation under a stream of nitrogen and subsequent hydration in salt media.

---

* These are indeed lipoprotein vesicles in the aqueous phase; the term "proteolipid" vesicles is used only to emphasize that they are formed from the complex after its partition into hexane, i.e. from the proteolipid.

Adolfo Martínez Palomo and I have investigated the freeze-fracture images of the proteolipid vesicles. As shown in Fig. 2, the fracture faces of the proteolipid vesicles show a homogeneous distribution of particles of 120 Å in diameter. In contrast, these particles are absent in vesicles prepared from equivalent phospholipids but without cytochrome oxidase. According to the currently accepted view (50,51), the plane of fracture during the freeze-fracture procedure occurs at the contact point of the terminal methyl groups of the hydrocarbon tails of the apposed monolayers. The absence of particles in pure lipid vesicles (see also 52) in contrast to their presence in proteolipid vesicles implies that the particles must be the cytochrome oxidase molecules themselves. Fig. 3 illustrates a comparison of the enzymatic specific activity of cytochrome oxidase as originally purified in detergent with that in the lipoprotein and proteolipid states. It is apparent the activity in all these interfaces is essentially the same. Furthermore, the first order character of the reaction and its reactivity to cyanide are conserved. This demonstrates that the extraction of cytochrome oxidase as a proteolipid into n-hexane exerts no deleterious effect on its activity. Thus, upon solvent evaporation and hydration, a smectic mesophase with incorporated functionally-active cytochrome oxidase is achieved.

*Proteolipid monolayer at the air-water interface.* It is well known that several proteins (not membranous) unfold at interfaces (53,54). Hence, survival of enzymatic activity following partition into hexane does not assure that surface denaturation will not occur upon spreading the proteolipid at an air-water interface. As the bilayer is to be formed from monolayers it must be shown that cytochrome oxidase is active in the monolayer and has not been denatured at the interface.

Fig. 4 illustrates the difference in the surface activity of the cytochrome oxidase proteolipid with that of the equivalent phospholipid.

The proteolipid exhibits an expanded $\pi/A$ (surface pressure/area molecule) behaviour with an area per phospholipid molecule of 82.5Å$^2$ at collapse pressure. In contrast, soybean lecithin is compressed monotonically and has an area per phospholipid molecule of 61 Å$^2$ at collapse pressure (see 55). The marked difference between the activity of lipid and proteolipid samples must be due to the protein since there is no other difference between them. The expanded profile of the proteolipid indicates the effective spatial occupation of the protein at the interface. Moreover, the difference in area/molecule at collapse pressure indicates that even at that pressure, the protein is not "extruded" from the monolayer into the aqueous phase but exists as an integral component of the monolayer. Thus, the amphipatic nature of both lipid and protein components, aided by hydrophobic interactions between them, must determine the coexistence of both elements at the organized interface.

Fig. 5 shows that cytochrome c oxidase proteolipid spread at an air-water interface over a subphase containing reduced cytochrome c can slowly catalyze the oxidation of cytochrome c. The oxidation reaction monitored spectrophotometrically (56), shows first-order kinetics and it is cyanide-sensitive. The oxidation of cytochrome c is indeed the result of enzymatic activity of the monolayer since increasing the amount of protein in the monolayer at first results in an enhancement of oxidation but as the amount of cytochrome oxidase is increased further its specific activity begins to decrease, reaching the maximum at a concentration close to that at which collapse of the monolayer occurs. Likewise, the first-order rate constant attains a plateau, indicating that once all the surface is covered by the monolayer the reaction proceeds at maximum velocity. This demonstrates that the cytochrome oxidase proteolipid retains its enzymatic activity after being spread at an interface.

*Proteolipid planar bilayers.* The necessary elements for the formation of a biologically active planar bilayer are now fulfilled. Thus, a planar bimolecular membrane can be assembled as previously described (23-25) by apposition of two monolayers; these monolayers can either be two proteolipid monolayers or one lipid and one proteolipid monolayers. Proteolipid bilayers formed by this method exhibit electrical properties which are distinct from those exhibited by the pure lipid membrane. As illustrated in Fig. 6, the electrical capacity of the proteolipid bilayer is lower than that of the equivalent lipid bilayer; the average values are 0.65 and 0.95 $\mu F \cdot cm^{-2}$, respectively. As a first approximation, the change in membrane capacity on incorporation of proteins can arise from changes in dielectric constant, dielectric thickness or both. There are no measurements of dielectric constant of membrane proteins available so far; therefore a variation in this parameter cannot, at present, be ruled out. It is clear, however, that incorporation of a partially hydrophilic protein into a membrane would, if anything, increase the dielectric constant, thus increasing the membrane capacity in comparison to that of equivalent lipid bilayers (57-59). Assuming no change in dielectric constant upon incorporation of cytochrome oxidase we can calculate a lower bound for the "effective" dielectric thickness of the cytochrome oxidase-containing membrane using the expression:

$$\ell = \varepsilon A / 4\pi C$$

where C is membrane capacity, A is area, $\varepsilon$ is the dielectric constant and $\ell$ is the effective dielectric thickness. Estimating 2.1, the dielectric constant of a long chained hydrocarbon, to be the dielectric constant of the membrane, the dielectric thickness of the lipid bilayer is 20 Å and that of the cytochrome oxidase membrane can be no smaller than 29 Å.

When the bilayers are formed in the absence of cytochrome c or in the presence of oxidized cytochrome c, there are no significant changes in membrane conductance. However, if the proteolipid is spread over a subphase containing oxidized cytochrome c only on one compartment, the resultant bilayer exhibits a conductance on reduction of cytochrome c by ascorbate. The magnitude of the conductance is of the order of $10^{-7}$mhos. $cm^{-2}$ and is stable, i.e. it does not correspond to a step in the breakdown of the membrane. These effects are not observed in the equivalent lipid membrane.

Following Dr. Racker's advice (43) that "where there is an assay, there is a will and there is a way" we searched for a redox-dependent membrane potential. This is shown in Fig. 7: a cytochrome oxidase proteolipid bilayer was formed in the presence of oxidized cytochrome c on only one compartment. On addition of 5mM ascorbate to both compartments, there developed following a lag of approximately 30 sec a membrane potential that reached a steady state value of 33mV after about 7 min. At this point, potassium cyanide was added to both compartments, and as can be seen in the record, in approximately 12 min the potential returned to the zero base line. The lower trace indicates the result of an experiment where the membrane was preincubated in the presence of cyanide. In this case no potential resulted from the reduction of cytochrome c by ascorbate. The sign of the potentials is such that the compartment where cytochrome c is located becomes positive with respect to the other one. Ascorbate and cyanide per se do not modify the conductance of either lipid or proteolipid bilayers and therefore the potential changes observed cannot be attributed to resistance changes. This is presented in the final portion of the record where the magnitude of the membrane resistance is tested by applying a constant current pulse and displaying the resultant membrane potential. It is found that the membrane conductance is essentially the same before and after the reaction, i.e. $10^{-7}$mho. $cm^{-2}$. The potential cannot be generated if cytochrome c is absent or when membranes are formed with proteolipid which has been heated to 50°C for one hour prior to the formation of the membrane. None of these effects have been observed with equivalent lipid bilayers under identical conditions.

We tentatively interpret this potential as being due to a transmembrane electron-transfer reaction which results in the production of $H^+$ in the cytochrome c side of the membrane and the consumption of $H^+$ in the opposite side consequent to oxygen reduction and water formation. It should be emphasized that the only assymmetry of the system is that due to cytochrome c. The following alternative considerations can be ruled out:

a. *Ionic diffusion potential.* The symmetric addition of tris-ascorbate and KCN to a medium of high ionic strength (0.2M) would make diffusion potentials of $tris^+$-$Ascorbate^-$ and $K^+CN^-$ negligible. The membranes were formed in 0.1M tris phosphate buffer; this system would buffer pH changes associated with the redox reactions to well below the detection level of the electrodes. Control experiments have demonstrated that no potential can be detected by adding an amount of $H^+$ that is 10-fold larger than that calculated to be produced by the reaction.

b. *Redox potential not due to transmembrane electron transfer.* Ascorbate reduces cytochrome c in one compartment and dehydroascorbate is formed. Hence a redox gradient of dehydro- and hydroascorbate could be formed across the membrane on reduction of cytochrome c. This potential however, would not be cyanide sensitive. The cyanide sensitivity of the potential measured thus rules out this possibility.

The membrane potential measured is at present poorly reproducible and even though the system potentially allows the study of as yet unexplored facets of the cytochrome oxidase reaction such as its modulation by the transmembrane electric field (60), investigation of the kinetics of a such a coupling, and the asymmetric topology of its components, a careful and methodic evaluation is required before the reconstitution can be accepted as a fact.

What factors could determine the lack of reproducibility?

In the hierarchy of complex factors that must be present in order to achieve a successful reconstitution, the cytochrome oxidase system occupies a high rank. The necessary enzyme-substrate interplay implies that strong protein-protein interactions must be occurring at the interface in such a way that the energies involved are not a challenge to the stability of the membrane; that the active-site of the enzyme is accessible to its substrate; that the orientation of the protein and its packing arrangement with respect to neighbouring lipids is analogous to the natural situation. These are among the important factors that must be recognized in all attempts to reconstitute membrane functions involving receptor-effector interactions. In such an operational scale the simplest system is that in which once the protein is incorporated into the bilayer the response is elicited by a physical and not a chemical stimulus. Such is the situation of rhodopsin.

## Rhodopsin

Rhodopsin is a lipoglycoprotein which constitutes over 80% of the membrane protein in the outer segment of rod photoreceptors (61-64).

As discussed in detail by Cone previously in this conference, the mechanism of action of rhodopsin in visual excitation still remains elusive (67-73), partly because natural rhodopsin-containing membranes amenable to the experimental methods required to investigate the transport properties of rhodopsin have not been successfully prepared.

Juan Korenbrot and I (58) succeeded in incorporating rhodopsin into planar bilayers. We have followed the same sequence outlined for cytochrome oxidase: a lipid vesicle containing purified native rhodopsin was prepared by the procedure of Hong & Hubbell (36,37) combining detergent solutions of lipid and rhodopsin followed by detergent removal on dialysis. In the presence of $Ca^{++}$, the partition of this lipoprotein into hexane is enhanced. In contrast to cytochrome oxidase, the rhodopsin-lipid complex in the non-aqueous phase is turbid. A possible explanation for this difference in optical appearance is a higher concomitant extraction of water into the solvent in the case of rhodopsin, the final state being that of an emulsion of the lipoprotein with water in hexane. The difference between these two membrane proteins can be accounted for by the presence in rhodopsin of carbohydrate residues covalently attached to it (63,64) that would drag considerable amount of water along with the protein into the solvent. Absorption spectra in the visible range of the organic solvent phase showed a peak with $\lambda max$ at 498 nm which changed upon illumination to a distinct yellow colour. The spectral similarity between the native and the extracted forms of rhodopsin is evidence that, despite the partition of rhodopsin between an aqueous and a non-aqueous phase, the protein remains native.

The rhodopsin "proteolipid" in hexane spreads into a monolayer at an air-water interface, and bilayers are formed either by apposing two proteolipid monolayers or under asymmetric conditions with pure lipid on one side. The latter situation results in the formation of stable membranes that exhibit an electrical capacity of about 0.65 $\mu F.cm^{-2}$. This result is analogous to that observed for cytochrome oxidase and for both cases the equivalent lipid bilayer has a capacity of about 0.95 $\mu F.cm^{-2}$. Thus, accepting for the time being the arguments advanced before on the variation of dielectric constant, the capacity data indicate an increase in the effective dielectric thickness of the bilayer on incorporation of rhodopsin.

In the dark, the membrane conductance of rhodopsin-proteolipid bilayers is essentially the same as that of the equivalent lipid membranes, i.e. $10^{-7}$-$10^{-8}$mhos.$cm^{-2}$ (Fig. 8A). Fig. 8B illustrates the membrane conductance changes upon dim continuous illumination. At this light intensity (200 $\mu$watt/$cm^2$) there is a slow development of conductance that reaches a stable value of 1.4 x $10^{-3}$ mhos.$cm^{-2}$ with a halftime of about 1 min. (please note scale calibrations). Fig. 8C presents the final stable state of the membrane with a capacity of 0.65 $\mu$F.$cm^{-2}$ and a conductance of 1.4 x $10^{-3}$ mhos.$cm^{-2}$. Thus, illumination does not affect the capacity of these membranes (within the resolution of our method) but does appear to increase membrane conductance by about five orders of magnitude. The steady-state current-voltage relations of the illuminated membrane are plotted in Fig. 9; the membrane exhibits ohmic behaviour up to its dielectric break-down voltage at approximately 200 mV; the slope conductance is 1.4 x $10^{-3}$mhos.$cm^{-2}$.

These preliminary results are the initial steps in our attempt to understand the molecular mechanism of rhodopsin action. We are just beginning to overcome some reproducibility problems and starting to explore the ionic specificity of this macroscopic conductance and its microscopic basis. The recent availability of techniques amenable to evaluating the formation of ion-channels across bilayers (cf. 9,13, 17) will allow us to test this possibility. Hagins (68) has proposed a mechanism for excitation in vertebrate rods (and cones) which assigns a central regulatory role to $Ca^{++}$ ions. In this mechanism $Ca^{++}$ ions are actively concentrated in the internal space of the discs; light transiently increases the disc membrane conductance to $Ca^{++}$, thus raising its cytoplasmic concentration. The $Na^{+}$ conductance of the plasma membrane is reduced by $Ca^{++}$ (68,70,72,74), presumably by reversibly blocking the sites at which the dark current enters (closing the plasma membrane $Na^{+}$ channels). In other words, $Ca^{++}$ acts as the connecting signal between discs and plasma membrane. This hypothesis may be tested with our experimental rhodopsin membrane.

Additionally, the so called "early receptor potential" has been attributed to charge displacements due to rhodopsin photolysis which contribute charge to the capacity of the plasma membrane (cf. 75). This fast photovoltage is proportional to the number of photopigment chromophores absorbing one photon and is direct manifestation of the light-induced changes in the protein itself. The experimental rhodopsin membrane developed allows, in principle, the direct measurement of these fast photovoltages and the associated action spectrum.

The approach outlined here is potentially powerful in the study of other membrane systems, in particular that of Halobacterium Halobium bacteriorhodopsin (76-78) recently shown by Racker and Stoeckenius (cf. 34) to be involved in light-induced $H^+$ transport; the "excitable vesicles" prepared from the electroplax of Electrophorus-Electricus by Kasai & Changeux (79) that possess not only the acetylcholine receptor, but also the components necessary for the chemical transduction process (80) and the $Ca^{++}$-ATPase of sarcoplasmic reticulum (SR) (cf. 81).

A unique feature of the principle of bilayer formation from lipid monolayers, which is not shared with either the black lipid films or the lipid vesicles, is the formation of asymmetric membranes (23-25). The importance of asymmetry and biological function is again manifested in the cytochrome-c-oxidase-and rhodopsin-containing membranes. Freeze-fracture studies have revealed that most membrane particles are preferentially associated with the inner, cytoplasmic monolayer of the bilayer (83). This is particularly the case with rhodopsin (36,37,58,82), bacteriorhodopsin (77) and the cytochrome c oxidase proteolipid vesicles shown in Fig. 2.

The asymmetric protein orientation and distribution may be accounted for by the synthesis of membrane proteins in the cytoplasm or on ribosomes bound to the cytoplasmic side of the bilayer that are subsequently incorporated into the membrane. In the case of yeast cytochrome oxidase, there is evidence for the asymmetric assembly of the subunits; the mitochondrially synthesized subunits are hydrophobic and are integrated into the inner mitochondrial membrane from the matrix side; "they are synthesized on mitochondrial ribosomes in close vicinity to their site of deposition" (84). The catalytic subunits are hydrophilic and synthesized in the cytoplasm; "they are imported from the outside and are situated close to the membrane surface" (84). In the case of rhodopsin, the available evidence indicates that the entire protein is synthesized in the inner segment cytoplasm, glycosylated in the Golgi apparatus, and subsequently exported (in a yet unknown form) through the connecting cilium to the outer segment (85). The asymmetric orientation arises from the "integration" of the protein into the plasma membrane which repeatedly infolds to form the disc membranes (86). The question of why protein molecules are predominantly on the inside of the discs is still open.

The use of proteolipids as a relevant and potentially informative model of membrane proteins can be foreseen. The recent observation that the major product of mitochondrial protein synthesis is a low molecular weight proteolipid (87,88) suggests that a concerted synthesis of lipid and protein is required for a hydrophobic protein to exist at polar-apolar interfaces in such a form that its insertion into a preformed membrane would be favored.

The combined use of spectroscopic and chemical techniques may shed some light into the kinetics and equilibrium characteristics of several processes occurring, now, in low dielectric constant media. The information botained in aqueous media may not be applicable to what occurs when the protein is located in the membrane environment (89). Our lack of information on the dielectric constant of membrane proteins (58) could be explained by the failure to obtain an adequate experimental system. The proteolipid may be such a model. Furthermore, the broad resolution power of dielectric relaxation techniques, that can follow molecular processes occurring in the $10^{-12}$ -$10^{-1}$ sec range may provide important information on the rate constants of charge and hydrophobic interactions, the extent of hydration and several other pertinent questions.

The operationally valid concept of proteolipid formation causes one reflect anew on the importance of ion-pairs in biological membrane function. The fact that many solubilized membranous enzymes, particularly ATP-ases, are predominantly reactivated by acidic phospholipids (cf. 90, 91) might imply that the optimum catalytic interface requires the formation of ion pairs between lipid and protein. The evidence that physiological ions allow the extraction of lipoproteins into hydrocarbon solvents suggests that analogously they may induce neutralization of the residual charge of a lipid-protein complex in a membrane and hence allow its penetration into the apolar core of the bilayer (28). Clowes established in 1916 (92) that an alteration of the ratio of univalent to divalent cations in favor of the divalent results in a phase inversion from an "oil in water" state to a "water in oil" state. Taking into consideration that the recombination rates of ion-pairs in liquids of low dielectric constant are in the nanosecond time-range (93,94), such transitions (lipoprotein → proteolipid) in location of the lipid-protein complex could be extremely fast. This would be in line with the reported rotational relaxation rates of rhodopsin (95,96) and cyto chrome oxidase (97) in their native membranes.

Following this reasoning Korenbrot and I (58) proposed that if excitation exposes $Ca^{++}$ binding sites which are not available in the dark, then charge neutralization of the lipoprotein would allow rhodopsin to sink deeper into the hydrocarbon region of the membrane. In the thin hydrocarbon region of photorreceptor membranes (∿ 28Å (98, 99) the large (∿ 45Å in diameter) (100) rhodopsin molecule need only sink a few Å to cross the low dielectric region. Thus, rhodopsin might translocate $Ca^{2+}$ ions by "bobbing" in the membrane. Indeed, Blasie (101) has interpreted the changes in X-ray scattering by photoreceptor membrane that follow illumination as resulting from rhodopsin molecules sinking by about 10 Å into the hydrophobic region of the membrane.

The proposed formation of ion-pairs and their existence in apolar media may be open to experimental validation, since optical and NMR spectroscopies have been successfully applied recently to detect and study the structure of ion pairs in solvents of low polarity (102).

The central role of $Ca^{++}$ in muscle (cf. 81), excitability (cf. 103), neurotransmitter release and action (103), hormone-receptor interaction (103), photoreception (cf. 68) oxidative phosphorylation (cf. 104), and in general in all bio-energy-transductions and its wide physiological importance make these ideas of ion-pair formation even more attractive.

The fact that biologically active bilayers can be formed from catalytically active proteolipid monolayers suggests that this principle of membrane formation might have played a role in prebiotic evolution, as proposed by Goldacre (105). This tempting speculation may find support in the successful laboratory synthesis of long chain fatty acids, porphyrins and protein analogues obtained under simulated abiotic atmospheres (106-107).

## Concluding Remarks

It is obvious that these findings give rise to as many -if not more- problems as those which they have attempted to solve, though at a different level of understanding. The encouraging results with the operationally valid concepts developed compeles me to propose this as an approach of furthering long-range projects of far-reaching consequence. I want to emphasize that this is just the beginning, but the consideration that this approach may allow us to go from the physiology and biochemistry of membrane-associated phenomena to what may be called Membrane Molecular Biology, is thought-provoking.

Acknowledgements

The author wishes to acknowledge with deepest thanks the kindness and helpfulness of Mrs. Silvia Almanza de Celis throughout this work, and the criticism of Drs. Carlos Gómez-Lojero and Juan Korenbrot.

References

1. Mueller, P., and Rudin, D.O. (1969) In; Current Topics in Bioenergetics. Vol. 3, pp. 157, ed. Sanadi, D.R. New York: Academic.
2. Chance, B. and Montal, M. (1971) In: Current Topics in Membrane and Transport. Vol. 2, pp. 99, eds. Bronner, F. and Kleinzeller, A. New York: Academic.
3. Hodgkin, A.L. and Huxley, A.F. J. Physiol. (London) 117 (1952) 500.
4. Singer, S.J. and Nicholson, G.L. Science 175 (1972) 720.
5. Gitler, C. Ann. Rev. Biophys. Bioengineer. 1 (1972) 51.
6. Mueller, P., Rudin, D.O.,Tien, H.T. and Wescott, W.C. Nature 194 (1962) 979.
7. Bangham, A.D., Standish, M.M. and Watkins, J.C. J. Mol. Biol. 13 (1965) 238.
8. Eisenman, G., Szabo, G., Ciani, S., McLaughlin, S. and Krasne, S. (1973) In: Progress in Surface and Membrane Science. Vol. 6, pp. 139. New York and London:Academic.
9. Haydon, D.A. and Hladkly, S.B. Quart. Rev. Biophys. 5 (1972) 187.
10. Stark, G., Ketterer, B., Benz, R., and Läuger, P. Biophys. J. 11 (1971) 98.

11. Jain, M. K., White, F.P., Strickholm, A., Williams, E. and Cordes, E.H. J. Membrane Biol. 8(1972) 363.

12. Jain, M.K., Mehl, L.E. and Cordes, E.H. Biochem. Biophys. Res. Commun. 51 (1973) 192.

13. Goodall, M.C., Ostroy, F., Sachs, G., Spenney, J.G. and Saccomani, G. (1973) In press.

14. Leuzinger, W. and Schneider, M. Experientia 28 (1972) 256.

15. Kemp, G., Dolly, J.O., Barnard, E.A. and Wenner, C.E. Biochem. Biophys. Res. Commun. 54 (1973) 607.

16. Redwood, W.R., Muldner, H. and Thompson, T. E. Proc. Nat. Acad. Sci. USA 64 (1969) 989.

17. Redwood, W.R., Gibbes, D.C. and Thompson, T. E. Biochim. Biophys. Acta 318 (1973) 10.

18. Vázquez, C., Parisi, M. and De Robertis, E. J. Membrane Biol. 6 (1971) 353.

19. Parisi, M., Reader, T.A. and De Robertis, E. J. Gen. Physiol. 60 (1972)454.

20. Ochoa, E., Fiszer de Plazas, S. and De Robertis, E. Molecular Pharmacol. 8 (1972) 215.

21. Storelli, C., Vogeli, H. and Semenza, G. FEBS Letters 24 (1972) 287.

22. Tosteson, M.T., Lau, F. and Tosteson, D.C. Nature New Biol. 243 (1973) 112.

23. Montal, M. and Mueller, P. Proc. Nat. Acad. Sci. USA 69 (1972) 3561.

24. Montal, M. Biochim. Biophys. Acta. 298 (1973) 750.

25. Montal, M. (1973) In: Methods in Enzymology. Biomembranes. eds.Fleischer, S., Packer, L. and Estabrook, R.W. New York: Academic. In press.

26. Kimelberg, H. K. and Papahadjopoulos, D. Biochim. Biophys. Acta 233 (1971) 805.

27. Gitler, C. and Montal, M. Biochem. Biophys. Res. Commun. 47 (1972) 1486.

28. Gitler, C. and Montal, M. FEBS Letters 28 (1972) 329.

29. Folch-Pi, J. and Lees, M.J. Biol. Chem. 191 (1951) 807.

30. Montal, M. J. Membrane Biol. 7 (1972) 245.

31. Razin, S. Biochim. Biophys. Acta 265 (1972) 241.

32. Kagawa, Y. and Racker, E. J. Biol. Chem. 246 (1971) 5477.

33. Hinkle, P.C., Kim, J.J. and Racker, E. J. Biol. Chem. 247 (1972) 1338.

34. Racker, E. Biochem. Biophys. Res. Commun. 55 (1973) 224.

35. Racker, E. and Eytan, E. Biochem. Biophys. Res. Commun. 55 (1973) 174.

36. Hong, K., and Hubbell, W.L. Proc. Nat. Acad. Sci. USA 69 (1972) 2617.

37. Hong, K. and Hubbell, W.L. Biochemistry 12 (1973) 4517.

38. Chen, Y.S. and Hubbell, W.L. Exp. Eye Res. 17 (1973) 517.

39. Appleburry, M.L., Zuckerman, D.M., Lamola, A. A. and Jovin, T.M. Biochemistry (1973) In press.

40. Chavre, M., Cavaggioni, A., Osborne, H.B. and Gulik-Krzywicki, T. FEBS Letters 26 (1972) 197.

41. Martonosi, A. J. Biol. Chem. 243 (1968) 71.

42. Lemberg, R. Physiol. Rev. 49 (1969) 48.

43. Racker, E. (1972) In: Membrane Research, pp. 97, ed. Fox, C.F. New York and London:Academic.

44. Mitchell, P. and Moyle, J. (1970) In; Electron-Transport and Energy Conservation. pp. 575. eds. Tager, J.M., Papa, S. Quagliarie-

llo, E. and Slater, E.C. Bari:Adriatica Editrice.

45. Mitchell, P. Fed. Proc. 26 (1967) 1370.

46. Mitchell, P. FEBS Letters 33 (1973) 267.

47. Hinkle, P.C. Fed. Proc. 32 (1973) 1988.

48. Racker, E., Kandrach, A. J. Biol. Chem. 248 (1973) 5841.

49. Jasaitis, A.A., Nemececk, I.B., Severina, I.I., Skulachev, V.P. and Smirnova, S.M. Biochim. Biophys. Acta 275 (1972) 485.

50. Branton, D. Proc. Nat. Acad. Sci. USA 55 (1966) 1048.

51. Pinto da Silva, P. and Branton, D. J. Cell Biol. 45 (1970) 598.

52. Deamer, D.W., Leonard, R., Tardieu, A. and Branton, D. Biochim. Biophys. Acta 219 (1970) 47.

53. Loeb, G.I. (1965) In: Surface Chemistry of Proteins and Polypeptides. U.S. Naval Research Laboratory, Washington, D.C.

54. Malcolm, B.R. (1973) In: Progress in Surface and Membrane Science. Vol. 7, pp. 183. eds. Danielli, J.F. Rosenberg, M.D. and Cadenhead, D.A. New York, London:Academic.

55. Shah, D.O. and Shulman, J.H. J. Colloid Interface Sci. 25 (1967) 107.

56. Smith, L. (1955) In: Methods of Biochemical Analysis, Vol. 2, pp. 427, ed. Glick, D. New York: Wiley (Interscience).

57. Bangham, A.D. (1968) In: Progress in Biophysics and Molecular Biology. Vol. 18, pp. 29, eds. Butler, J.A.Y. and Noble, D. New York: Pergamon Press.

58. Montal, M. and Korenbrot, J.I. Nature 246 (1973) 219.

59. Hanai, T., Haydon, D.A. and Taylor, J. J. Theoret. Biol. 9 (1965) 422.

60. Hinkle, P.C. and Mitchell, P. J. Bioenergetics 1 (1970) 45.

61. Robinson, W.E., Gordon-Walker, A. and Bownds D. Nature New Biol. 235 (1972) 112.

62. Heitzman, H. Nature New Biol. 235 (1972)114.

63. Heller, J. and Lawrence, M.A. Biochemistry 9 (1970) 864.

64. Steineman, A., and Stryer, L. Biochemistry 12 (1973) 1499.

65. Abrahamson, E.W. and Fager, R.S. (1973) In: Current Topics in Bioenergetics, Vol. 5, pp. 125, ed. Sanadi, D.R. New York, Academic.

66. Hubbard, R. and Kropf. A. Proc. Nat. Acad. Sci. USA 44 (1958) 130.

67. Wald, G., Brown, P. K. and Gibbons, I.R. J. Opt. Soc. Am. 53 (1963) 20.

68. Hagins, W.A. Ann. Rev. Biophys. Bioengineer. 1 (1972) 131.

69. Hagins, W.A., Penn, R.D. and Yoshikami, S. Biophys. J. 10 (1970) 380.

70. Korenbrot, J.I. and Cone, R.A. J. Gen. Physiol. 60 (1972) 20.

71. Korenbrot, J.I., Brown, D.T. and Cone, R.A. J. Cell Biol. 56 (1973) 389.

72. Korenbrot, J.I. Exp. Eye Res. 16 (1973) 343.

73. Tomita, T. Quart. Rev. Biophys. 3 (1970) 179.

74. Brown, H.M., Hagiwara, S. Koike, H. and Meech, R.W. Fed. Proc. 30 (1971) 69.

75. Cone, R.A. and Pak, W.L. (1971) In: Principles of Receptor Physiology. Vol. I, Handbook of Sensory Physiology pp. 345. ed. Loewenstein, W.R. Berlin:Springer-Verlag.

76. Oesterhelt, D. and Stoeckenius, W. Nature New Biol. 233 (1971) 149.

77. Blaurock, A.E. and Stoeckenius, W. Nature New Biol. 233 (1971) 152.

78. Oesterhelt, D. and Stoeckenius, W. Proc. Nat. Acad. Sci. USA 70 (1973) 2853.

79. Kasai, M. and Changeux, J.P. J. Membrane Biol. 6 (1971) 1-88 (1,24,58).

80. Katz, B. and Miledi, R. J. Physiol. (London) 244 (1972) 665.

81. Inesi, G. Ann. Rev. Biophys. Bioengineer. 1 (1972) 191.

82. Clark, A.W. and Branton, D. Z. Zellforsch. 91 (1968) 586.

83. Branton, D. and Deamer, D. (1972) In: Membrane Structure. New York, N.Y., Springer-Verlag.

84. Ebner, E., Mason, T.L. and Schatz, G. J. Biol. Chem. 248 (1973) 5369.

85. Hall, M.O., Bok, D. and Bacharach, A.D.E. J. Mol. Biol. 45 (1969) 397.

86. Young, R.W. (1969) In: The Retina: Morphology, Function and Clinical Characteristics, p. 177. ed. Straatman, B.R. Hall, M.O. Allen, R.A. and Crescitelli, F Los Angeles: University of California.

87. Kadenbach, B. Biochem. Biophys. Res. Commun. 44 (1971) 724.

88. Tzagoloff, A. and Akai, A. J. Biol. Chem. 247 (1972) 6517.

89. Kassner, R.J. Proc. Nat. Acad. Sci. USA. 69 (1972) 2263.

90. Kagawa, Y. Biochim. Biophys. Acta. Rev. Biomemgranes 265 (1972) 297.

91. Triggle, D.J. (1970) In: Recent Progress in Surface Science. Vol. 3, pp. 273. New York: Academic.

92. Clowes, G.H.A. J. Phys. Chem. 20 (1916) 407.

93. Ludwig, P.K. and Huque, M.M. Ber. Bunseges Physik. Chem. 72 (1968) 352.

94. Ludwig, P.K. J. Chem. Phys. 50 (1969) 1787.

95. Brown, P.K. Nature New Biology 236 (1972) 35.

96. Cone, R.A. Nature New Biology 236 (1972) 39.

97. Junge, W. FEBS Letters 25 (1972) 109.

98. Blaurock, A.E. and Wilkins, M.H.F. Nature 223 (1969) 206.

99. Blaurock, A.E. and Wilkins, M.H.F. Nature 236 (1972) 313.

100. Blasie, J.K., Worthington, C.R. and Dewey, M.M. J. Mol. Biol. 39 (1969) 407.

101. Blasie, J.K. Biophys. J. 12 (1972) 191.

102. Smid, J. Angew, Chem. Internat. Edt. 11 (1972) 112.

103. Triggle, D.J. (1972) In: Progress in Surface and Membrane Science, Vol. 5, pp. 267. New York, and London:Academic.

104. Lehninger, A.L., Carafoli, E. and Rossi, C. S. Advan. Enzymol. Relat. Areas Mol. Biol. 29 (1967) 259.

105. Goldacre, R.J. (1958) In: Surface Phenomena in Chemistry and Biology, pp. 278, eds. Danielli, J.F., Pankhurst, K.G.A. and Riddiford, A.D. Oxford: Pergamon Press.

106. Ponnamperuma, C. Quart. Rev. Biophys. 4 (1971) 77.

107. Fox, S.W. Naturwissenschaften 56 (1969) 1.

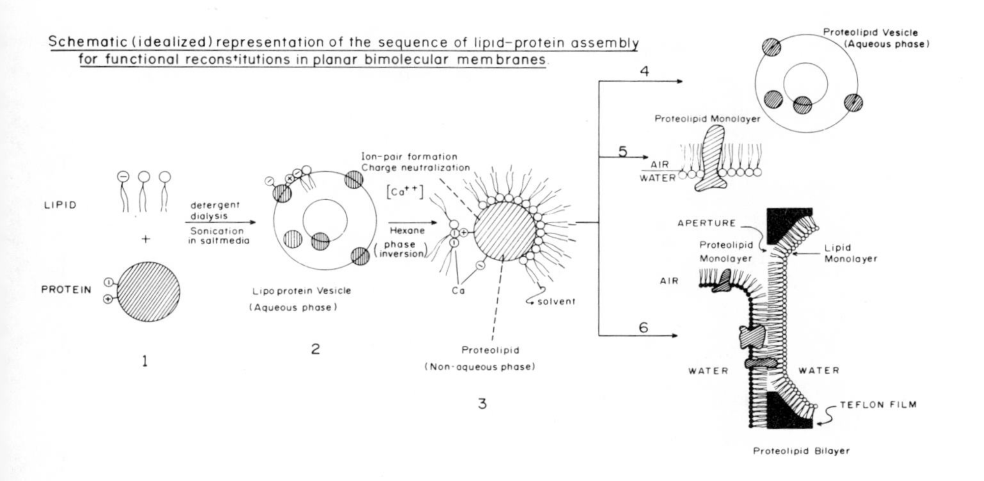


Figure 1. Schematic (idealized) representation of the sequence of lipid-protein assembly for functional reconstitutions in planar bimolecular membranes. (Not drawn to scale).

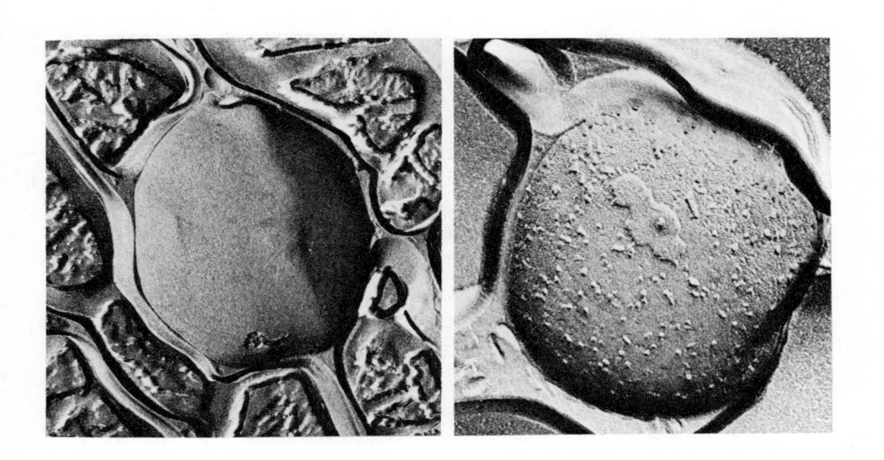

Figure 2. Freeze-etch replicas of lipid vesicles (left) and cytochrome c oxidase proteolipid vesicles (right). The vesicles were frozen in 25% glycerol; x 43,000.

ENZYMATIC SPECIFIC ACTIVITY OF MEMBRANOUS CYTOCHROME C OXIDASE BEFORE AND AFTER EXTRACTION AS PROTEOLIPID INTO n-HEXANE

| PREPARATION | SPECIFIC ACTIVITY $SEC^{-1}$/MG PROTEIN (Δ) |
|---|---|
| CYTOCHROME C OXIDASE (+) | 1.5 |
| LIPOPROTEIN | 1.5 |
| PROTEOLIPID (‡) | 1.0 |

(Δ) ASSAYED IN THE PRESENCE OF 15μM ASCORBATE REDUCED-CYTOCHROME C (DIALYZED) IN 0.01M PHOSPHATE BUFFER CONTAINING 0.2% TWEEN 80 AT 25°C. AVERAGE OF THREE DIFFERENT PREPARATIONS.

(+) PREPARATION IN 2% TWEEN 80.

(‡) ASSAYED AFTER EVAPORATION OF HEXANE AND DISPERSION IN 0.25M SUCROSE -0.01M PHOSPHATE BUFFER pH 7.0. THE EFFICIENCY OF PROTEIN EXTRACTION INTO THE HEXANE PHASE IS OVER 94%.

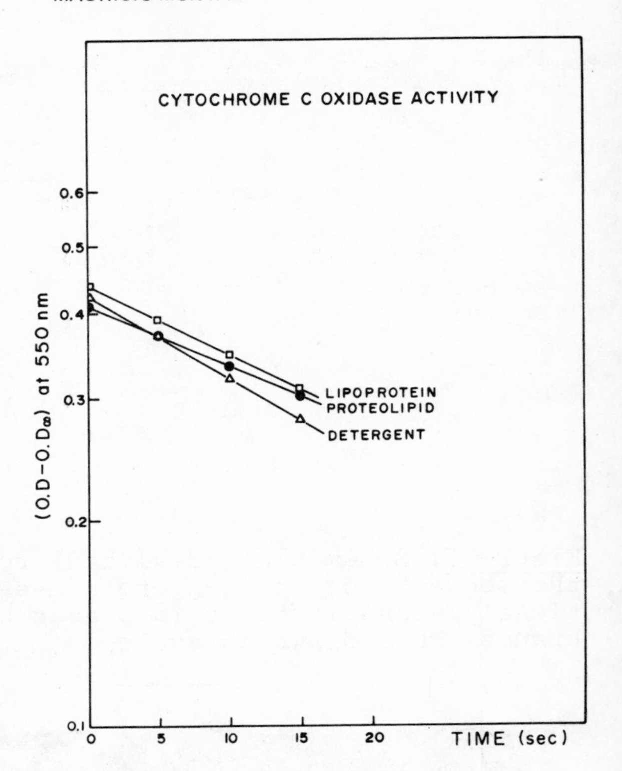


Figure 3. Cytochrome c oxidase activity in detergent, lipoprotein and proteolipid interfaces. Cytochrome oxidase was prepared and the lipoprotein vesicles reconstituted according to reference 34. Briefly, soybean lecithin (20 mg) was dispersed in 0.5 ml of 50 mM phosphate buffer pH 7.0. Next, 250 μg of cytochrome oxidase in 0.1 ml of 0.1 M phosphate buffer pH 7.4 containing 2% Na-cholate, pH 7.4 were added. The mixture was sonicated by immersion of the test tube in a Bransonic.(Heat systems ultrasonics, Inc. Plainview, N.Y., ultrasonic cleaner, tank dimensions: $3^1/_2$ x $3^1/_2$ x $2^5/_8$ in.; power output; 40 watts, for 25 min. The clear vesicles were brought to a final volume of 2.0 ml with water. Then, 0.2 ml of a 100 mM $CaCl_2$ solution and 1.0 ml of hexane were added to the suspension in rapid sucession. The tube was vigorously mixed for 5 min and the two immiscible

phases were then allowed to separate. After extraction, the hexane had a characteristic green colour. The proteolipid vesicle was prepared by evaporating the hexane under a stream of nitrogen and dispersing the residue in 0.25 M sucrose containing 50 mM phosphate buffer, pH 7.0. Activity was assayed according to Smith (56) following the decrease in absorbance of reduced cytochrome c at 550 nm in a Cary 14 spectrophotometer with the use of 1 cm path length cuvettes at 25º ± 2º C. The incubation medium contained 15 μM ascorbate-reduced cytochrome-c (dialyzed) in 0.01 M phosphate buffer pH 7.0 containing 0.2% Tween-80.

---

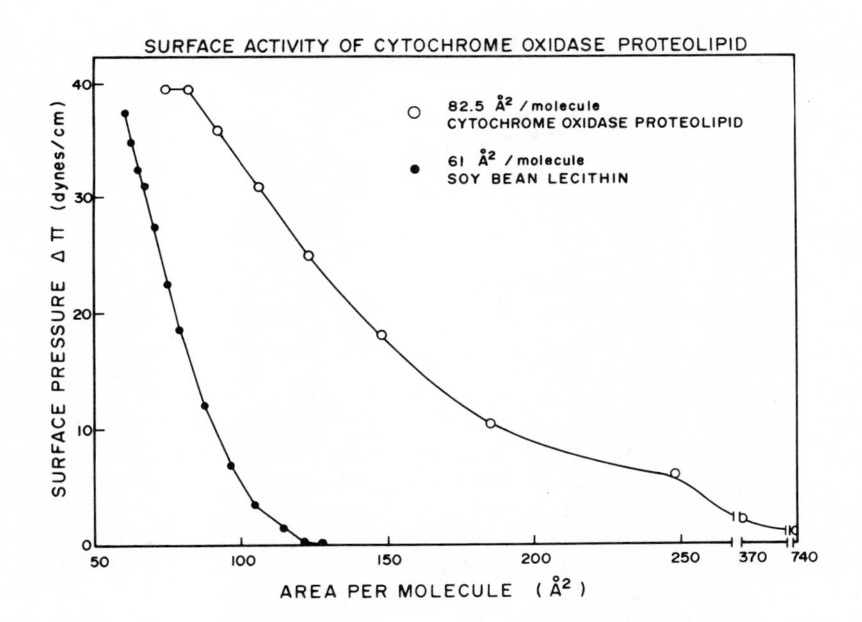


Figure 4. Surface pressure-area curves of soybean lecithin and cytochrome oxidase proteolipid. The interfacial tension was measured with a Cenco-DuNouy interfacial tensiometer by the "ring-detachment" method. The surface pressure ($\pi$) was calculated according to the formula: $\pi = \gamma o - \gamma$, where

$\gamma_o$ is the surface tension of the clean subphase and $\gamma$ the surface tension of the film-covered surface. A circular glass-through with an internal diameter of 10 cm and a depth of 1.0 cm was used. The subphase consisted of 0.01 M phosphate buffer pH 7.0. The lipid and proteolipid solutions in hexane (prepared as described in Fig. 2 caption) were spread on a clean surface of subsolution (25). The measurements were taken five minutes after spreading. The area per molecule was calculated on the basis of direct chemical determination of phospholipid phosphate in both samples. Temperature = 20±2°C.

---

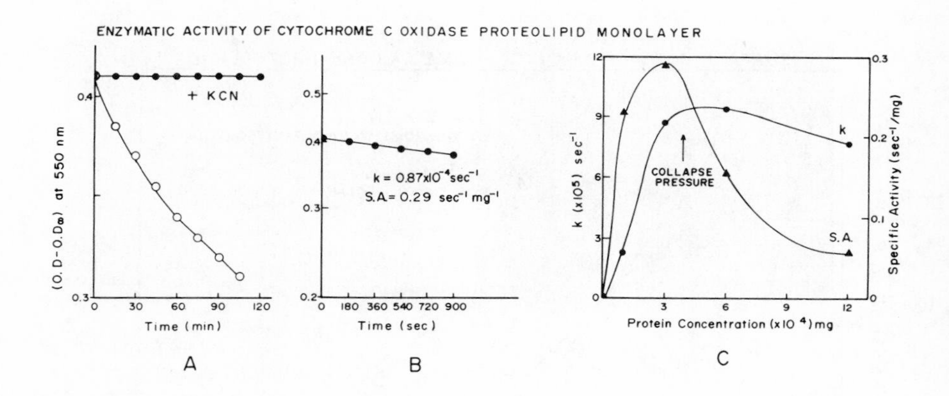


Figure 5. Enzymatic activity of cytochrome oxidase proteolipid monolayer at the air-water interface.

A. Oxidation of Reduced cytochrome c by a cytochrome oxidase proteolipid monolayer. Fifteen ml of 0.25 M sucrose - 0.05 M phosphate buffer pH 7.0 solution containing cytochrome c, (50 µg/ml) reduced with ascorbate and dialyzed overnight in 50 mM phosphate buffer pH 7.0, were pipetted into a 10 cm path length cuvette, placed inside a Cary 14 spectrophotometer. After cleaning the resultant surface (17.8 $cm^2$) the proteolipid, containing 0.4 µg of protein, was spread from a hexane solution. The decrease in absorbance at 550 nm was continuously monitored. At the end of the measuring interval, ferricyanide was added to oxidize chemically all the cytochrome c present. The concentration of KCN was 5 mM when present. Temperature = 20°± 2° C.

B. Semilogarithmic plot of data shown in A. The calculated first-order rate constant, (κ) is $0.87 \times 10^{-4}\ sec^{-1}$, and the specific activity, $0.29\ sec^{-1} \cdot mg^{-1}$.

C. Dependence of κ and of the specific activity on the amount of proteolipid protein. The experiments were performed as in A.

---

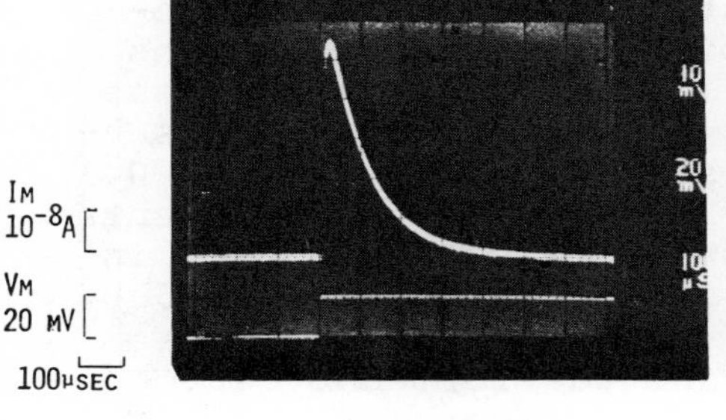


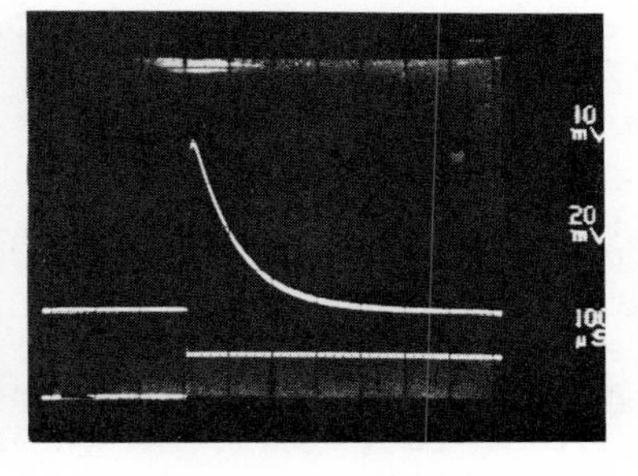

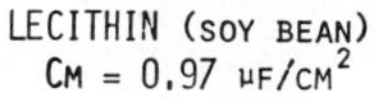


CYTOCHROME OXIDASE PROTEOLIPID
$C_M = 0.65\ \mu F/cm^2$

Figure 6. Electrical capacity of bimolecular membranes. Bilayers were formed from monolayers and their electrical properties studied as described in detail in (ref. 23-25). The records illustrate the membrane current (I) in response to a 20 mV voltage pulse (V) across symmetric bilayers. The upper trace in each record is current and the lower trace is voltage. The left hand record corresponds to a soybean lecithin membrane, which is the lipid present in the proteolipid; the other record corresponds to a membrane formed by apposition of two cytochrome oxidase proteolipid monolayers. The aqueous subphase for both membranes was 0.25 M sucrose-0.05 M phosphate buffer pH 7.0. Both membranes were formed across the same aperture ($3.5 \times 10^{-4}.cm^2$). The membrane capacity (C), calculated from the time (t) integral of the current records illustrated (23-25) according to the formula $C=\int_0^\infty Idt/V$, is: soybean lecithin, 0.95 $\mu F.cm^{-2}$ and cytochrome oxidase proteolipid, 0.65 $\mu F.cm^{-2}$. Temperature 20±2°C.

---

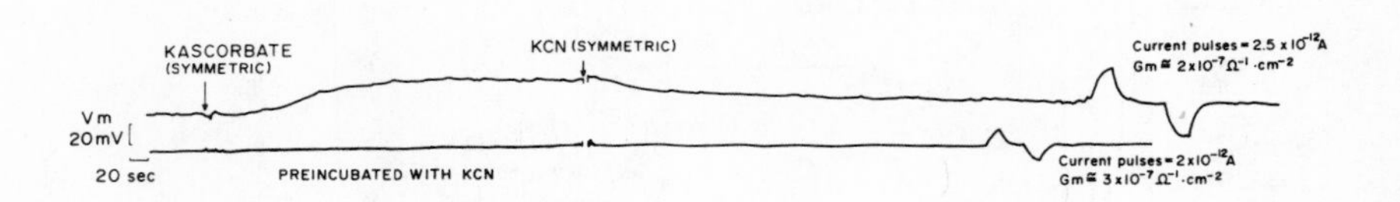


Figure 7. Membrane potential changes of a cytochrome oxidase proteolipid bilayer to additions of ascorbate and cyanide. The cytochrome oxidase proteolipid in hexane is spread over a clean subphase containing 0.25 M sucrose 0.1 M KCl and 0.1 M tris-phosphate pH 7.0. One of the compartments contained 250 µg/ml of oxidized cytochrome c, in addition. The volume of each compartment was 5.0 ml and the area 3.5 $cm^2$. The aperture area was $3.5 \times 10^{-4}$ $cm^2$. A membrane was formed by apposing the two monolayers. Tris-ascorbate and KCN both at pH 7.0 and at a final concentration of 5 mM were added to both compartments where indicated. At the end of each record the membrane resistance was measured by applying constant current pulses of the indicated amplitude and from the steady state value of the resultant potential, the membrane resitance (Rm) and conductance (Gm=1/Rm) were calculated. Temperature = 20±2°C.

---

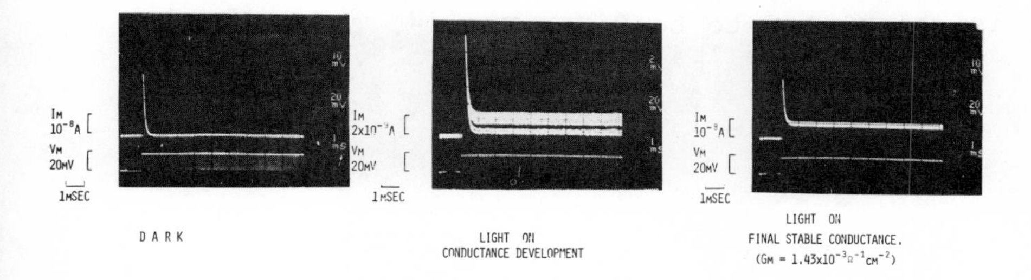


Figure 8. Electrical properties of Rhodopsin-proteolipid bilayers in the presence and absence of illumination. The Rhodopsin-proteolipid was prepared as described in reference 58. Asymmetric membranes were formed by apposing a monolayer of rhodopsin proteolipid with a monolayer of soybean lecithin. Both monolayers were spread over a clean subphase containing 20 mM NaCl, 10 mM KCl, 0.5 mM $CaCl_2$ and 10 mM Imidazole pH 7.0. The temperature was 20±2°C. The dimensions of the chamber and aperture used are identical to those described in Figure 7 caption. The records present membrane currents in response to 20 mV voltage pulses across the bilayers. The upper trace in each record is current (please note the different scale calibration in B) and the lower trace is voltage.

A. Characteristics of the membrane in the absence of illumination. The membrane capacity as determined by integration of the membrane current record is 0.65 $\mu F.cm^{-2}$. The membrane conductance is of the order of $10^{-7}$-$10^{-8}$ $mho.cm^{-2}$.

B. Development of conductance upon illumination of the membrane. Voltage pulses were repeatedly applied (1/5 sec) and the current traces were superimposed on a storage oscilloscope. The membranes were continuously illuminated with white light at an energy of 200 $\mu watt/cm^2$. The current scale is expanded and therefore the capacitative transient is out of scale. The conductance develops slowly reaching a stable value of $1.4 \times 10^{-3}$ $mho.cm^{-2}$ with a half time of about 1 min.

C. Characteristics of the membrane in the presence of illumination. The membrane capacity is calculated to be 0.65 μF.cm$^{-2}$. The membrane conductance is 1.4 x 10$^{-3}$ mho.cm$^{-2}$.

---

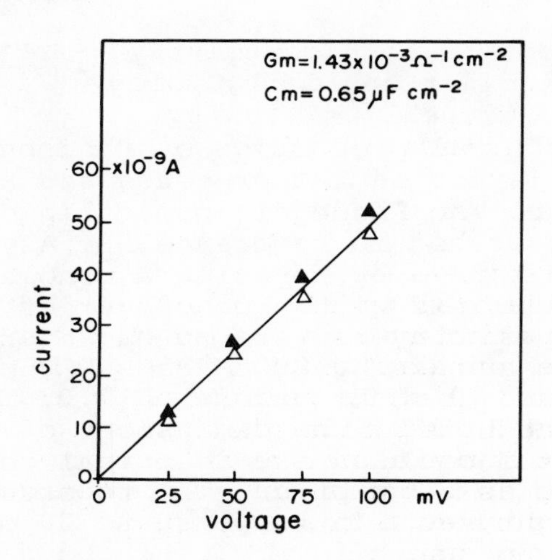


Figure 9. Steady-state current-voltage relations of the illuminated rhodopsin-proteolipid membrane. Solid and open triangles are positive and negative pulses, respectively. The behaviour is ohmic up to the breakdown voltage at about 200 mV. The slope conductance is 1.4 x 10$^{-3}$ mho.cm$^{-2}$.

---

# FROM MEMBRANES TO CANCER

E. Racker

Section of Biochemistry,
Molecular and Cell Biology
Cornell University
Ithaca, N. Y. 14850
U.S.A.

## Introduction

We were charged by the organizers of this interesting symposium not to "release recent data" or "review the subject" but to engage in "serious speculation" and draw up perspectives for the future. I have accepted this challenge but must apologize that I could not avoid to present some facts to serve as a basis for the speculations. Perhaps we might see a forest without knowing the trees but how can we tell that we are not dealing with a mirage?

The major two themes of my talk are a) the mechanism of ion transport and b) control of ion transport. While the first subject is very popular and has engaged many investigators, the second has been remarkably neglected.

I have recently proposed (1) that the basic problem of phosphorylation coupled to oxidation has been solved. The chemiosmotic hypothesis of Mitchell (2) has laid the foundation to experiments (3) which showed that there is a vectorial translocation of proton a) via an asymmetrically organized oxidation chain and b) via an asymmetrical oligomycin-sensitive ATPase complex. Experiments with reconstituted vesicles that catalyze oxidative phosphorylation at Site III and Site I (cf 1) have convinced us that the basic concepts of the chemiosmotic hypothesis are correct, thus bearing out the prediction (4) that the principle of oxidative phosphorylation will be solved before 1969. It is now a question how long it will take the experts in the field to recognize that the problem indeed has been solved. On this point I am not willing to make any prediction, since it belongs into the area of psychology and I cannot indulge in serious speculation in that field.

By accepting the vectorial translocation of protons via the oxidation chain and the formation of a membrane potential as the primary processes in oxidative phosphorylation, we have transformed the problem of oxidative phosphorylation into a more general one, namely, the mechanism of ATP-drive ion pumps. It was shown that the $Na^{+}K^{+}$ pump (5) and the $Ca^{++}$ pump (6,7) operating in reverse catalyze ATP formation. Thus the crucial question is how ion gradients can accomplish this. The fact that the mitochondrial pump utilizes a proton gradient is incidental and significant only in its relation to the oxidation chain, which creates the proton gradient.

## Mechanism of ATP-Driven Ion Pumps

How does an ion gradient or a membrane potential participate in the generation of ATP when a pump operates in reverse? In 1965, I had lunch with Peter Mitchell and he scribbled on a napkin a mechanism how the ATPase may be involved in such a reaction. I regret that I did not keep this napkin for historical reasons, particularly since it showed that Mitchell has broader views that he is credited for. He proposed that ATP formation may take place with a soluble ATPase provided that substrates and the ATPase are used stoichiometrically and that a proper conformational change can be induced, perhaps by alterations of the pH. Thus Mitchell was an early proponent that a conformational change may take place in the ATPase in response to the formation of a membrane potential. This conformational change concept is however fundamentally different from proposals (8,9) that a conformational change takes place in an oxido-reduction catalyst as the primary step in oxidative phosphorylation.

We took Mitchell's conformational speculations serious enough to conduct some experiments with crude and purified preparation of $CF_1$, the coupling factor of chloroplasts. We chose $CF_1$ over $F_1$, the mitochondrial ATPase, because in $CF_1$ the hydrolytic cleavage of ATP is masked so that any ATP formed would not be degraded. Moreover, we were aware at the time that an acid-base catalyzed formation of ATP takes place in chloroplasts (10). Unfortunately, our experiments were negative. We have repeated them more recently with greater sophistication. The results are still negative but of course in a more sophisticated way. Yet I still think it is a good hypothesis and we are encouraged to continue our attempts because of the following four experimental observations.

i) *Formation of phosphoproteins with inorganic phosphate*. It was shown ten years ago (1) that a phosphatase can be induced to take up inorganic phosphate and incorporate it into a serine residue. Since this takes place without an added substrate, the energy for the reaction must have been derived from the protein itself. Preparations of sarcoplasmic reticulum vesicles incorporate $^{32}P_i$ in the absence of an added energy source (12). In fact, ATP competes with $P_i$ for the phosphorylation site in a reaction which is sensitive to hydroxylamine. $Mg^{++}$ is required and very low $Ca^{++}$ concentrations ($K_i$ of 10 μM) inhibit. $^{32}P_i$ was also incorporated into the $Na^+K^+$-ATPase but in this case ATP-$Mg^{++}$ as well as ouabain and $Na^+$ were present (13, 14).

ii) *Formation of phosphorylated pump proteins with ATP*. The $Na^+K^+$-ATPase is embedded in the plasma membrane. $Na^+$ and $K^+$ interact with the enzyme on both sides but $Na^+$ has a higher affinity at the inside, $K^+$ at the outside. I shall henceforth refer to these sites as $Na^+$- and $K^+$-sites and describe the reaction as formulated by Sen et al (13). As shown in Fig. 1, $E_1$ is the form of the enzyme which interacts on the $Na^+$-site with ATP-$Mg^{++}$ to yield a phosphorylated protein which undergoes the transformational change required for the translocation of $Na^+$ to the outside. This reaction yields $E_2$-P which is characterized by a high reactivity with $K^+$ and ouabain. On interaction with $K^+$ the protein becomes dephosphorylated and after deposition of $K^+$ on the inside reverts to $E_1$. It is of particular interest that ouabain sensitizes the phosphoenzyme to slow hydrolysis and $P_i$ exchange, but renders it insensitive to the rapid dephosphorylation by $K^+$ (13). In the absence of ATP but in the presence of $Mg^{++}$ the enzyme reacts slowly with ouabain. This form of the enzyme, which is considered (13) to have little physiological significance, catalyzes a $K^+$ dependent $H_2{}^{18}O$-$P_i$ exchange (15). This and other similarities with oxidative phosphorylation have stimulated interesting speculations on the mechanism of action of the two systems (15). The interaction of the enzyme with ouabain appears to be complex (13). The steroid interacts only at the $K^+$-site and inhibits the ATPase activity as well as ion translocation. When ouabain interacts with the native enzyme in the absence of monovalent cations (or below 16 mM) but in the presence of $Mg^{++}$ the formation of phosphoenzyme by ATP is inhibited. It is therefore clear that the order of addition of the reagents may determine the response of the enzyme to ATP.

Although two separate phosphoenzyme intermediates ($E_1$ and $E_2$) are listed in Fig. 1, they are only conformational variants of the same phosphoenzyme species, the phosphoryl group being attached to an aspartyl residue (17,18). Whether the conformational changes include rotational motion or movement from one side of the phospholipid to the other or whether they take place within a channel is still unknown. Although there are several examples for mobile carriers such as valinomycin and nigericin, I am more inclined to think of a channel mechanism when we are dealing with proteins with a molecular weight of 50,000 or more. My prejudice against the popular carousel model is probably the result of the fact that I get easily seasick. Moreover, rapid motions of a heavy protein or even part of a protein through both layers of the phospholipid bilayer do not appeal to me, but I agree that it is somewhat easier to visualize a 3 $Na^+$:2 $K^+$ exchange by a carousel than by a channel.

In the case of the $Ca^{++}$ pump of sarcoplasmic reticulum a channel mechanism appears to be operative. Dr. A. Knowles observed that the pump is still active at low temperatures when the ionophore 23187 is completely immobilized. Moreover, she found that during $Ca^{++}$ translocation in reconstituted vesicles, the terminal P of ATP is released on the outside and not, as assumed by the carrier-type hypothesis, on the inside of the vesicles.

Although phospholipids are needed for the hydrolysis of ATP by sarcoplasmic reticulum fragments they do not seem to be required for the formation of the phosphoenzyme (cf 19). These findings are reminiscent of the phospholipid dependency of the oligomycin-sensitive ATPase (20).

It is thus apparent that in the $Na^+K^+$-pump as well as in the $Ca^{++}$-pump the steps of phosphorylation and dephosphorylation are functionally separable under appropriate experimental conditions.

iii) *A chemical model system.* In a model system we observed an ATP transfer reaction with rather remarkable properties. In the presence of 70% dimethylsulfoxide, $Mg^{++}$, sodium maleate, ATP and $P_i$, pyrophosphate was formed (21). If instead of $P_i$ arsenate was added, ATP was hydrolyzed. The dependency of the transfer and ATPase reactions in this model system on $Mg^{++}$ and on a dicarboxylic acid is particularly interesting because the $Mg^{++}$-dependent ATPase of spinach chloroplast $CF_1$ exhibits the same properties. Studies of the phosphotransfer reaction in dimethylsulfoxide by laser Raman spectroscopy suggest that $Mg^{++}$ binds to the α and β phosphate of ATP and that in the presence of maleate and dimethylsulfoxide a relatively stable intermediate is formed (22). The intermediate has a Raman spectrum similar to that of ADP, but is formed long before significant amounts of the end product accumulate. It is conceivable that the role of dimethylsulfoxide is to provide the proper hydrophobic environment for the formation of the intermediate and the labilization of the γ phosphate of ATP, and that similar conditions prevail in a crevice of $CF_1$ where the active center is located. Unfortunately, limitation in sensitivity of R man spectroscopy do not at present allow for a direct test of this possibility. The observation that hydrophobic reagents such as iodine (23), NBD-chloride (24) or quercetin (25) inhibit $F_1$ or $CF_1$ while hydrophilic reagents are quite ineffective, is consistent with the notion that the active center is in a hydrophobic environment.

iv) *Conformational changes in $CF_1$ and $F_1$.* Membrane-bound $CF_1$ molecules take up tritium during illumination of chloroplasts (26). Exposure of chloroplasts to light in the presence of DTT permits the isolation of a $Ca^{++}$-dependent unmasked ATPase (27) and a light-dependent interaction of N-ethylmaleimide with the γ-subunit of $CF_1$ was observed (28). In unpublished experiments Kagawa (1972) has observed tritium uptake into $F_1$ during oxidation of succinate by submitochondrial particles. It therefore appears that the energization of the membrane by either light or substrate oxidation induces changes in the coupling factors. This is particularly significant in view of recent findings that have convinced us that $F_1$ and $CF_1$ are in vivo attached to the membrane by a stalk. It has been repeatedly suggested that the morphological appearance of the protein outside the membrane is an artifact of the negative staining procedure resulting from a displacement of the protein from the membrane (cf 29). This possibility seems now remote since 85 Å spheres can be seen outside the membrane after fixation and staining with osmium tetroxide (30) and in freeze-etch preparations (31).

In view of these findings we are considering the possibility that proton movements take place via the stem and through the coupling factor and that this process is responsible for the conformational change and the formation of ATP from ADP and $P_i$. This formulation almost demands that it should be possible to induce the isolated coupling factor to form ATP stoichiometrically if the appropriate conformational alteration and local proton changes could be induced. We are using these considerations to design experiments with purified coupling factors with and without inhibitors to induce either the formation of a phosphoenzyme from $P_i$ or ATP or the formation of ATP from $P_i$ and ADP in amounts stoichiometric with the enzyme.

## Mechanism of ATP-Generating Proton Pumps

*Proton translocation in mitochondria and chloroplasts via the ATPase complex.* Those in the audience who are not chemiosmotically oriented should be reminded of some of the properties of the mitochondrial ATP-driven proton pump. By taking advantage of the fact that at pH 6.2 the hydrolysis of ATP does not give rise to a change in pH, Thayer and Hinkle have shown (32) that in submitochondrial particles close to 2 $H^+$ are translocated per ATP hydrolyzed. The process is sensitive to oligomycin and uncouplers of oxidative phosphorylation. It was reported, however, that in chloroplasts 4 $H^+$ are translocated per ATP produced (33). It is rather important to establish the true value of this ratio particularly in view of calculations which have been used to challenge the validity of the chemiosmotic hypothesis on thermodynamic grounds (34). For those who consider the possibility that the mitochondrial proton pump is on a side-path of oxidative phosphorylation, I would like to point out that in reconstituted vesicles which catalyze oxidative phosphorylation (35) the oligomycin-sensitive ATPase is part of the coupling device and that proton translocation is observed on addition of ATP (36).

We are now working on the hypothesis that the hydrophobic component ($CF_o$) of the oligomycin -sensitive ATPase is responsible for proton translocation through the membrane while the soluble coupling factors are concerned with the proper attachment and function of the soluble ATPase. This hypothesis is supported by the observation that particles that are depleted in coupling factors, e.g. ASU or STA particles (37) exhibit respiratory control in the presence of oligomycin. Moreover, DCCD which inhibits, like oligomycin, proton translocation, interacts at low concentrations only with the hydrophobic component and not with the soluble protein factors.

Based on these considerations we have designed an assay for the proton transporter of mitochondria. We incorporate fractions of the hydrophobic preparation of the oligomycin-sensitive ATPase during the reconstitution of cytochrome oxidase vesicles (38) which exhibit respiratory control and test for a) activation of respiration and b) sensitivity of respiration to oligomycin. Only the latter provides a reliable assay, since many hydrophobic proteins, e.g. denatured proteins, make the membrane leaky to protons.

Dr. Chien in our laboratory has isolated a protein fraction which is about 10 times more active than $CF_o$ in this assay. It contains one major protein component with a molecular weight of about 30,000 when analyzed in SDS-acrylamide gels. However, several minor components will have to be removed before we can state that only one polypeptide chain is involved in proton translocation.

*Proton translocation via the oxidation chain.* In view of the fixed asymmetry of the respiratory catalysts (39) proton translocation via the oxidation chain proceeds by a channel mechanism as outlined by the tentative scheme in Fig. 2. Although in principle this scheme is similar to that of Mitchell (3), the assignment of the proton carriers in Site I and II are different, but no less speculative. The first site as drawn requires that the FMN site of the DPNH dehydrogenase is on the C-side of the membrane. I have inserted a crevice mechanism so that DPNH can reach the FMN. Since Site III is really a half-loop we need two proton donors at the C-side of the membrane during operation of Site II. Perhaps the new oxidation factor (40) serves as a proton carrier.

*Proton translocation catalyzed by bacterial rhodopsin.* Proton translocation in H. halobium catalyzed by bacterial rhodopsin (41) is associated with the generation of ATP (42). Our study of the light-driven proton pump reconstituted from bacterial rhodopsin and phospholipids appears to yield answers to several questions (43, 44). The first is whether this pump can substitute for the electron transport chain in the reconstitution of vesicles that catalyze ATP formation via the oligomycin-sensitive ATPase. The answer to that question is yes (43). The fact that ATP formation was linear over the tested time period of 20 minutes eliminates the possibility of a stoichiometric reaction with rhodopsin. Althoug it has not been possible to determine the exact $H^+$/ATP ratio because of the high concentration of phosphate buffer required for the efficient operation of the coupling device, extrapolation of data obtained in the absence of buffer suggests an approximate ratio of 5.

The second question is whether the pump operates as a channel or as a carrier (44). Reconstitution was performed with synthetic phospholipids with known transition temperatures, dimyristoyl phosphatidylcholine (24°) and dipalmitoyl phosphatidylcholine (42°). As shown in Fig. 3 the pump is operative over a wide range of temperatures with only small differences in the extent of proton uptake. Although the rate of $H^+$ uptake was slower at lower temperatures it was compensated by a slower proton leakage with the result that the extent of proton uptake was actually higher at 10° than at 20° In the presence of an ionophore like nigericin which collapses the proton gradient by $K^+/H^+$ exchange, proton uptake was eliminated at temperatures above the transition temperature, but was either normal or slightly faster than the control without nigericin when the membrane was frozen (Fig. 4). In contrast to nigericin, which is a mobile carrier, gramicidin which forms a channel (45,46,47) effectively eliminated proton uptake at all temperatures provided it was added to the flu fluid membrane at a high temperature. When added to the frozen membrane it was inactive.

On the basis of these data we have concluded that the rhodopsin pump operates by a channel mechanism. What could be the mechanism of a vectorial $H^+$ translocation via a channel? Two possibilities come to mind. One is that proton conduction takes place via water "frozen" along the polypeptide chain of a protein channel. This would be in line with the known mobility of $H^+$ in ice (48) but is somewhat difficult to visualize to take place vectorially. Alternatively, proton movement could take place along the $NH_2$ groups of the amino acid residues by a domino mechanism inhibited on the inside of the membrane by the photochemical release of a proton from the protonated Schiff base of the retinal-lysine chromophore of rhodopsin. This formulation requires that rhodopsin is transmembranous with the chromophore located on the inside of the vesicles.

What are the factors responsible for the asymmetric orientation of a protein within the phospholipid bilayer? It is particularly puzzling that in the reconstituted vesicles with bacterial rhodopsin protons are pumped in, while in intact _H. halobium_ bacteria protons are pumped out (41). It does not seem that the composition of the phospholipids is a deciding factor since a variety of phospholipids including those isolated from _halobacteria_ yield vesicles which pumped protons in (49). If the curvature of the vesicles were a decisive factor in the orientation, the natural membrane should reveal cristae or concavities which have not been observed. A third possibility is that the conditions of reconstitution play an important role. We have recently encountered a striking example illustrating this point. We have prepared vesicles with trapped ADP inside which catalyze an atractyloside and bongkrekate-sensitive ADP-ATP exchange translocation (50). When submitochondrial particles were prepared by sonication of mitochondria in the presence of ADP they are all inside-out and were inhibited only when atractyloside was added before sonication. When the vesicles were reconstituted from an insoluble protein fraction obtained after extraction of the membrane with cholate, they exhibited the same sidedness as intact mitochondria in that they were sensitive to external atractyloside. Unpublished experiments by Dr. Ian Ragan also indicate that the orientation of complex I in reconstituted vesicles varies with the conditions of reconstitution.

A fourth possibility is that in the case of rhodopsin the ratio of protein to phospholipids is very critical and that the large excess of phospholipids used in our experiments favor inside-out conformation, while the protein to phospholipid ratio of 3 in the natural membrane (51) favors the conformation resulting in proton excretion. The protein:phospholipid ratio controls some of the properties of the reconstituted $Ca^{++}$ pump, e.g. the response to externally added oxalate, which increases when the ratio becomes larger (52). I have recently examined the purple membrane fragments with and without added phospholipids. Reducing the phospholipids from 25 μmoles per mg rhodopsin did not change the direction of proton movements. Without any added phospholipids the purple membrane showed essentially no response to light at room temperature (43) but a proton release was noted when the temperature of the assay mixture was lowered (Fig. 5). However, while with reconstituted vesicles $H^+$ uptake was catalytic, with the membrane fragments proton release was somewhat less than stoichiometric with rhodopsin even at the lowest temperature tested. In contrast to the proton uptake in reconstituted vesicles this reaction has a relatively sharp pH optimum (Fig. 5). Since the reaction was insensitive to gramicidin we are probably dealing with the stoichiometric photochemical event and by lowering the temperature have slowed down the return of protons to the Schiff-base.

## Control of Ion Pumps

*Control of a calcium pump.* A calcium pump reconstituted with the ATPase of sarcoplasmic reticulum and phospholipids by the cholate-dialysis procedure (53) catalyzed a rapid ATP dependent uptake of $Ca^{++}$. Instead of a $Ca^{++}$/ATP ratio of 2.0 observed in sarcoplasmic reticulum (54) ratios below 0.1 were observed. Even vesicles that were reconstituted without detergent (52) were at first much less efficient than the sarcoplasmic reticulum fragments, but manipulations of the phospholipid composition and extention of the sonication time greatly increased the $Ca^{++}$/ATP ratio to values of about 1. In the course of these studies we noted that on aging the solubilized ATPase preparation became turbid. After centrifugation the clear supernatant, which contained the ATPase activity, yielded vesicles with very low efficiency. The insoluble precipitate contained a heat-stable factor which increased the efficiency of the pump by inhibiting the ATPase activity without interfering with $Ca^{++}$ translocation. We know little about this heat-stable factor, but it may be similar to the heat-stable regulatory ε subunit of $CF_1$ or $F_1$ (55, 56).

*Control of the proton pump of chloroplasts and mitochondria.* $CF_1$ from chloroplasts contains 5 subunits (55). The smallest ε subunit has a molecular weight of about 13,000 and is a regulatory protein which inhibits the ATPase activity without interfering with photophosphorylation (57). In mitochondria a somewhat smaller protein of 10,500 MW acts in exactly the same manner, but appears to be more readily dissociable (56). The dissociation of the inhibitor protein from $F_1$ is influenced by the state of the membrane and oxidoreduction processes (58). Submitochondrial particles which have been partially depleted of the inhibitor protein catalyze ATP hydrolysis much faster than ATP generation. They therefore do not accumulate ATP during oxidation and an efficient ATP trapping system (hexokinase plus glucose) is required for measurements of P:O ratios. Only when inhibitor protein is added can ADP itself be used as phosphate acceptor (56). We have recently found that some bioflavonoids such as quercetin substitute for the protein inhibitor (25). Within an appropriate range of concentration quercetin inhibits ATPase activity but not oxidative phosphorylation. Similar observations were made with $CF_1$ and photophosphorylation, the ATPase activity being much more sensitive to quercetin than photophosphorylation (24).

*Control of the $Na^+K^+$ ATPase.* The ion pump of the plasma membrane translocates 3 $Na^+$ out and 2 $K^+$ into the cell for each ATP hydrolyzed to ADP and $P_i$ (13). Since glycolysis requires 1 ADP and 1 $P_i$ for each lactate that is formed from glucose, one can calculate the efficiency of the pump in cells that utilize the $Na^+K^+$-ATPase to regenerate ADP and $P_i$, provided there is no other pathway of either ATP generation or ATP hydrolysis. This can be accomplished in ascites tumor cells by the addition of bongkrekate which inhibits mitochondrial ATP generation as well as hydrolysis. It was found (59) that under these conditions the transport of each $Rb^+$ molecule was associated with the hydrolysis of 3 to 4 molecules of ATP. At appropriate concentrations quercetin inhibited the excess ATP hydrolysis without interfering with the pump mechanism thereby lowering the ATP/$Rb^+$ close to 0.5. This results in a corresponding inhibition of the aerobic lactate formation without change in the transport of $Rb^+$.

We are not concerned at present whether the loosely coupled pump in the tumor cells is a primary lesion; what we would like to know is whether the production of lactic acid and the resulting lowering of the pH both inside the cell and in the tissues surrounding the tumor cell have any bearing on the degree of malignancy. One could imagine how an altered pH may affect important controls of DNA, RNA and protein metabolism and how the excreted lactic acid may damage some of the neighboring cells and thus provide room for expansion for the cancer tissue.

These are indeed serious speculations and we would like to test their validity. We hope to find the right chemical that specifically repairs the lesion responsible for the increased lactic acid production. Actually quercetin has some of the required properties but it combines readily with serum albumin which is abundant in the blood of animals. We are therefore searching for analogous compounds with greater specificity. If we find one we would like to see it in animals with spontaneous and transplanted tumors. If we can inhibit glycolysis *in vivo* we could settle the old question whether or not the high aerobic glycolysis of tumors is any way associated with malignancy.

## References

1. Racker, E. *Biochim. Biophys. Acta* (1974) In press.
2. Mitchell, P. *Nature* 191 (1961) 144.
3. Mitchell, P. *Biol. Rev. Cambridge, Phil. Soc.* 41 (1966) 445.
4. Racker, E. 1965. *Mechanisms in Bioenergetics*. Academic Press, Inc., New York.
5. Glynn, I. M. and Lew, V.L. 1969. In: *Membrane Proteins (New York Heart Association Symposium)* pp. 289, Little, Brown & Co., Boston.
6. Makinose, M. and Hasselbach, W. *FEBS Letters* 12 (1971) 271.
7. Panet, R. and Selinger, Z. *Biochim. Biophys. Acta*. 255 (1972) 34.
8. Boyer, P.D. 1965 In: *Oxidases and Related Redox Systems*. Vol. 2, pp. 994, eds. King, T.E. Mason, H.S. and Morrison, M. New York: John Wiley.

9. King, T.E., Kuboyama, M. and Takemori, S. 1965. In: Oxidases and Related Redox Systems. Vol. 2, pp. 707, eds. King, T.E., Mason, H.S. and Morrison, M. New York: John Wiley.

10. Jagendorf, A.T. and Uribe, E. Proc. Nat. Acad. Sci. U.S.A. 55 (1966) 170.

11. Schwartz, J.H. Proc. Nat. Acad. Sci. U.S.A. 49 (1963) 871.

12. Masuda, H. and deMeis, L. Biochemistry 12 (1973) 4581.

13. Sen, A.K., Tobin, T. and Post, R.L. J. Biol. Chem. 244 (1969) 6596.

14. Post, R.L. and Toda, G. IUM 9th. International Congress of Biochemistry. Abstracts (1973) p. 254.

15. Dahms, A.S. and Boyer, P.D. J. Biol. Chem. 248 (1973) 3155.

16. Boyer, P.D. Cross, R.L. and Momsen, W. Proc. Nat. Acad. Sci. 70 (1973) 2837.

17. Post, R.L. and Orcutt, B. 1973. In: Organiza tion of Energy-transducing Membranes. pp. 35, eds. Nakao, M. and Packer, L. University of Tokyo Press.

18. Degani, C. and Boyer, P.D. J. Biol. Chem. 248 (1973) 8222.

19. Martonosi, A. 1972. In: Current Topics in Membranes and Transport. Vol. 3, pp. 83, eds. Bronner, F. and Kleinzeller, A. Academic Press:New York and London.

20. Kagawa, Y. and Racker, E. J. Biol. Chem. 241 (1966) 2467.

21. Nelson, N. and Racker, E. Biochemistry 12 (1973) 563.

22. Lewis, A., Nelson, N. and Racker, E. In preparation.

23. Penefsky, H. S. J. Biol. Chem. 242 (1967) 5789.

24. Deters, D., Nelson, N. and Racker, E. (1974) In preparation.

25. Lang, D. and Racker, E. Biochim. Biophys. Acta (1974) In press.

26. Ryrie, I. and Jagendorf, A.T. J. Biol. Chem. 246 (1971) 3771.

27. McCarty, R. and Racker, E. J. Biol. Chem. 243 (1968) 129.

28. McCarty, R.E. and Fagan, J. Biochemistry 12 (1973) 1503.

29. Wrigglesworth, J.M., Packer, L. and Branton, D. Biochim. Biophys. Acta 205 (1970) 125.

30. Telford, J.N. and Racker, E. J. Cell Biol. 57 (1973) 580.

31. Garber, M.P. and Steponkus, P.L. J. Cell Biol. (1974) In press.

32. Thayer, W.S. and Hinkle, P.C. Fed. Proc. Abstracts (1973) 2568.

33. Rumberg, B. and Schroder, H. 1972. In: Proc. Internatl. Congress of Photobiology, Bochum ed. G.O. Schenk. In press.

34. Avron, M., Bamberger, E.S., Rottengerg, H. and Shuldiner, S. IUB International Congress of Biochemistry (1973) 215.

35. Racker, E. and Kandrach, A. J. Biol. Chem. 248 (1973) 5841.

36. Kagawa, Y. and Racker, E. J. Biol. Chem. 246 (1971) 5477.

37. Racker, E. and Horstman, L.L. 1972. In: Energy Metabolism and the Regulation of Metabolic Processes in Mitochondria, pp. 1, eds. Mehlman, M.A. and Hanson, R.W. Academic Press Inc., New York.

38. Hinkle, P.C., Kim, J.J. and Racker, E. J. Biol. Chem. 247 (1972) 1338.

39. Racker, E. 1970 In:Essays in Biochemistry Vol. 6 p. 1, eds. Campbell, P.N. and Dickens, F.

40. Nishibayashi-Yamashita, H., Cunningham, C. and Racker, E. J. Biol. Chem. 247 (1972) 698.

41. Oesterhelt, D. and Stoeckenius, W. Proc. Nat. Acad. Sci. U.S.A. 70 (1973) 2853.

42. Danon, A. and Stoeckenius, W. Proc. Nat. Acad. Sci. U.S.A. (1974) In press.

43. Racker, E. and Stoeckenius, W. J. Biol. Chem. (1974) In press.

44. Racker, E. and Hinkle, P.C. Membrane Biology (1974) In press.

45. Hladky, S.B. and Haydon, D.A. Nature (London) 225 (1970) 451.

46. Krasne, S., Eisenman, G. and Szabo, G. Science 174 (1971) 412.

47. Mueller, P. and Rudin, D.O. 1969. In: Current Topics in Bioenergetics, Vol. 3 pp. 157.

48. Eigen, M. and de Mayer, L. Z. Electrochem. 60 (1956) 1037.

49. Racker, E. Biochem. Biophys. Res. Commun. 55 (1973) 224.

50. Shertzer, H.G. and Racker, E. J. Biol. Chem. (1974) In press.

51. Oesterhelt, D. and Stoeckenius, W. Nature New Biol. 233 (1971) 149.

52. Racker, E. and Eytan, E. Biochem. Biophys. Res. Commun. 55 (1973) 174.

53. Racker, E. J. Biol. Chem. 247 (1972) 8198.

54. Hasselbach, W. and Makinose, M. Biochem. Z. 339 (1963) 94.

55. Nelson, N., Deters, D.W. Nelson, H. and Racker, E. J. Biol. Chem. 248 (1973) 2049.

56. Pullman, M.E. and Monroy, G.C. J. Biol. Chem. 238 (1963) 3762.

57. Nelson, N., Nelson, H. and Racker, E. (1972) J. Biol. Chem. 247 (1972) 7657.

58. Van Den Stadt, R.J., De Boer, B.L. and Van Dam, K. Biochim. Biophys. Acta 292 (1973) 338.

59. Suolinna, E., Lang, D. and Racker, E. J. Natl. Cancer Inst. (1974) In press.

---

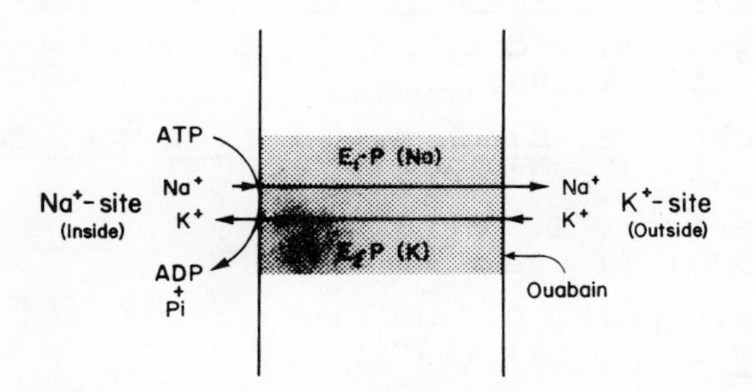


Figure 1. The $Na^+$ $K^+$-ATPase of plasma membrane (channel mechanism).

---

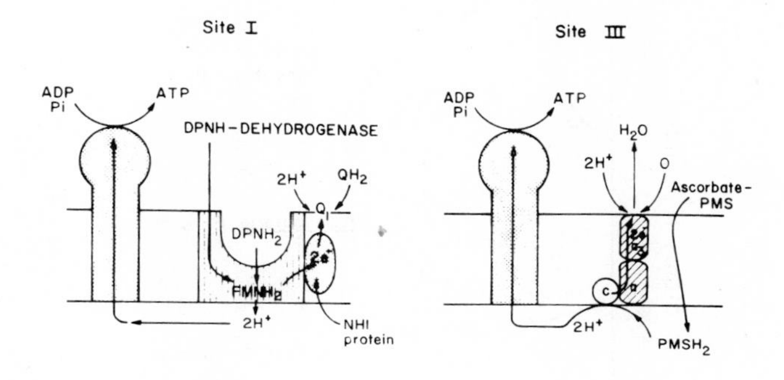


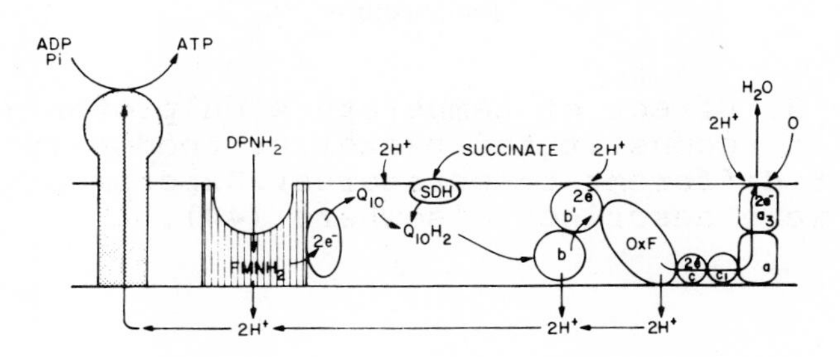


Figure 2. Proton movements in oxidative phosphorylation.

---

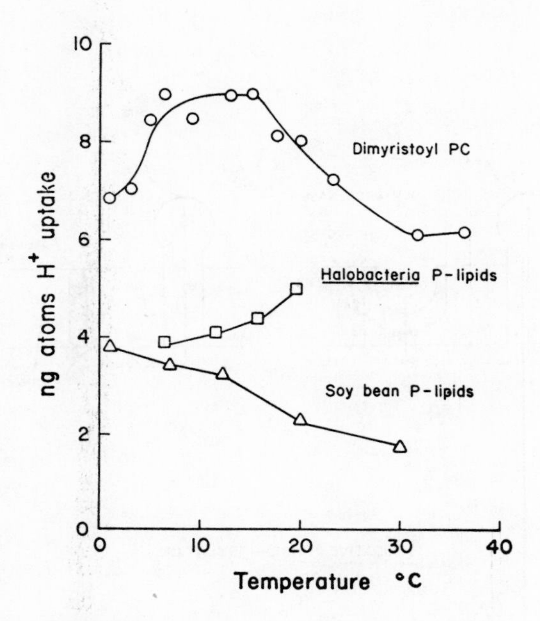


Figure 3. Effect of temperature on proton movements in reconstituted bacterial rhodopsin vesicles(at different temperatures).Experimental details were described elsewhere (44).

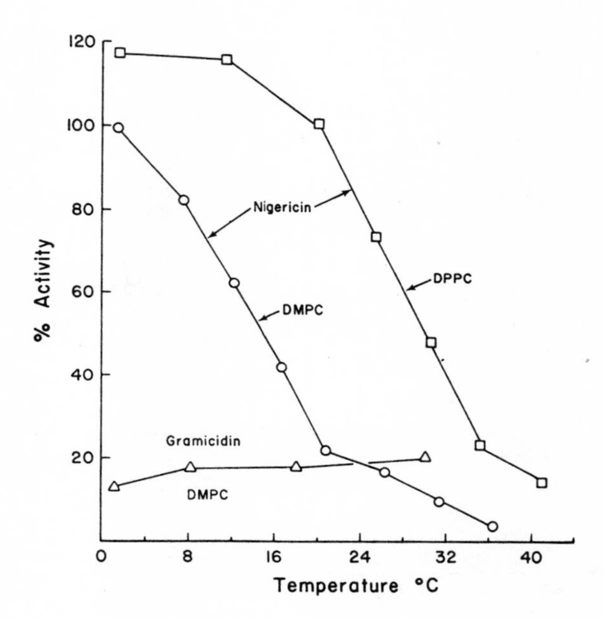


Figure 4. Effect of nigericin and gramicidin on proton movements in reconstituted bacterial rhodopsin vesicles at different temperatures. Experimental details were described elsewhere (44).

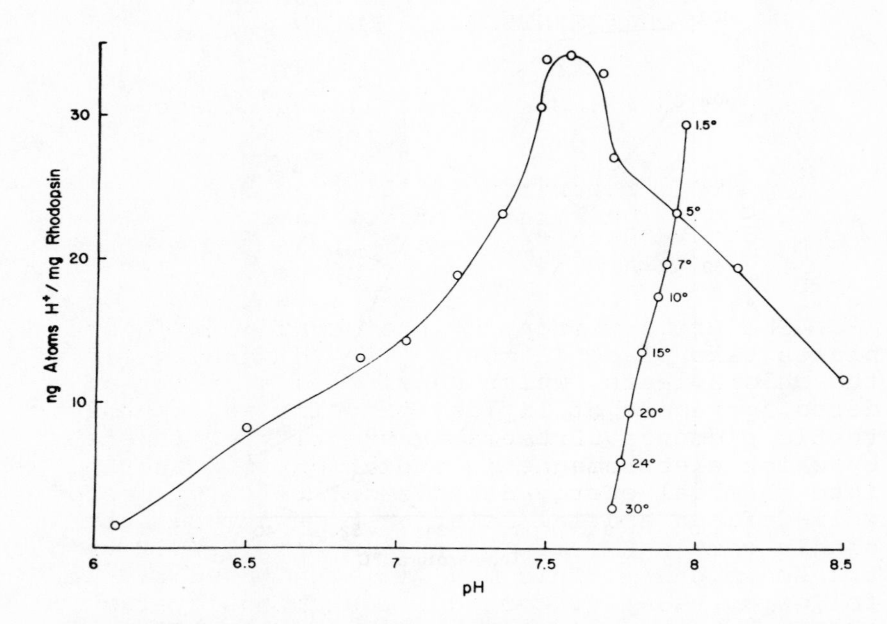


Figure 5. Effect of temperature and pH on proton release from purple membrane preparations. The assay system was as described previously (43) with 100 μg of purple membrane preparation in 1 ml of 0.15 M KCl. The pH was adjusted by addition of KOH or HCl.

# ARTIFICIAL LIPID MEMBRANES AS POSSIBLE TOOLS FOR THE STUDY OF ELEMENTARY PHOTOSYNTHETIC REACTIONS.

P. Läuger, G.W. Pohl, A. Steinemann
and H.-W. Trissl

Department of Biology,
University of Konstanz,
7750 Konstanz, Germany

The primary steps of photosynthesis in green plants take place in the thylakoid membrane of the chloroplasts, which consist of a highly ordered aggregate of lipids, proteins and photosynthetic pigments. Presumably, the high efficiency by which electromagnetic radiation is converted into chemical energy, depends on the existence of well-defined spatial relationships between the single components of the photosynthetic electron transport chain. This may be illustrated by the following example. Consider a chemical system in which a certain reaction may take place if components A, B and C come together in a linear collisional complex of structure ABC. If A, B and C are present in dissolved state, the spatial distribution is completely random, and the occurrence of a configuration ABC is a rare event. In this case the reaction will proceed with a low efficiency. A much more favourable situation arises, however, if A, B, and C form an ordered two-dimensional aggregate in which the components have fixed positions with respect to each other (1). Such an arrangement may be achieved in various ways if A, B, and C are building blocks of a membrane.

Most of the chemical components of the thylakoid membrane have been identified. Lipids (main ly galactolipids) presumably act as a supporting matrix. Besides the main photosynthetic pigment, chlorophyll a, there are a number of accessory pigments in the membrane, such as chlorophyll b and different carotinoids. Another constituent of lipoic nature is plastochinone which acts as a redox carrier. Most other components of the electron transport chain are proteins (cytochrome b, cytochrome f, plastocyanine, etc).The precise geometrical arrangement of these components in the thylakoid membrane is not known with any certainty, but various models for the membrane architecture have been proposed (2).

For the physical chemist the existence of a structure such as the thylakoid membrane presents a challenge to built up artificial systems in which the single components have a similar order. The ultimate goal of such an attempt would be to reconstruct the photosynthetic apparatus from its separate components. At present, this goal is indeed very remote, but nevertheless it is worthwhile to go a few steps into this direction. There is also another, more modest reason for the study of artificial model systems. In the past several constituents of the photosynthetic electron transport chain have been isolated in pure form and their physicochemical properties have been studied in solution, i.e., in an isotropic medium. On the other hand we know that the photosynthetic membrane is a highly anisotropic structure in which most molecules become oriented in such a way that part of the molecule may be accessible from the aqueous phase adjacent to the membrane while the other part remains in an hydrophobic environment. Therefore, if we want to examine the properties of photosynthetic pigments and other functional components of the thylakoid membrane more closely, we have to introduce these molecules again into an anisotropic medium such as a lipid bilayer membrane.

Several different techniques for the formation of artificial lipid bilayer membranes have been described in the past years. A first method consists in the ultrasonication of a suspension of lipid in water and leads to small, single-shelled lipid vesicles of a diameter of a few hundred Angstroms in diameter (3). These vesicles may be easily obtained in large quantities. They enclose an internal aqueous space and are therefore suitable for transport studies. A disadvantage is the small radius of curvature which leads to an appreciable distortion of the lipid structure. This disadvantage is not present in a second method which has been introduced by Takagi, Azuma and Kishimoto (4) and by Montal and Mueller (5). With the Montal-Mueller technique planar bilayers may be formed starting from two lipid monolayers at the air-water interface. This new method has great potentialities, mainly because asymmetric membranes may be obtained in this way. Since up to now only membranes of rather small area could be formed so that optical studies have not yet been carried out with this method. Most experiments on photosynthetic pigments in lipid systems have been done with the well-known Mueller-Rudin membranes which are formed on a plastic diaphragm in aqueous phase using a solution of lipid in an organic solvent (6). In the following is a description of the manner in which these membranes may be used to study the interaction of chlorophyll with lipids and of the behaviour of this pigment in an anisotropic environment.

Chlorophyll is an amphiphilic molecule consisting of a strongly hydrophobic tail, the phytyl chain, and the porphyrin ring which has a somewhat more hydrophilic character. A molecule of this type is well adapted to interact in a specific way with a lipid bilayer. We may expect that the phytyl chain becomes anchored in the hydrocarbon core of the membrane, whereas the porphyrin ring presumably is located near the interface. It is therefore an obvious experiment to form an artificial lipid membrane from a mixture of chlorophyll *a* and a lipid such as lecithin. The composition of such a membrane is not simply given by the ratio of chlorophyll to lipid in the membrane-forming solution because the bimolecular membrane is in equilibrium with the bulk lipid solution in the torus surrounding the membrane. The composition of the membrane may be studied with optical methods. A sensitive method consists in measuring the fluorescence of the membrane (7, 8). If the fluorescence signal of the membrane is compared with the signal obtained from a chlorophyll solution of known concentration, the number of chlorophyll molecules per $cm^2$ of the membrane may be determined. The values which are calculated in this way agree with the results of measurements of the optical absorption of chlorophyll-lecithin membranes using a sensitive double-beam spectrophotometer (9). It is found that up to $3.10^{13}$ chlorophyll molecules per $cm^2$ can be incorporated into the membrane, corresponding to a mean distance of about 20 Å between the porphyrin rings. This is close to the estimated means spacing of about 15 Å between chlorophyll molecules in vivo (10). It is well known that this high packing density leads to energetic interactions between porphyrin rings in the thylakoid membranes, and to corresponding spectral shifts. The same is found with artificial chlorophyll membranes. At high chlorophyll concentration a shift of about 5 nm towards longer wavelengths occurs both in the blue and in the red absorption band (9, 11). Similar (but larger) shifts are observed with crystalline chlorophyll and with chlorophyll in vivo.

The dense packing of porphyrin rings in the artificial membranes gives rise to still another interesting phenomenon. If we measure the fluorescence intensity of the membrane as a function of chlorophyll *a* concentration, it is found that the fluorescence goes through a maximum and becomes small at high concentrations (7). Similar observations have been made with chlorophyll *a* in monolayers at the air-water interface (12, 13). At least two different mechanisms may be responsible for this self-quenching of fluorescence. The simplest mechanism is the so-called static quenching. This term implies that an equilibrium exists between fluorescent monomers and non-fluorescent dimers (or higher aggregates). When the pigment concentration is increased, more and more pigment molecules are present in the form of non-fluorescent aggregates. In many cases, however, quenching is greatly enhanced by energy transfer, whereby the excitation energy migrates among the assembly of pigment molecules and may eventually reach a nonfluorescent molecule. There is strong evidence that this process takes place with high efficiency in the chloroplasts of green plants (10, 14). The most probable mechanism for energy migration is the resonance transfer or Förster mechanism, which leads to high transfer rates if the distance between pigment molecules becomes smaller than about 50 Å. For instance, the distance at which emission and intermolecular energy transfer become equally probable is 80 Å for chlorophyll a in ethyl ether solution and 65 Å in lipid vesicles (10). In contrast to static quenching, energy migration leads to depolarization of fluorescence. It is therefore possible to discriminate between the two mechanisms by experiments in which the exciting light is linearly polarized and the emitted radiation is analyzed with a second polarization filter (9). At low chlorophyll concentrations where quenching is small, the degree of polarization, P, is found to be 0.065. (P is defined as $P = (I_p - I_s)/(I_p + I_s)$ where $I_p$ and $I_s$ are the measured fluorescence intensities when the analyzer is parallel and perpendicular to the polarizer, respectively). At high chlorophyll concentration, however, where quenching is appreciable the polarization is practically zero. This result indicates that energy transfer takes place in the artificial membrane at the higher chlorophyll concentrations.

In photosynthetic organisms energy migration not only occurs between pigment molecules of the same kind, but also between different pigments. The evolutionary advantage of this process is obvious: the organism may utilize for photosynthesis also those parts of the sunlight spectrum which are not normally absorbed by chlorophyll a. The so-called accessory pigments in the thylakoid membrane presumably have, at least in part, such a light-harvesting function. One prominent example is chlorophyll b which accompanies chlorophyll a in the higher plants and which differs from chlorophyll a only in that the methyl group at position 3 is replaced by a formyl group. The blue absorption peak of chlorophyll b lies near 450 mm where the chlorophyll a absorption is rather weak.

Energy transfer from a donor pigment D to an acceptor pigment A is possible if the emission band of D overlaps with the absorption band of A. For a demonstration of energy transfer, again fluorescence experiments may be used. For this purpose the system containing either A or D alone, or both together is irradiated with light which is absorbed only by D. Under these circumstances no fluorescence occurs if only A is present. If, however, pigment D is present in addition to A, energy which is absorbed by D may be transfered to A by the resonance mechanism so that now the fluorescence of A is observed. Such an experiment may be performed with an artificial bilayer membrane in the presence of chlorophyll a and b (15). The peak positions which are observed from bilayers containing either chlorophyll a or b alone are the following:

| Chlorophyll | excitation | emission |
|---|---|---|
| a | 430 nm | 685 nm |
| b | 470 nm | 665 nm |

Ideally, energy transfer could be demonstrated if on exciting chlorophyll b with light of wavelenght 470 nm the fluorescence of chlorophyll a at 685 nm is observed. The situation is somewhat complicated, however, by the poor separation of the excitation and emission bands of a and b, and therefore the fluorescence spectrum obtained from a membrane containing both chlorophyll a and b has to be corrected for the emission which would be observed in the absence of energy transfer. Nevertheless, the experiments demonstrate that efficient energy transfer from chlorophyll b to a takes place in the membrane. Similar results have been reported for carotinoids (donor) and chlorophyll a (acceptor) in artificial lipid membranes (16).

The theory of energy transfer in an assembly of pigment molecules has been developed in great detail in the past years, but the application of this theory to the photosynthetic membrane is still difficult. The efficiency of energy transfer between two pigment molecules not only depends on the distance but also on the precise mutual orientation of the two transition moments. Unfortunately, the orientation of the chlorophyll in the thylakoid membrane is not known with certainty (17). For this reason a structural analysis of artificial model systems may be helpful; in particular such a study may provide information how chlorophyll molecules are built into a lipid bilayer.

The first question we may ask is whether chlorophyll is randomly distributed in the hydrocarbon core of the bilayer (as in a solution) or whether it is localized in the membrane-solution interfaces. These two cases may be distinguished if a nonpermeating reagent which changes the optical properties of chlorophyll is added to the aqueous phase on one side of the membrane. A suitable reagent is potassium peroxidisulfate ($K_2S_2O_8$) which quenches the chlorophyll fluorescence (9, 18). From electrical conductances measurements it may be informed that the membrane is practically impermeable to $S_2O_8^{--}$. If $K_2S_2O_8$ is added to only one external phase of a chlorophyll-lecithin bilayer membrane, the fluorescence drops to nearly half the original value and remains then constant for at least 30 min. If the reagent is added also to the second aqueous phase, the fluorescence intensity goes to zero. The finding that half of the chlorophyll in the film reacts after the addition of $S_2O_8^{--}$ to one side excludes the solubility model and gives evidence that the chlorophyll is localized in the membrane interfaces. Furthermore, we may conclude that if there is an exchange of chlorophyll between the two interfaces, it must be rather slow. These conclusions are in accordance with the amphiphilic character of the chlorophyll molecule.

An important problem of course is the question whether the porphyrin ring has a preferential orientation with respect to the plane of the bilayer. This problem may be studied by measuring the linear dichroism of the membrane (19, 20). If linearly polarized light passes through the membrane and if the transition moments of the pigment molecules are oriented, the light absorption will depend on the angle between the plane of polarization and the plane of the membrane.

A rather sensitive measurement of the dichroism becomes possible if the electrical vector of the plane-polarized light is rotated at a certain frequency. The photomultiplier signal monitoring the transmitted light then consists of a d-c signal with a small superimposed a-c component. This a-c signal which depends on the dichroism of the membrane is further processed in lock-in amplifier system. It is found that an artificial bilayer membrane made from lecithin and chlorophyll *a* or *b* shows a relatively strong dichroism in the blue absorption band and a weaker dichroism in the red band. From these data it is possible calculate the orientation of the "blue" and "red" transition moments and furthermore, as the angle between the two transition moments in the porphyrin plane is known, also the angle of tilt between the porphyrin ring and the membrane. This angle turns out to be about 45-50° for chlorophyll *a* in a lecithin bilayer (19, 20). Even if the angle of tilt is known there are still several possibilities how the porphyrin ring may be inserted into the bilayer. The porphyrin ring may be located in the aqueous phase outside the polar groups of the lipid molecules or may be present in the hydrocarbon layer of the membrane. A third possibility would be a location between the polar groups. Evidence against the first case is given by optical reflectance measurements (21). Also the second case seems to be rather unlikely in view of the observed reaction of chlorophyll with the impermeable ion $S_2O_8^{--}$ and in view of the fact that the electrical capacitance of a lipid membrane does not appreciably change by the introduction of chlorophyll. The most probable location of the porphyrin ring is therefore between the polar groups of the lipid molecules. This conclusion is consistent with the fact that the porphyrin ring is partly hydrophobic but possesses a hydrophilic edge in the vicinity of the cyclopentanone ring and the phytyl ester bond. It is tempting to assume that the porphyrin ring makes contact to the aqueous phase with this hydrophilic region.

One of the reasons for studying artificial chlorophyll membranes is the hope that these model systems will become useful in understanding the primary reactions of photosynthesis. The events taking place in the thylakoid membrane immediately after the absorption of a light quantum by a chlorophyll molecule are still poorly understood. There is some evidence, however, that the primary chemical event is a redox reaction in which the excited chlorophyll is directly involved. One possibility would be that the excited chlorophyll gives off an electron to an acceptor in or near the membrane. In a second step the oxidized chlorophyll returns back to the ground state by accepting an electron from a donor molecule. The electromagnetic energy is then stored in the oxidized-donor, reduced-acceptor pair. In order to prevent the immediate back reaction, the oxidized donor and the reduced acceptor must be separated by some barrier which may be represented simply by the hydrophobic core of the membrane. As the overall process involves the transport of charge across the membrane, an electric field should appear as an early event in photosynthesis. Indeed, spectroscopic experiments give rather compelling evidence that a change in the electric potential difference across the thylakoid membrane occurs shortly after the absorption of light (22).

In the last years observations on photoelectric phenomena in lipid bilayer membranes containing chlorophyll *a* or related compounds have been reported from several laboratories (23-30). Unfortunately, most of these experiments have been carried out under poorly defined conditions. In many cases chloroplast extracts have been used yielding bilayer membranes of unknown composition. Moreover, there is spectroscopic evidence that in the chloroplast the pigment which participates in the primary chemical reaction is not ordinary chlorophyll a but a chlorophyll which shows a large shift of the red absorption band towards larger wavelengths and which presumably represents chlorophyll a in a special environment. Nevertheless, the photoelectric experiments carried out so far with lipid model membranes are encouraging and may eventually contribute to a better understanding of the primary reactions in photosynthesis.

The basic observation consists in the following. If a lipid-chlorophyll membrane separating two different aqueous redox systems is illuminated, a change in the voltage across the membrane occurs under open-circuit conditions. Conversely if the aqueous solutions are electrically short-circuited, a photocurrent flows through the membrane upon illumination. Depending on the system, the photovoltage and the photocurrent will be of transient nature or will persist as long as light is falling on the membrane. An example of a redox system which gives large photoelectric effects is N, N, N', N'-tetramethyl-p-phenylenediamine( TMPD ). This compound readily gives off an electron yielding the radical cation $TMPD^+$ (Wurster's blue). If TMPD is added to the aqueous phase on one side of a lecithin membrane containing chlorophyll a, steady photovoltages of the order of 40mV and steady photocurrents of several $nA/cm^2$ are observed upon continous illumination (24, 29). The action spectrum of the photoeffects is similar, but not identical, to the absorption spectrum of chlorophyll a. This suggests that a derivative of chlorophyll a, which is formed in the light is the active compound. The photoeffects which are observed in this system may be explained if it is assumed that the excited pigment P* is reduced by a TMPD molecule at one membrane-solution interface:

$$P^* + TMPD \longrightarrow P^- + TMPD^+$$

In this way the mobile, lipid-soluble cation $TMPD^+$ is formed. The reduced pigment $P^-$ remains fixed in the interface and presumably is reoxidized by an unspecified component of the system in a subsequent reaction. The cations $TMPD^+$ which are generated in the interface may either jump into the adjacent aqueous phase, or they may migrate through the membrane towards the opposite aqueous phase. The first process does not contribute to the steady current which is measured externally under short -circuit but the second process gives rise to a stationary photocurrent flowing from the TMPD side to the TMPD-free side. On the other hand, if the external solutions are connected to a high-impedance voltmeter, a photovoltage (TMPD side negative with respect to the TMPD-free side) is generated by the transport of $TMPD^+$ through the membrane.

At least two different mechanisms may lead to photoelectric phenomena at a lipid bilayer membrane. One is the continous production of membrane-permeable charge carriers at one interface by the action of light. An example of this mechanism is presumably represented by the TMPD system. A second type of photoprocess occurs when neither the pigment molecule nor the redox component with which the excited pigment exchanges an electron are able to cross the membrane (29, 31). For instance, it may be imagined that the pigment is fixed at the interface (as it is the case with chlorophyll) and that a water-soluble protein which acts as the redox component becomes adsorbed to the interface on one side of the membrane. In this case the absorption of light in the pigment and the subsequent electron exchange between pigment and protein generate an array of dipoles in the membrane-solution interface. An alternative possibility would be that the impermeable redox component is simply dissolved in the aqueous solution and reacts during a collision with the excited pigment in the interface. Under these circumstances only one half of the electric double layer is fixed, the other half being represented by a diffuse Gouy-Chapman layer in the aqueous solution near the interface. Both mechanisms, the production of permeable charge-carriers at the interface and the generation of a dipole layer, can be distinguished by the time-behaviour of the photocurrent. In the first case a steady photocurrent is observed under continous illumination, whereas in the second case the photocurrent is of a transient nature, because here a current can only flow until the membrane capacitance has been charged up.

The photogeneration of an electric double-layer may be observed if a protein such as cytochrome c is adsorbed at the interface of a lipid-chlorophyll membrane (29). It is known from other experiments that cytochrome c binds electrostatically to the membrane surface at low ionic strength (32). In the presence of cytochrome c in one aqueous phase and at ionic strength below $10^{-3}$M transient photocurrents are observed upon illumination of the membrane. The time constants of the rise and the fall of both photocurrent and photovoltage agree with the theoretical predictions for the dipole-layer model. The observed effects may be explained by the assumption that an electron is transfered from the cytochrome to the excited pigment. A change of the redox state of the heme group may be excluded, however, because it is found that both reduced and oxidized cytochrome c give almost the same photoeffects. It is therefore likely that the protein moiety of cytochrome c is involved in the redox reaction.

From the foregoing it becomes clear that the experiments with photosynthetic model systems which have been done so far are in a rather early stage. On the other hand these studies may indicate the possible routes for further investigations. The optical experiments have shown that a bilayer membrane formed from chlorophyll and lipid has a well-defined structure in which the chromophore has a non-random orientation with respect to the plane of the membrane, which presumably is similar to the orientation in vivo.

Such systems may therefore be used for a more detailed analysis of the mechanism of energy transfer between photosynthetic pigments under conditions which are much closer to the physiological state than it was the case in the previous experiments with dissolved pigments. For this purpose fluorescence studies with lipid vesicles containing photosynthetic pigments seem to be suitable (33, 34). Flat bilayer membranes, on the other hand, offer the advantage that both sides of the membrane are accesible from the aqueous solutions and that the electric potential difference across the membrane may be easily measured or controlled. It seems possible, for instance, to use flat bilayer membranes for a closer study of the role of plastoquinone in the photosynthetic electron transport chain. It is known that many plastoquinone molecules together form a kind of "pod", accepting electrons from excited chlorophyll a in photosystem II and donating electrons *via* intermediates to chlorophyll a in photosystem I (35). It would be highly desirable to know more about the physicochemical basis of this process and of its coupling to proton translocation. Of particular interest, of course, are bilayer membranes of the Montal-Mueller type which may be built up in an asymmetric fashion. For instance it should be possible to introduce chlorophyll only to one half of the bilayer and to have a second redox component such as plastoquinone only in the other half (36). The structure of the membrane may then be further modified by asymmetric adsorption of proteins (plastocyanine, cytochromes) from the aqueous phases. In this way it seems feasible to construct step by step functional systems of increasing complexity.

## References

1. Bücher, H., Drexhage, K.H., Fleck, M., Kuhn,H. Möbius, D., Schäfer, F.P., Sonderma n, J. Sperling, W., Tillmann, P. and Wiegand, J. J. Mol. Cryst. (1967) 199.

2. Kreutz, W. Angew. Chem. Internat. Edit. 11 (1972) 551.

3. Huang, C. Biochemistry 8 (1969) 344.
4. Takagi, M., Azuma, K. and Kishimoto, U. Ann. Rep. Works Fac. Sci. Osaka Univ. 13 (1965)107.
5. Montal, M. and Mueller, P. Proc. Nat. Acad. Sci. U.S.A. 69(1972) 3561.
6. Mueller, P., Rudin, D. O., Tien M. Ti and Wescott, W.C. Nature 194 (1962) 979.
7. Alamuti, N. and Läuger, P. Biochem. Biophys. Acta 211 (1970) 362.
8. Trosper, T. J. Membrane Biol. 8(1972) 133.
9. Steinemann, A., Alamuti, N., Brodmann, W., Marschall, D. and Läuger, P. J. Membrane Biol. 4 (1071) 284.
10. Colbow, K. Biochem. Biophys. Acta 314 (1973) 320.
11. Cherry, R.J., Hsu, K. and Chapman, D. Biochem. Biophys. Res. Comm. 43 (1971) 351.
12. Tweet, A.G., Gaines, G.L.Jr. and Bellamy, W. D. J. Chem. Phys. 40 (1964) 2596.
13. Trosper, T., Park, R.B. and Sauer, K. Photochem. Photobiol. 7 (1968) 451.
14. Robinson, G.W. Brookhaven Symp. Biol. 19 (1966) 16.
15. Pohl, G.W., Biochem. Biophys. Acta 288 (1972) 248.
16. Strauss, G. and Tien, H.T. Photochem. Photobiol. 17 (1973) 425.
17. Kreutz, W. Z. Naturforsch 23b (1968) 520.
18. Huebner, I.S. J. Membrane Biol. 8 (1972) 403.
19. Steinemann, A., Stark, G. and Läuger, P. J. Membrane Biol. 9 (1972) 177.
20. Cherry, R.J., Hsu, K. and Chapman, D. Biochim. Biophys. Acta 267 (1972) 512.
21. Cherry, R.J., Hsu, K. and Chapman, D. Biochim. Biophys. Acta 288 (1972) 12.

22. Junge, W. and Witt, H.T. Z. Naturforsch. 23b (1968) 244.

23. Tien, H.T. and Verma, S.P. Nature 227 (1970) 1232.

24. Trissl, H.-W. and Läuger, P. Z. Naturforsch. 25b (1970)1059.

25. Tien, H.T. Photochem. Photobiol. 16(1972)271.

26. Hong, F.T. and Manzerall, D. Biochim. Biophys. Acta 275 (1972) 479.

27. Hong, F.T. and Manzerall, D. Nature New Biol. 240 (1972) 154.

28. Ilani, A. and Berns, D.S. J. Membrane Biol. 8 (1972) 333.

29. Trissl, H.-W. and Läuger, P. Biochim. Biophys. Acta 282 (1972) 40.

30. Tien, H.T. and Hebner, I.S. J. Membrane Biol. 11 (1973) 57.

31. Ullrich, H.M. and Kuhn, H. Biochim. Biophys. Acta 266 (1972) 584.

32. Steinemann, A. and Läuger, P. J. Membrane Biol. 4 (1971) 74.

33. Trsoper, T., Raveed,D. and Ke, B. Biochim. Biophys. Acta. 223 (1970) 463.

34. Colbow, K. Biochim. Biophys. Acta 318 (1973) 4.

35. Stiehl, H.H. and Witt, H.T. Z. Naturforsch. 24b (1969) 1588.

36. Montal, M. Personal communication.

A 4
B 5
C 6
D 7
E 8
F 9
G 0
H 1
I 2
J 3